Health Maintenance and Principal Microbial Diseases of Cultured Fishes

Health Maintenance and Principal Microbial Diseases of Cultured Fishes

John A. Plumb

IOWA STATE UNIVERSITY / AMES

John A. Plumb is Professor Emeritus in the Department of Fisheries and Allied Aquacultures, Auburn University, Alabama, where he taught graduate courses in microbial diseases and disease diagnosis of fish. His research includes investigations of viral and bacterial diseases of fish. Plumb has served as president of the Fish Health Section of the American Fisheries Society. Widely published, he leads the Southeastern Cooperative Fish Disease project and advises internationally.

Originally published as *Health Maintenance of Cultured Fishes: Principal Microbial Diseases* by CRC Press, Boca Raton, Florida 1994

Iowa State University Press
2121 South State Avenue, Ames, Iowa 50014

Orders:	1-800-862-6657
Office:	1-515-292-0140
Fax:	1-515-292-3348
Web site:	www.isupress.edu

♾ Printed on acid-free paper in the United States of America

First ISUP edition, 1999

International Standard Book Number: 0-8138-2298-X

Library of Congress Cataloging-in-Publication Data
Plumb, John A.
[Health maintenance of cultured fishes]
Health maintenance and principal microbial diseases of cultured fishes / John A. Plumb. — 1st ISUP ed.
p. cm.
Originally published: Health maintenance of cultured fishes. Boca Raton : CRC Press, c1994.
Includes bibliographical references and index.
ISBN 0-8138-2298-X (alk. paper)
1. Fishes—Infections. 2. Fish—culture. I. Title.
SH171.P66 1999
639.3—dc21 98-38489

The last digit is the print number: 9 8 7 6 5 4 3 2 1

This book is dedicated to the 1966–1967 teaching and research staff at the U.S. Fish and Wildlife Service, Eastern Fish Disease Laboratory (National Fish Health Research Laboratory), Leetown, West Virginia: S. F. Snieszko, Ken Wolf, Glenn Hoffman, Pete Bullock, Bob Putz, Roger Herman, and Ed Dunbar.

Contents

Preface

Infectious diseases of cultured fish are among the greatest constraints to expansion and realization of aquaculture's full potential. Viral, bacterial, and parasitic agents infect many wild and all cultured fish species. Most pathogenic agents are endemic to natural waters, where, under normal conditions, they cause no great problem. When these same viruses, bacteria, and parasites are present in an aquacultural environment, they may cause significant disease and mortality. Fish are often held in environments to which they are not biologically accustomed, a circumstance that often increases susceptibility to infectious disease. It is virtually impossible to separate the relationship of infectious disease from problems associated with environmental quality. In the following pages, the objective will be to emphasize salient points of host–pathogen–environment relationships, elucidate important aspects of infectious diseases, and explore how management can be used to help reduce effects of fish diseases.

The text is divided into three sections: Part I emphasizes the principles of fish health maintenance and diagnosis and management of infectious fish diseases. Parts II and III concentrate on viral and bacterial diseases, respectively, that are most important to aquaculture and to wild fish populations, where applicable. They include geographical distribution and species susceptibility, clinical signs, etiological agents and descriptions, epizootiology, pathological manifestations, significance of disease, and management of diseases within the fish groups. I have tried to create a balance between diseases of warmwater, coolwater, and coldwater fishes. Although much of the following information has been derived from North America, important disease problems from other parts of the world are included.

Diseases are organized into fish groups or families that are most extensively cultured. In the event a disease affects members of more than one fish family, emphasis is placed on the family most commonly or severely affected. Although some viral and bacterial diseases occur in both marine and freshwater fish, no specific distinction is made between the two environments.

It is not my intent to list every reported disease or all published papers on each disease or subject matter mentioned. Only those publications that are pertinent to the discussion have been cited. This book is intended for students and scientists who are interested in health maintenance of aquatic animals, aquatic pathobiology, and infectious fish diseases, as well as for practicing pathologists, aquaculturists, and fishery managers, biologists, and aquatic veterinarians.

Acknowledgments

I thank the following individuals who reviewed portions of this book and provided valuable suggestions: Kevin Amos, Thomas Bell, Frank Hetrick, John Grizzle, John Grover, Emmitt Shotts, Rudy Schmittou, and especially my wife, Peggy, without whose unselfish help this revision would not have been completed.

Common and Scientific Names

Latin names are based on *World Fishes Important to North Americans,* American Fisheries Society, Special Publication 21, American Fisheries Society, Bethesda, Maryland, USA.

COMMON NAME	SCIENTIFIC NAME
Alewife	*Alosa pseudoharengus*
Amberjack	*Seriola dumerili*
Amberjack (yellowtail)	*Serola lalandi*
American eel	*Anguilla rostrata*
Arctic char	*Salvelinus alpinus*
Arctic grayling	*Thymallus arcticus*
Atlantic cod (Baltic)	*Gadus morhua*
Atlantic croaker (Baltic)	*Micropogonius undulatus*
Atlantic herring	*Clupea harengus*
Atlantic halibut	*Hippoglossus hippoglossus*
Atlantic mackerel	*Scomber scombrus*
Atlantic menhaden	*Brevoortia tyrannus*
Atlantic salmon	*Salmo salar*
Atlantic seasnail	*Liparis atlanticus*
Atlantic tomcod	*Microgadus tomcod*
Australian pilchard	*Sardinops sagax neopilchardus*
Ayu	*Plecoglossus altivelis*
Baramundi perch	*Lates calcarifer*
Bighead carp	*Aristichthys nobilis*
Black bullhead	*Ameiurus melas*
Black carp	*Mylopharyngodon piceus*
Black seabream (porgy)	*Hypophthalmichthys schlegeli*
Bleak	*Alburnus alburnus*
Blue catfish	*Ictalurus furcatus*
Bluegill	*Lepomis macrochirus*
Blue tilapia	*Oreochromus aureus*
Bream (common)	*Abramis brama*
Brook trout	*Salvelinus fontinalis*
Brown bullheads	*Ameiurus nebulosus*
Brown-spotted grouper	*Epinephelus lovina*
Brown trout	*Salmo trutta*
Channel catfish	*Ictalurus punctatus*
Chinook salmon	*Oncorhynchus tshawytscha*
Chum salmon	*Oncorhynchus keta*
Coho salmon	*Oncorhynchus kisutch*
Common carp	*Cyprinus carpio*
Common minnow (Eurasian)	*Phoxinus phoxinus*
Common shiner	*Lulilus cornutus*
Cutthroat trout	*Oncorhynchus clarki*
Dab (North Sea)	*Pleuronectus limanda*
Damselfish	*Chrysiptera* sp.

COMMON NAME	SCIENTIFIC NAME
Danio (sind)	*Danio devario*
Doctorfish	*Labroides dimidatus*
Emerald shiner	*Notropis atherinoides*
Estuarine grouper	*Epinephelus tauvina*
European catfish (sheatfish or wels)	*Silurus glanis*
European eel	*Anquilla anquilla*
European flounder	*Platichthys flesus*
European seabass	*Morone (Decentrarchus) labrax*
European smelt	*Osmerus eperlanus*
Fathead minnow	*Pimephales promelas*
Formosa snakehead	*Channa maculata*
Gourami	*Osphronemus goramy*
Gilthead seabream	*Sparus auratus*
Gizzard shad	*Dorosoma cepedianum*
Glass knifefish	*Eigenmannia virescens*
Golden shiner	*Notemigonus crysoleucas*
Goldfish (Crucian carp)	*Carassius auratus*
Grass carp	*Ctenopharyngodon idella*
Grayling	*Thymallus thymallus*
Greenback flounder	*Rhombosolea tapirina*
Gudgeon (topmouth)	*Gobio gobio*
Gulf killifish	*Fundulus grandis*
Gulf menhaden	*Brevoortia patronus*
Guppy	*Poecilia reticulata*
Hardhead catfish	*Arius felis*
Ide	*Leuciscus idus*
Indian glassfish	*Chanda ranga*
Itipa mojarras	*Diapterus rhombeus*
Japanese catfish	*Silurus asotus*
Japanese eel	*Anguilla japonica*
Japanese striped knife jaw	*Oplegnathus faciatus*
Kelp (red) grouper	*Epinephelus moora*
Lake sturgeon	*Acinpenser fulvescens*
Lake trout	*Salvelinus namaycush*
Macquarie perch	*Macquaria australasica*
Masu (yamame, cherry salmon)	*Oncorynchus masou*
Mosquitofish	*Gambusia affinis*
Mountain galaxias	*Galaxias olidus*
Mozambique tilapia	*Oreochromis mossambicus*
Muskellunge	*Esox masquinongy*
Neon tetra	*Paracheirodon innesi*
Nile tilapia	*Oreochromis niloticus*
Northern anchovy	*Engraulis mordax*
Northern pike	*Esox lucius*
Olive flounder	*Paralichthys olivaceus*
Pacific cod	*Gadus macrocephalus*
Pacific halibut	*Hippoglosus stenolepis*
Pacific herring	*Clupea pallasi*
Pacific sardine	*Sardinops sagax*
Pearl danio	*Brachydanio albolineatus*

COMMON NAME	SCIENTIFIC NAME
Pejerrey	*Odonthestes banariensis*
Pinfish	*Lagodon rhomboides*
Pink salmon	*Oncorhynchus gorbuscha*
Plaice	*Pleuronectes platessa*
Rainbow smelt	*Osmerus mordax*
Rainbow trout	*Oncorhynchus mykiss*
Rare minnow	*Gobiocypris rarus*
Red drum	*Sciaenops ocellatus*
Redfin perch (yellow perch)	*Perca fluviatilis*
Red seabream (Asia)	*Pagrus major*
Red seabream (New Zealand)	*Chrysophrys major*
Roach	*Rutilus rutilus*
Rudd	*Scardinius erythrophthalmus*
Sablefish	*Anoplopoma fimbria*
Seabream	*Sparus aurata*
Sea raven	*Hemitripterus americanus*
Shiner perch	*Cymatogaster aggregata*
Shorthorn sculpin	*Myoxocephalus scorpius*
Silver carp	*Hypophthalmichthys molitrix*
Silver perch	*Bidanus* spp.
Silver perch (North America)	*Bairdiella chrysura*
Silver seatrout	*Cynoscion nothus*
Sockeye (kokanee)	*Oncorhynchus nerka*
Snakehead (Chevron)	*Channa striata*
Sole (Dover)	*Solea solea*
Spotted grouper	*Epinephelus akaara*
Striped bass	*Morone saxatilis*
Striped jack (White trevaly)	*Caranx dentex*
Striped mullet	*Mugil cephalus*
Striped trumpeter	*Latris lineata*
Tench	*Tinca tinca*
Threespot gourami	*Trichogaster trichopteru*
Turbot	*Scophthalmus maximus*
Walking catfish	*Clarias batrachus*
Walleye	*Stizostedion vitreum*
White bass	*Morone chrysops*
White bream (silver)	*Blicca bioerkna*
White catfish	*Ameiurus catus*
Whitefishes (ciscos)	*Coregonus* spp.
White perch	*Morone americana*
White seabass	*Atractoscion nobilis*
White sturgeon	*Acipenser transmontanus*
White sucker	*Catostoma commersoni*
Willow shiner	*Gnathopogon elongatus*
Winter flounder	*Pleuronectus americanus*
Yellow bullhead	*Ameiurus natalis*
Zebra danio	*Danio rerio*

I
Health Maintenance

Fish health maintenance emphasizes many areas that affect the health of cultured fishes. It requires continuous efforts, which include the location and construction of a culture facility; selection and introduction of culture species; and reproduction, culture, and harvesting of the final product. The aquatic habitat—a dynamic and continuously changing environment—is affected by structural material, facility design, soil quality and type, volume and quality of water, fish species present, amount and quality of nutrients introduced into the system, climate, and daily human activities.

Health maintenance involves a series of principles that apply to most farm-raised animals. Fish, however, tend to react more quickly to environmental change than terrestrial animals. Because of their homothermic nature, most terrestrial animals respond comparatively slowly to unfavorable environmental conditions, whereas fish—being poikilothermic—respond quickly and often fatally to handling, temperature change, excessive or insufficient dissolved gasses in the water, metabolites, or chemical additives, and so forth, to which they are unable to adapt. These factors also increase fish susceptibility to infectious agents and compromise their immune response.

One objective of health maintenance is to help control environmental fluctuations through management practices, thus reducing the magnitude of change and producing a more economical, healthier, and better quality product.

Specific areas of concern addressed include principles of health maintenance, epizootiology and pathology of fish diseases, disease recognition, basic concepts in disease diagnosis, and prevention and control of infectious fish diseases. Aquatic animal health management encompasses the entire production process, including disease diagnosis and treatment.

The ultimate goals of health management are (1) disease prevention, (2) reduction of infectious disease incidence, and (3) reduction of disease severity when it occurs. Successful health maintenance and disease prevention and/or control do not depend on any single procedure but are the culmination of the application of integrated concepts and exercising management options.

1 Principles of Health Maintenance

"An ounce of prevention is worth a pound of cure" is a familiar phrase that describes one approach to the culture of food animal resources. Health maintenance actually is a concept in which animals are reared under conditions that optimize growth rate, feed conversion efficiency, reproduction, and survival while minimizing problems related to infectious, nutritional, and environmental diseases, all within an economic context. "Health maintenance" encompasses the entire production management plan for food animals, whether they be swine, cattle, poultry, or fish either in the public (governmental) or private sector. Aquaculture involves human intervention in the growth process of fish and other organisms in an aquatic environment. The degree of intervention is progressive, ranging from extensive (few fish per unit of water volume) to increasingly intensive (greater numbers of fish per unit of water volume) in ponds, raceways, cages, and recirculating systems where higher fish densities are maintained. As culture becomes more intense, need for intervention increases accordingly, and principles of health maintenance become of greater importance. These principles apply to aquaculture anywhere in the world, regardless of fish species or culture method.

Fish health management is not a new approach to aquaculture. Snieszko (1958) recognized the need for health maintenance in fish culture when he stated, "We are beginning to realize that among animals (including fish) there are populations, strains, or individuals which are not susceptible all of the time, or even temporarily, to some of the infectious diseases." He proposed the theory that fish possess a certain level of natural resistance to infectious diseases that can be enhanced through proper management and that environmental stressors and/or fish cultural practices can adversely affect that natural resistance. Another contributor to a health maintenance concept for aquatic animals is Klontz (1973), who established a course in fish health management at Texas A & M University. This course combined the studies of fish culture and infectious diseases into a health management concept. The Great Lakes Fishery Commission published the *Guide to Integrated Fish Health Management in the Great Lakes Basin,* which was a regional concept for fish health management (Meyer et al. 1983). These publications deal with the improvement of aquatic animal health through management. The most useful contribution to maintaining health of domestic (cultured) animals was made by Schnurrenberger and Sharman (1983) when they set forth a series of principles for animal health maintenance that apply in a general sense to all domesticated food animals. In the following pages, these principles are applied to aquaculture. Theoretically, if these principles are used in daily, monthly, yearly, and long-term management of an aquatic culture facility, there will be fewer environmental and disease problems and optimum production will be more readily obtained.

MAINTAINING HEALTH

In an aquatic environment, there is a profound and inverse relationship between environmental quality and disease status of fish. As environmental conditions deteriorate, severity of infectious diseases increases; therefore, sound health maintenance practices can play a major role in maintaining a suitable environment where healthy fish can be grown. The aquatic environment is a dynamic ecosystem that changes during a 24-hour period and seasonally, particularly in ponds with limited water exchange. Tucker and van der Pfloeg (1993) noted that in static catfish ponds, periods of poorest water quality occurred during summer months, when feeding, temperature, and standing crops were at a maximum but rainfall and available water were at a minimum, thus producing a higher potential for stressful conditions requiring health management.

Fish health management is a positive concept that aids in disease prevention, emphasizes interruption of a disease cycle, deals with multiple segments of health maintenance, and results in more efficient production. Health maintenance does not simply target infectious diseases, but it emphasizes proper utilization of physical facilities, use of genetically improved fish and certified "specific pathogen free" (SPF) stocks whenever available and/or feasible, environmental control, prophylactic therapy, feed quality and quantity, pond, cage, raceway, tank, or recirculating system management, control of vegetation, aeration and use of other water quality maintenance practices, and a management commitment to provide an optimum habitat in terms of water quality for fish being cultured. Its goal is to improve the health and well-being of animals that appear to be generally healthy. If sound health maintenance principles are followed, production will be more efficient and result in a healthier product. Obviously, all activities, policies, and improvements must be based on sound economic criteria.

STRESS

"Stress" is difficult to define because it is used to describe many adverse situations that affect the well-being of individuals, but generally it is the reaction of an animal to a physical, physiological, or chemical insult (Barton 1997). Stress may also produce a nonspecific response to factors that are perceived as harmful; however, stress in fish is usually related to handling, transport, environmental quality, or fright. For clarity in this text, "stressors" are factors that cause a "stress response," which is the sum of physiological changes that occur as fish react to physical, chemical, or biological stressors as they attempt to compensate for changes that result from these stressors (Wedemeyer 1996). The corticosteroid level in plasma is the usual quantitative measure for stress; however, amounts of glucose, lactic acid, and ions will also increase during stressful conditions (McDonald and Milligan 1997).

The aquatic environment is in a continuous state of flux and because fish are poikilotherms and body functions are controlled by temperature, oxygen concentration, and many other water quality parameters, they must continually adapt physiologically to environmental changes. An inability to adjust to these changes may be manifested in lower productivity, reduced weight gain, increased feed conversion, decreased immunity, reduced natural disease resistance, increase in infectious disease, lowered hardiness in general, death, reduced profits for the commercial fish farmer, and underproduction for public hatcheries.

Some commonly known stressors in the aquatic environment are unionized ammonia, nitrites, chronic exposure to low concentrations of pesticides or heavy metals, insufficient oxygen, high concentrations of carbon dioxide, rapidly changing or extremes in pH or water temperature, external salinities, nutrition, and fish density (Barton 1997). Low alkalinity and hardness are also not conducive to good fish health or performance (Boyd 1990). Many of these factors are exacerbated by type, quality, and quantity of feed put in a pond and by waste accumulation. Sensitivity to these conditions will vary with fish species. Successful and efficient health maintenance programs for aquacultural facilities will include measures to reduce and modify stressful conditions that may be present in a fish population.

HAZARD REDUCTION THROUGH MANAGEMENT

Experience has shown that a wide variety of bacterial, parasitic, and other fish diseases will cause mortality if cultured fish are held in unfavorable environmental conditions (Wedemeyer 1996). Health and environmental management decisions are not independent, and a change in one area should not be made without evaluating its effect in other areas. Notable stressor-related fish diseases that result from a culmination of management and biological factors (Figure 1.1) are furunculosis, enteric redmouth, motile *Aeromonas* septicemia, columnaris, vibriosis, bacterial gill disease, streptococcus, external fungal infections, and some protozoan parasites (Table 1.1).

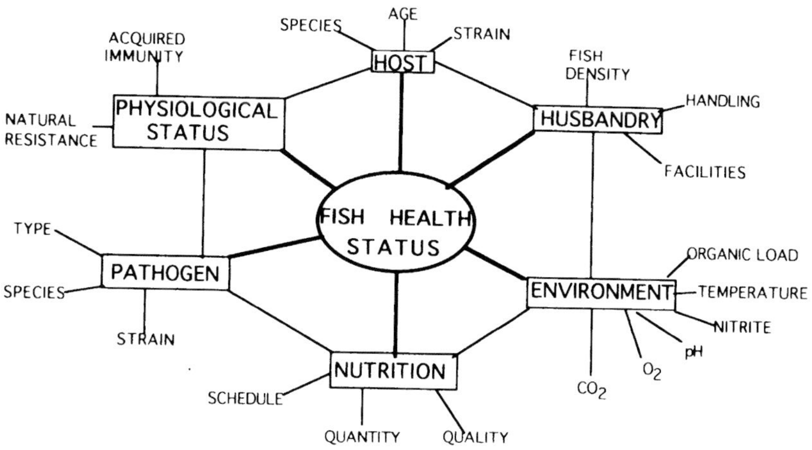

FIGURE 1.1. The relationship of environmental conditions, biological factors, and management practices in aquaculture that influences health and infectious diseases of fish.

TABLE 1.1. MICROBIAL DISEASES OF FISH COMMONLY CONSIDERED STRESS MEDIATED

Disease	Predisposing environmental factor
Spring viremia of carp	Handling after overwintering
Bacterial gill disease	Crowding, poor water quality, elevated ammonia, particulate material in water, presence of causative bacteria
Columnaris	Crowding, poor water quality, handling, seining, adverse temperature, physical injury
Cold-water disease	Temperature decrease from >10 to <10°C
Enteric redmouth	High stocking density, elevated water temperature, handling, transport, poor water quality
Furunculosis	Low oxygen, handling, environmental stress
Motile *Aeromonas* septicemia	Injury to skin, transport, improper handling, temperature stress, poor water quality, other parasites
Ulcer disease of goldfish and carp erythrodermatitis	Handling and stocking in late winter or early spring
Vibriosis	Handling, poor environmental conditions, moving from freshwater to salt water.
Streptococcosis	Handling, poor water quality, parasites

Sources: Walters and Plumb (1980); Piper et al. (1982); Roberts (1989); Wedemeyer (1996).

Stress on fish increases when environmental conditions approach the host's limit of tolerance (Snieszko 1973). For example, if water temperature is critically high and oxygen concentration is adequate, fish may survive, and if oxygen is critically low and water temperature is normal, fish may also adjust and survive. When multiple parameters are at stressful levels but are not individually lethal, however, the problem is compounded synergistically, and if the animal is unable to adapt, death may result. An example of environmental stressors and their synergistic effect on fish involves dissolved oxygen (DO) and carbon dioxide (CO_2) concentrations. Channel catfish can adapt to an elevated level of CO_2 (20 to 30 mg/L) if the DO concentration is optimal (Boyd 1990). If, however, the CO_2 level is critically high and DO is critically low, fish cannot eliminate CO_2 and will become listless (narcotized) and may die. If fish do adapt to these environmental stressors and survive, pathogen resistance is often compromised.

A theory of host-pathogen-environment relationship was applied to fish with regard to development of infectious diseases by Snieszko (1973) (Figure 1.2). This theory is based on the premise that to have an infectious disease, a host and pathogen are required but an unfavorable environmental condition often acts as a trigger for disease to develop. Potential pathogens are often endemic in surface waters, especially in warmwater fish culture, and

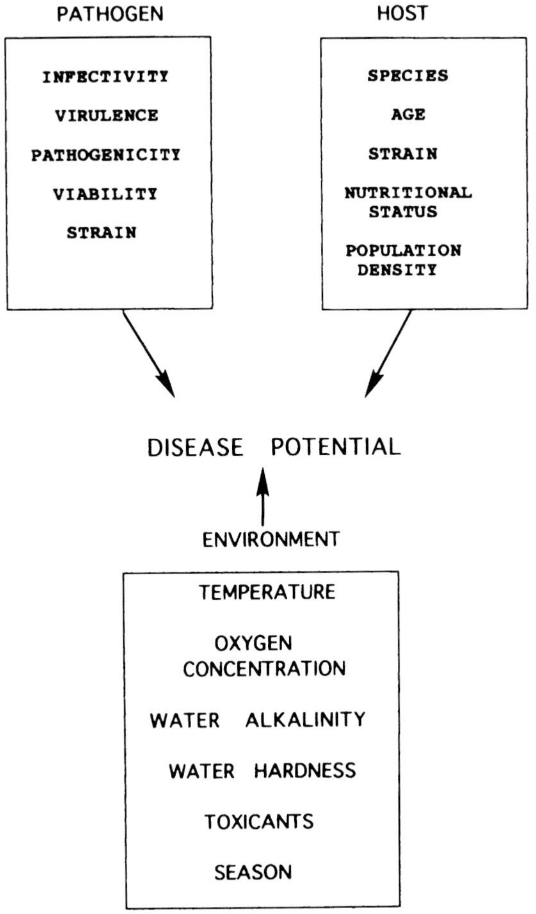

FIGURE 1.2. Some variables of the infectious agent, host, and environment that influence the potential for disease occurrence. Adapted from Snieszko (1973) and Schnurrenberger (1983a).

only environmental conditions and/or the host's natural resistance can dictate onset of the disease process. The interaction of these factors is expressed in the equation:

$$H(A + S^2) = D$$

where: H = Species or strain of host (natural resistance)
A = Etiological agent
S = Environmental stressors
D = Disease

Environmental stressors are squared because as fish approach adaptation limits, stressors increase accumulatively rather than additively. Also, when more than one stressor is involved (oxygen, ammonia, carbon dioxide, temperature, and so forth), detrimental factors act synergistically.

The relationship between water quality deterioration and bacterial infection was shown by Plumb et al. (1976). A sudden die-off of cyanobacteria (blue–green algae) in a channel catfish pond was followed by reduced DO production, decreased pH, and increased CO_2 and NH_4. These water quality changes in the pond resulted in "oxygen depletion" and a fish kill (Figure 1.3). This phenomenon has since been described by R. Schmittou (Department of Fisheries and Allied Aquacultures, Auburn University, Alabama, personal communication) as "low dissolved oxygen syndrome" (LODOS), which refers to the fact that low dissolved oxygen is part of an environmental condition that includes a variety of separate but interrelated elements. When fish first began to die as DO dropped to less than 1 mg/L, no bacteria or other significant pathogens were found during necropsy. However, 4 days after oxygen depletion, channel catfish with hemorrhaged and depigmented skin and muscle lesions were found (Figure 1.4). When first observed, no bacteria were isolated from internal organs or skin-muscle lesions of these fish, but 2 days later and for several days thereafter, *Aeromonas hydrophila* was isolated from both. When fresh water was added to the pond and remedial aeration provided, mortality ceased and clinical signs of infectious disease abated. It was theorized that while the water was in a state of hypoxia, some muscle areas in the fish also became hypoxic, which led to tissue necrosis, hemorrhaging, and skin depigmentation. When protective epithelium integrity was lost, naturally occurring *A. hydrophila* invaded the muscle beneath areas where skin had been injured and focal infections were established that progressed into septicemia. Walters and Plumb (1980) demonstrated that either low oxygen, low pH, high ammonia, or high $C0_2$ alone were not particularly stressful and did not lead to bacterial disease; however, if two or more of these adverse environmental conditions occurred simultaneously, infection was much more likely to occur (Figure 1.5).

Intensively reared fish provide unique, but manageable, cultural problems (Boyd 1990; Piper et al. 1982; Wedemeyer 1996). All fish require adequate water maintained at a suitable temperature and oxygen concentration level for proper growth and reproduction according to the species being cultured. Water temperature requirements will vary from species to species (Table 1.2). For example, channel catfish require a water temperature of 20

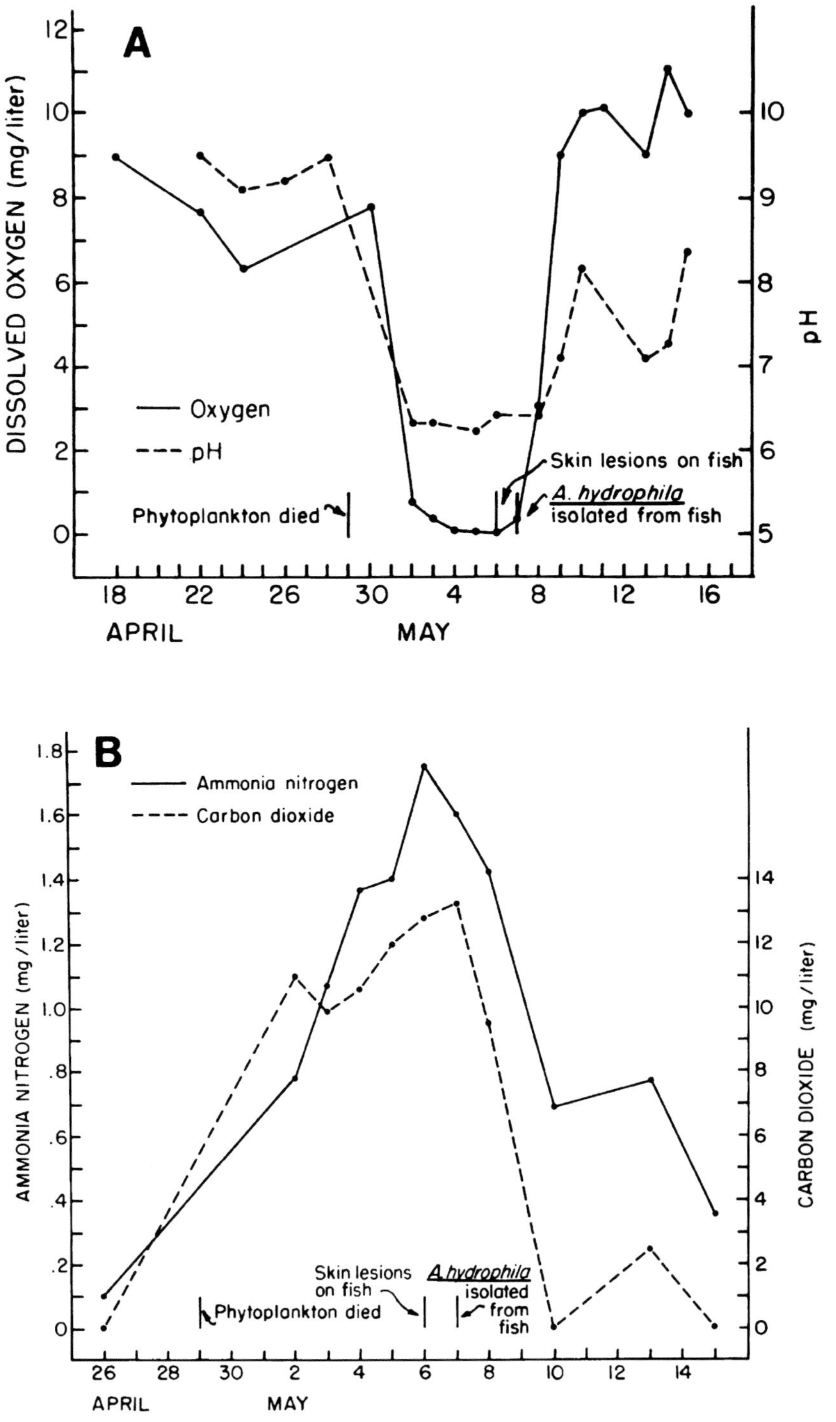

FIGURE 1.3. Water quality parameters before, during, and after a channel catfish mortality and subsequent Aeromonas hydrophila *infection (Plumb et al. 1976). (Reprinted with permission of* Journal of Wildlife Diseases.*)*

to 30°C for growth (optimum of 28°C); however, they spawn at 27°C and prefer oxygen concentrations greater than 5 mg/L. Channel catfish will survive at 1 to 5 mg of oxygen per liter, but prolonged exposure at levels less than 1 mg/L is lethal. If oxygen levels drop below a critical concentration, supplemental aeration is necessary as either a routine practice or as emergency management (Figure 1.6). Rainbow trout require water temperatures of 8 to 18°C for growth, spawn at temperatures as low as

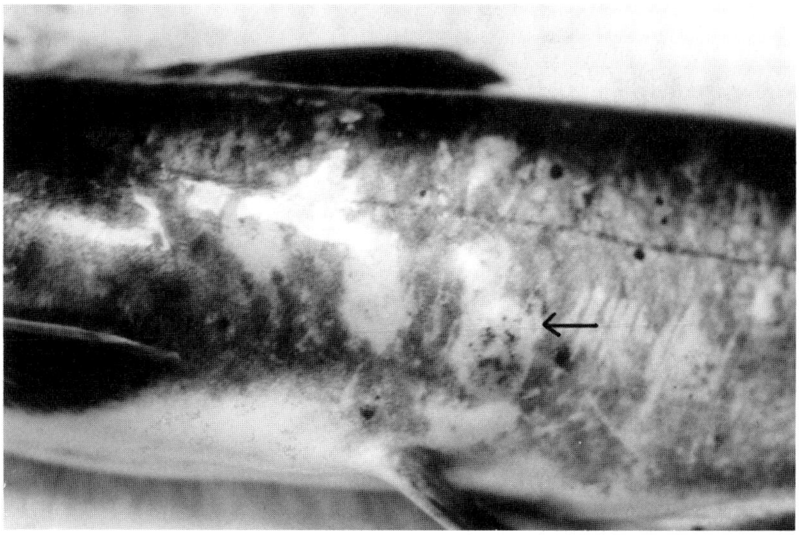

FIGURE 1.4. Channel catfish with depigmented, hemorrhaged, and necrotic lesions that appeared 6 days after oxygen depletion (Plumb et al. 1976). (Reprinted with permission of Journal of Wildlife Diseases.*)*

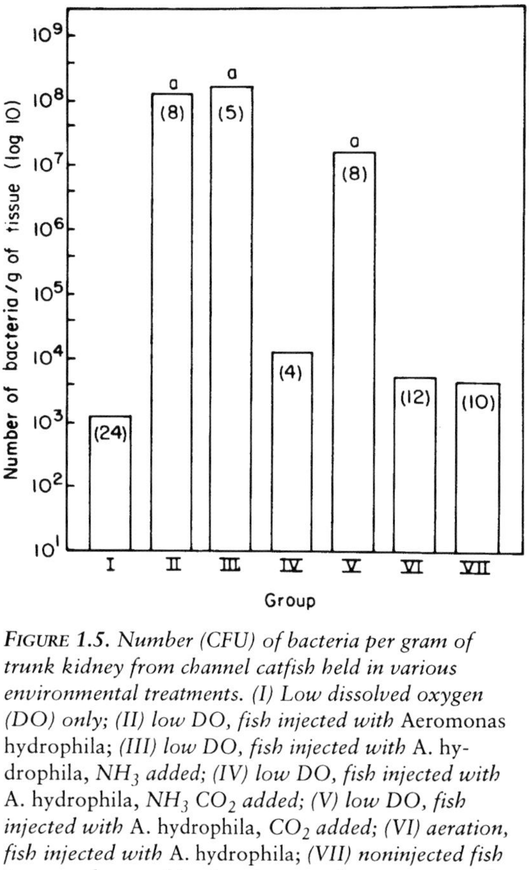

FIGURE 1.5. Number (CFU) of bacteria per gram of trunk kidney from channel catfish held in various environmental treatments. (I) Low dissolved oxygen (DO) only; (II) low DO, fish injected with Aeromonas hydrophila; *(III) low DO, fish injected with* A. hydrophila, *NH_3 added; (IV) low DO, fish injected with* A. hydrophila, *NH_3 CO_2 added; (V) low DO, fish injected with* A. hydrophila, *CO_2 added; (VI) aeration, fish injected with* A. hydrophila; *(VII) noninjected fish in aerated water. Numbers in parentheses represent the numbers of fish samples; significantly higher numbers of bacteria are designated by "a" (Walters and Plumb 1980). (Reprinted with permission of the Fisheries Society of the British Isles.)*

TABLE 1.2. RANGE OF TEMPERATURE TOLERANCE, OPTIMUM TEMPERATURE FOR GROWTH, AND SPAWNING TEMPERATURE OF SELECTED CULTURED FISHES

	Temperature		
Common name	Range	Optimum	Spawning
Atlantic salmon	1–24	10–17	7–10
Brook trout	1–22	8–13	8–13
Brown trout	1–25	9–17	9–13
Channel catfish	4–35	28–30	25–27
Chinook salmon	1–25	10–14	8–13
Coho salmon	1–25	9–14	8–13
Common carp	4–35	23–30	13–27
Eel	4–35	25–28	16–17
Grass carp	4–35	22–30	22–27
Lake trout	1–21	7–14	9–11
Largemouth bass	4–35	13–27	16–20
Milkfish	10–35	25–35	23–32
Northern pike	1–27	4–18	4–9
Rainbow trout	1–25	10–17	10–13
Sockeye salmon	1–21	10–15	8–12
Striped bass	2–32	13–24	13–22
Tilapias	15–35	23–32	23–32
Walleye	1–27	8–16	9–13
Walking catfish	13–38	20–30	20–30

Sources: Ney (1978); Piper et al. (1982); Synopsis of potential aquaculture species (1984).

6°C, and require oxygen concentrations above 5 mg/L. The walking catfish of Southeast Asia can survive water temperatures higher than 35°C and very low oxygen concentrations. These fish, however, have an auxiliary gill that allows them to extract oxygen directly from the air. Extreme high or low water temperatures may result in greater fish

FIGURE 1.6. Routine and emergency aeration: (A) *with water pump attached to tractor power takeoff,* (B) *with electric paddle wheel in a catfish pond (could also be powered by tractor power takeoff), and* (C) *by flowing water over a column of perforated plates.*

susceptibility to viral, bacterial, and parasitic diseases (Austin and Austin 1987).

For a fisheries manager, biologist, or diagnostician to understand infectious fish diseases, he or she must also understand how and to what extent the environment affects the host and disease. It is not enough simply to culture and/or identify a pathogen or parasite; it is equally important to identify and understand environmental stressors that predispose fish to disease. When stressors are known, corrective and preventive measures can often be initiated to help prevent or minimize a reoccurrence of the condition. In the aquatic world, maintaining an optimal environment is essential to good animal health, and extreme deviations will result in stress manifested by reduced feeding, higher feed conversions, poor growth, disease, or death.

LOCATION, SOIL, AND WATER

Choosing a proper site for an aquaculture facility is paramount to its success. Land topography, soil quality, water quality and abundance, and proximity to market are primary factors to be considered when choosing a location. To maintain healthy fish populations, soil, water, and fish species to be grown must be compatible. Sometimes, environmental modifications can be made to accommodate a species not indigenous to a specific area, but these modifications must be cost-effective. For example, channel catfish are not normally grown in northern latitudes of the United States because of a short growing season and lack of warm water. If, however, warm water from power plants or geothermal supplies are available, channel catfish or other warmwater fish can be grown successfully, although seldom economically.

TABLE 1.3. WATER QUALITY CRITERIA FOR OPTIMUM FISH HEALTH MANAGEMENT OF WARMWATER AND COLDWATER SPECIES OF FISH (MG/L EXCEPT FOR pH)

Characteristic	Coldwater	Warmwater
Oxygen	5–saturation	5–saturation
pH	6.5–8	6.5–9
Ammonia (un-ionized)	0–0.0125	0–0.02
Calcium	4–160	10–160
Carbon dioxide	0–10	0–15
Hydrogen sulfide	0–0.002	0–0.002
Iron (total)	0–0.15	0–0.5
Manganese	0–0.01	0–0.01
Nitrate	0–3.0	0–3.0
Phosphorus	0.01–3.0	0.01–3.0
Zinc	0–0.05	0–0.05
Total hardness ($CaCO_3$)	10–400	10–200
Total alkalinity ($CaCO_3$)	10–400	10–400
Nitrogen (gas saturation)	<100%	<100%
Total solids	0–80	50–500

Sources: Piper et al. (1982); Boyd (1990); Hajek and Boyd (1994).

Some soil requirements for culture ponds are very specific. Soil must have a high clay content to prevent ponds from leaking, as leaky ponds are unstable, do not hold a suitable plankton bloom to shade rooted vegetation, and require continuous replenishment of precious water. Conversely, in an analysis of chemicals naturally present in soil, Boyd (1990) stated that it is possible to rear fish in ponds built on soils that have wide ranges of chemical properties.

Certain regions, particularly coastal areas, have acid-sulfate soils containing high levels of iron pyrite (Boyd 1995). As long as these soils are submerged, the iron pyrite is stable and usually causes no problem. When the pond is drained and the bottom exposed to air, however, the iron pyrite is oxidized and, upon dehydration, sulfuric acid is produced. When the pond is refilled, the water becomes highly acid (pH may be as low as 3.5), rendering an unproductive environment. Generally, such areas should be avoided for aquaculture sites, but the acidity level can be corrected to some degree by addition of huge quantities of lime.

It is important that pond soil is free of chemical and pesticide residues. If toxicants, usually of anthropogenic origin, are present, they can leach into the water and kill fish. Concern is not only with soil used in pond construction but also with soil in the watershed; therefore, before construction of a culture facility, watersheds should be inspected for toxicants that could contaminate the water supply.

The most singularly important ingredient in successful fish health management is water quality and its availability. Water quality as described by Boyd (1990) varies, but most water can be made suitable for aquaculture, except under unusual circumstances. Water characteristics, such as hardness, alkalinity, pH, presence of toxicants, or undesirable gases, are important in aquaculture (Table 1.3). Before a fish farm is built, water quality, volume, and reliability of the water source must be determined.

Water sources for aquaculture may be lakes, rivers, springs, pumped or artesian wells, surface runoff, or irrigation canals. From a fish disease management standpoint, water from springs or wells is preferred because these sources are free of wild fish that may be carriers of infectious disease agents. From a water quality standpoint, however, these sources may not be usable without modification because of acidity, low dissolved oxygen, and/or high concentrations of carbon dioxide or nitrogen gases that require removal. These waters may be very soft (less than 50 mg/L $CaCO_3$) or contain high concentrations of iron, sulfur, or manganese (Table 1.3). Although well and spring water may have a constant temperature, it may not be optimum for the aquaculture species to be cultured, thus requiring heating or cooling in the case of geothermal sources. If pumping is necessary to use a water source, it can be an expensive procedure prone to mechanical breakdowns and interrupted water flow. Most water quality problems can, however, be overcome by proper planning and management.

When streams or reservoirs are used as a water supply, there are inherent problems involved. Wild fish, including fry and eggs that harbor parasites, pathogenic bacteria, or viruses, may be indigenous to these sources. These waters can be disinfected with ozone or ultraviolet light, but the efficacy of treatment depends on the physical nature of the water. For example, ultraviolet treatment is ineffective in silt or particulate-laden water. Also, surface waters including streams, reservoirs, and irrigation canals may be prone to wide seasonal temperature fluctuations, variable oxygen levels, silt loads, increased organic loads, volume fluctuations, and pollutants from municipal, industrial, or agricultural sources. Installation of sand and gravel filters or use of porous (saran) socks on inlet pipes are some options to prevent fish contamination.

AVOIDING EXPOSURE

The ideal way to control infectious fish diseases is to prevent fish exposure to pathogenic agents whenever possible, thus avoiding most devastating health problems. When dealing with the aquatic environment, however, it is virtually impossible to define all disease-causing agents and to keep them isolated from the fish host. Water provides an excellent medium for transfer of many communicable agents from fish to fish or from locality to locality. Moreover, many disease-causing organisms are endemic to the aquatic environment and are opportunistic, facultative pathogens that remain viable under various conditions.

Since the mid-1960s, some governments, fish hatcheries, and privately owned fish farms have made great strides in avoiding fish exposure to certain infectious diseases by using fish certified SPF, quarantine, routine water disinfection, and destruction of populations infected with specific disease organisms when treatment is not feasible. The pros and cons of the latter approach should be carefully weighed before making a decision to destroy or treat. To paraphrase a statement by S. F. Snieszko, a disease management practice should not destroy more than it saves.

The earliest attempt to prevent fish exposure to infectious pathogens occurred when trout stocks known to be positive for infectious pancreatic necrosis virus (IPNV) were not used as egg sources. This practice is now applied to other diseases. It is impossible to declare a fish group, or population, simply "disease free," implying they are free of all disease agents; therefore, the term "disease free" is limited to a specific pathogen. Currently, SPF certification is the best way to prevent introduction of unwanted pathogens into a "clean" facility, but it must be understood that testing procedures are not infallible because they are based on a statistically determined sample number of individuals taken from a larger population (Thoesen 1994). To detect a given pathogen with 95% confidence, the appropriate sample number for a 2% prevalence is 120 fish per 100,000 and for a 5% prevalence, 60 fish per 100,000 (Simon and Schill 1984). Thorburn (1996) indicated, however, that these numbers may not be uniform for all diseases and all fish species. In some instances, a larger number than recommended should be sampled to improve accuracy.

Many states have now established fish health protection programs that specify that incoming fish must be accompanied by a document certifying them to be SPF. California has one of the oldest and most rigorous fish health regulatory programs in the United States. Regional fish health plans have also been established to prevent introduction of unwanted pathogens into geographically defined fish populations. The Colorado River Basin Council has a strong inspection program to prevent introduction of fish that are infected with certain disease agents into natural or hatchery waters. The Great Lakes Fisheries Commission has a regional fish health protection program for the states and Canadian provinces contiguous to the Great Lakes. The major problem with state and regional fish health regulations is statutory inconsistency and degree of implementation.

International fish disease control concepts are gaining support to keep pace with a growing, worldwide aquacultural industry. An increasing number of countries now require health certificates of some type, verifying that certain imported fish products (alive or dead) are free of specific diseases. Although not foolproof, these methods have generally been successful in inhibiting the spread of many infectious disease agents. Rohovec (1979) reported that Great Britain, the European Union, Canada, and the United States all have regulations limiting fish movement, with emphasis on fish health. Great Britain introduced what was probably the first fish health regulation by passing the Diseases of Fish Act of 1937 (Hill 1996), with several subsequent amendments. It is believed that its continuous implementation has had a positive effect on the United Kingdom's fish industry. In the United States, Title 50 applies to injurious wildlife and requires that fish and fish eggs imported into the United States be certified free of certain pathogens (Salmonid Import Regulations 1993). The European Union has directives that provide guidelines that must be met before aquaculture products are shipped into their sovereign territories to ensure protection against introduction of exotic diseases (Daelman 1996). The European Union also has an operative fish disease control service that requires that all aquaculture facilities be inspected and registered. Some countries, such as Japan, consider current aquaculture practices to be too diverse and extensive for implementation of an effective national disease control program (Wakabayashi 1996).

Quarantine is an approach to disease avoidance when fish are moved from one area to another. Fish should be isolated for a specific period before contact with a resident population. If disease develops in newly arrived animals, it can be dealt with more effectively and without exposing resident stocks.

Drastic measures such as eradication of an entire fish population are sometimes necessary to avoid exposing healthy fish to highly infectious, nonendemic disease agents. When contemplating eradication, the following factors should be considered: the economic and biological significance of the disease, whether the disease is indigenous to an area, and whether it can be adequately controlled through management or chemotherapeutics. In other words, is eradication worth long-term potential savings in fish and can the area be maintained free of the disease organism?

In 1989, an attempt was made to prevent establishment of the viral hemorrhagic septicemia virus (VHSV) in the United States by destroying adult salmonids as they returned to spawn at two sites in the Pacific Northwest where the virus had been found (Hooper 1989). This constituted the first confirmed VHSV occurrence outside Europe, so these drastic measures appeared to be justified. In subsequent years, however, the virus was found in different areas of the northwestern United States, so eradication did not solve the problem (Winton et al. 1991).

Destruction of lake trout and disinfection of contaminated hatcheries in the Great Lakes region of the United States and Canada in the late 1980s was apparently successful in eliminating epidermal epitheliotropic disease, a severe herpesvirus infection of juvenile lake trout (R. Horner, Illinois Department of Natural Resources, personal communication). Another large-scale fish destruction occurred in 1990 when trout brood stock at Jackson National Fish Hatchery (Wyoming) (Anderson 1991) and White Sulfur Springs National Fish Hatchery (West Virginia) (Cipriano et al. 1991) were killed because they were subclinically infected with *Renibacterium salmoninarum* (bacterial kidney disease [BKD]). These two facilities were major trout egg suppliers for many federal and state agencies, and the potential was present to spread BKD by egg distribution. Therefore, an administrative decision was made to destroy large numbers of valuable rainbow, lake, cutthroat, and brook trout. This decision was based on a highly sensitive enzyme-linked immunosorbent assay (ELISA) method for detecting very low numbers of *R. salmoninarum*. The fish were killed in spite of the fact that neither facility demonstrated any evidence of clinical BKD. Snieszko's theory on the application of fish health management practices comes to mind: "Was the cost and consequence of implementation greater than the value of what was saved?

Infectious disease prevention is best accomplished through avoidance whenever possible, but there are times when one must look beyond the host for a disease source. Some disease agents (helminths, for example) have complex life cycles that involve fish-eating birds, snails, or copepods, and the pathogen life cycle is broken by controlling nonfish vectors, a nearly impossible task. Another example of a disease agent that may have a nonfish vector is *Edwardsiella ictaluri,* the causative agent of enteric septicemia of catfish. This organism was found in intestines of 53% of cormorants and other fish-eating birds examined (Taylor 1992). It would be extremely difficult to prevent exposure of warmwater fish to these contamination sources, and, furthermore, it is not fully known whether birds actually contribute to dissemination of fish diseases.

Each disease outbreak must be considered individually, and rarely will a general policy be applicable; therefore, "avoiding exposure" decisions must be made using biological facts and common sense, with an eye to the overall good of the aquatic environment.

THE EXPOSING DOSE

As previously noted, total avoidance of infectious agents is the best way to prevent disease; however, this is not always possible or practical because many organisms that infect fish are normally free-living, facultative, and opportunistic pathogens. Fish and some pathogens can coexist without disease unless the animal's immune system or other defensive mechanisms are compromised.

The bacterium *Aeromonas hydrophila,* causative agent of motile *Aeromonas* septicemia (MAS), is a ubiquitous organism that occurs naturally in most fresh waters of the world. It is capable of living and proliferating in any water that has organic enrichment, usually causing disease only after a fish's resistance has been compromised by environmental stressors or when fish have suffered some mechanical or biophysical injury. The best way to prevent MAS is to reduce organic load and bacterial numbers in the water and to keep fish at a high state of resistance by environmental management. Prophylactic treatments after fish are handled will aid in healing superficial wounds and in reducing bacterial populations on the skin. In a normal, healthy fish population, a certain percentage of fish may have systemic *A. hydrophila;* however, disease occurs only when bacterial numbers overwhelm the fish's resistance.

Most fish, wild or cultured, normally have some parasites present on their gills or skin. As long as these parasites, either protozoans or monogenetic trematodes, are present in low numbers, there generally is no health problem; however, if water quality deteriorates to a point that fish are stressed, parasite numbers will increase. Parasite populations can be reduced with prophylactic chemotherapy, and good health reestablished with restoration of environmental quality.

Culling old brood stock and using younger fish for reproduction before they become heavily parasitized can reduce the effects of metazoan parasites. Using young largemouth bass as brood stock has proved successful in the southeastern United States, where bass tapeworm (*Proteocephalus ambloplitis*) larvae can be a problem if large numbers invade ovaries of adult fish, causing a reduction in egg production (W. A. Rogers, Auburn University, Alabama, personal communication).

Most infectious agents require specific levels of infective units before adversely affecting a host's health status. Therefore, if infectious organisms can be maintained at levels below disease threshold, losses will be reduced.

EXTENT OF CONTACT

Diseases that involve opportunistic (facultative) organisms are not considered in this particular discussion. The following remarks pertain to obligate pathogen–induced fish diseases (viruses, some bacteria, and a few parasites). In managing such diseases, initial consideration is given to the pathogen source as it exists in the culture system.

Pathogens must have a route of transmission from source to susceptible host, which may be by direct contact from fish to fish, via the food chain, through water, on nets, buckets, and other gear, or via reproductive products (Figure 1.7). Infected

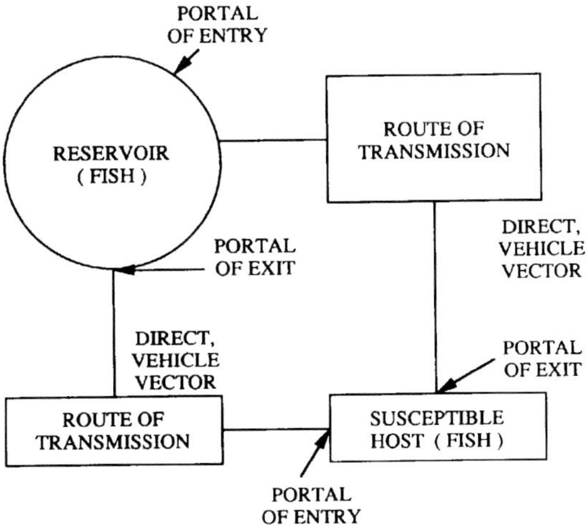

FIGURE 1.7. Typical transmission cycle for communicable disease agents. The direct, vehicle vector may be water, food, contaminated equipment, eggs, and so forth. (Adapted from Swan 1983.)

fish that survive can become pathogen carriers and a reservoir for disease agents. There are numerous examples of fish diseases (particularly viruses) in which epizootics are associated with a reservoir host. A pathogen must have a way to escape the host; in fish, this can be accomplished by shedding virus or bacteria across the gill membranes, through skin mucus, from skin lesions, in feces, urine, or via reproductive products. Internal metazoan parasites often must wait until the host fish is eaten by an intermediate or final host to continue its life cycle. Nematode and trematode parasite larvae that live in the eyes of some fish are released when the eye is destroyed.

Some pathogens, especially viruses, are released from a reservoir host during spawning via eggs, ovarian fluid, or milt. In some viral diseases, the number of virus-shedding fish actually increases during spawning time (infectious hematopoietic necrosis virus [IHNV]), which aids transmission (Mulcahy and Pascho 1985).

Bacteria and parasites can be transmitted to cultured fish populations through feed that contains raw wild fish flesh. Marine fish are an especially good reservoir source for these pathogens. Mycobacteria (acid fast staining bacteria) are transmitted to uninfected fish when uncooked fish flesh or viscera is incorporated into their feed. Transmission of mycobacteria was common in Pacific salmon hatcheries in the 1950s and 1960s when ground raw marine fish was incorporated into hatchery feeds. Another example of a parasite being transmitted by feeding contaminated fish flesh occurred in Florida when chopped Atlantic menhaden, infected with *Goezia* (a nematode that inhabits the stomach and intestinal wall) was fed to cultured striped bass fingerlings. The infected striped bass were then stocked into Florida's inland lakes, where the parasite was transmitted to resident tilapia and largemouth bass (W. A. Rogers, personal communication). Although virus transmission resulting from feeding raw fish has not been reported, most virulent fish viruses have been transmitted orally under laboratory conditions.

Although infectious disease-induced mortalities in wild fish populations are rare, high losses are much more common in cultured fish populations. Like other animals, fish are more susceptible to infection when crowded. As fish density increases, the rate of contact and pathogen transmission increases accordingly. In high-density fish populations, many epizootics have a potential for reaching catastrophic proportions. Most fish disease organisms do not manifest themselves in clinical disease until infected fish are placed in stressful conditions.

PROTECTION THROUGH SEGREGATION

Segregating fish according to species and age can reduce transmission of disease-producing agents because these organisms are not equally pathogenic to all fish species, all strains within a species, or to all ages within a susceptible species. Some fish disease–producing organisms lack host specificity, whereas others are more host specific. Non–host-specific organisms are generally facultative or free-living opportunistic parasites or bacteria. Some diseases, such as furunculosis (*Aeromonas salmonicida*), and most viral agents are generally more host specific. Brook trout are considered highly susceptible to *A. salmonicida* and should be reared completely segregated from carrier populations. Rainbow trout and brown trout are also susceptible to *A. salmonicida,* but to a lesser degree than brook trout, and may be considered for culture where this disease is endemic. Rainbow trout and some species of Pacific salmon are highly susceptible to IHNV; therefore, culture of a less susceptible salmonid species should be considered where this virus occurs.

Channel catfish are the most susceptible species to channel catfish virus disease (CCVD). Within channel catfish stocks, inbred strains are more susceptible to CCVD than are outbred strains (Plumb et al. 1975). In areas where CCVD is endemic, less-susceptible catfish species, blue catfish, can be cultured even though they are not totally CCV resistant. The same catfish susceptibility pattern exists for *E. ictaluri,* the bacterial causative agent of enteric septicemia of catfish (ESC) (Wolters et al. 1996).

Generally, young fish have greater pathogen susceptibility than do older fish, particularly with regard to viruses and some parasites. In the case of IPNV and IHNV of salmonids and CCV of catfish, fish less than 3 months old are highly susceptible, but disease effects diminish with increasing age. If these fish species can be reared in noncontaminated water until past the critical age of susceptibility, severe losses can be avoided. The same is true with *Myxobolus cerebralis,* a parasite that infects young trout when cartilage is still predominate in the skeleton. Susceptible fish should be reared in well or spring water free of wild fish. After fish have grown past a susceptible size or age, they can be transferred to less-protected waters.

Water temperature affects the disease-causing ability of some fish pathogens. Infectious hematopoietic necrosis virus normally occurs at a temperature of 10 to 17°C; thus, if fish can be reared in water warmer than 17°C, disease can be avoided. At elevated temperatures, however, trout may be more susceptible to other disease organisms and growth rate will be reduced. Channel catfish virus is most severe at temperatures higher than 25°C; therefore, if catfish are held at cooler temperatures, CCV can be avoided but growth rate will be reduced. Unfortunately, in most instances, optimum temperature for pathogens is also optimum for hosts.

THE PROBLEM OF NEW ARRIVALS

Newly arrived fish may bring pathogenic organisms with them, either in an active or carrier state, to which the resident population has not been exposed. Therefore, no new animals should be introduced into an existing population until reasonably certain they will not be detrimental. This rule applies to home aquaria, large aquaculture facilities, movement of fish from farm to farm, and interstate and international fish shipments.

Several precautions that can be taken to reduce potential for introducing disease with new arrivals are use of SPF eggs or familiarity with the health history of fish to be introduced: know if any serious disease problems occurred and, if so, what type treatment was given. If not treated before transport, newly arrived fish or eggs should be treated prophylactically with appropriate drugs to remove any external pathogens. If possible, new fish should be segregated (quarantined) from the resident population until shown to be disease free. Fish farmers must be extremely careful when replenishing brood stocks with outside sources to ensure that no new pathogens are introduced into their facility. This danger is especially prevalent if fish from wild populations with unknown disease histories are used to replace brood stocks.

Disease-free certification of fish or eggs, although not 100% reliable, is applicable to IPNV, IHNV, VHSV, and some salmonid bacterial diseases, but it is less applicable for warmwater fish diseases. Currently, there is no economical way to certify channel catfish free of CCV, although methods for detecting CCV DNA in fish tissue are emerging. When considering specific disease-free certification, one should refer to the *Blue Book: Suggested Procedures for the Detection and Identification of Certain Finfish and Shellfish Pathogens* for information about which diseases are certifiable in the United States (Thoesen 1994). To stop movement of disease agents, one must stop movement of infected animals. The prudent fish farmer must take every precaution available when acquiring new stock.

BREEDING AND CULLING

Domestication of wild animals has been the foundation of successful animal husbandry throughout the ages. Selection, culling, and cross-breeding has been vital in developing today's herds and flocks of domesticated animals. Culturing fish for food has been practiced for thousands of years in some areas of the world, especially in China. Comparatively few studies, however, have dealt with development of disease-resistant fish through selective breeding and genetic manipulation. In most genetic fish improvement experiments, objectives have been to increase growth rates and fecundity or to improve feed conversions; therefore, any disease resistance has resulted from happenstance.

Lack of available scientifically developed disease-resistant brood fish should not prevent the aquaculturist from improving his or her stocks. This can be gradually accomplished by selecting fish that are less affected by specific disease outbreaks. If a particular lot of brood fish routinely produces offspring that develop a specific disease year after year (for example, CCVD), replacing those brood fish should be strongly considered. Routine culling of inferior individual brood fish will also improve performance.

There is much interest in genetically improving fish for cultural purposes by gene manipulation, hybridization, and genetic engineering. Such studies may produce valuable culture animals in the future, but in the meantime, the aquaculturists should continue to improve present stocks by culling and selective breeding.

THE NUTRITIONAL BASIS OF HEALTH MAINTENANCE

Proper nutrition is essential for survival, growth, and reproduction of any animal species, and nutrient requirements of fish are similar to those of other animals (Lovell 1988). This applies to wild or cultured fish populations, although nutritional management in the two systems is entirely different. For wild fish populations, proper nutrition depends on the natural food chain and availability of primary nutrients. Nutrients can be supplemented in these populations by addition of organic or inorganic fertilizers to stimulate growth of macroflora (vascular plants), microflora (phytoplankton), zooplankton, and other invertebrate fish food organisms. Forage fish use these organisms as food and in turn are fed upon by predator species. A fish population's health depends on an available food source at any given position in the food chain.

In many parts of the world, traditional aquaculturists continue to use manure and/or ingredient-based nutrient sources to feed cultured fish. This practice causes increased levels of organic matter in semiliquid layers of unoxygenated sediments, greater water quality problems, poorer fish growth, higher feed conversion ratios, and higher incidences of infectious disease, especially stress-induced epizootics. More advanced aquaculture management practices advocate the use of manufactured feeds that contain proteins, fats, carbohydrates, fiber, vitamins, amino acids, and minerals.

A deficiency in any one of these nutrients can result in nutritionally related disease or lowered disease resistance. The amount of each nutrient needed for proper growth depends on the species and age of the fish being fed and the intensity of the culture system. Fresh feed of the highest quality available should always be used.

It has been shown that certain nutritional deficiencies can result in specific pathological conditions (Halver 1972). Injury to the spinal column of channel catfish (broken-back syndrome) results from a vitamin C deficiency. Lack of sufficient riboflavin can cause eye cataracts in rainbow trout and possibly in channel catfish. A niacin deficiency increases sensitivity to ultraviolet irradiation and "sun burn." Nutritional gill disease of trout is caused by a pantothenic acid deficiency and can progress into bacterial gill disease if the deficiency is not corrected. Also, severe anemia in channel catfish is linked to dietary deficiency of folic acid or presence of the folic acid antagonist pteroic acid (Butterworth et al. 1986).

Several researchers have shown that a deficiency of specific elements in a fish's diet can increase disease susceptibility and that megadoses may increase immune response. Channel catfish fed a vitamin C–deficient diet were more susceptible to *Edwardsiella tarda* and *E. ictaluri* (Durve and Lovell 1982), and megadoses of vitamin C (five times the recommended 60 mg/kg of diet per day) enhanced immune response (Li and Lovell 1985). These studies were conducted in aquaria, however, and when the same principles were applied to pond-reared channel catfish, vitamin C–deficient diets did not affect susceptibility, nor did megadoses of the vitamin enhance immunity (Liu et al. 1989).

The attributes of organic zinc sulfate versus inorganic zinc sulfate in a fish diet has received significant attention recently. Paripatananont and Lovell (1995) reported that channel catfish required less organic than inorganic zinc sulfate and that resistance to *E. ictaluri* increased with addition of organic zinc sulfate in the diet. Subsequently, Lim et al. (1996) reported that the zinc source was unimportant and did not change disease resistance of channel catfish.

Fish in poor nutritional condition (starved or lightly fed) may be more resistant to some infectious agents than are well-fed fish. Anecdotal observations indicate that channel catfish susceptibility to CCV is decreased following starvation for 2 weeks (unpublished). More recently, Kim and Lovell (1995) showed that channel catfish that were not fed at all during winter were significantly less susceptible to *E. ictaluri* than were fish fed every day or less frequently during the winter. Taking channel catfish off feed when ESC occurs has become a management practice throughout the catfish industry. Fish that continue to be fed suffer higher mortality than do either medicated or nonfed populations (D. Wise, Delta Fish Farming Center, Stoneville, Mississippi, personal communication). To what degree nutrition affects disease resistance in fish remains unclear.

Subtle effects of a nutritionally deficient diet are reduced growth, poorer feed conversion, and reduced fecundity, and all dietary factors have the potential to affect health status and disease susceptibility of fish (Lovell 1988). When water quality becomes critical with high organic fertility and low DO, reduced feeding can hasten a reversal of this trend; therefore, flexibility in feeding can be a valuable tool in aquaculture management.

ERADICATION, PREVENTION, AND CONTROL

The terms *eradication, prevention,* and *control* are defined as follows: *Eradication* is the complete elimination of a disease-causing agent from a facility or specific geographical region, and practical eradication is elimination of an agent from its reservoir of practical importance. *Prevention* is the avoidance of introduction of a disease-producing agent into a region or facility and/or the stopping of a disease process before it can become a problem. *Control* involves reduction of a problem to a level that is economically and biologically manageable and/or confinement of a problem to a defined area.

Eradication of a fish disease from a facility, watershed, or region is desirable but difficult to accomplish. To date, there are no reported examples of a fish disease agent being totally eradicated from a large geographical region (epizootic epitheliatropic disease of lake trout may be one exception). Practical eradication of IPNV, furunculosis, and bacterial kidney disease from individual fish hatcheries and farms in the United States have been reported. This was accomplished by removal of all fish and sterilization of facilities with chlorine or formaldehyde gas. These select facilities had closed water supplies with no indigenous fish populations to serve as disease reservoirs, and precautions were taken to prevent reintroduction of disease. Only SPF eggs were used to repopulate the facilities, and strict sanitation policies were established that included disinfection of equipment and fish transport trucks that accessed the culture area. A commitment to eradicate a specific disease must be balanced with an equal commitment to keep the facility disease free.

Disease prevention is primarily a farm management approach and should begin with construction of a culture facility. Disease prevention considerations should include site selection, development of water supply, and selection of fish to be stocked. Original fish stocks, as well as any subsequent fish brought onto the facility, should come from populations known to be free of obligate pathogens. When available, vaccination should be considered as a disease prevention tool. Sanitation practices are extremely important, and disinfection of nets, buckets, boots, and other equipment should be a routine practice whenever used in different culture units. This precaution will also help prevent spread of disease within a facility once it occurs.

Control is appropriate when nonobligate pathogens are involved or when effective chemotherapeutic and management practices are available. The objective of control is to reduce pathogen levels in an environment or host so that acute disease will not occur. This approach dictates that some fish will die while attempting to keep losses at a minimum. Chemotherapeutics are most successful when used in conjunction with elimination of environmental stressors and good health management practices.

Biological and economic constraints usually dictate what approach will be used when dealing with fish disease problems. Eradication may be the best approach when isolated cases of obligate pathogen infections are involved; however, destruction of entire animal populations remains controversial. Eradication verses control should be determined on a case-by-case basis, with strong consideration being given to economic consequences. Eradication is only feasible if a facility can be maintained free of a specific disease agent after sterilization.

VARIABLE CAUSES REQUIRE VARIABLE SOLUTIONS

Epizootiology is the study of disease patterns in animals other than humans and implies that the presence or absence of overt disease is not necessarily indicative of the presence or absence of a disease-causing agent. As previously discussed, many infectious fish diseases are caused by the complex interaction of host, pathogen, and environment. Although a host and pathogen must be present to cause infectious disease, environmental influences are often the trigger that elevates a subclinical infection to disease. Most pathogenic organisms can be identified, but environmental influences that precipitate disease are far more complex and difficult. Each disease situation is different and must be evaluated and dealt with on an individual basis.

When a fish disease problem arises, an environmental evaluation before and during the time of disease outbreak must be part of the diagnostic process (Schnurrenberger 1983a). Facts pertinent to establishing an accurate diagnosis are whether there had been a recent oxygen depletion or any other water quality problem noted, whether there was any color change in the water, whether heavy rains had occurred recently, whether mortality occurred suddenly or increased gradually, whether other fish had recently been introduced, and the nature of the current water quality characteristics. Determination of these factors will influence the type of management procedures or treatments to be used.

At the time of disease outbreak, it is imperative that a complete disease examination (necropsy) be performed to determine whether a single species of bacterium, parasite, or virus is involved or multiple pathogens are present. It must be emphasized that even after a pathogen has been found, necropsy must be completed or other contributing pathogens may go undetected. Different disease organisms, requiring different treatments, can produce very similar clinical signs; therefore, an incomplete necropsy may result in a premature and incorrect diagnosis and improper treatment. The key to treating multiple infections is first to identify and treat the primary disease-causing agent and then to deal with lesser disease agents.

DON'T JUST CURE, PREVENT

A health maintenance program should be in place when an aquacultural facility begins production, and health management issues should be addressed on a daily basis, not just when disease outbreaks occur. When disease does occur, managers often expect an immediate diagnosis and treatment prescription from a fish disease specialist. This can put pressure on a diagnostician to recommend a treatment that will correct the immediate problem without addressing long-range problems. Often, a chemical can be added to water and/or an antibiotic can be incorporated into feed, which will result in a positive response and cessation of mortality. The effect of these treatments may be temporary, however, and disease may reoccur unless a commitment is made to seek and eliminate all predisposing disease factors. Even though a fish farmer is more likely to accept chronic disease losses because they are not dramatic and occur over an extended period, these losses can be minimized through proper health maintenance. If a subacute disease occurs and great numbers of fish die in a few days, the farmer will be more inclined to take preventive steps.

Although preventive health maintenance is important in controlling fish disease, it is not a "cure all." Some fish diseases are less preventable through environmental management than are others, and some have no known prevention or treatment. No health maintenance program is perfect, but one that emphasizes disease prevention is well worth the effort economically, productively, and for the pure satisfaction of knowing that the best possible aquaculture techniques are being used.

THE LAW OF LIMITING FACTORS

The law of limiting factors plays an important role in our ecology as well as in health management of fish. Schnurrenberger (1983b) likened the law to the idea that a "chain is no stronger than its weakest link." Also, each major link consists of its own set of limiting factors, and a weak spot within a subset of factors can cause failure in the overall chain. If one considers the fish production cycle a chain, with one end being spawning adults and the other harvestable-size fish, many potential weak links exist, any of which could be considered a limiting factor. Water is essential for fish production, and components present in a given body of water make it either suitable or unsuitable to support a healthy fish population. Volume of water, temperature, pH, and other parameters must all be compatible with the fish species being cultured. If adequate water is available but its temperature is not suitable, temperature then becomes a weak link and water becomes a limiting factor in the production cycle.

Food quantity and quality are often limiting factors in unmanaged fish ponds. With fertilization, the base of the food chain in these ponds can be extended to support an increased standing crop, but if culture procedures become too intensive, other limiting factors may become involved. As a standing crop increases, supplemental feeding is required so that food will not again become a limiting factor. Supplemental feeding will increase the organic water load, and pond fertility will increase as a result of uneaten feed and the accumulation of metabolic wastes (feces, urine, ammonia, and so on). Decomposition of dead plants and uneaten feed remove oxygen from the water, and this oxygen loss may become a limiting factor. Accumulation of nitrite and CO_2 may also become limiting factors.

The law of limiting factors becomes acute as a fish culture system becomes more intensive. As fish density increases, more feed is required; more water is necessary to remove, dilute, and neutralize metabolic wastes; more efficient supplemental aeration is required; and fish stressors become more critical. At what point these limiting factors become detrimental to effective production must be determined and appropriate corrective measures taken. A limiting factor does not always have to be biological. It can be financial or dependent on a manager's ability to operate a facility properly and effectively.

Each fish species has its own set of factors that are necessary for reproduction, growth, and survival. Factors for fish survival lie between minimum requirements and maximum tolerances. Consideration must also be given to the synergistic effect one limiting factor has on another and how altering one factor to improve health may illuminate a previously unrecognized factor.

STAYING ON TOP OF THE OPERATION

Many people go into aquaculture without fully understanding what is required to initiate and maintain a successful operation. Fish farming, especially of warmwater species, is possibly the most demanding of all agricultural enterprises during certain seasons of the year. During critical periods in the production cycle, a 24-hour schedule must be maintained to stay on top of the operation. It is as important in aquaculture for a fish farmer to observe and evaluate routinely his or her culture procedures and facility as it is for cattle or swine farmers to inspect herds daily.

Although casual scrutiny of ponds, tanks, or raceways can be helpful in detecting possible health changes, specifics must also be considered. Operational procedures should be observed in such detail that any departure from good health management is immediately detected. The most obvious indicator of deteriorating fish health is a change (reduction) in feeding behavior, which can be affected by adverse environmental conditions and/or by infectious disease. If, on a given day, fish are eating actively, and the next day there is a noticeable reduction in feed consumption, the cause should be investigated. Water quality should be checked to determine dissolved oxygen, nitrite, or carbon dioxide concentrations, and a sample of fish should be examined for presence of infectious or parasitic disease. Some fish, such as channel catfish, normally reduce feed consumption in autumn as water temperatures drop, but when this happens at other times of the year, it may indicate a health problem.

Color changes in the water may indicate a potential water quality problem. When a pond turns from green (indicating a viable phytoplankton bloom) to dingy brown or suddenly clarifies, it could indicate algae are dying, or a pond "turn over" is occurring (good upper water mixes with unoxygenated deep water) and an oxygen depletion is likely. If the cause can be detected and corrected while occurring, stress on fish can be minimized.

Dissolved oxygen concentrations should be measured on a daily and nightly basis during critical periods, either by an individual or automatic DO recording device. Computerized oxygen monitoring systems have been developed that will register oxygen concentrations 24 hours a day and automatically turn on aeration when needed. Monitoring of oxygen is particularly critical in heavily stocked ponds that are receiving large amounts of feed. If, during a period of days and nights, a downward trend in oxygen concentration becomes evident, the dissolved oxygen may be ready to "crash" as a result of pond respiration, and nightly aeration should be initiated. Also, when fish swim erratically or lethargically just beneath the surface, gasp (pipe) at the surface, or move into shallow water, a manager should look for causes for this abnormal behavior, which could be attributed to poor water quality, toxicants, infectious disease, or parasitic infestation. When fish surface, aeration or introduction of fresh water should be initiated as soon as possible because a delay may cause fish to become so stressed they will not move to the aerated water.

Staying on top of the operation can prevent catastrophes from occurring in a fish population, but it requires diligence and regularly scheduled visits to the entire culture facility. During critical times, if several days or nights are allowed to pass without checking water quality, high fish losses can occur in a very short period. Also, sublethal stressful conditions may develop that can lead to infectious disease.

EARLY DIAGNOSIS

Because an early and accurate diagnosis when disease occurs is very important in maintaining healthy fish, it is imperative that an aquaculturist know what is normal. It is important to be able to recognize clinical signs of disease and to have affected fish examined as soon as possible. It is also important to be able to distinguish signs of oxygen depletion and chemical toxicants from those of infectious disease.

Few specific infectious fish diseases can be diagnosed solely on the basis of clinical signs; therefore, whenever possible, fish should be examined by a professional fish pathologist. In most cases of infectious disease, only a few fish die during early stages of an outbreak, followed by a gradual increase in the daily mortality rate. When the first sick or dead fish appear, suitable specimens consisting of moribund fish showing clinical signs typifying the condition should be examined.

Although an early and accurate diagnosis is important in nearly all infectious fish diseases, it is absolutely crucial in bacterial infections when use of medicated feed is indicated. If infection progresses to a degree that fish are no longer feeding, oral treatment is no longer an option. In any disease situation, the earlier the diagnosis, the sooner corrective measures or treatment can begin, thus reducing losses.

A DYNAMIC TEAM EFFORT

A dynamic fish health maintenance program must be flexible and coordinated, not a mixture of unrelated, unchanging procedures. Although objectives and principles for fish health maintenance programs can follow a general outline, each facility needs to design and implement a health plan that best meets the needs of the facility it serves.

There are three principal times when an animal food resource producer should be ready to discuss the establishment of a health maintenance program (Hudson 1983): (1) when experiencing a disease crisis, (2) when starting a new unit, and (3) when considering expansion with borrowed capital. The broad scope of basic knowledge and technology that applies to health maintenance of aquatic animals is usually beyond the expertise of any one individual. Therefore, a team approach, which uses input from a variety of specialists, is often indicated. In most geographical areas, State Cooperative Extension Services or state, federal, or university laboratories have professional, qualified fishery specialists available to assist farm managers in developing a good health maintenance program that meets their specific needs.

KEEPING CURRENT

All health management decisions should be based on the most current and accurate information available. Shell (1993) quoted an old folk-saying, "It's not what I don't know that hurts me. It's what I know that's not true that hurts me." In the past 25 years, major advances in aquaculture have been made in breeding, genetics, nutrition, disease identification, treatment, and processing. During this time, however, few new drugs or chemicals have been developed to aid in disease treatment and/or prevention. Because of constraints imposed by the U.S. Food and Drug Administration, there has actually been a reduction in number of drugs that can be used on food fish in the United States. This trend has also been observed in many regions of the world. The reduction in approved chemotherapeutics is in part due to technological advances that have been made in identifying hazardous characteristics of drugs previously thought to be safe. It is, therefore, imperative that a producer stay current about which drugs and chemicals are approved for use on food fish.

A fish farmer can stay current in aquaculture management techniques by being active in professional or trade organizations on a local, regional, state, national, and international level and by reading literature published by these organizations. Extension fishery specialists and agricultural agents have access to pamphlets, brochures, and newsletters with accurate, up-to-date information. If modern fishery managers are to maximize production facilities, they must keep pace with technological advances being made in the field.

MAINTAINING A CLEAN ENVIRONMENT

Everyone recognizes that an accumulation of trash, excrement, decaying animals, and so forth in a terrestrial environment is associated with unhealthy conditions. Why should we expect it to be different in the aquatic environment? Clearly it is not. Generally, when visiting private and governmental aquacultural facilities around the world, it is often possible to identify facilities that have the most disease problems on the basis of their appearance. This is not to say that a "spit and polish" facility will never have disease problems, but they tend to have fewer and less severe disease outbreaks. Water is an excellent medium for transmitting disease agents; therefore, sick or dead fish should be removed to reduce pathogen reservoirs. In *E. ictaluri*–infected catfish ponds, areas where dead fish were allowed to accumulate had significantly higher pathogen concentrations in the water than where carcasses had not accumulated (Earlix 1995). Maintaining clean areas around ponds and controlling vegetation at water's edge will not allow sick or dead fish to go unnoticed. If dead fish are allowed to remain in a culture unit, there is also the possibility that scavengers will carry infected carcasses to other ponds, thus spreading the disease. Accumulation of feces, uneaten feed, and other organic detritus is a major problem in ponds, raceways, and recirculating systems and contributes significantly to water quality degradation and can serve as a substrate for facultative pathogens. Removal of excreta and bottom detritus from water will help improve and maintain water quality. These problems can be avoided and/or corrected by improving feed quality and by regular cleaning and periodic draining, drying, and when necessary, disinfecting of holding facilities. Removal of accumulated sediment can be accelerated by drying the pond bottom and adding lime to the surface and tilling (Boyd and Pippopinyo 1994). Increasing soil pH to 7.5 to 8.0 with calcium hydroxide or calcium carbonate will enhance respiration and reduce pond sediment.

Cleaning and disinfection of seines, nets, buckets, tanks, and other equipment will reduce transmission of many pathogens. Maintaining a clean aquaculture facility will pay dividends by providing a safer workplace, reduced disease incidence, increased production, and generally healthier fish, all of which translates into more efficient production.

THE HIGH-RISK CONCEPT

Aquaculture is a high-risk endeavor because of its dynamic relationship to environmental change. Because some aspects of fish culture present higher risk factors than others, management efforts should concentrate on areas that include production and financial losses; namely disease, growth rate, feed quality and conversion, water quality, and product marketability. Water quality is a major factor in rearing healthy fish, and if not suitable, fish will not feed or grow well regardless of feed quality, and potential for disease becomes more prevalent. Consequently, major emphasis must be placed on water quality in any aquaculture health maintenance program.

Infectious disease can also be classified as a major risk factor, but risks will differ from farm to farm depending on water type and quality, species and strain of fish being reared, and general management practices, and so forth. Disease problems also change from year to year. The aquaculturist should be aware of high-risk factors, practice preventive health maintenance that addresses these areas, and at the same time not lose sight of lesser risk factors and their potential for causing problems.

RECORD KEEPING AND COST ANALYSIS

Well-organized records are basic to a successful aquaculture business. In addition to production records that include stocking and feeding rates, feed conversion, and harvest totals, each fish production facility should maintain a set of health records that include mortality patterns and dates, infectious disease incidences and treatments, and treatment results. Size, water volume, and water quality characteristics of each production unit should also be included in health records. After several growing seasons, a manager may be able to ascertain from these records whether a certain disease is likely to occur and what conditions predispose fish to a specific disease. This information may help prevent disease outbreaks by timely prophylactic treatments or other remedial actions. Detail and extent of record keeping may vary with size and intensity of an operation; however, records should be kept by every facility regardless of size.

Accurate records are important when evaluating health maintenance procedures because without them, an economic feasibility and cost-effective evaluation is impossible. When considering the economic feasibility of a procedure, cost, efficacy, and benefits must be considered, and the most cost-effective (not necessarily the cheapest) procedure selected.

When initiating fish health management practices, a long-range, cost-effective analysis should be made. Initial financial outlay may be high, but over time, good health management will pay for itself in reduced mortality and optimal growth. A cost-to-benefit ratio (C:B) should be calculated for each health management procedure, and benefits must outweigh cost except under unusual circumstances.

REFERENCES

Anderson, D. E. 1991. BK episode (epitaph), Jackson National Fish Hatchery, a 1988 case study [abstract]. In: *16th Annual Eastern Fish Health Workshop*. Martinsburg, WV, June 11–13.

Austin, B., and D. A. Austin. 1987. *Bacterial Fish Pathogens; Diseases in Farmed and Wild Fish*. Chichester, UK: Ellis Horwood Ltd.

Barton, B. A. 1997. Stress in finfish: past, present and future—a historical perspective. In: *Fish Stress and Health in Aquaculture*, edited by G. K. Iwama, A. D. Pickering, J. P. Sumpter, and C. B. Schreck, 1–33. Society for Experimental Biology Seminar Series 62. Cambridge, UK: Cambridge University Press.

Boyd, C. E. 1990. *Water Quality in Ponds for Aquaculture*. Auburn, AL: Alabama Agricultural Experiment Station, Auburn University.

Boyd, C. E. 1995. *Bottom Soils, Sediment, and Pond Aquaculture.* New York: Chapman & Hall.

Boyd, C. E., and S. Pippopinyo. 1994. Factors affecting respiration in dry pond bottom soils. *Aquaculture* 120:283–293.

Butterworth, C. E., Jr., J. A. Plumb, and J. M. Grizzle. 1986. Abnormal folate metabolism in feed-related anemia of cultured channel catfish. *Proceedings of the Society for Experimental Biology and Medicine* 181:49–58.

Cipriano, R. C., C. E. Starliper, and J. D. Teska. 1991. Case study on the investigation of salmonids at the White Sulfur Springs National Fish Hatchery for *Renibacterium salmoninarum* [abstract]. *16th Annual Eastern Fish Health Workshop,* Martinsburg, WV, June 11–13.

Daelman, W. 1996. Animal health and the trade in aquatic animals within and to the European Union. *Review in Scientific Technologies* 15(2):711–722.

Durve, V. S., and R. T. Lovell. 1982. Vitamin C and disease resistance in channel catfish (*Ictalurus punctatus*). *Canadian Journal of Fisheries and Aquatic Science* 39:948–951.

Earlix, D. 1995. Host, pathogen, and environmental interactions of enteric septicemia of catfish. Auburn, AL: Auburn University. Doctoral dissertation.

Hajek, B. F., and C. E. Boyd. 1994. Rating soil and water information for aquaculture. *Aquacultural Engineering* 13:115–128.

Halver, J. 1972. *Fish Nutrition.* New York: Academic Press.

Hill, B. J. 1996. National legislation in Great Britain for the control of fish diseases. *Reviews in Scientific Technology* 15(2):633–645.

Hooper, K. 1989. The isolation of VHSV from chinook salmon at Glennwood Springs, Orcas Island, Washington. *Fish Health Section/American Fisheries Society News Letter* 17(2):1.

Hudson, R. S. 1983. A dynamic, coordinated entity. In: *Principles of Health Maintenance,* edited by P. R. Schnurrenberger and R. S. Sharman, 12–20. New York: Praeger Press.

Kim, M. K., and R. T. Lovell. 1995. Effect of overwinter feeding regimen on body weight, body composition and resistance to *Edwardsiella ictaluri* in channel catfish, *Ictalurus punctatus. Aquaculture* 38:237–246.

Klontz, G. W. 1973. *Fish Health Management.* College Station, TX: Texas A & M Sea Grant Program, Texas A & M University.

Li, Y., and R. T. Lovell. 1985. Elevated levels of dietary ascorbic acid increase immune response in channel catfish. *Journal of Nutrition* 115:123–131.

Lim, C., P. H. Klesius, and P. L. Duncan. 1996. Immune response and resistance of channel catfish to *Edwardsiella ictaluri* challenge when fed various dietary levels of zinc methionine and zinc sulfate. *Journal of Aquatic Animal Health* 8:302–307.

Liu, P. R., J. A. Plumb, M. Guerin, and R. T. Lovell. 1989. Effect of megalevels of dietary vitamin C on the immune response of channel catfish *Ictalurus punctatus* in ponds. *Diseases of Aquatic Organisms* 7:191–194.

Lovell, T. 1988. *Nutrition and Feeding of Fish.* New York: AVI, Van Nostrand Reinhold.

McDonald, G., and L. Milligan. 1997. Ionic, osmotic, and acid-base regulation in stress. In: *Fish Stress and Health in Aquaculture,* edited by G. K. Iwama, A. D. Pickering, J. P. Sumpter and C. B. Schreck, 119–144. Society for Experimental Biology Seminar Series 62. Cambridge, UK: Cambridge University Press.

Meyer, F. P., J. W. Warren, and T. G. Carey. 1983. *A Guide to Integrated Fish Health Management in the Great Lakes Basin.* Special publication no. 83-2. Ann Arbor, MI: Great Lakes Fishery Commission.

Mulcahy, D., and R. J. Pascho. 1985. Vertical transmission of infectious hematopoietic necrosis virus in sockeye salmon *Oncorhynchus nerka* (Walbaum): isolation of virus from dead eggs and fry. *Journal of Fish Diseases* 8:393–396.

Ney, J. J. 1978. A synoptic review of yellow perch and walleye biology. In: *Selected Coolwater Fishes of North America,* edited by R. L. Kendall. Special publication no. 11. Bethesda, MD: American Fisheries Society.

Paripatananont, T., and R. T. Lovell. 1995. Responses of channel catfish fed organic and inorganic sources of zinc to *Edwardsiella ictaluri* challenge. *Journal of Aquatic Animal Health* 7:147–154.

Piper, R. G., I. B. McElwain, L. E. Orme, J. P. McCraren, L. G. Gowler, and J. R. Leonard. 1982. *Fish Hatchery Management.* Washington, DC: U.S. Department of the Interior, Fish and Wildlife Service.

Plumb, J. A., O. L. Green, R. O. Smitherman, and G. B. Pardue. 1975. Channel catfish virus experiments with different strains of channel catfish. *Transactions of the American Fisheries Society* 104:140–143.

Plumb, J. A., J. M. Grizzle, and J. deFigueiredo. 1976. Necrosis and bacterial infection in channel catfish (Ictalurus punctatus) following hypoxia. *Journal of Wildlife Diseases* 12:247–253.

Roberts, R. J. 1989. Fish Pathology. London: Baillière Tindall.

Rohovec, J. 1979. Review of international regulations concerning fish health. In: *Proceedings from a Conference on Disease Inspection and Certification of Fish and Fish Eggs.* Publication No. ORESU-W-79-001. Corvallis, OR: Oregon State University Sea Grant College Program.

Salmonid Import Regulations. 1993. Injurious wildlife: Import of fish and fish eggs. Department of the Interior, Fish and Wildlife Service. *Federal Register 50* C.F.R. Part 16.

Schnurrenberger, P. R. 1983a. Variable causes require variable solutions. In: *Principles of Health Maintenance,* edited by P. R. Schnurrenberger and R. S. Sharman, 54–63. New York: Praeger Publishers.

Schnurrenberger, P. R. 1983b. The law of limiting factors. In: *Principles of Health Maintenance.* Edited by P. R. Schnurrenberger and R. S. Sharman, 23–34. New York: Praeger Publishers.

Schnurrenberger, P. R., and R. S. Sharmon, Editors. 1983. *Principles of Health Maintenance.* New York: Praeger Publishers.

Shell, E. W. 1993. *The Development of Aquaculture: An Ecosystems Perspective.* Auburn, AL: Alabama Agricultural Experiment Station, Auburn University.

Simon, R. C., and W. B. Schill. 1984. Tables of sample size requirements for detection of fish infected by pathogens: three confidence levels for different infection prevalence and various population sizes. *Journal of Fish Diseases* 7:515–520.

Snieszko, S. F. 1958. Natural resistance and susceptibility to infections. *The Progressive Fish-Culturists* 20: 133–136.

Snieszko, S. F. 1973. Recent advances of scientific knowledge and development pertaining to diseases of fishes. *Advances in Veterinary Science and Comparative Medicine* 17:291–314.

Swan, A. I. 1983. The extent of contact. In: *Principles of Health Maintenance,* edited by P. R. Schnurrenberger and R. S. Sharman, 35–42. New York: Praeger Publishers.

Synopsis of potential aquaculture species. 1984. Auburn University, AL: Department of Fisheries and Allied Aquacultures. Xeroxed.

Taylor, P. W. 1992. Fish-eating birds as potential vectors of *Edwardsiella ictaluri. Journal of Aquatic Animal Health* 4:240–243.

Thoesen, J. C., Editor. 1994. *Blue Book: Suggested Procedures for the Detection and Identification of Certain Finfish and Shellfish Pathogens.* Bethesda, MD: Fish Health Section, American Fisheries Society.

Thorburn, M. A. 1996. Apparent prevalence of fish pathogens in asymptomatic salmonid populations and its effect on misclassifying population infection status. *Journal of Aquatic Animal Health* 8:271–277.

Tucker, C. S., and M. van der Pfloeg. 1993. Seasonal changes in water quality in commercial channel catfish ponds in Mississippi. *Journal of the World Aquaculture Society* 24:473–481.

Walters, G. R., and J. A. Plumb. 1980. Environmental stress and bacterial infection in channel catfish, *Ictalurus punctatus* Rafinesque. *Journal of Fish Diseases* 17:177–185.

Wakabayashi, H. 1996. Importation of aquaculture seedlings to Japan. *Reviews in Scientific Techniques* 15:409–422.

Wedemeyer, G. A. 1996. *Physiology of Fish in Intensive Culture Systems.* New York: Chapman & Hall.

Winton, J. R., W. N. Batts, R. E. Deering, R. Brunson, K. Hooper, T. Nischzoua, and C. Stehr. 1991. Characteristics of the first North American isolates of viral hemorrhagic septicemia virus [abstract]. *Proceedings of the Second International Symposium on Viruses of Lower Vertebrates,* Corvallis, OR.

Wolters, W. R., D. J. Wise, and P. H. Klesius. 1996. Survival and antibody response of channel catfish, blue catfish, and channel catfish female X blue catfish male hybrids after exposure to *Edwardsiella ictaluri. Journal of Aquatic Animal Health* 8:249–254.

2 Epizootiology of Fish Diseases

Understanding the dynamic aquatic environment and its role in fish health is imperative to management of infectious diseases. Fish respond in concert with, and quickly to, environmental changes that affect disease susceptibility and overall general health. An effective fish health maintenance program will continually take into account interaction between a fish population and its environment.

EPIZOOTIOLOGY

In a broad sense, *epizootiology* refers to the study of infectious diseases of animals including spread of pathogens, mode of infection, and control. Relative to epizootiology, the following terms are important:

Disease: Disease is a deviation from normal or good health; not necessarily implying cause of deviation. Disease may be a condition resulting from infectious agents, nutritional deficiencies, toxicants, or environmental factors or may be genetic.

Communicable disease: A communicable disease results from multiplication, replication, or reproduction of the causative organism in a host and the likelihood that the organism can be transmitted (communicated) to other hosts.

Epizootic (epidemic): An epizootic is a disease outbreak among animals, other than humans, in a specified and localized area. To be classified as an epizootic, an abnormal number of animals must be infected.

Enzootic (endemic): Enzootic indicates the continuing presence of a causative disease organism in a localized area, such as a pond, raceway, hatchery, river, or fish population but is not necessarily obvious as clinical disease.

Infection: Infection is the presence of a pathogen in a host that might or might not be diseased. Fish may be infected and, thus, carriers of the pathogen, without becoming diseased. Many fish disease organisms are enzootic in the aquatic environment and will only cause disease when conditions are favorable.

There are two basic types of organisms involved in communicable disease in terms of host relationship: *obligate pathogens* and *nonobligate (facultative) pathogens*. Obligate pathogens must have a host or intermediate host to reproduce and survive indefinitely in nature. Examples of fish obligate pathogens are all viruses, bacterial pathogens *Mycobacterium marinum* (Mycobacteriosis) and *Renibacterium salmoninarum,* and the parasitic protozoan *Ichthyophthirius multifiliis*. Facultative or nonobligate pathogens can live and multiply in a host or live freely, deriving nutrients from organic matter in water, mud, or living host. *Aeromonas hydrophila* (motile Aeromonas septicemia) is an example of a facultative pathogen. The fungus *Saprolegnia* spp. is saprophytic and derives nutrients from living or dead organic material.

SEASONAL TRENDS OF FISH DISEASES

Fish diseases generally occur seasonally and tend to fluctuate with temperature changes, presence of young susceptible fish in a population, and environmental conditions that affect immunity and natural resistance (Meyer 1970; Plumb 1976; Warren 1991). In tropical climates, only minor differences are evident in numbers of fish disease outbreaks throughout the year; however, in temperate and colder climates, seasonal fluctuations can be noted, particularly in warmwater species (Figure 2.1). Disease incidence in cultured and wild fish populations in the southern United States is low from November through February, increases during March through June as waters begin to warm, decreases during the warm summer months, and rises again in autumn. A similar disease pattern can be noted as one moves northward, but an increase in disease occurs later in the spring and earlier in the fall. Fish that live in waters with more constant temperatures, such as trout or tropical fish, appear to have seasonal diseases when, in reality, disease is more related to age class or stage of development.

Incidence of disease rises in spring because natural resistance and fish immunity are lower in late winter and early spring. Also, naturally occurring antimicrobial substances in the blood are low at this time, making fish more susceptible to infection by facultative disease-causing organisms (Snieszko 1958). Bly and Clem (1991) demonstrated that catfish immunity may be compromised during winter and that fish will not respond positively until water temperatures rise to 18 to 28°C in late spring or early summer. However, the immunologically favorable water temperature also coincides with optimum growth temperature for many fish pathogens.

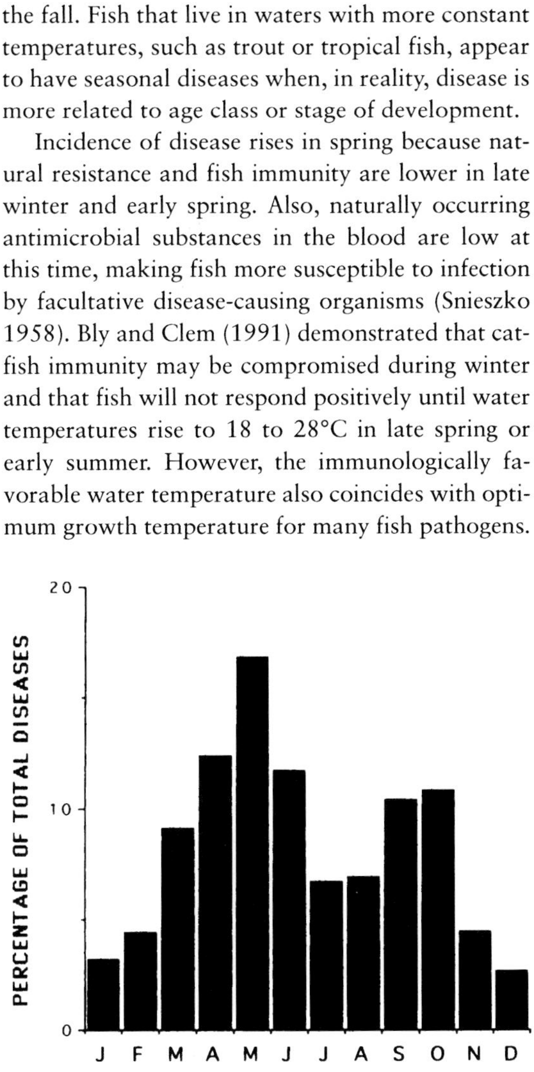

FIGURE 2.1. Monthly distribution of 6698 disease outbreaks in Alabama from 1991 through 1996. Data compiled from diagnostic records at the Southeastern Cooperative Fish Disease Laboratory, Auburn University, Alabama, and the Alabama Fish Farming Center, Greensboro, Alabama. (Data from Alabama Fish Farming Center, courtesy of William Henstreet.)

Changes in most water quality variables appear to be related to seasonal periodicity of phytoplankton abundance in channel catfish ponds (Tucker and van der Pfloeg 1993). Most water quality problems occur during summer and can be correlated with elevated water temperatures, solar radiation, accumulation of organic matter, increased nutrient input, and peak primary pond production and standing crop. Stressful conditions precipitated by these events during summer months may actually be partially responsible for a delayed disease increase in autumn.

FACTORS IN DISEASE DEVELOPMENT

When an infectious outbreak occurs, factors that dictate disease severity are (a) source of infection, (b) mode of transmission, (c) portal of entry, (d) virulence of the pathogenic organism, and (e) resistance of the host. If one of these factors can be altered through management, disease may be eliminated or its impact greatly reduced.

Source of Infection

An infectious disease requires a pathogen source, which may include dead or moribund fish, carrier fish that show no clinical signs of disease, contaminated eggs from infected brood stock, or infected water supplies. Contaminated feed can be a source of bacterial and parasitic agents, especially if it contains uncooked fish flesh or viscera. Transmission of many pathogens can be prevented by disinfection of equipment, filtration of water, adjustment of stocking densities in infected waters, removal of dead and moribund fish, use of seed fish from specific pathogen free (SPF) brood stocks, and use of proper feed.

Mode of Transmission

The mode by which fish disease is transmitted is closely related to, and some times inseparable from, the source of infection. Water, because it contains fish metabolites and waste products, often provides an ideal high nutrient environment in which potential fish pathogens can grow and proliferate, thus serving as a pathogen source as well as a primary mode of transmission.

Eggs produced by disease-carrying brood stock are often responsible for vertical transmission of viral and some bacterial agents from one generation to another. Also, intermediate hosts such as birds, snails, and crustaceans are important in the life cycles of some fish parasites and can serve as vehicles for disease organisms. Humans often contribute to transmission of fish diseases by mechanically transporting infective organisms in water or on fish transport equipment, nets, boots, and other utensils. Management practices can help disrupt disease transmission by use of prophylactic treatments, control of intermediate hosts, use of water filtration, water treatment with ozone or ultraviolet light, or use of a fish-free water source.

Portal of Entry

Each disease organism has an optimal point of entry to the host, and if entry points are identified, management may help prevent infection. Common entry points are the intestine, gills, and skin. In fish, the mucous layer that covers the epithelium provides protection against some pathogens. If this mucous layer is compromised, underlying epithelium is exposed to bacteria or parasites present in the water and, as a result, the skin, gills, or intestines become portals of entry for the pathogen. Also, parasites may penetrate the mucous layer and epithelium, making fish vulnerable to invasion by viruses and bacteria. Proper handling and use of prophylactic treatments will often reduce or prevent epithelial infections.

Virulence of the Pathogenic Organism

Virulence is a measure of an organism's ability to cause disease and may range from low to high. Facultative bacteria can vary greatly in virulence from location to location. *A. hydrophila,* a facultative bacterium in water, can be highly virulent to fish in one area, whereas an isolate from another body of water can be avirulent (not pathogenic). When *A. hydrophila* is isolated from diseased fish, it is usually more virulent than when isolated from water (de Figueiredo and Plumb 1977). It is more likely that a severe disease outbreak will occur when highly virulent pathogen strains are present. Although it is difficult to manipulate virulence of pathogenic organisms in nature, management can help prevent severe disease outbreaks by ensuring that environmental conditions are more favorable to the host than to the pathogen.

Natural Resistance of the Host

Natural resistance occurs when a host has an inherent ability to subdue a pathogen to the point that clinical disease will not occur. A fish's natural resistance to disease is extremely important in deterring infection (Chevassus and Dorson 1990). Intrinsic factors that influence a fish's natural resistance to disease are nonspecific phagocytic activity of neutrophils and macrophages, tissue integrity, and nonspecific serum components (interferon, complement, and so on). Extrinsic influences such as nutritional well-being and environmental conditions can, however, adversely affect these traits. Also, natural resistance is related to species, strain, and age of fish. Natural resistance of adult fish is often reduced during, or before, the spawning season, when some fish stop feeding and energy is diverted into reproductive activities. In temperate zones, older fish tend to have reduced resistance in spring owing to overwintering and presence of young fish that have not yet acquired natural resistance. Taking advantage of natural resistance is an integral part of fish health management and can be enhanced by culturing more disease-resistant strains.

HOST–PATHOGEN RELATIONSHIP

That a basic host–pathogen relationship exists is a fundamental concept in the epizootiology of fish diseases. A pathogen may be present in the environment or host, but as long as the host remains in good physical condition and its natural resistance remains high, clinical disease is less likely to occur. Disease organisms tend to cause infections in fish when there is an environmentally induced host–pathogen imbalance. This imbalance can be manifested by long-term nutritional problems and/or deterioration of the aquatic environment to the degree that a fish's natural resistance is compromised and "infection" progresses into "disease." Through good management, a proper host–pathogen balance can be better maintained and clinical disease outbreaks kept to a minimum, resulting in better growth and higher survival. There are, however, some pathogens that cause fish disease even though there is a balance in the host–pathogen relationship.

DEGREE OF INFECTION

Fish diseases are categorized according to disease severity, mortality and morbidity rates, and clinical signs present and are described as either acute, subacute, chronic, or latent (Figure 2.2). The mortality pattern is related to the severity of the causative agent or degree of infection and is helpful in determining whether the cause of death is due to toxicants, adverse water quality, infectious agents, or nutritional deficiencies.

Acute

Acute mortality is marked by a high death rate in a short period (hours to 1–2 days). In most instances, truly acute mortalities are associated with events such as water quality problems, oxygen depletions, or toxicants. In rare instances when infectious agents are involved, overt clinical signs (ulcerative lesions, severe hemorrhage, and so forth) may be minimal or totally lacking.

Subacute

A subacute infection is less severe than an acute one, and mortality will accelerate during a 3- to 4-day period, lasting for several weeks with clinical signs present. Some highly virulent organisms that are capable of causing acute or subacute mortality include channel catfish virus (CCV), infectious hematopoietic necrosis virus (IHNV), and infectious pancreatic necrosis virus (IPNV). On rare occasions, bacterial pathogens (*Flavobacterium columnare* [*Flexibacter columnaris*]) (columnaris) and some parasites (*I. multifiliis*) can cause subacute mortality. Environmental conditions may also be associated with subacute mortality.

Chronic

Many infectious fish diseases exhibit chronic characteristics that include maximum clinical signs that can persist for weeks. In these infections, the mortality rate is gradual with an undetectable peak,

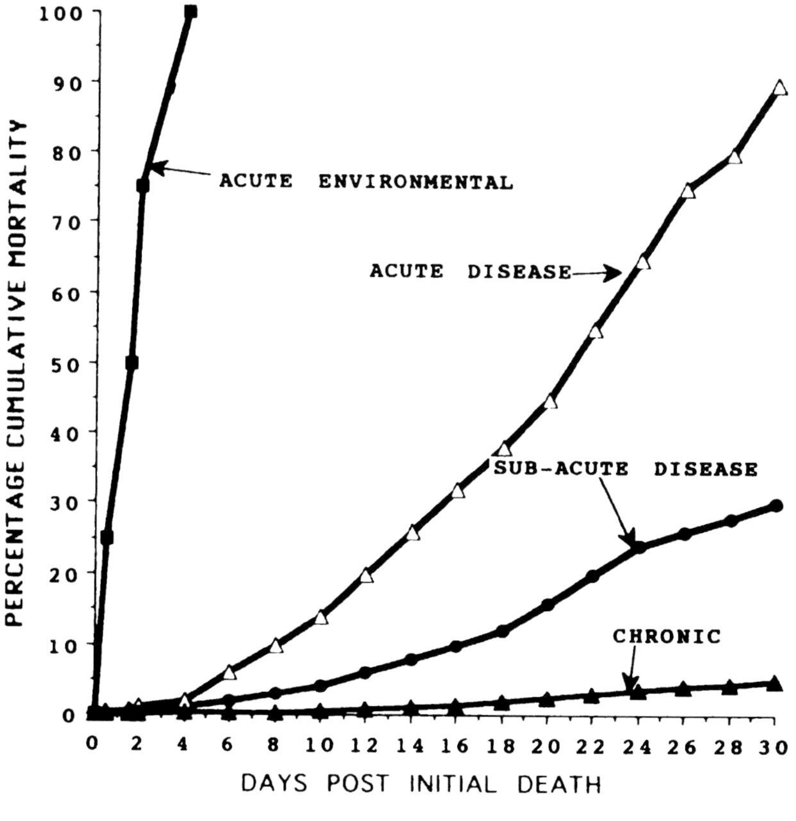

FIGURE 2.2. Theoretical patterns of mortality in fish. Acute patterns are usually associated with environmental problems and, rarely, infectious agents. Subacute mortalities may be associated with environmental problems or infectious agents. Chronic mortalities are frequently associated with infectious agents.

but cumulative mortality can be high. Examples of chronic fish infections are motile *Aeromonas* septicemia (*Aeromonas* spp.), columnaris, furunculosis (*Aeromonas salmonicida*), bacterial kidney disease (*R. salmoninarum*), and most infections caused by protozoan parasites.

Latent

In a latent condition, a disease organism is present but its host exhibits no overt clinical signs of disease and little or no mortality occurs. Latency is usually associated with viral infections (for example, IPNV and IHNV), obligate bacterial pathogens (*A. salmonicida, R. salmoninarum,* and protozoans *Myxobolus cerebralis* [whirling disease]).

PRIMARY VERSUS SECONDARY INFECTION

Fish diseases may result from a *primary* or *secondary infection.* Primary infections are those caused by primary pathogens, usually obligate pathogens, and can result in morbidity or death without the presence of additional factors. Primary pathogens are capable of causing disease in healthy fish even under ideal environmental conditions; however, any adverse environmental factor can synergize infection to a more serious level (Wedemeyer 1996 and 1997). Channel catfish virus, *Edwardsiella ictaluri,* and *R. salmoninarum* are examples of primary pathogens that infect fish.

Generally, a secondary pathogen is a free-living, facultative, saprophytic, opportunistic organism that infects a host when its defenses have been compromised by other pathogens or stressors, thus causing a secondary infection. These infections usually result from mechanical or physiological injury, disease that renders the host weakened, poor management practices, or environmental stressors such as low oxygen levels, temperature shock, and excessive or improper handling. There are many secondary pathogens that affect fish, including *A. hydrophila, F. columnare,* and *Saprolegnia* spp.

A minor disease can progress to a severe condition if good management procedures are not practiced during disease outbreaks. For example, if diseased fish in a pond are moved to a confined holding facility for treatment, the act of moving the fish will probably be more detrimental and stressful than any benefit derived from treatment.

REFERENCES

Bly, J. E., and L. W. Clem. 1991. Temperature-mediated processes in teleost immunity: in vitro immunosuppression induced by in vivo low temperature in channel catfish. *Veterinary Immunology and Immunopathology* 28:365–377.

Chevassus, B., and M. Dorson. 1990. Genetics of resistance to disease in fishes. *Aquaculture* 85:83–107.

de Figueiredo, J., and J. A. Plumb. 1977. Virulence of different isolates of *Aeromonas hydrophila* in channel catfish. *Aquaculture* 11:349–354.

Meyer, F. P. 1970. Seasonal fluctuations in the incidence of disease in fish farms. In: *A Symposium on Diseases of Fishes and Shellfishes,* edited by S. F. Snieszko, 21–29. Publication no. 5. Bethesda, MD: American Fisheries Society.

Plumb, J. A. 1976. An 11-year summary of fish disease cases at the Southeastern Cooperative Fish Disease Laboratory. *Proceedings Annual Conference of the Southeastern Association of Game and Fish Commissioners* 29:254–260.

Snieszko, S. F. 1958. Natural resistance and susceptibility to infections. *The Progressive Fish-Culturist* 20:133–136.

Tucker, C. S., and M. van der Pfloeg. 1993. Seasonal changes in water quality in commercial channel catfish ponds in Mississippi. *Journal of the World Aquaculture Society* 24:473–481.

Warren, J. W. 1991. *Diseases of Hatchery Fish.* 6th ed. Washington, DC: U.S. Fish and Wildlife Service.

Wedemeyer, G. A. 1996. *Physiology of Fish in Intensive Culture Systems.* New York: Chapman & Hall.

Wedemeyer, G. A. 1997. Effects of rearing conditions on the health and physiological quality of fish in intensive culture. In: *Fish Stress and Health in Aquaculture,* edited by G. K. Iwama, A. D. Pickering, J. P. Sumpter and C. B. Shreck, 35–71. Society for Experimental Biology Seminar Series 62. Cambridge, UK: Cambridge University Press.

3 Pathology

Pathology, the study of disease, includes functional and morphological alterations and reactions that develop in a living organism as a result of injurious agents, nutritional deficiencies, or inherited characteristics. Pathology terms used in description and interpretation are essentially the same for fish as for higher vertebrates (Roberts 1989; Ferguson 1989; Cotran et al. 1989). Fish pathology is too vast a subject to cover thoroughly in one chapter; therefore, the objective of this chapter is to introduce basic pathology terms and present examples of the most common pathological changes that occur in fish.

PATHOLOGY TERMS

The following terms are often used when describing a disease or considering the disease process.

Etiology: Etiology is the study of the cause of disease.

Etiological agent: An etiological agent (pathogen, nutritional deficiency, and so on) is that which causes a particular disease; for example, channel catfish virus (CCV) is the etiological agent of channel catfish virus disease (CCVD).

Pathogenesis: Pathogenesis is the sequence of events by which a disease develops, including method of infection, establishment of the agent in the host, and injury inflicted on the host.

Pathogenicity: This term refers to a pathogen's characteristic ability to gain access to susceptible tissue, become established, proliferate, avoid host defense mechanisms, and cause injury and disease to the host.

Virulence: Virulence refers to degree of pathogenicity of a disease agent: avirulent (not virulent), low, medium, or highly virulent. These terms are usually related to a specific group, species, or strain of pathogenic organism.

Lesion: A lesion is any gross, microscopic, or biochemical tissue abnormality. The lesion can be either functional or morphological.

Clinical Signs: Clinical signs are characteristics or conditions associated with disease and can be seen during a macroscopic external examination of the affected animal. Clinical signs include behavioral and gross morphological changes but do not include microscopic, biochemical, or internal changes.

Histology: Histology is the study of tissue on a microscopic level. Representative species from several important fish groups have been studied and their normal histology described. These include channel catfish (Grizzle and Rogers 1976), striped bass (Groman 1982), salmonids (Yasutake and Wales 1983), Atlantic cod (Morrison 1987–1993), and walking catfish (Chinabut et al. 1991).

Histopathology: Histopathology is the microscopic study of injuries to tissues and cells.

CAUSE OF DISEASE

Disease can be caused by an etiological (specific cause) or a nonetiological (contributing cause) agent. Etiological agents can be classified as either *inanimate* or *animate*. Inanimate etiological agents are factors without life of their own and can originate within a host (*endogenous*) or outside of a host (*exogenous*). Endogenous, inanimate factors are those associated with genetic and/or metabolic disorders of the host. Exogenous, inanimate agents include trauma, temperature shock, electrical shock, chemical toxicity, and dietary deficiencies. These etiological agents may serve as sublethal stressors that predispose fish to infectious disease. Animate etiologies are living communicable infectious agents, which include viruses, bacteria, fungi, protozoa, helminths, and copepods.

Nonetiological causes of disease are characterized as *extrinsic* (from outside the body) or *intrinsic* (within the body). Extrinsic factors are usually associated with environmental conditions or dietary problems, and intrinsic factors include age, gender, heredity, and fish species. Both fish species and strain of fish are important because all are not equally susceptible to a specific disease organism. Feed quality, water quality factors, and water temperature extremes can be classified as either etiological or nonetiological extrinsic factors and can contribute to infectious disease.

PATHOLOGICAL CHANGE

Pathological changes in individual fish can aid in recognition and identification of disease. Using definitions from Roberts (1989) and Ferguson (1989), the following are brief descriptions of some pathological changes that can occur in diseased fish.

Circulatory Disturbances

Circulatory disturbances are those abnormalities (lesions) that reflect injury to the vascular system. These disturbances may be anemia, hemorrhage, edema, hyperemia, congestion, emboli, or telangiectasis.

• Anemia

Anemia is a reduction of red blood cells that results from cell lyses during infection, presence of toxicants, or lack of erythrocyte production. Anemia may result in packed red blood cell volumes (hematocrit) falling from a normal range of 25% to 40%, depending on fish species, to less than 15%.

• Hemorrhage

Hemorrhage is the escape of blood from vessels and can occur in the skin or mucous membranes, within serous cavities, or between cells of any tissue or organ. Hemorrhages are generally classified according to size or degree of area affected. *Petechiae* are hemorrhages that measure less than 2 mm in diameter. They are common in bacterial and viral infections and typically occur in the skin or on the surface of visceral organs. *Ecchymotic* hemorrhages are larger and more diffuse than petechiae and measure 3 mm or larger in diameter. *Paintbrush* hemorrhage refers to large affected areas that appear as though they were splashed with red paint. *Hematomas* (seldom seen in fish) are localized hemorrhages that result in a blood-filled swelling.

Hemorrhages can be caused by ruptured blood vessels or increased vessel porosity resulting from the presence of infectious bacteria, viruses, or toxicants. In fish, overall effects of hemorrhaging are mild if the process is gradual because hematopoiesis is able to compensate for loss of blood cells. If, however, hemorrhaging is acute, as often is the case in infectious disease, anemia can occur and will be characterized by pale gills and internal organs and a low hematocrit (percent packed red blood cells).

• Edema

Edema is an abnormal accumulation of fluids (not whole blood) in body cavities, interstitial spaces of tissues, and organs that results in swelling. Four categories of edema occur in fish: (1) *Ascites* is fluid accumulation in the body cavity caused by dysfunction of the heart, liver, or kidney and gives affected fish a "pot-bellied" appearance. This fluid may range from colorless to pale yellow, cloudy, or bloody depending on the etiology. Turbidity is caused by high numbers of bacteria. (2) *Hydropsy* results from fluid in muscle, interstitial, and connective tissues and gives fish a swollen appearance. When pressure is applied to the skin, an indentation will remain (pitting on pressure). (3) *Lepidorthosis* is the accumulation of fluid in scale pockets, causing scales to be pushed away from the body, making its surface rough. (4) *Exophthalmia* is the accumulation of fluid, either behind or in the eye, causing a protrusion of the eyeball that gives affected fish a "popeyed" appearance.

Edema indicates an imbalance of hydrostatic pressure or improper osmolarity of the blood, increased capillary permeability, vascular obstruction, or organ dysfunction. These conditions are normally associated with chemical toxicants; viral, bacterial, or parasitic agents; or mechanical injury. Edema caused by either injury or disease may predispose fish to further infection, as edematous fluids provide a medium for bacterial growth. The effects of edema can be harmful and long lasting.

• HYPEREMIA AND CONGESTION

Hyperemia and congestion involve an excessive volume of blood within the vessels; however, mechanisms of development differ. Hyperemia is an active engorgement of blood in the vascular bed and is often associated with inflammation. In contrast, congestion is a passive accumulation of blood in vessels caused by an impairment of blood flow. Hyperemia and congestion cause affected tissue to appear red with possible swelling, and blood vessels in the fins and mesenteries become more prominent.

• TELANGIECTASIS

Telangiectasis is the bulging of a blood vessel in fish gills that is equivalent to an aneurysm in higher vertebrates. Telangiectasis is passive and reversible, whereas an aneurysm is the permanent swelling of an artery. Telangiectasis may be caused by mechanical injury, toxicants, viruses, bacteria, bacterial toxins, parasites, and in some cases, nutritional deficiencies.

• EMBOLISM

An embolism is an obstruction, often gas bubbles, in the vascular system that causes an interruption of blood flow. Gas bubbles in blood vessels caused by a supersaturation of gas (usually oxygen or nitrogen) in the water are common in fish (Bouck 1980). When dissolved gas in water exceeds 100% saturation (supersaturation) for an extended period, gas bubbles may form in the blood, producing "gas bubble disease." Fish affected by this disease may have microscopic or macroscopic gas bubbles in the operculum, gills, eyes, and fins and around the head. Gas bubble disease in cultured fishes occurs when deep wells are used as a water supply or when cool water is heated just before it flows to a culture unit. Cool water has a higher gas saturation point, and when water is warmed, it will become supersaturated. Wild fish populations living just below high dams may be affected by gas bubble disease as a result of water tumbling over spillways and entrapment of gas in the plunge pool.

Cellular Degeneration

Degeneration is a broad term that refers to a process in which cells or tissues deteriorate, usually with a corresponding degree of functional inhibition. In the process of cellular degeneration, cells go through biochemical alterations and functional abnormalities that finally result in morphological change. The degree and progress of degeneration vary with type and number of cells affected, nature of the injurious agent, intensity, and duration of injury. Degeneration of cells and tissues can be reversible, changing from a regressive to a progressive process, and affected cells can be restored to normal if the injurious cause is removed before cell death occurs.

Degeneration can result from: (1) deficiency of a critical material (oxygen or a vital nutrient); (2) lack of an energy source that affects metabolism; (3) mechanical, thermal, or electrical injury; or (4) accumulation of an abnormal substance in cells caused by viral, bacterial, or parasitic pathogens and their toxins, toxic chemicals, nutritional imbalance, mild irritants, and others. Types of cellular degeneration found in fish include cloudy swelling, fatty change, and mucoid degeneration.

• CLOUDY SWELLING

Cloudy swelling or cellular swelling often results from bacterial toxemia and is the earliest microscopic indication of cellular degeneration. Cells affected with cloudy swelling are enlarged, and cytoplasm has a homogeneous, hazy appearance identified by histology.

• FATTY DEGENERATION

Fatty degeneration (or fatty change) is the result of an accumulation of lipids, most commonly in the liver. This change is usually caused by infectious disease, nutritional imbalance, hypoxia, anemia, and some toxicants in which the liver is characterized by a pale yellowish to gray appearance. Although macroscopic evidence of fatty degeneration may eventually be noticed, early indications are microscopic.

• NECROSIS

Necrosis is the death of cells following cell degeneration in a living animal and is the final stage in irreversible degeneration. Characteristics of necrotic tissue can include the following: (1) paler than normal color, (2) loss of tensile strength (tissue becomes friable and easily torn), (3) cheesy or pasty consistency, and (4) an unpleasant odor. Necrosis should not be confused with postmortem change, which is cellular deterioration after death of an animal.

Necrosis can be caused by trauma, biological agents (viruses, bacteria, fungi, and parasites), chemicals, or an interrupted blood supply to a specific area. Three types of necrosis are (1) liquefaction, (2) coagulation, and (3) caseation. When a cell dies, enzymes within the cell cause *liquefaction necrosis,* which appears as a semisolid or fluid mass. When this occurs in the epidermis or exposed muscle, the necrotized tissue is sloughed off, leaving an open ulcerative lesion that can be invaded by facultative pathogens present in the water. Liquefaction necrosis is the most common type of fish necrosis. *Coagulation necrosis* refers to an area in which the gross and microscopic nature of affected tissue is still recognizable and is associated with injury caused by several types of toxicants. *Caseation necrosis* results when a pathogenic organism produces cheesy, whitish material in a lesion, but this type of necrosis is uncommon in fish.

Disturbances of Development and Growth

Several growth changes (excessive, deficient, or abnormal growth in cells, tissues, or organs) can occur in fish in response to an infection, toxicant, or other irritant. Generally, growth disturbances involve either an abnormal number of cells, cells of abnormal size in an organ or tissue, or a combination of the two.

• Atrophy

Atrophy refers to a decrease in size of a mature body part or organ due to decreased size or number of cells. Atrophy is a slow process and results from starvation or malnutrition (most common cause), lack of adequate blood supply, or chronic infection.

• Hypertrophy

Hypertrophy is an increase in size or volume of a body part or organ due to an increase in size of individual cells. Hypertrophy usually results from an increased demand for function but can also be initiated by an infectious agent such as lymphocystis virus.

• Hyperplasia

Hyperplasia refers to an increase in size of a body part or organ due to an increase in number of cells. One form of hyperplasia is characterized by an increased thickness of gill lamellae epithelium because of infection or exposure to a continuous mild irritant. Chemical water pollutants, as well as some fish viruses or bacteria, cause formation of hyperplastic lesions, particularly of the integument. Enlargement of a thyroid gland to form a goiter in fish is an example of hyperplasia.

Inflammation

Inflammation, which can be acute or chronic, is an aggressive vascular and cellular defensive response of living animal tissue to a sublethal injury that helps to minimize the effect of an irritant or pathogen on tissue. When injury or infection occurs, fluids from the vascular system and lymphocytes, neutrophils, macrophages, and other blood components migrate to the injured area. The accumulated fluids and cells attempt to dilute, localize, destroy, and remove any irritant or infectious agent and to stimulate replacement and repair of injured tissue. Inflammation can be caused by viruses, bacteria, parasites, trauma, heat, irradiation, and toxicants. The "cardinal signs" of inflammation in higher vertebrates are (1) redness (hyperemia), (2) swelling, (3) heat, (4) pain, and (5) loss of function. How exactly these apply to fish is not clearly understood, but in most instances, the process is similar to that of other animals.

Acute inflammation is classified according to type of exudate involved: serous, fibrinous, hemorrhagic, catarrhal, purulent, or granulomatous. The exudates are named according to histological appearance, and all types are seen in fish to varying degrees.

• Serous

Serous inflammation is exudation of clear fluid from the vascular system and serosa surfaces in response to a mild irritant or infection. A blister is a good example of serous inflammation, although it is seldom seen in fish.

• Fibrinous

Fibrinous inflammation occurs when large amounts of fibrin escape from blood vessels and form a clear clot when exposed to air. This type of inflammation usually occurs in the peritoneal cavity of fish and is associated with hyperemia. Fibrinous inflammation, along with hemorrhage, has been observed in body cavities of tilapia with systemic amoeba infections (unpublished).

• PURULENT

Purulent inflammation typically contains a combination of necrotic cells, neutrophils, and proteolytic enzymes in the exudate. Generally, fish do not produce pus, although in some bacterial diseases focal infections do produce purulent exudate. Purulent inflammation does occur in furunculosis of trout and has occasionally been found in motile *Aeromonas* septicemia.

• CATARRHAL

Catarrhal inflammation is the excessive production of mucus on the epitelium surfaces of the skin, gills, and digestive tract. The exudate can be clear, cloudy, or pink with a consistency of fluid to mucoid. The fecal cast of trout infected with infectious hematopoietic necrosis virus contains catarrhal exudate. Catarrhal exudate forms on the gills and skin of fish infected with *Ichthyophthirius multifiliis.*

• HEMORRHAGIC

Hemorrhagic inflammation is usually characterized by the presence of large numbers of erythrocytes and other blood components on organ surfaces or in exudate. Hemorrhagic inflammation will generally be diffused on serous or mucous membranes and can be caused by viruses, bacteria, parasites, or toxicants. Erythema, a reddish appearing skin condition, is usually associated with hemorrhagic inflammation.

• GRANULOMATOUS

Granulomatous inflammation is usually a chronic condition associated with several microbial fish infections and is characterized by a predominance of macrophages or related cell types. *Edwardsiella ictaluri* causes granulomatous inflammation in channel catfish, but tissue does not form a typical granuloma, although the spongy, soft tissue surrounding skull lesions is granulomatous. Other pathogens, *Mycobacterium* spp. and *Photobacterium damsella* subsp. *piscicida* (bacteria) and *Ichthyophonus hoferi* (fungus), cause granulomatous inflammation, but these infections often progress into macroscopic granulomas. Acute granulomatous inflammation occurs initially in internal organs and muscle tissue. Macrophages and other inflammatory cells form a layer resembling epithelium around the irritant, and as the lesion ages, increased numbers of fibroblasts and lymphocytes appear. The resulting granulomas are white to yellow and have a cheesy to hard consistency. Tissue affected with granulomatous inflammation will have small white spots, and the surface may have a rough sandpaper texture. In some cases, several granulomas may be connected by fibrous connective tissue to form a large encapsulated hard nodule. Once inflammation has disappeared, a central necrotic area remains that may retain the causative agent.

REFERENCES

Bouck, G. R. 1980. Etiology of gas bubble disease. *Transactions of the American Fisheries Society* 109:703–707.

Chinabut, S., C. Limsuwan, and P. Kitsawat. 1991. *Histology of the Walking Catfish, Clarias batrachus.* Canada: International Development Research Center.

Cotran, R. S., V. Kumar, and S. L. Robbins. 1989. *Robbins Pathologic Basis of Disease.* 4th ed. Philadelphia: W. B. Saunders Company.

Ferguson, H. W. 1989. *Systematic Pathology of Fish: A Text and Atlas of Comparative Tissue Responses in Diseases of Teleosts.* Ames, IA: Iowa State University Press.

Grizzle, J. M., and W. A. Rogers. 1976. *Anatomy and Histology of the Channel Catfish.* Auburn, AL: Auburn University, Alabama Agricultural Experiment Station.

Groman, D. B. 1982. *Histology of the Striped Bass.* Bethesda, MD: American Fisheries Society.

Morrison, C. E. 1987–1993. *Histology of the Atlantic Cod, Gadus morhua: An Atlas,* vols. I–IV. Ottawa, Canada: Department of Fisheries and Oceans.

Roberts, R. J. 1989. *Fish Pathology,* 2nd ed. London: Baillière Tindall.

Yasutake, W. T., and J. H. Wales. 1983. *Microscopic Anatomy of Salmonids: An Atlas.* Resource Publication 150. Washington, DC: U.S. Fish and Wildlife Service, Department of the Interior.

4 Disease Recognition and Diagnosis

An accurate fish disease diagnosis involves recognizing the presence of an infectious agent or other etiology, its role in the disease process, environmental factors that may have influenced the process, and how the problem can be corrected. Aquaculturists may be capable of recognizing some infectious diseases, but usually it is necessary to consult a fish health professional for positive identification.

DISEASE RECOGNITION

Although infectious agents play a major role in fish health, other important factors must be identified and considered when a disease diagnosis is made (Meyer and Barclay 1990; Noga 1996). Actual identification of a parasitic, bacterial, or viral agent causing disease may not be difficult; however, knowing how to deal with the problem and preventing it from reoccurring are often more difficult. History, mortality pattern, and water quality characteristics provide vital information for an accurate diagnosis and treatment regimen (Lasee 1995). Once an etiological agent is identified and a treatment strategy established, steps must be taken to prevent a reoccurrence of disease.

History

History of a fish population exhibiting morbidity or mortality should include origin, size, age, number, and species of fish involved, previous disease history, type of holding facility, how long fish with clinical signs of illness have been noted, behavior and feeding patterns, and type of treatment administered. It is also essential to collect as much environmental and water quality data as possible, which should include water temperature; dissolved oxygen, carbon dioxide, ammonia, and nitrite concentrations; any recent change in water color, clarity, or flow; stocking density; weather conditions just prior to mortality; agricultural activities in the area; and size of culture unit. All of these data can be vital in diagnosing and treating disease and, more importantly, in preventing reoccurrence.

Mortality Pattern

A mortality pattern may indicate whether cause of death was due to an infectious agent, poor water quality, or presence of a toxicant. Generally, infectious disease outbreaks are marked by a gradual acceleration in morbidity and mortality, unless an extremely virulent pathogen is involved (see Figure 2.2). Infectious disease agents seldom, if ever, result in "overnight" mass mortality but most often produce subacute or chronic mortality patterns resulting in cumulative losses of 30% to 40% or less. They seldom reach 100%; however, when highly virulent agents (some viruses and bacteria) infect highly susceptible fish, 90% to 100% cumulative mortality will occasionally occur. When the majority of a fish population dies overnight or in a 24-hour period, oxygen depletion and other water quality changes, chemical toxicants, or other environmental causes should be strongly considered. Mortalities due to nutritional deficiencies are usually very protracted.

Clinical Signs

Clinical signs is a term used to describe clinical findings that include behavioral, physical, or other pathological changes that can be observed. When combined with gross internal pathological changes, these findings can be helpful in diagnosing fish diseases. In most cases, it is not possible to diagnose a specific disease or determine a specific etiological agent based solely on clinical signs because very few are specific (pathognomonic) for a particular disease (Table 4.1). In this text, emphasis is placed on recognition of clinical signs that are readily apparent by gross inspection.

• Behavior

Initial reaction of fish to many diseases is a cessation of feeding activity; therefore, any sudden drop in feed consumption should be investigated. Stress of any type can cause fish to "go off feed (Table 4.1)." Diseased fish may swim into shallow water, swim lethargically at the surface, lie listlessly on the pond or tank bottom, float downstream, swim erratically, or rub against underwater structures as if trying to eliminate a skin irritant. Some disease agents cause afflicted fish to swim in a longitudinal spiral or "tail chase" in a circle. Crowding around a water inlet to take advantage of oxygenated water may be a sign of diseased gills or low dissolved oxygen concentrations. Sick fish may gather in a tight school or ride high in the water.

• External

External clinical signs of infectious fish diseases are varied and generally not distinctive for a specific disease; however, in a few instances, infection is characterized by specific types of lesions (Table 4.1). Most obvious external clinical signs are in-

TABLE 4.1. Summary of behavioral and external clinical signs and internal lesions of fish and etiological agent that could be associated with them

Clinical sign or lesion	Etiologic agent
Behavior	
Reduce or stop feeding	Viral, bacterial, parasitic, or environmental factor
Lethargic swimming	Viral, bacterial, parasitic, fungal, or water quality
Crowding to freshwater	Bacteria or parasites on gill, or low oxygen
Spinning, erratic swimming	Viral, parasitic, or toxicants
External	
Gills necrotic	Bacterial, parasitic, or fungal
Gills with excess mucus	Bacterial, parasitic, environmental, or nutritional
Gills are pale	Viral, bacterial, or nutritional
Skin with excess mucus	Parasitic or environmental
Depigmented areas in skin	Bacterial or parasitic
Dark skin pigmentation	Viral, bacterial, nutritional, or eye parasites
Hemorrhage, erythemia	Viral, bacterial, parasitic, or toxicants
Frayed, eroded, erythemia in fins	Bacterial, parasitic, or mechanical or physiological disorder
Exophthalmia, hemorrhaged opaque eyes	Viral, bacterial, parasitic, or gas supersaturation
Ulcerative, necrotic lesions	Bacterial or parasitic
Hydropsy	Bacterial, viral, or metazoan parasites
Lepidorthosis	Bacterial, viral, or parasitic
Enlarged abdomen (fluid)	Viral, bacterial, or parasitic
Growths, nodules, raised white spots on skin	Viral, parasitic, neoplasms, or fungal
Internal gross pathology	
Clear, yellow fluid in peritoneal cavity	Viral; bacterial rarely
Cloudy, bloody fluid in peritoneal cavity	Bacterial
Viscera generally hyperemic	Viral or bacterial
Liver pale, friable	Viral, bacterial, or nutritional
Liver mottled, hyperemic; petechiae present	Viral or bacterial
Liver, kidney, spleen with white nodules	Parasitic, bacterial, or neoplasms
Spleen dark red, swollen	Viral or bacterial
Kidney dark red, pale, or soft	Viral or bacterial
Intestine erythemic, flaccid, void of food	Viral or bacterial
Kidney with large white, irregular nodules	Bacterial, parasitic, or fungal
Muscle with petechiae; necrotic pustules	Viral or bacterial
Muscle with white, yellow nodules	Metazoan parasites

flammation and/or hemorrhage of fins, skin, or head; frayed fins; open necrotic and ulcerative lesions at any location on the body; lepidorthosis; and excessive or total lack of mucus. Excessive mucus production on the skin will give fish a grayish appearance. Edema, an enlarged abdomen, growths on the body surface, presence of yellow or black spots on the skin, prolapsed anus, and exophthalmia are all clinical signs of disease. Varying degrees of erythema, hyperemia, or hemorrhage may also be observed on the body and fins.

Gills are adversely affected by viruses, bacteria, and parasites. They can become frayed and necrotic (gray or white), pale (anemic), swollen as a result of hyperplasia, or can produce excessive mucus that interferes with oxygen absorption from the water and causes affected fish to crowd around water inflow.

Skin lesions of diseased fish are diverse and usually nonspecific. Initially, they often appear as slightly raised or swollen, depigmented areas where the epithelium has not been totally necrotized or sloughed. In advanced stages of disease, or when pathogens severely affect the skin, the epithelium and dermis can be totally missing (necrotized and sloughed into the water), leaving exposed musculature. Necrotic lesions accompanied by areas of deep ulceration can occur in muscle or on opercles, head, or mouth. Margins of ulcerative lesions are often white (necrotic) and surrounded by inflammation, hemorrhage, or erythema in the epithelium. In the center of such lesions, muscle bundles (myomeres) are often distinguishable.

Lesions caused by infectious agents often appear first on fins, which may become extensively necrotic, hyperemic, or pale. Injury may progress until soft tissue of the fins is destroyed, leaving only hard spines, or fins can be completely missing. Scale pockets may show lepidorthosis, or the body may appear swollen as a result of muscle edema (hydropsy) and presence of fluid in the abdominal cavity (ascites). Growths on the body and fins may appear as cottony material (fungus); solid tumors of various sizes, colors, and textures; or as parasites embedded in or beneath the skin. Embedded helminthic parasites will appear as yellow, white, or black spots in the skin, muscle, or organs.

In fish that have systemic bacterial infections, the anus is often prolapsed, swollen, and red; however, this condition should not be confused with the swollen vent of a gravid female that is ready to spawn. Protruding (exophthalmic), hemorrhaged, or opaque eyes may be associated with an accumulation of fluid (edema) or inflammatory exudate behind the eye caused by bacterial infection, helminth parasites, or supersaturation of gas in the water. Opaqueness of the eye may also be due to a trematode or nematode infestation, dietary deficiency, or bacterial infection.

Mortalities caused by most acute water quality changes or toxicants generally affect all species of fish present and will occur during a short period. These fish generally show no external lesions or clinical signs other than flared gill covers and gaping mouth at death. However, lesions such as brown blood (methemoglobin anemia) caused by nitrites can occur. Some chemical toxicants will cause hemorrhagic inflammation of the skin and fins. Fish suffering from oxygen depletion will gasp at the surface and have dark red gills.

Fish deformities are common and usually involve the head and/or spine. Juvenile trout infected with the parasite *Myxobolus cerebralis* (whirling disease) can develop head deformities and/or curvature of the spine. Spinal deformities may result in scoliosis (lateral curvature) or lordosis (forward or vertical curvature), which may be congenital, a result of dietary related (vitamin C deficiency) factors, or infections (mycobacteriosis and infectious hematopoietic necrosis virus [IHNV]).

Gross Internal Lesions

After observing and recording behavioral and clinical signs of affected fish, they must undergo necropsy (postmortem examination). Internal gross lesions may help determine whether an infection is viral, bacterial, or parasitic (Table 4.1). Clear, straw-colored fluid in the abdominal cavity usually, but not always, indicates a viral infection. Bloody and cloudy fluid and/or large white pustular or variable size granulomatous lesions on visceral organs usually indicate a bacterial infection. Some bacterial infections will cause the visceral cavity to have an intense putrid odor even in fresh fish. The presence of small white or yellow uniform-sized cysts in internal organs is indicative of metazoan parasites.

A uniform hyperemia or hemorrhage in the viscera is indicative of a viremia or bacterial septicemia, in which case the spleen will usually be dark red and enlarged and the liver may be friable, soft and pale, or mottled with petechiae. The kidney is often swollen and soft. Intestines are usually devoid of food and may contain white or bloody mucoid material, and the intestinal wall is often flaccid and erythemic. Internal organs may appear brown owing to presence of methemoglobin anemia caused by nitrite toxicity. Some toxicants cause necrosis and chronic exposure to them can result in several types of lesions.

DISEASE DIAGNOSIS

In diagnosing fish diseases, a small number of animals are examined and information derived from these individuals is used to determine the health status of an entire population; therefore, it is essential that fish being examined are representative of the affected population. Moribund specimens exhibiting clinical signs are best for examination and yield most dependable results. Fish that have been dead more than a few minutes should not be used for diagnosis unless they are put on ice immediately upon death, especially when water temperatures are warm. Dead fish that have lost normal skin color, have pale and soft gills, cloudy eyes, and an offensive odor have most likely been dead too long for proper necropsy. When a fish dies, parasites tend to leave the gills and skin, bacteria escape the gut and invade the visceral organs and body cavity, and any bacteria present in the water may invade the skin. Also, apparently healthy fish caught by hook and line are usually not suitable for necropsy.

Etiological agents of fish diseases should not be determined from clinical signs alone because many viral, bacterial, or parasitic diseases have similar clinical signs. Conversely, clinical signs of ichthyophthiriasis (parasitic), enteric septicemia of catfish, and columnaris (bacterial) can provide reasonably accurate clues to the etiological agent. In either scenario, a complete laboratory examination (necropsy) of affected fish should be carried out. A diagnosis without proper examination often leads to an erroneous diagnosis; improper treatment; and a waste of time, money, and fish. Diagnostic procedures should include examination for external parasites, attempted isolation of possible viruses and/or bacteria, and pathogen identification. If a pathogen is found on initial examination, all phases of the necropsy should be completed because multiple infections and successful therapy will require that all pathogens be addressed.

Parasitic Diseases

This book does not directly address parasitic fish diseases, but a necropsy is not complete without considering these pathogens (Thoesen 1994; Lasee 1995; Mitchum 1995; Noga 1996; Hoffman 1998). It is essential to use alcohol or some other disinfectant when isolating bacteria; therefore, a parasitic examination should be made before aseptic necropsy. It is better to kill fish by pithing rather than an overdose of anesthetic, which may affect external parasites. When examining fish for external parasites, material is taken from gills and skin including fins and placed on a glass slide, a drop of water added, and a cover slip applied. The slide is examined first under light microscopy at low magnification (40–100×) and then at higher magnification (400–440×). Most protozoa, myxosporidians, helminths (monogenetic trematodes), and parasitic copepods are identifiable within these magnifications. Phase contrast is beneficial in detecting and identifying parasites.

Internal fish parasites are detected by dissection and examination of internal organs and tissues including the muscle, gut, liver, spleen, and kidney. Many metazoan parasites are visible as small white, yellow, or black cysts, and their presence can be verified in wet mounts. Tissues may contain larval or adult nematodes, trematodes, or cestodes. Muscle dwelling parasites, primarily larval trematodes, can be detected by dissection of cysts that are examined in wet mounts. Histopathological examination is valuable in detecting many tissue-dwelling parasites. Infestation severity and extent of injury to the host can be ascertained by this method.

Viral Diseases

Detection and diagnosis of virus diseases that occur in fish require special expertise (Wolf 1988; Sanz and Coll 1992; Thoesen 1994). Detection methods include isolation of the virus in tissue culture, electron microscopic examination of tissue, and use of serological and molecular procedures. In spite of available biotechnology and the fact that some viruses are known only via electron microscopy, isolation of viral pathogens in tissue culture continues to be the usual method for fish virus diagnosis. Type of tissue used for virus assay is based on size and age of fish: when assaying fry, the entire fish is homogenized;

only visceral organs are excised for virus assay of fingerlings; and parts or whole organs are used from larger fish. When checking brood fish for viruses during egg collection, ovarian fluids are assayed. Collected tissue is homogenized in a stomacher or tissue grinder, diluted in phosphate-buffered saline at 1:5 or 1:10, centrifuged, and the supernatants decontaminated by passing them through a filter membrane with 0.22 to 0.45 µm porosity. An alternative method for sanitizing homogenates is to incubate them in a solution of phosphate-buffered saline containing a high concentration of antibiotics for 24 hours (Thoesen 1994).

Viruses cannot be detected by light microscopy because they are generally less than 300 nm in size, and many are less than 100 nm; therefore, living cells grown in vitro are used to detect their presence. Viruses infect cells and cause cell injury known as cytopathic effect (CPE), which can be seen with light microscopy. There are many permanent fish cell lines that have been developed for virus diagnosis and research. A cell line derived from the same species or one phylogenetically close to the fish species being assayed for virus should be used when available. However, many known fish viruses replicate in a variety of fish cell lines.

Once a cell line is chosen, filtrates or antibiotic-sanitized materials are inoculated onto cell cultures. Depending on cell line and virus assay samples, inoculated cells are incubated at 15 to 30°C for a few days to several weeks. If a virus is present and infectious to the inoculated cells, CPE will usually occur in 1 to 5 days depending on virus and incubation temperature. Some viruses will require a longer incubation period to develop CPE. The type of CPE is often specific for a certain virus group, but differences are usually too subtle for positive identification. Once viruses are isolated, they can be positively identified by a variety of biotechnical procedures, including serum neutralization, fluorescent antibody and enzyme linked immunosorbent assay (ELISA) techniques, plaquing, radiolabeling, biotin-avidin, immunoperoxidase, and polymerase chain reaction techniques. Some of these techniques are also used to detect viral antigen in infected fish tissue, which provides a more rapid diagnosis.

Bacterial Diseases

Diagnosis of most bacterial fish diseases is accomplished by isolation of the pathogen on laboratory media and then identifying the organism biochemically, serologically, or by molecular methods. Some bacterial organisms are fastidious, however, and are difficult to isolate and grow on culture media. Basic equipment needed for isolating bacteria includes dissecting kit, method for sterilizing instruments to be used, inoculating loop, sterilized bacterial culture agar, and incubation temperatures of 15 to 30°C (Plumb and Bowser 1983; Austin and Austin 1989; Lasee 1995; Noga 1996).

Scrapings taken from body, fin, or gill lesions before disinfection should be examined in wet mounts for presence of external bacteria and fungi. Skin and muscle lesions are disinfected with alcohol or hot scalpel before an incision is made and an inoculum taken. Internal organs can be accessed through the abdominal region by disinfecting the skin surface and removing the muscle flap from the left side of the body cavity via three cuts (1—medial from anterior of the rectum to the isthmus, 2—from anterior of the rectum along the dorsal line of the coelomic cavity, and 3—from isthmus to above the pectoral fin) to yield access to the peritoneal cavity, liver, spleen, or kidney. A second approach to accessing the kidney is to disinfect the dorsal area and cut through the back and spinal column just posterior to the dorsal fin, thus penetrating the dorsal musculature to reach the kidney.

Most bacterial organisms that infect fish grow on general laboratory media such as brain heart infusion (BHI), tryptocase soy agar (TSA), nutrient, or blood agar; however, some specialized media are required for isolating flavobacteria, flexibacteria, bacterial kidney disease, and mycobacteria organisms. Incubation temperatures are not critical for most fish bacterial pathogens, but generally those from warmwater fishes are cultured at 25 to 30°C and those from salmonids at 15 to 20°C.

Initial criteria for making a presumptive identification of fish pathogenic bacteria are that bacteria be isolated from clinically diseased fish. The presumptive bacterial identification can be made based on relatively few biophysical and biochemical tests using conventional bacteriological procedures and tube media (Plumb and Bowser 1983; Shotts and Teska 1989; Shotts 1994). When making presumptive bacterial identification, the following criteria should be included: Determine the bacteria's cell morphology and motility; Gram stain reaction; production of cytochrome oxidase; reaction on carbohydrates (primarily glucose); indole, gas, and hydrogen sulfide production; temperature sensitivity; and sensitivity to certain growth inhibitors (0/129 or Novobiocin). Dichotomous keys are helpful in identifying many fish pathogens and can be used to classify most to genus and, by using a variety of additional tests, can be further classified to species (Shotts and Teska 1989; Shotts 1994).

A battery of conventional tube media tests can be used to identify bacteria but are labor intensive, and final identity may take several days. Numerous commercially prepared systems can be used to shorten identification time, but these systems are not always accurate for fish pathogens. The most popular of these systems are API, Minitek, Biolog GN Microplate, Abbott Diagnostics, and Vitek Systems, all of which were originally developed for human or environmental microbiological application, and attempts to adapt them to fish pathogens have not been completely successful (Shotts and Teska 1989; Teska et al. 1989; Taylor et al. 1995; Robohm 1997). Most fish pathogens do not occur in these systems' numerical database because the systems require 35 to 37°C incubation, a temperature that is higher than normally used to incubate fish pathogens.

The BIONOR Mono-kit, which is based on latex agglutination of bacteria, has proved to be quick and accurate in detecting most strains of *Vibrio anguillarum, Yersinia ruckeri,* and *Renibacterium salmoninarum* (Romalde et al. 1995). However, reagents used to identify *Vibrio splendidus* and motile *Aeromonas* isolates have showed some cross-reactivity and false positives. The kits are considered cost-effective and suitable for screening large numbers of samples when used in fish health laboratories but may not be universally applicable. Other serological systems are being developed specifically for some fish pathogens.

To speed up identification and increase accuracy, serology and biotechnical procedures are being used to detect some pathogens in situ and in vitro after bacteria are isolated and cultured. Serological procedures use direct and indirect fluorescent antibody, several variations of ELISAs, radiolabeling, and polymerase chain reactions (Schill et al. 1989). These rapid, highly sensitive procedures can be beneficial in diagnosing infectious fish diseases, especially those associated with bacteria. Most of these procedures require specialized equipment and expertise; therefore, they are not currently available in all fish disease diagnostic laboratories.

REFERENCES

Austin, B., and D. A. Austin. 1989. *Methods for the Microbiological Examination of Fish and Shellfish.* Chichester, UK: Ellis Horwood Ltd.

Hoffman, G. L. 1998. *Parasites of North American Freshwater Fishes.* Ithaca, NY: Cornell University Press.

Lasee, B. A., Editor. 1995. *Introduction to Fish Health Management.* Washington, DC: U.S. Fish and Wildlife Service, Department of the Interior.

Meyer, F. P., and L. A. Barclay, Editors. 1990. *Field Manual for the Investigation of Fish Kills.* Resource publication no. 177. Washington, DC: U.S. Fish and Wildlife Service.

Mitchum, D. L. 1995. *Parasites of Fishes in Wyoming.* Cheyenne, WY: Wyoming Game and Fish Department.

Noga, E. J. 1996. *Fish Disease Diagnosis and Treatment.* St. Louis, MO: Mosby-Year Book, Inc.

Plumb, J. A., and P. R. Bowser. 1983. *Microbial Fish Disease Laboratory Manual.* Auburn, AL: Auburn University, Alabama Agricultural Experiment Station.

Robohm, R. A. 1997. An evaluation of the use of Biolog GN Microplate reactions in constructing taxonomic trees for classification of bacterial fish pathogens [abstract]. *22nd Annual Eastern Fish Health Workshop.* Atlantic Beach, NC, March 18–20.

Romalde, J. L., B. Magariños, B. Fouz, I. Bandín, S. Núñez, and A. E. Toranzo. 1995. Evaluation of BIONOR Mono-kits for rapid detection of bacterial fish pathogens. *Diseases of Aquatic Organisms* 21:25–34.

Sanz, F., and J. Coll. 1992. Techniques for diagnosing viral diseases of salmonid fish. *Diseases of Aquatic Organisms* 13:211–223.

Schill, W. B., G. L. Bullock, and D. P. Anderson. 1989. Serology. In: *Methods for the Microbiological Examination of Fish and Shellfish,* edited by B. Austin and D. A. Austin, 98–140. Chichester, UK: Ellis Horwood Ltd.

Shotts, E. B., Jr. 1994. II. Flow chart for the presumptive identification of selected bacteria from fish. In: *Bacterial Diseases of Fish. Bluebook: Suggested Procedures for the Detection and Identification of Certain Finfish and Shellfish Pathogens,* 4th ed., edited by J. C. Thoesen. Bethesda, MD: Fish Health Section/American Fisheries Society.

Shotts, E. B., Jr., and J. D. Teska. 1989. Bacterial pathogens of aquatic vertebrates. In: *Methods for the Microbiological Examination of Fish and Shellfish,* edited by B. Austin and D. A. Austin, 167–186. Chichester, UK: Ellis Horwood Ltd.

Taylor, P. W., J. E. Crawford, and E. B. Shotts, Jr. 1995. Comparison of two biochemical test systems with conventional methods for the identification of bacteria pathogenic to warmwater fish. *Journal of Aquatic Animal Health* 7:312–317.

Teska, J. H., E. B. Shotts, and T. C. Hsu. 1989. Automated biochemical identification of bacterial fish pathogens using the Abbott Quantum II. *Journal of Wildlife Diseases* 25:103–107.

Thoesen, J. C., Editor. 1994. *Bluebook: Suggested Procedures for the Detection and Identification of Certain Finfish and Shellfish Pathogens,* 4th ed. Bethesda, MD: Fish Health Section/American Fisheries Society.

Wolf, K. 1988. *Fish Viruses and Fish Viral Diseases.* Ithaca, NY: Cornell University Press.

5 Disease Management

In addition to practicing good fish husbandry, a fish disease control plan includes health management, judicial use of chemotherapeutics either as prophylaxis or treatment, and the use of vaccines when available. A "team effort" is important to good fish health management, and the central figure of that team should be a fish health professional. The individual can be a trained fish health specialist or veterinarian as long as he or she is familiar with good fish husbandry practices and understands the aquatic environment and its relationship to fish health.

FISH HEALTH MANAGEMENT

Fish health management can be classified as "applied art" or a "science" depending on one's perspective; in reality, it is probably a combination of the two, with science combined with "good ol' common sense." A number of encompassing fish health management principles have been outlined in earlier chapters. This chapter emphasizes aspects that relate more directly to disease prevention and control. Fish health management through environmental manipulation may not totally eliminate infectious disease in cultured fish, but when disease does occur its impact will be minimized. There are, however, numerous examples of fish diseases that are directly associated with poor environmental conditions (see Table 1.1), many of which can be addressed through environmental management.

Fish physiology is important in health management because as intensity of the culture system increases, physiological response increases proportionally. Wedemeyer (1996) discussed in detail the relationship between an aquatic environment and homeostasis of fish, particularly in intensive systems; the practical application of water management blended with science; and how these affect disease susceptibility.

Extreme deviations from optimal environmental conditions are stressful and can result in reduced feed consumption, higher feed conversion ratios, poor growth, disease, and/or death. Management procedures that help reduce stress as it relates to increased disease susceptibility can be divided into several areas: (a) fish handling and stocking, (b) feed management, (c) water flow and temperature management, (d) aeration management, (e) other environmental problems, and (f) waste management. When discussing these concerns, however, it is difficult to separate them because they do not function independently.

Fish Handling and Stocking

As one old saying goes, "fish are not potatoes" and should not be handled as such. Fish should be handled gently, whether as individuals or groups. Careful handling during transport, sorting, spawning, and stocking is critical to disease susceptibility because production of mucus can be adversely affected and injury to the skin and gills will provide an opportunity for pathogens to become established. A good time to treat prophylactically to help prevent secondary infections is while fish are still in hauling or holding tanks following handling. It is usually counterproductive to move fish strictly to treat them.

Slowly changing water temperature from that of the current environment into one in which fish will be moved (tempering) is advisable for most fish. Tempering enables the fish's physiology to adapt to its new environment; time required varies depending on fish species and temperature differential. Some fish species (for example, trout) require several hours to be properly tempered, whereas other species (for instance, carp) require less time. The greater the temperature differential, the longer tempering time required. Fish generally can adapt to lower temperatures more quickly than to higher ones. For this reason, if possible, fish should not be handled during hot weather.

Crowding is one of the most common factors that adversely affects the general health of fish (Wedemeyer 1996). Carrying capacity in raceways can be increased by increasing water flow rates, but according to R. Schmittou (personal communication) overcrowding is almost never a limiting factor in pond or cage culture. Increased fish density depends on the quality and quantity of feed required and can be adversely affected by any associated water quality problem that relates to feed management.

Carrying capacity of most fertilized bodies of water is usually 25 to 150 kg of fish per hectare (about 30 to 175 lb/acre) depending on fertility. Under these conditions, there is sufficient natural food, oxygen, and other life-sustaining substances to support a fish population adequately, but not one that is economical from a commercial standpoint. Therefore, an aquaculturist in the business of rearing fish for profit must maintain a high fish population density. When nutrients are added to support higher stocking densities (fish numbers and weight per unit volume), water quality can be adversely affected and the potential for disease increases. In an effort to improve productivity, some catfish farmers elevate fish densities from approximately 10,000 fish per hectare (4050/acre) to 20,000 fish per hectare (8100/acre). A catfish pond stocked with fingerlings to yield as much as 10,000 kg of food size fish per hectare is a potentially deteriorating environment because of the quantity of feed required to sustain growth for that biomass. Uneaten feed and fecal and metabolic waste contribute an enormous organic load to the water, providing an abundance of nutrients to support plant and bacterial growth, which in turn consume oxygen essential for fish survival.

As stocking densities increase, disease becomes more critical, but not necessarily as a direct result of increased numbers of fish per volume of water. Supplemental feeding is required to support the greater numbers and weight of fish, and this additional nutrient input proportionally increases environmental instability and can cause extreme fluctuations in critical water quality factors. Consequently, a change in one management technique (for instance, addition of feed) often must be matched by another management procedure to offset any detrimental effects of the first (for instance, addition of aeration).

Tucker et al. (1992) reported that catfish ponds stocked at 11,120 fish per hectare produced 5200 kg/ha per year and that ponds stocked at 19,500 fish per hectare produced 6600 kg/ha per year (Table 5.1). Although the higher stocking density resulted in a slightly greater weight of fish, they did not grow as well, had greater size variation, reduced survival, higher feed conversion ratios, increased need for aeration, and increased waste. The addition of nutrients to sustain the higher densities caused deterioration in water quality and increased potential for low dissolved oxygen syndrome, which lead to reduced disease resistance and higher mortality rates in the higher-density ponds. Actual net profit in each culture system was about the same.

TABLE 5.1. PROJECTED PRODUCTION IN A 131 HA (288 ACRE) CATFISH FARM STOCKED AT TWO FISH DENSITIES AND HARVESTED ANNUALLY FOR 3 YEARS (4 PONDS/TREATMENT) BASED ON EXPERIMENTAL DATA FROM STUDIES ON A SMALLER SCALE

Criteria	Treatment A[a]	Treatment B[a]
Stocking density (no./ha)	11,120	19,770
Average no. harvested/year	8800	13,500
Mean survival	79%	68%
Average net harvest (kg/yr)	5200	6600
Average wt./fish (kg)	0.59	0.49
Feed conversion ratio	1.37	1.56
Aeration hours	1606	2120
Net revenue/ha ($US)	$3560	$3550

Source: Tucker et al. (1992).
[a]Each pond was harvested annually and the number of fish removed was replaced with equal number of seed stock.

Feed Management

Quality of feed should be the primary consideration in feed management. Feed should always be fresh, of the highest quality, and consistent with nutrient requirements of a particular species, size of fish, and type of culture unit involved (Lovell

1989). A second consideration in feed management, as it relates to disease, is the amount and frequency of feed applied per unit of surface area. Accumulated metabolic wastes in water creates a demand for oxygen and serves as a source of nutrients for phytoplankton, which in turn exerts its own oxygen demand. As feed rates increase, phytoplankton growth also increases, thus effecting a decrease in water quality and a need for management procedures (such as aeration, reduction in feeding rates, and phytoplankton bloom control).

Water Flow Management

In some culture systems (such as trout raceways), water quality problems associated with high fish densities are overcome by increasing water flow volume in conjunction with maintaining adequate dissolved oxygen levels. In large volume culture ponds, however, routine continuous fresh water exchange is usually limited or nonexistent and actually may be harmful (Lorio 1994). Recirculating water for 8 hours at night in catfish ponds increased yield by a factor of 2.5, but these data may be misleading because of a lack of adequate controls in the study. Indications are that excessive water flow in ponds can adversely affect the environment by removing more nutrients than desirable and that continuous water flow has not proved to be a deterrent to infectious diseases. During periods of poor water quality, particularly oxygen depletion, addition of good quality water can, however, be beneficial and may prevent severe stress and reduce disease susceptibility.

Aeration Management

Mechanical aeration is a major reason for the rapid growth of aquaculture. A variety of aerators (paddle wheels, agitators, or sprayers) (see Figure 1.6) are available; each having unique advantages (Boyd and Wattson 1989). Energy sources are tractor power–take-off, propane gas, or electricity. Natural oxygenation is a result of photosynthesis driven by sunlight, consequently oxygen levels fluctuate diurnally, being highest in the afternoon and lowest during the night and, therefore, should be monitored 24 hours a day during warm weather. To calculate a pond's oxygen concentration at dawn, measure the oxygen concentration at dark and 3 hours later and extrapolate a straight

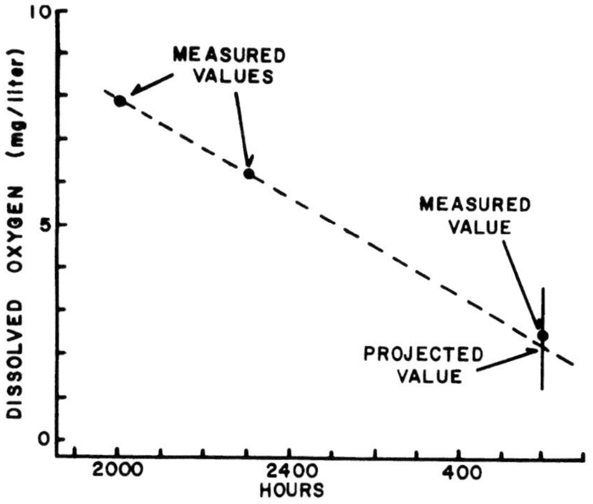

FIGURE 5.1. Projection technique for estimating nighttime dissolved oxygen (DO). By measuring oxygen concentrations at 8:00 PM and 12:00 midnight, the DO at daylight can be estimated by straight line projection (Boyd et al. 1978). (Reprinted with permission of the American Fisheries Society.)

line to determine the predicted oxygen concentration at dawn (Figure 5.1) (Boyd et al. 1978). If the predicted oxygen concentration at dawn is below a minimum level (that is, 3 mg/L), night aeration should be initiated.

Nightly aeration during warm weather prevents stressful oxygen depletion that can lead to increased disease incidence in pond cultured fish. Lai-Fa and Boyd (1988) showed that during the growing season, 6 hours of aeration at night almost doubled channel catfish production; feed conversion ratios for fish in the aerated ponds were 1.32 versus 1.75 in unaerated ponds.

Successive cloudy days have a significant detrimental effect on water quality in aquaculture ponds. Piedrahita (1991) found by computer modeling that short-term management schemes (lowering pH, raising alkalinity, and increasing nitrogen or phosphorus) were not effective methods for alleviating oxygen loss or depletion. The author concluded that the best way to counter cloudy-day water quality syndrome is to reduce water depth and flush fresh water through the pond at a rate of at least 20% per 24-hour period. It is essential, however, that oxygen concentrations in the incoming water be higher than those of the pond water.

Aeration in closed or recirculating systems can be achieved by using compressed air, liquid oxygen, pure oxygen, or by various mechanical agitators. Bubbling air or oxygen directly into the culture system or pipeline manifold is probably not as efficient as introducing oxygen or air by injection into a sealed column of water (Doulos et al. 1994).

Other Environmental Problems

Oxygen concentration is not the only environmental water quality parameter important to health of cultured fish. Maintaining low levels of carbon dioxide, nitrite, and ammonia, as well as proper pH, alkalinity, and water hardness are also important (see Table 1.3). Carbon dioxide is normally elevated during pond respiration and decomposition of organic matter but seldom poses a problem because algae and plants utilize it almost as fast as it is produced. High CO_2 can be a problem in closed systems. Water acidity, which depends on its buffering capacity (alkalinity), should be pH 6.5 to 9. For aquaculture purposes, total alkalinity and total hardness should be higher than 20 mg/L ($CaCO_3$) with values of more than 50 mg/L preferred (Boyd 1990). Low or widely fluctuating pH, low alkalinity, or hardness in ponds can be partially corrected by adding agricultural lime and in recirculating or flow-through systems by dripping lime into the water or passing water through a bed of oyster shells. Ammonia is seldom a problem in open ponds, but it does accumulate in closed systems and can be controlled with biological filters. Some of these water quality problems result from high fish density and can be partially corrected using manipulative management; reducing the standing crop, limiting quantity of feed, or increasing aeration. In the long term, routine but judicial application of Environmental Protection Agency (EPA) registered herbicides, such as copper sulfate or aquazine, can control massive blooms of phytoplankton or filamentous algae.

In geographical areas where water temperatures normally fall below optimal levels for growth and health of cultured fish, water from warm underground aquifers or heated water from power plants can be used to increase water temperatures in a rearing system. Heat-exchange units or solar heating can also be used to warm water if the amount of water required is relatively small.

Waste Management

Uneaten feed, accumulation of fecal waste, and dead plant matter in the culture system will elevate the water's organic load. In trout raceways, with high water-exchange rates, most of this waste material is flushed out, whereas in ponds it settles on the bottom. Contrary to popular belief, old pond soils do not necessarily contain high quantities of organic matter (Boyd et al. 1994). Bottom soil of new shrimp ponds in southern Thailand contained 1.1% organic matter, whereas sediment in recently drained established shrimp ponds contained only 1.9%. Boyd and Pippopinyo (1994) found that optimum respiration and oxidation of organic material occurred when soil moisture was 12% to 20% and at pH 7.5 to 8.0. Addition of either calcium hydroxide or calcium carbonate will enhance respiration of acidic soils. Pulverizing the hard surface crust of pond soil by tilling will also accelerate respiration. In closed recirculating systems, organic material must be removed by mechanical scrubbers and biological filter devices.

Herbivorous fish can help control vegetation in environments with high fertility and an overabundance of phytoplankton or vascular plants. Grass carp eat vascular plants, and silver carp and other filter feeders eat the phytoplankton. When practical, controlling aquatic vegetation with herbivorous fish is a better alternative than chemicals. Benefits are usually longer lasting, more economical, and more user friendly. Precautions must be taken to avoid escape of exotic species into natural waters, where they could be ecologically detrimental. Because of this possibility, exotic carp are prohibited by law in some states; however, in some instances stocking sterile triploid grass carp is allowed.

DRUGS AND CHEMICALS

Aquaculture has expanded globally during the past 30 years, aided by the use of drugs and chemicals for treating fish diseases. Chemotherapeutics are, however, only a small part of a comprehensive management plan and should not be relied on exclusively to solve all aquacultural health problems. Realistically, use of such compounds has in part allowed farmers to raise carrying capacity and reach new food production levels, but in some instances drugs have been detrimental. Overdependency on drugs, which can be harmful to the animal being produced, is ill-advised in today's environmentally conscious society. Before any drug can be used or developed for fish, regulatory, public health, and biological constraints must be considered. A drug must be registered as safe for use on fish, and in the United States this is a major obstacle.

Drugs can be used to prevent a disease from occurring (prophylactic) or to treat an existing infection (therapeutic). Advantages of drug therapy include availability; that it can be purchased in advance for use when needed; that it can be administered in several ways under a variety of conditions; that it may affect multiple disease organisms; and that it is effective for diseases that require treatment by immersion.

On the other hand, drug effectiveness is often short term and once withdrawn, disease may reoccur, especially if all pathogens are not eliminated or predisposing disease factors are not corrected. Bacterial kidney disease may not develop in trout or salmon as long as erythromycin is incorporated into feed, but when treatment ceases clinical infection reappears (Groman and Klontz 1983). Enteric septicemia of catfish often behaves in a similar manner, particularly when water temperatures remain optimum for the pathogen for an extended period. By the time chemotherapy is applied, the pathogen being treated has already taken a toll on the fish population either in mortalities, growth reduction, or productivity or a combination thereof. Also, chemotherapy presents a variable cost problem because it cannot be predetermined how often treatment will be required. Bacteria may develop resistance to antimicrobials if used too often or for an extended period or if applied improperly (Chaslus-Dancla 1992; Dixon 1994). Some drugs used on fish may be environmentally hazardous or toxic to treated animals, the applicator, or consumer. Accumulation of drugs in aquaculture products is an area of concern and one reason why drugs are closely controlled in most countries.

Alderman and Michel (1992) reviewed drug use in aquaculture and noted that in spite of the industries' huge global expansion, the number of approved chemotherapeutics available for food fishes is limited and the potential for pharmaceutical products in aquaculture is still small when compared with those for beef, swine, or poultry. In a review of drugs that can be used for fish, Stoffregen et al. (1996) emphasized a lack of drug availability for aquaculture in the United States. They identified several specific antimicrobials that have potential, but the registration process, as governed by the U.S. Food and Drug Administration (FDA), appears to be counter to approval of new drugs. The approval process is so arduous, time consuming, and expensive and the outcome so uncertain that it is difficult to persuade pharmaceutical companies to invest in drug research and development for aquaculture. Also, regulatory constraints further limit drugs from being used or developed in many other countries (Schnick 1992; Schlotfeldt 1992). Drug regulations vary among countries, but with increasing international trade of aquaculture products, regulations between countries have become more interrelated.

Treatment Process

Infectious diseases will be minimized if the principles of health maintenance outlined in Chapter 1 are followed and incorporated into individual management programs. Even the highest level of health maintenance will not completely eliminate infectious diseases; therefore, the question of control will eventually arise. Treatment of most aquaculture diseases must be considered on a population basis because it is impractical, and often impossible, to treat diseased fish individually unless only small numbers are involved. In a confined aquatic environment, if one fish in a population has an infectious disease, it is assumed that the entire population has at least been exposed and all fish must be treated.

Disease treatments and procedures described here are designed for cultured fish populations. In most wild populations, diseases are not treatable for either practical or economic reasons. Because drugs seldom completely eradicate all pathogens, their application is often simply "buying time" until fish can overcome infection. A therapeutant can reduce infection, temporarily prevent pathogen reproduction, or retard pathogen growth, thus giving the host's natural or acquired defensive mechanisms time to develop and overcome disease.

Wellborn (1985) proposed critical questions to be considered before treatment is applied: What is the prognosis if treatment is applied or withheld? and Does potential mortality justify treatment? If treatment is still indicated after answering these questions, four additional criteria should be considered: (1) know the water, (2) know the fish, (3) know the chemical, and (4) know the disease. Ignoring any one of these factors can result in an ineffective treatment and, over the long term, may prove detrimental to affected fish.

• **Know the Water**
Before applying a treatment, water volume to be treated must be accurately calculated to prevent a lethal or ineffectual treatment. Total alkalinity and hardness, pH, organic load, and temperature of the water should be known, as these factors influence efficacy and toxicity of some drugs.

• **Know the Fish**
In fish, chemical toxicity varies from species to species, strains within species, and between different age groups. If the drug of choice has never been used in a particular water supply to treat a specific fish species or age class of that species, toxicity should be tested. This involves treating a few fish in a small vessel containing water to be treated and observing chemical effects on the fish.

• **Know the Chemical**
Toxicity and percent active ingredient of a drug must be known to calculate correctly the proper amount to be used. Some drugs are affected by sunlight, pH, temperature, organic water load, and alkalinity. Some drugs are toxic to plants and will contribute to oxygen depletion by chemically removing oxygen from the water. All of these factors should be considered before applying drugs into an aquatic environment.

• **Know the Disease**
The disease must be accurately diagnosed by a complete necropsy. Multiple infections involving different pathogens often occur and will require multiple treatments beginning with the most serious pathogen. An incorrect diagnosis and treatment can prove disastrous to the fish.

Therapeutic Applications

The most effective and economical method of drug application is determined on a case-by-case basis and depends on disease involved, drugs prescribed, type of unit to be treated, and age and species of fish. Six basic methods for treating fish diseases are (1) dip, (2) flush, (3) prolonged bath, (4) indefinite, (5) orally in the feed, and (6) injection.

• **Dip**
Fish to be treated are immersed in a drug solution for a short time, usually 15 to 60 seconds. Drug concentrations can be expressed as percentage of material, milligrams per liter (mg/L), microliters per liter (μL/L) (both equivalent of parts per million [ppm]), or a ratio of chemical to volume of water (for example, 1:5000). Dipping fish requires caution because of strong drug concentrations used and the potential for toxic response. Time of immersion is critical; therefore, it is best to treat a few fish to determine their reaction before treating an entire population. The dip method is usually used to treat small numbers of easily confined fish infected with external parasites or bacteria. A disadvantage of this procedure is the necessity of handling fish.

• **Flush**
A flush treatment involves adding specific amounts of drug to a trough, tank, or raceway and allowing it to flush through without interrupting water flow. Drug concentrations are usually expressed as mg/L, μL/L, or as a ratio of drug to volume of water and can last from a few minutes up to 1 hour. In the latter, a stock solution of drug is introduced by a continuous flow delivery system, which ensures a desired concentration throughout the entire treatment period. This method is popular at trout and salmon hatcheries for treating eggs and fish in raceways for external parasites and bacteria. The flush treatment is seldom used on warm water farms or hatcheries except for egg treatment.

• **Prolonged Bath**
A prolonged bath treatment exposes fish to a drug for a specified period. Concentrations are normally in mg/L or μL/L, or as a ratio of drug to volume of water. The drug is added to a trough or holding unit at the desired concentration and left for a predetermined time, usually 1 hour, under static conditions with aeration. When treatment is terminated, water flow is resumed and the drug is flushed out as quickly as possible. Fish should be observed continuously during treatment, and if any signs of discomfort, such as gasping, "flashing," or loss of equilibrium are noted, the drug should be flushed immediately regardless of exposure time. Adequate amounts of water must be available at all times for rapid flushing if needed. Prolonged treatments use high drug concentrations, and caution is required to ensure equal distribution to avoid "hot spots" that can be toxic to fish. Prolonged treatment can be used for either external parasitic or bacterial infections.

• INDEFINITE

Indefinite treatment means that a drug is introduced into a pond or tank at comparatively low concentrations for an undetermined length of time and allowed to break down and dissipate naturally. Drug concentrations are usually mg/L or µL/L; this method is generally considered safe for treating fish. Indefinite treatments often require large quantities of drug that are difficult to apply, especially in large ponds, but can be applied to small bodies of water by hand or sprayers. A boat is required for treating larger areas. Easily dissolved dry chemicals, such as potassium permanganate, can be dissolved in water and then dispensed with a boat bailer or siphoned depending on kind and quantity of drug being applied. Indefinite treatments can be used for parasitic or bacterial infections; however, it is necessary to apply the drug uniformly to avoid hot spots, and supplemental aeration should always be available.

• ORAL

Treatment of systemic bacterial infections and intestinal parasites requires incorporation of antimicrobial or anthelminthic drugs into feed. The only legal way to use medicated feed in the United States is to purchase commercially prepared feed. Drugs incorporated into floating feeds must be heat resistant; otherwise, medicated sinking pellets are used. Spraying or coating feed with antibiotics suspended in oil is usually not satisfactory, as the drug is quickly washed off when pellets come into contact with water. Medicated feed is administered to fish on the basis of body weight to be treated. Standard units of treatment are given in grams of active ingredient per 45 kg (100 lb) of fish or in milligrams active ingredient per kilogram (mg/kg) of body weight per day for a defined number of days. Drugs are incorporated into feed at a concentration that delivers a desired dose per unit of fish weight per day, and fish are fed at a specific feeding rate (percentage of body weight divided into a defined number of feedings per day). Prophylactic use of drugs by feeding them for short periods or continuous feeding at low dosage is not advised because these practices enhance drug resistance of bacteria. A "withdrawal time," which is defined as the time (days) that must elapse between the last day drug is fed and day of slaughter (or in the case of fish released into public waters, the time when they are legally catchable), is usually required when drugs are fed to food animals, to ensure that no drug residues are present in the animal's flesh when eaten. Withdrawal time varies with drug, fish species, and temperature when drug is administered.

• INJECTION

Broodfish or small numbers of other valuable fishes can be treated for some diseases by drug injection. Doses to be injected are measured in either international units (IU) or milligrams (mg) of active drug per kilogram (or pound) of fish and are administered intraperitoneally (IP) or intramuscularly (IM). Intraperitoneal injections are given in the posterior area of the body cavity by inserting a needle through the peritoneal wall at a 45° angle at the base of the pelvic fin, taking care not to injure internal organs. Intramuscular injections are given slowly in the thick dorsal musculature near the dorsal fin. The IM method is slower to administer, and the quantity of injectable material is reduced, because of potential drug loss through the puncture wound. Injection is usually used to treat bacterial infections, but the need to handle each fish may increase stress if fish are already in poor health.

Calculations

There are two basic methods of mass drug application for fish: bath/immersion and medicated feed. Once the drug of choice is established for bath/immersion, it is essential to know the volume of water to be treated, concentration of drug to be used, and percent active ingredient in order to calculate how much drug to apply. The following formula can be used to calculate the amount of drug required to treat static tanks, raceways or ponds.

$$\begin{matrix}\text{Wt. or} \\ \text{vol. water}\end{matrix} \times \frac{100}{\%\ \text{Activity}} \times \begin{matrix}\text{Desired} \\ \text{conc.}\end{matrix} = \begin{matrix}\text{Wt. or volume} \\ \text{of drug to use}\end{matrix} \qquad (5.1)$$

Example No. 1: To treat 1000 L of water at 2 mg/L (ppm) with a drug that is 100% active, the calculation is as follows:

$$1000\ (\text{L}) \times \frac{100}{100} \times 2\ (\text{mg/L}) = \text{Wt./vol. of drug required}$$

Therefore: 1000 × 1 × 2 = 2000 mg (2 g) of drug required.

Example No. 2: To treat a 5-acre (2.02 ha) pond that averages 4 ft. (1.22 m) at 4 mg/L (ppm) with a drug that is 50% active, the calculation is as follows:

One acre foot is 1 surface acre (43,560 ft^2) that

is 1 foot deep; therefore, the pond contains 20 acre feet of water. Also, 1 acre ft. weighs 2.7 million pounds; therefore 2.7 pounds of anything in 1 acre foot = 1 mg/L (ppm), and 1 acre foot contains 1,238,230 L of water. Using the above formula, the calculation is:

$$1{,}238{,}230\ (\text{L}) \times \frac{100}{50} \times 4\ (\text{mg/L}) = \frac{\text{mg of drug required}}{\text{per acre ft. of water}}$$

$$\begin{aligned}\text{Therefore; } 1{,}238{,}230 \times 2 \times 4 &= 9{,}905{,}840 \text{ mg of drug}\\ &= 9.9 \text{ kg/acre ft. of water}\\ &\times 20 \text{ acre ft.}\\ \hline &198.0 \text{ kg (436 lb) of drug required to treat the 5-acre pond.}\end{aligned}$$

When antibiotics are fed to fish, it is essential to know the weight of fish to be treated, treatment rate (weight of drug fed per unit weight per day), and concentration of active ingredient in the drug. The following formula is used to calculate how much drug to feed per day.

$$\begin{matrix}\text{Wt. of}\\ \text{Fish}\end{matrix} \times \frac{100}{\%\ \text{Activity}} \times \begin{matrix}\text{Treat. Level}\\ \text{(mg/kg/day)}\end{matrix} = \begin{matrix}\text{Wt. of}\\ \text{Drug}\end{matrix} \qquad (5.2)$$

Example No. 3: To treat 100 kg of fish with a drug that is 25% active at a rate of 100 mg/kg of fish per day, the calculation is:

$$\begin{matrix}100\ (\text{kg})\\ \text{of Fish}\end{matrix} \times \frac{100}{25} \times 100\ (\text{mg/kg}) = \text{Wt of drug/day}$$

Therefore: 100 × 4 × 100 = 40,000 mg (40g) of drug mixed in a 1-day ration.

Drugs for Fish

Drugs were used with no restrictions to prevent or treat fish diseases in the United States until the mid-1960s when a more cautious attitude began to develop concerning their indiscriminate use. Herwig (1979) listed approximately 275 different drugs and chemicals that had at one time been used in aquaculture, many of which were ineffective, uneconomical, known carcinogens, or remained in fish flesh for long periods. As the potential for drug residue to remain in fish flesh and the possibility of it being passed on to consumers became apparent, steps were taken worldwide to restrict use of unsafe drugs for treating fish diseases. Not including drugs that are already registered, Stoffregen et al. (1996) listed nine antimicrobials that are currently being used experimentally for fish disease control in the United States.

The FDA's Center for Veterinary Medicine (FDA CVM) has primary responsibility for approving drugs and chemicals used for disease control in food animals, including fish. Because fish are a "minor" agricultural product, they fall under the Minor Use Guidance Document (draft) written by the FDA CVM (Minor Use Guidance Document 1996), but drugs to treat fish must still meet certain guidelines before they can be used. To accomplish this, drugs must go through an Investigational New Animal Drug (INAD) application process during which requisite data are developed. Once toxicity, product safety, efficacy, residue in target animals, and environmental impact data are accumulated, the sponsor (manufacturer of the drug) may then file a New Animal Drug Approval (NADA) application. After critical review of submitted data, CVM rules whether the drug can be used on the target animal or if more data are necessary. As a result of this lengthy process, only a few drugs are currently approved for use on fish (Table 5.2) in the United States. The list includes only one new drug (Romet-30 in 1986) approved by CVM for aquaculture since Terramycin's approval in the early 1970s.

The CVM classifies drugs as (1) approved drugs that are the subject of an NADA, (2) drugs registered by the United States Environmental Protection Agency (EPA) as pesticides for use in aquatic systems, (3) unapproved drugs for which data are being generated under an INAD exemption, (4) drugs of low regulatory priority (LRP), (5) drugs of high regulatory priority, (6) drugs that are the subject of case-specific regulatory discretion, (7) drugs legally administered to species for indications not listed on the approved drug label by a licensed veterinarian as defined in the Animal Medicinal Drug Use Clarification Act (AMDUCA) (extralabel use), and (8) illegal drugs (Guide to Drug, Vaccine, and Pesticide Use in Aquaculture 1994; Extralabel Drug Use in Animals 1996).

Drugs currently being used in aquaculture for disease treatment are discussed here, including status with the FDA, methods of application, concentration and dosage, length of application, limitations, and diseases for which they are used. Some of these drugs are dangerous, especially in concentrated form, and safety precautions, which include wearing protective clothing, rubber gloves, breathing masks, and safety glasses, should be taken

TABLE 5.2. DRUGS AND CHEMICALS APPROVED BY THE U.S. FOOD AND DRUG ADMINISTRATION FOR DISEASE CONTROL IN FOOD FISH

Drug/Chemical	Species	Indication	Dosage	Withdrawal
Finquel[a]	All species	Anesthetic	50–100 mg/L	21 d
Formalin				
Formalin F[b]	All fish	External protozoa and monogenetic trematodes	Raceways, tanks:	NA
Paracide-F[b]	Fish eggs	External protozoa and monogenetic trematodes	>10°C 170 µg/L for 1 h <10°C 250 µg/L for 1 h	
Parasite-S[b]	Penaeid shrimp	Fungus	Ponds: 15 to 25 µl indefinitely 1000 to 2000 uL/L for 15 min Tank: 50–100 uL/L for 4 h	
Romet 30[c]	Salmonids	Furunculosis, enteric redmouth	50 mg/kg fish per day for 5 d	42 d
	Catfish	Enteric septicemia	50 mg/kg fish per day for 5 d	3 d
Terramycin[d]	Salmonids	Furunculosis, enteric redmouth, vibriosis	50–75 mg/kg fish per day in feed for 10 d	21 d
	Catfish	Motile *Aeromonas* septicemia, *Pseudomonas* septicemia, enteric septicemia	50–75 mg/kg fish per day in feed for 10 d	21 d
Sulfamerazine in Fish Grade[e]	Rainbow, brook, brown trout	Furunculosis	220 mg/kg/day for 14 d	21 d

Source: Schnick et al. (1989); Guide to Drug, Vaccine, and Pesticide Use in Aquaculture 1994.
[a]Tricaine Methanesulfonate—"tricaine."
[b]37–36% formaldehyde gas.
[c]Sulfadimethoxine–ormetoprim in 5:1 ratio.
[d]Oxytetracycline hydrochloride.
[e]According to manufacturer, this product is no longer available.

when weighing, measuring, and applying them. If the skin comes into contact with any drug, it should be thoroughly washed immediately.

Registered Drugs For Fish

In the United States, manufacturers have received FDA approval and registration of four drugs for fish disease control and one anesthetic (Guide to Drug, Vaccine, and Pesticide Use in Aquaculture 1994) (Table 5.2). These drugs are registered for a specified application method for a specified disease of a specific fish species, and technically their use for other diseases or other fish species is illegal unless used under an INAD or via AMDUCA.

• FORMALIN

There are three registered formalin sources that can be used as a parasiticide for fish: Formalin F (Natchez Animal Supply), Paracide-F (Argent Chemical Laboratories Inc.), and Parasite-S (Western Chemical Inc.). Formalin is a clear liquid that contains 37% to 40% formaldehyde gas, but in treatment calculations, it is considered 100% active. Formalin is approved as a parasiticide for salmon, trout, catfish, bluegill, and largemouth bass and as a fungicide for salmon and esocid eggs (Table 5.2). Approved formalin concentrations as a parasiticide in tanks, troughs, and raceways are 170 µL/L (1:6000) for 1 hour as a prolonged treatment at temperatures above 10°C and 250 µL/L (1:4000) for 1 hour at temperatures below 10°C. In ponds, formalin is used as an indefinite treatment at 15 to 25 µL/L, but care must be taken during warm weather because a critical oxygen depletion can occur several days after application that will require aeration. Formalin is also approved for application as a flush treatment for eggs at 1000 to 2000 µL/L for 15 minutes to control fungus. In penaeid shrimp, formalin is used as an external protozoan treatment at 50 to 100 µL/L in tanks for up to 4 hours and in ponds at 25 µL/L indefinitely.

• TERRAMYCIN FOR FISH

Terramycin (oxytetracycline) (Pfizer Inc.) is a bacteriostatic antibiotic in powder form, which is added to feed to control motile *Aeromonas* septicemia, *Pseudomonas* septicemia, and enteric septicemia in catfish and furunculosis, enteric redmouth, and vibriosis in salmonids. It is also approved for marking skeletal tissue of juvenile Pacific salmon for identification when they return to spawn and for controlling *Gafkemia* in lobster (Guide to Drug, Vaccine, and Pesticide Use in Aquaculture 1994). Terramycin is manufactured in various concentrations of active ingredient, which must be taken into consideration when calculating dosage. Oxytetracycline is fed at a rate of 50 to 75 mg/kg or 2.5 to 3.75 g/45 kg (100 lb) of fish per day for 10 consecutive days. This dosage equates to the addition of about 1 to 1.5 g of active antibiotic per 0.45 kg (1 pound) of feed and fed at 2% of body weight per day. Withdrawal time for the drug is 21 days. Terramycin has been used so extensively in aquaculture (sometimes improperly) that a high percentage of target organisms, especially motile *Aeromonads* spp. and typical *Aeromonas salmonicida,* have become resistant to it. Another concern is that Terramycin is only available in sinking pellets because the amount of heat required to extrude floating pellets during manufacture destroys the drug.

• ROMET-30

Romet-30 (Hoffmann-La Roche Inc.) is a bactericidal potentiated sulfonamide in powder form. The drug is a combination of sulfadimethoxine and ormetoprim in a ratio of 5:1 and is approved for treating enteric septicemia of catfish (ESC) and furunculosis in salmonids. Romet-30 is incorporated into feed to provide 50 mg/kg of fish per day for 5 days. At higher concentrations, fish tend to refuse feed; therefore, it is better to use a reduced drug concentration and increase feeding rate accordingly. Romet-30 is heat resistant and can be used in extruded floating pellets. Withdrawal time is 42 days for salmonids and 3 days for catfish. Mortalities are reduced quickly when fish consume feed containing Romet-30; however, a low incidence of drug resistance has been encountered with frequent infection recrudescence (Plumb et al. 1995).

• SULFAMERAZINE IN FISH GRADE

Sulfamerazine in Fish Grade (American Cyanamide Company) is approved for treating furunculosis at a feeding rate of 220 mg/kg of fish per day for 14 days followed by a 21-day withdrawal period. This drug is no longer manufactured in the United States, but it or similar sulfa-drugs may be available elsewhere.

• FINQUEL

Although not a disease treatment drug, Finquel (tricaine methanesulfonate, MS-222, tricaine) (Argent Chemical Laboratories Inc.) is an anesthetic used in a variety of aquaculture procedures that require tranquilizing and handling individual fish. It is registered by the FDA for catfish, salmonids, esocids, and percids, and if applied to other fish, it is suggested that it be done only in hatcheries or laboratories.

Chemicals Registered by the United States Environmental Protection Agency

Three chemicals, copper sulfate ($CuSO_4$), Diquat, and potassium permanganate ($KMnO_4$) have been registered by the EPA for use as herbicides or oxidizing agents in aquaculture (Schnick et al. 1989). The FDA has deferred regulatory action on potassium permanganate and copper sulfate while information is gathered on environmental impact and bioaccumulation of copper and manganese in edible tissue. The FDA has indicated that as long as these chemicals are used per EPA registered labels, no objections will be made if the products have an incidental effect on fish diseases (Guide to Drug, Vaccine, and Pesticide Use in Aquaculture 1994).

• COPPER SULFATE

Copper sulfate ($CuSO_4$), also known as bluestone, comes in two formulations, "crystal" or "snow," both of which are considered 100% active. Copper sulfate is registered by the EPA as an herbicide and can be used in water harboring food fish. The chemical is a moderately effective therapeutic for external bacterial and/or protozoan infections and because of availability is one of the most widely used drugs in the fish industry worldwide. Copper sulfate has a narrow margin of safety between an effective treatment concentration and toxicity to fish; therefore, it is most often used in low concentrations as an indefinite treatment. Toxicity of copper sulfate varies with the copper formulation, water pH, total alkalinity, and water hardness (Straus and Tucker 1993). These authors also reported that chelated copper is significantly less toxic than copper in sulfate salt. These data, and general literature, support the argument that copper sulfate should never be used unless the aforementioned water chemistry characteristics are known. Gener-

TABLE 5.3. UNAPPROVED DRUGS BY THE U.S. FOOD AND DRUG ADMINISTRATION BUT THAT ARE UNDER AN INVESTIGATIONAL NEW ANIMAL DRUG EXEMPTION AND ARE CONSIDERED EXPERIMENTAL

Drug	Experimental use	Application concentration/method
Amoxycillin	*Streptococcus* spp. in tilapia	50–80 mg/kg of fish in feed; 10 d
Chloramine-T	External bacterial infections (Columnaris, bacterial gill disease)	8–10 mg/L for 1 h; 3 d
Erythromycin	Bacterial kidney disease (BKD) in brood salmonids; control of BKD	2 or 3 10–20 mg/kg by injection, 10–60 d before spawning; 100 mg/kg of fish in feed, 28 d
Enrofloxacin	Furunculosis in trout	10 mg/kg of fish in feed; 10 d
Nalidixic acid	Furunculosis in trout	40 mg/kg of fish in feed; 5–10 d
Oxolinic acid	Furunculosis and enteric redmouth in trout	12 mg/kg of fish in feed; 10 d
Sarafloxacin	Enteric septicemia of catfish	10 to 15 mg/kg in feed; 5–10 d

Source: R. S. Schnick, National Aquaculture INAD Coordinator, personal communication; and Guide to Drug, Vaccine, and Pesticide Use in Aquaculture (1994).

ally, 1.0 mg of copper sulfate per liter can be used in water for each 100 mg/L total alkalinity. More specifically, when TA is 0 to 49 mg/L, a toxicity test should be run before use; when TA is from 50 to 99 mg/L, use 0.5 to 1 mg/L of $CuSO_4$; 100 to 149 mg/L, use 1 to 2 mg/L of $CuSO_4$; and 150 to 300 mg/L, use 2 to 3 mg/L of $CuSO_4$. In salt water or if TA of fresh water is above 300 mg/L, copper sulfate precipitates rapidly and will most likely be ineffective.

• DIQUAT

Diquat is an EPA-approved herbicide for use in food fish ponds and has some efficacy against external bacterial infections, especially bacterial gill disease and columnaris. It is applied at 0.25 to 2.5 mg/L indefinitely or 2 to 4 mg/L for 1 hour as a prolonged bath. Diquat can only be used in culture systems where it cannot escape and where terrestrial animals have no access to it for 14 days.

• POTASSIUM PERMANGANATE

Potassium permanganate ($KMnO_4$), considered 100% active, is a purple crystalline material previously registered by the EPA for use in oxidizing organic matter in ponds. The EPA recently required all sponsors of registered pesticides and water treatment compounds to resubmit data for reregistering their products. The sponsor of $KMnO_4$ chose not to reregister it, thus technically negating any legal use.

Potassium permanganate is effective for some external protozoa and bacterial infections, especially columnaris, and can be applied in tanks at 5 to 10 mg/L for 1 hour as a prolonged bath. Fish should be observed continuously, and at the first hint of discomfort, fresh water should be introduced. Pond application rate is 2 to 4 mg/L indefinitely depending on organic content of the water. The $KMnO_4$ concentration must be 2 mg/L over the oxidizing demand of water to be treated. This demand can be determined by the concentration of $KMnO_4$ that is necessary to turn water from reddish purple (active) to brown (inactive) within 15 minutes (Tucker and Boyd 1977). The burgundy color should remain for 12 hours after application to be therapeutically effective. The toxicity margin of $KMnO_4$ is narrow (1 to 3 mg/L); therefore, caution is required at higher application rates. Also, if $KMnO_4$ is used too often on the same fish, severe gill injury will occur; therefore, a minimum of 5 days should elapse between treatments. Potassium permanganate reduces phytoplankton and may temporarily have a negative effect on oxygen levels. Cost is another distinct disadvantage of $KMnO_4$ if treating large volumes of water.

Unapproved Drugs (Some under Investigational New Animal Drug Exemption)

Several new drugs have been investigated for potential application in treatment of fish diseases and are currently in some stage of consideration for INAD exemption (Table 5.3). Most of these antimicrobials, if approved by the FDA, would likely be available only through prescription by veterinarians.

• AMOXYCILLIN

Amoxycillin is a penicillin-type antimicrobial that has shown promise for treating *Streptococcus* spp. infections in tilapia when fed at 50 to 80 mg/kg of fish for 10 days. It is also efficacious against the Gram-negative bacteria *A. salmonicida*. This drug was recently approved in the United Kingdom for use against furunculosis.

• CHLORAMINE-T

Chloramine-T is an antimicrobial used as a prolonged bath to treat bacterial gill disease in trout. It is applied in water at 10 to 20 mg/L for 1 hour for 3 consecutive days. Although not approved by the FDA, it has been used experimentally in the United States and extensively in Europe.

• ERYTHROMYCIN

Erythromycin (a macrolide) is a drug that is active against Gram-positive and some Gram-negative bacteria. It is not approved by the FDA for use on fish but is under INAD exemption. It has been administered experimentally by injection to prevent vertical transmission of bacterial kidney disease in Pacific salmon (Moffitt 1992). Adult salmon are injected with erythromycin at a rate of 10 to 20 mg/kg of body weight two or three times prior to spawning, with the final injection being administered 10 days before spawning. To control bacterial kidney disease (BKD) in salmonids, erythromycin is fed at a rate of 100 mg/kg of body weight for 28 days. Erythromycin may also be effective against *Streptococcus* spp. in tilapia when fed at the same rate.

• FLUOROQUINOLONES AND QUINOLONES

The fluoroquinolones, second-generation derivatives of synthetic quinolones, are a relatively new group of antimicrobials for possible use in aquaculture. Several members of this group (sarafloxacin, enrofloxacin, nalidixic acid, and oxolinic acid) have received significant attention as potential drugs for fish. Approval for use in aquatic systems in the United States may be difficult to obtain because of concern that bacteria may easily develop resistance to these drugs. If bacteria are resistant to one of these compounds, they are often resistant to all others.

• ENROFLOXACIN has been studied extensively as a control for furunculosis, BKD, enteric redmouth, and vibriosis in trout and salmon. An effective concentration of 10 mg/kg of fish per day for 10 days has been experimentally established (Bowser et al. 1994).

• NALIDIXIC ACID is used as a feed additive for furunculosis in trout and salmon at 40 mg/kg of body weight for 5 to 10 days (Uno et al. 1992).

• OXOLINIC ACID is registered in Japan and several countries in Europe for use in treating furunculosis and enteric redmouth in trout and has shown potential efficacy against other fish pathogenic bacteria. The drug is fed at a rate of 12 mg/kg of fish per day for 10 days (Austin et al. 1983).

• SARAFLOXACIN has been investigated as an oral treatment for enteric septicemia in channel catfish and for furunculosis, cold water vibriosis, and enteric redmouth disease in salmonids. Sarafloxacin is fed for 5 or 10 days at 10 to 15 mg/kg of fish per day (Plumb and Vinitnantharat 1990). The sponsor of sarafloxacin has fulfilled all FDA requirements, but approval has not been granted for use in the United States.

Drugs of Low Regulatory Priority

Low regulatory priority (LRP) compounds, some of which are not used for treating fish diseases, are acetic acid, calcium chloride, calcium oxide, carbon dioxide, Fuller's earth, garlic (whole), hydrogen peroxide, ice, magnesium sulfate (epsom salt), onion (whole), papain, potassium chloride, povidone iodine compounds, sodium bicarbonate (baking soda), sodium chloride (salt), sodium sulfite, urea, and tannic acid (Table 5.4). These drugs, chemicals, and other household compounds are substances used in our everyday lives and, amazingly, are not approved for use on fish. The FDA is unlikely to object to their use, however, if the following conditions are met (Guide to Drug, Vaccine, and Pesticide Use in Aquaculture 1994): (1) that they are used for prescribed indications, including species and life stage where specified, (2) that they are used at prescribed dosages, (3) that they are used according to good management practices, (4) that the product is of an appropriate grade for use in food animals, and (5) that an adverse effect on the environment is unlikely. The position of the FDA on the use of these substances should not be considered an approval or endorsement of their safety or efficacy. The LRP compounds that may have value in fish health are briefly discussed.

TABLE 5.4. UNAPPROVED DRUGS USED IN FISH HEALTH BUT CLASSIFIED BY THE U.S. FOOD AND DRUG ADMINISTRATION AS LOW REGULATORY PRIORITY

Drug	Indication	Dosage/treatment
Acetic acid (vinegar)	External parasites	Dip at 1000–2000 mg/L for 1–10 min
Calcium oxide	External parasites	Dip at 2000 mg/L for 5 s
Garlic (whole)	Helminths, sea lice	Unknown
Hydrogen peroxide	Fungus on fish/eggs	250–500 µL/L 15 min
Magnesium sulfate (epsom salt)	Monogenetic trematodes	Dip 3000 mg/L plus 7000 mg/L NaCl for 5–10 min
Onion (whole)	Crustacean parasites	Unknown
Povodine iodine	Egg disinfection	50 mg/L, 30 min during water hardening
Sodium chloride (salt)	Osmoregulatory, prophylaxis	0.5% indefinitely; 1–3% for seconds to 30 min or until loss of equilibrium

Source: Guide to Drug, Vaccine, and Pesticide Use in Aquaculture 1994.

• ACETIC ACID (VINEGAR)
Acetic acid is used as a parasiticide dip at 1000 to 2000 mg/L for 1 to 10 minutes; efficacy is unclear.

• CALCIUM OXIDE
This compound is used as an external protozoan treatment at 2000 mg/L for a 5 second dip with unknown efficacy.

• GARLIC (WHOLE FORM)
Garlic is used to control external helminths (monogenetic trematodes) and sea lice of marine salmonids. The efficacy of garlic is unknown.

• HYDROGEN PEROXIDE (H_2O_2)
Hydrogen peroxide is used at 250 to 500 µL/L (based on 100% activity) to control fungi on all fish species and at all life stages, including eggs. Efficacy studies in controlling fish egg fungus have been inconclusive.

• MAGNESIUM SULFATE (EPSOM SALT)
This compound is used to treat external monogenetic trematode and crustacean infestations on freshwater fish. Fish are dipped in 30,000 mg/L of $MgSO_4$ plus 7000 mg NaCl/L for 5 to 10 minutes.

• ONION (WHOLE FORM)
Onion is used to treat external crustacean parasites and to deter sea lice infestations on the skin and gills of salmonids.

• POVIDONE IODINE
Povidone iodophores (PVP) are solutions of iodine known as Betadine, Wiscodyne, and Novadine-iodine that are used for egg disinfection. The iodine compounds are applied at concentrations of 50 mg/L during water hardening for 30 minutes or 100 mg/L after water hardening for 10 minutes.

• SODIUM CHLORIDE (SALT)
Salt (NaCl) is used as a parasiticide, prophylaxis, and for stress reduction on freshwater fish following handling. Depending on fish species, fish may be held for a short period or indefinitely in 0.5% to 1% NaCl for relief of stress and to prevent shock, or they may be held in 3% NaCl for 30 seconds or until loss of equilibrium occurs.

Animal Medicinal Drug Use Clarification Act

Extralabel use of drugs for animals was redefined in the AMDUCA (Extralabel Drug Use in Animals 1996). This act allows veterinarians to prescribe extralabel uses of certain approved food animal and human drugs, as well as over-the-counter drugs, for aquatic animals. However, this same extralabel use and AMDUCA disallow the use of a medicated feed labeled for a specific disease in one grouping of fish (i.e., furunculosis in trout) to be applied for a different diseases in a different grouping of fish (i.e., enteric septicemia in catfish). Veterinarians exercising extralabel use for fish should be knowledgeable in aquatic animal health.

Other Chemicals and Drugs Used in Aquaculture

Several chemicals and drugs of value in fish health are used in other countries but are not approved for use in the United States or for use in aquaculture products destined for U.S. markets. The discussion of these drugs should not be construed as an endorsement or recommendation for use where not approved.

• ACRIFLAVINE
Acriflavine is used as a fish egg disinfectant, an antibacterial, or for external protozoa and is usually administered as a prolonged prophylactic bath. The recommended treatment rate is 10 mg/L for 1 hour or at 2 mg/L indefinitely. Because of cost, it is seldom used in large culture units.

• CHLORINE
Chlorine, primarily in the form of sodium hypochlorite granules, has become a popular fish treatment in ponds for a variety of ailments, but there are very little research data to support its value as a disease treatment.

• MALACHITE GREEN
Malachite green, also known as aniline green and Victoria green, is an organic dye that has for years been used as a protozoan parasiticide, especially for treatment of *Ichthyophthirius multifiliis* (Ich), and as a fungicide for fish and fish eggs. Although it is strictly banned in the United States, it is still incorporated in treatment formulations for ornamental fish. Because of potential teratogenicity, malachite green is gradually losing popularity everywhere it has been approved. For treating parasites, recommended prolonged bath treatment concentrations in troughs and tanks are 0.15 mg/L for 1 hour, and at 0.1 mg/L indefinitely in ponds. Malachite green (0.1 mg/L) in combination with formalin (25 mg/L) is an excellent indefinite treatment for Ich infections. Eggs are treated at 0.5 mg/L for 1 hour or 5 mg/L for 15 minutes for fungus prevention and control. Malachite green has a very narrow toxicity range for fish; therefore, accurate calculations are mandatory and extra precautions should be taken when handling it.

• NITROFURANS
Nitrofurans are a group of antimicrobials that are not approved for food animals in the United States because of their carcinogenic potential. These compounds (nitrofurazone, furacin, furazolidone, and others) are used to treat systemic bacterial infections at 200 mg/kg of body weight in feed for 14 days. As a prolonged external bacterial bath, nitrofurazone and furacin are applied at 20 mg/L for 1 hour or indefinitely at 20 mg/L. Furanace (furpyridinol) is more toxic and is applied to ornamental fish at 1 mg/L for 1 hour.

• QUATERNARY AMMONIA COMPOUNDS
Hyamine and Roccal, with various concentrations of active quaternary ammonia, are disinfectants used for treating bacterial gill disease in salmonids. The recommended concentration of active ingredient is 1 to 2 mg/L for a 1-hour flush to be repeated on 3 consecutive days. Quaternary ammonia is also a good net and utensil disinfectant at 200 mg/L.

Future Outlook

Although many countries have a more diverse list of "legal drugs" for use in fish than does the United States, before any of the aforementioned compounds are used they must be approved by the governing agency of the country where they are to be used. Also, if aquaculture products are destined for international trade, any compound being used must be acceptable to the country of destination.

In the European Union (EU) and other areas around the world, drugs and chemicals used in aquaculture and regulations controlling them are diverse (Schlotfeldt 1992). The subject of chemotherapy is dynamic, and most regulatory changes have resulted in more restrictive drug use and/or decline in drug availability for aquaculture. As the world population becomes more environmentally conscious and concerned with the role of drugs in human health problems, regulation and use of chemotherapeutics will become even more restrictive. In the final analysis, these restrictions will result in a more complicated and expensive drug registration process, thus discouraging pharmaceutical companies from investing in research and development of new aquaculture drugs. On the positive side, these restrictions have brought about a significant shift in attitude away from chemotherapeutic fish disease control to one of prevention by management. Chemotherapy, however, will continue to be important in aquaculture because chemicals and drugs are key components in health management, disease prevention, and control of cultured animals.

VACCINATION

The first experimental fish vaccination was reported by Duff (1942) when he fed a killed *A.*

salmonicida preparation to cutthroat trout to prevent furunculosis. From the early 1940s to the mid-1960s, little effort was made to advance the potential for vaccine use in preventing and/or controlling fish diseases; this period might be characterized as the "golden age of antibiotics." It was thought that chemotherapy could correct all ills, but it became obvious that drugs were not a panacea, and alternative methods of disease control were sought. A renewed interest in development of fish vaccines occurred in the early 1970s, and vaccination of cultured aquatic animals has now become a viable fish health management tool for some diseases. Like drugs, vaccines cannot solve all fish disease problems but, when available, can be an aid to the industry.

An acceptable vaccine for aquaculture must have the following attributes and characteristics (Leong and Fryer 1993):

1. Provide adequate immunoprotection for a specific disease under intensive rearing conditions.
2. Provide protection when the animal is most susceptible to disease.
3. Provide protection of long duration.
4. Protect against all serotypic variants of the disease agent.
5. Easily administered; preferably orally, immersion, or spray and application should minimally disrupt the normal management routine.
6. Safe for the vaccinated animal.
7. Inexpensive to produce and license and be cost-effective.

Implementing a vaccination program to prevent infectious fish disease is a "proactive" approach as opposed to a "reactive" approach when using drugs. Vaccines present distinct advantages over drugs: they reduce disease impact, decrease the need for drugs, provide long-term protection that often continues until slaughter, and provide a more fixed-cost disease-prevention expenditure; and in most instances, vaccinated fish grow better than unvaccinated fish. Perhaps the greatest value of vaccines is that no residue is left in edible flesh of vaccinated animals, except possibly in the case of some adjuvants.

Vaccines cannot completely eliminate disease organisms or prevent the target organism from being present in vaccinated populations because of individual immunological variances. Vaccinated fish can become carriers and, consequently, reservoirs of the disease organism. Vaccines are expensive to make and apply, but in some instances, a low percentage of improved survival justifies vaccination cost. For example, a 5% improved survival in vaccinated channel catfish for ESC can economically justify vaccination.

The first commercially licensed fish vaccine became available in 1976 to treat trout and salmon for enteric redmouth (ERM) (*Yersinia ruckeri*) and was soon followed by a vaccine for vibriosis (*Vibrio anguillarum* and *V. ordalii*). Since the introduction of these vaccines, they have been used internationally wherever ERM and vibriosis are a problem and are also occasionally used in treatment of nonsalmonids. In 1994, there were 15 licensed fish vaccines available in the United States for use against five fish pathogens (Guide to Drug, Vaccine, and Pesticide Use in Aquaculture 1994) and more are now available globally (Table 5.5). Autogenous bacterins made from specific isolates from individual aquaculture sites are also available from some biological companies.

Fish vaccines are used extensively in Norway, where at least six commercial biological manufacturers have marketed up to 24 different products. The Norwegians vaccinate salmon and trout against infectious pancreatic necrosis virus (IPNV), *V. anguillarum*, *V. ordalii*, *V. salmonicida*, *A. salmonicida*, and *Y. ruckeri*. Vaccines are available as univalent or multivalent preparations and are applied by immersion, injection, spray, and orally. Vaccines have improved survival and production of salmon and rainbow trout populations in Scandinavia and other areas where used. According to Press and Lillehaug (1995), in 1986, 3500 L of fish vaccine was used in Norway; 76,250 L in 1988; and in 1992, vaccine quantity had decreased to 24,780 L. Early vaccines were monovalent and effective against only a single disease organism, but more recently, multivalent vaccines containing antigens of two or more pathogens (viral and bacterial) or strains have been developed, resulting in reduced volumes of vaccine being used.

Antigens

Most bacterial vaccine research, development, and application have targeted salmonid diseases (*A. salmonicida*, *Y. ruckeri*, *V. anguillarum*, and *V. salmonicida*). Research and development of immunological products for nonsalmonids include vaccines for the most serious bacterial pathogens

TABLE 5.5. A PARTIAL LIST OF VACCINES CURRENTLY AVAILABLE FOR FINFISH AQUACULTURE[a]

Trade name	Antigen	Type vaccine	Company[b]	Fish species	Application
Biovax 1150	*Yersinia ruckeri*	Bacterin	Alpharma	Trout/salmon	Immersion
Biovax 1300	*Vibrio anguillarum, V. ordalii*	Bacterin	Alpharma	Trout/salmon	Immersion
Biojec 1500	*Aeromonas salmonicida*	Bacterin	Alpharma	Trout/salmon	Injection
Biojec 1800	*A. salmonicida* *V. anguillarum*	Bacterin	Alpharma	Trout/salmon	Injection
Biojec	*A. salmonicida* *V. anguillarum* *V. salmonicida*	Bacterin	Alpharma	Trout/salmon	Injection
Autogenous vaccine	Various bacterial antigen	Bacterin	Alpharma	Cultured fish	Immersion, injection, oral
Norvax Vibriose	*V. anguillarum*	Bacterin	Intervet	Trout/salmon	Immersion, injection
Norvax Protect-IPN	*V. anguillarum*+ Infectious pancreatic necrosis virus	Bacterin + Recombinant Infectious pancreatic necrosis virus protein	Intervet	Trout/salmon	Immersion, injection
Norvax Protect	*A. salmonicida* *V. anguillarum* *V. salmonicida*	Bacterin with double adjuvant	Intervet	Trout/salmon	Immersion, injection
Furogen	*A. salmonicida*	Bacterin	Aqua Health	Trout/salmon, Koi carp	Immersion, injection
Furogen 2	*A. salmonicida*	Bacterin, oil adjuvant	Aqua Health	Trout/salmon	Injection
Lipogen Triple, Lipogen Forte	*A. salmonicida* *V. anguillarum* *V. ordalii* *V. salmonicida*	Bacterin, oil adjuvant	Aqua Health	Trout/salmon	Injection
Escogen	*Ewardsiella ictaluri*	Encapsulated bacterin	Aqua Health	Catfish	Oral
Ermogen	*Y. ruckeri*	Bacterin	Aqua Health	Trout/salmon	Immersion, injection
Vibrogen	*V. anguillarum*	Bacterin	Aqua Health	Trout/salmon, sea bream, sea bass	Immersion, injection
AQUAVAC ERM	*Y. ruckeri*	Bacterin	AVL	Trout/salmon	Immersion, oral
AQUAVAC Vibrio	*V. anguillarum*	Bacterin	AVL	Trout/salmon	Immersion, oral
AQUAVAC Furvac 5	*A. salmonicida*	Bacterin	AVL	Trout/salmon	Immersion, injection, oral
AQUAVAC Furvac 5, Vibrio	*A. salmonicida* *V. anguillarum*	Bacterin	AVL	Trout/salmon	Injection
AQUAVAC FHV	*A. salmonicida,* *V. anguillarum,* *V. salmonicida*	Bacterin	AVL	Trout/salmon	Injection
AQUAVAC Cyprivac CE	*A. salmonicida*	Bacterin	AVL	Koy carp	Immersion
MICROViB	*V. anguillarum*	Bacterin	Microtek	Trout/salmon	Immersion, injection
MICROSaL$^{imm,\ inj}$	*A. salmonicida*	Bacterin	Microtek	Trout/salmon	Immersion, injection
MICROSaL^{oral}	*A. salmonicida*	Bacterin	Microtek	Trout/salmon	Oral
MICROCwV^{oral}	*Y. salmonicida*	Bacterin	Microtek	Trout/salmon	Oral

MICROViB^{oral}	*V. anguillarum*	Bacterin	Microtek	Trout/salmon	Oral
MULTIVaCC^{3}	*V. anguillarum*, *A. salmonicida*	Bacterin	Microtek	Trout/salmon	Injection
MULTIVaCC^{4}	*V. anguillarum*, *V. salmonicida*, *A. salmonicida*	Bacterin	Microtek	Trout/salmon	Injection
MICROPaS	*Phot. damsela*	Bacterin	Microtek	Sea bream/bass	Immersion
MICROErM	*Y. ruckeri*	Bacterin	Microtek	Trout/salmon	Immersion, oral
MICRO-SrS	*P. salmonis*	Bacterin	Microtek	Trout/salmon	Immersion
MICROSaL^{Atten}	*A. salmonicida*	Attenuated	Microtek	Trout/salmon	Immersion
MICROViR^{IHN}	IHNV	Adjuvanted	Microtek	Trout/salmon	Injection
MICROViR^{IPN}	IPNV	Adjuvanted	Microtek	Salmon	Injection

[a]Compiled from information supplied by the manufacturers. Other vaccine manufacturers did not supply requested information.
[b]Alpharma, Redmond, Washington, USA; Intervet Norbio, Bergen, Norway; Aqua Health Ltd., Charlottetown, Prince Edward Island, Canada; AVL, Aquaculture Vaccine Ltd. Essex, United Kingdom; Microtek International Limited, Saanichton, British Columbia, Canada.

that affect them; *Flavobacterium columnare* (*Flexibacter columnaris*), *Edwardsiella ictaluri*, *E. tarda*, *Photobacterium damsela* subsp. *piscicida* (*Pasteurella piscicida*), and *Streptococcus* sp.

Vaccine developed to protect against a specific pathogen for a specific species or fish group (multivalent vibriosis and furunculosis vaccines for trout and salmonids) can also be used for nontarget species. Rogers and Xu (1992) used a commercial *V. anguillarum–V. ordalii* bacterin developed for salmonids to protect striped bass against vibriosis.

Fish virus vaccine research for aquaculture has lagged behind that for bacterial pathogens; however, some experimental vaccines are now being developed with several attenuated, chemically inactivated, and subunit (recombinant DNA technology) preparations being produced (Leong and Fryer 1993). These include vaccines for IPNV, viral hemorrhagic septicemia virus (VHSV), infectious hematopoietic necrosis virus (IHNV), channel catfish virus (CCV), and spring viremia of carp virus (SVCV). Most recently, a company in Norway has developed a recombinant vaccine against IPNV that is marketed in a tetravalent preparation that also includes antigens for furunculosis, vibriosis, and coldwater vibriosis (Intervet Norbio, Bergen, Norway). According to some reports, 50 million Atlantic salmon smolts have been vaccinated with this multivalent vaccine, resulting in a 96% relative percent survival.

Vaccine Preparations

Most early fish vaccines, as well as many currently on the market, are simple bacterins to which some substance (generally formalin) has been added to kill the pathogen. Although bacterins are relatively easy to make, formalin or other additives can alter the organisms antigenic quality, thus reducing its efficacy. Attenuated bacterial or viral vaccines are potentially more effective because they produce a cell-mediated immune response more readily than do other preparations (Shoemaker et al. 1997). It is, however, possible that conventionally attenuated live pathogens can revert to a pathogenic form when released into an aquatic environment where water disposition cannot be controlled, a condition that is of great concern to regulatory agencies and aquaculturists. In some channel catfish diseases (such as *E. ictaluri*), and possibly other organisms, live bacterial cells may be essential to stimulate a cell mediated response and elicit protection (Shoemaker and Klesius 1997). In fact, Lawrence et al. (1997) and Thune et al. (1997) have shown that immersion of channel catfish in attenuated *E. ictaluri* baths conveys immunological protection. Also, Klesius and Shoemaker (1998) described immunological protection against ESC as a result of vaccinating channel catfish with a modified live *E. ictaluri* vaccine by immersion.

To avoid pathogen virulence, Vallejo et al. (1992) proposed that synthetic peptides be used as fish vaccines. Vaughan et al. (1993) described a nonpathogenic aromatic-dependent preparation of *A. salmonicida* (*aero*A) that protects trout against furunculosis. The advantage of aromatic-dependent mutants is that they will not revert to pathogenic form but retain their ability to produce antigenic components such as the A layer. Attenuation can also be linked to a bacterium's inability to synthesize *p*-aminobenzoic acid, which is important in pathogenesis. Cooper et al. (1996) used a nonpathogenic chondroitinase negative *E. ictaluri* to protect channel catfish against ESC. Cloning, mapping, and sequencing genes for individual proteins of *A. salmonicida* may also yield mechanisms by which immunogenically protective products will be pathogen specific (Bennett et al. 1992).

Biotechnology used in vaccine production for humans and veterinary animals is now being applied to vaccines for aquaculture. Genetic recombination, attenuation, protein engineering, and subunit vaccines have proven to be promising, and the possibility for reversion to pathogenicity is eliminated. According to reports from Norway, recombinant viral vaccine preparations are economically feasible, especially when combined with bacterins (Press and Lillehaug 1995).

Development strategies for fish virus vaccines emphasize genetic engineering and molecular biology. Most of this work has targeted IPNV, VHSV, and IHNV of salmonids (Leong and Fryer 1993). As already noted, a commercial DNA recombinant vaccine for IPNV is available in Norway. The recombinant subunit vaccine, in the form of a crude bacterial lysate, protected fish against IPNV via immersion vaccination. A subunit VHSV vaccine has been produced by using the viruses surface glycoprotein and inserting the gene into the bacterium *Escherichia coli,* making it possible for large quantities of viral glycoprotein to be inexpensively produced (Thiery et al. 1990). An experimental IHNV vaccine has been constructed using this same technology; a crude lysate of *E. coli* containing fusion IHNV protein was used to immunize several salmonid species from challenge with a lethal dose of IHNV (Xu et al. 1991). In a field trial, fish were immersed in IHNV subunit vaccine and mortality to virus was reduced from 27% to about 4% (Leong et al. 1992). For virus vaccines to provide protection against most serotypes of a specific target virus, it is important to identify different strains of the particular virus (Novoa et al. 1995).

Some research has been directed at producing a vaccine for the protozoan *I. multifiliis* (Goven et al. 1981; Dickerson et al. 1984). Although research in this area continues, no practical protozoan vaccine has been developed.

Adjuvants

Immunostimulants or adjuvants in fish vaccine preparations have recently received attention. These substances include glucans and other yeast extracts, lipopolysaccharides (LPS), extracts of abalone, rough mutants, natural or synthetic oil-based materials, and prevaccination salt baths. Adjuvants are added to oral, injection, or immersion vaccines to enhance immune response, increase longevity, and/or broaden effects of the vaccines, while decreasing pathogen specificity. Anderson (1992) listed 19 different immunostimulants, adjuvants, or vaccine carriers, but not all were equally suitable or effective for fish. Press and Lillehaug (1995) identified bovine lactoferrin given orally, or levamisole in an immersion preparation, as showing promise for fish vaccination. Saponin (Quil-A) administered orally or anally increases absorption of human gamma globulin in tilapia.

Methods of Vaccine Application

Fish vaccines are applied by injection, immersion, spraying (Figure 5.2), or orally in feed. Injectable vaccines, either IP, IM, or subcutaneously provide the highest level of protection and require relatively small amounts of vaccine. They are economical for use with larger and highly valuable fish, and multivalent vaccine preparations and adjuvants can be conveniently included as needed. Vaccination by injection against multiple disease organisms, or different serological strains of the same

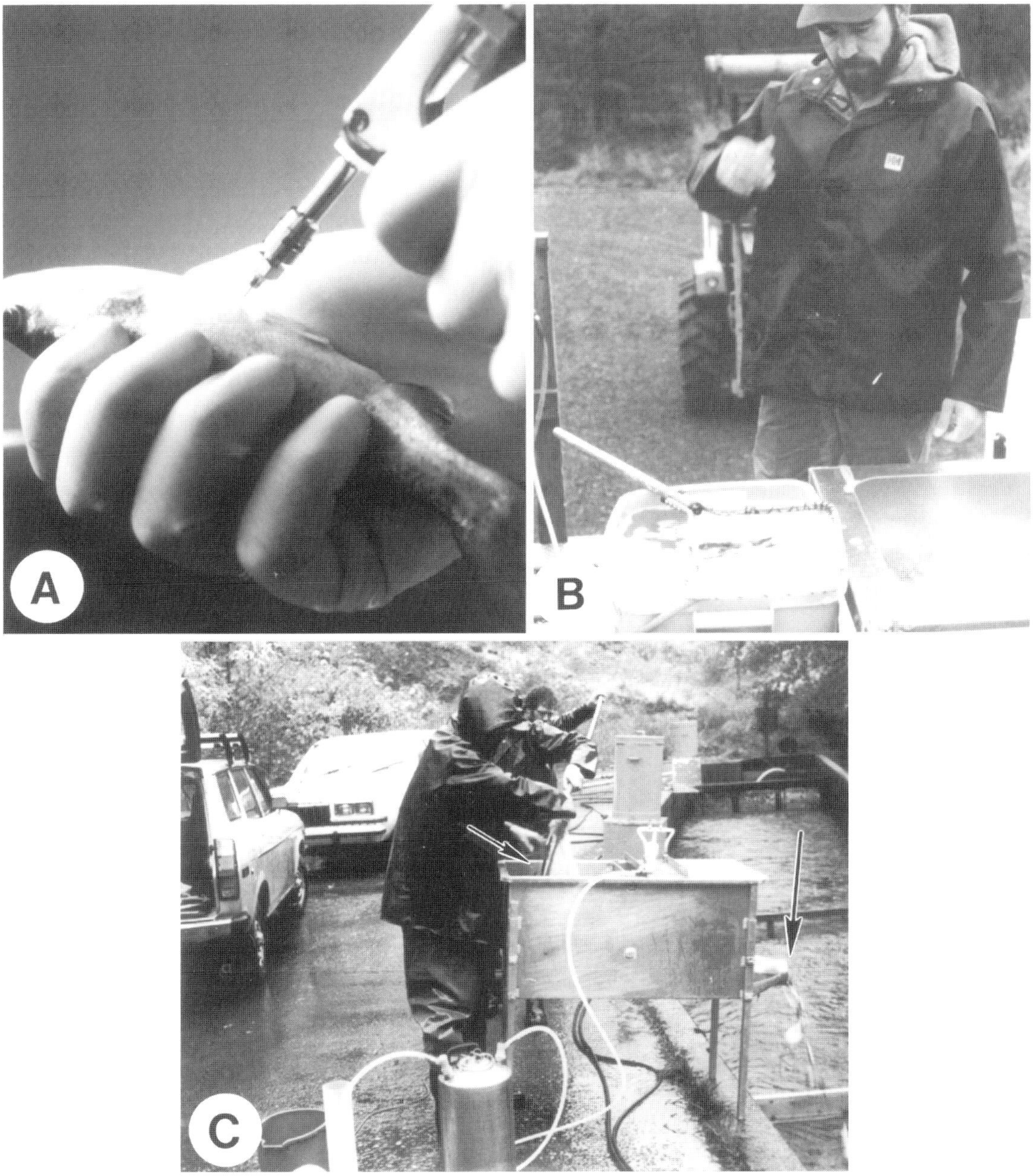

FIGURE 5.2. Vaccinating fish. (A) *Vaccinating trout by intraperitoneal injection.* (B) *Vaccination by immersion for specified period.* (C) *Vaccination by spraying the vaccine onto fish as they pass through the shower apparatus. Unvaccinated fish netted into top* (small arrow) *and vaccinated fish coming out of tube* (large arrow). *(Photographs courtesy of David Powell, ALPHARMA, Bellevue, Washington.)*

species simultaneously, is economically superior to other procedures and also provides a higher degree of protection (Press and Lillehaug 1995). Although semiautomated injection equipment is available, any vaccination by injection is labor intensive and requires handling of individual fish. Injection is impractical for devastating diseases of juvenile fish in their first few months of life, and there is also a potential safety hazard to persons administering the vaccine.

Immersion vaccination constitutes dipping fish in a solution of diluted vaccine for seconds to minutes and allowing them to absorb the antigen through the skin, across the gill membrane, and/or by ingestion (Figure 5.2). Immersion is widely used for vaccinating fish and can be easily incorporated into the normal culture routine. Immersion often produces an acceptable level of protection, and fish are less stressed than when injected. Disadvantages include the need to handle fish, that vaccination of larger fish is not economical, and that the procedure can be labor intensive, although use of semiautomated equipment will expedite the process.

Spray or shower is a modification of immersion vaccination that provides a high level of protection and can accommodate a higher weight of fish per unit of vaccine volume (Figure 5.2). Disadvantages include a need to handle fish, that it is labor intensive, and that specialized equipment is required. A semiautomated system has been developed to expedite spray vaccination.

Oral vaccine is logistically easy to apply, fish do not have to be handled, and protection is variable to moderate; however, the large quantity of vaccine required and the need for antigen encapsulation are disadvantages. Oral vaccination may best be suited to secondary or booster vaccinations. The efficacy of orally delivered fish vaccines is hindered by the potential for proteins (antigens) to be denatured in the fish's acidic stomach before they can be absorbed by the intestine and gain access to immunologically competent tissue. The problem can be circumvented by encapsulating or coating the antigen; however, this process is expensive. Plumb et al. (1994) demonstrated that protective immunity was obtained in channel catfish when a coated vaccine preparation of *E. ictaluri* was incorporated into their diet. A titration of the oral vaccine indicated that 0.5%, 1.0%, and 10% of vaccine (w/w) in the diet for two 5-day treatments, with a 5-day window between, significantly increased survival. Vaccines can also be bioencapsulated in fish food organisms such as *Artemia* nauplii, which are then fed to fish (Joosten et al. 1995).

Problems

When contemplating fish vaccination, several criteria must be considered. Newman (1993) noted 17 factors that influence protective immunity in fish, including vaccine characteristics, biological nature of fish and their nutritional well-being, and environmental conditions. Three important aspects of vaccination are age or size of fish, effect of temperature on immunity, and level and duration of protection. Most fish have a minimum age or size at which they become immunocompetent. Rainbow trout are immunocompetent at 0.5 g, but when vaccinated at a larger size, they will develop a stronger immunity (Johnson et al. 1982a). Duration of immunity also depends on fish size when vaccinated. Rainbow trout vaccinated at 1 g will retain immunity for about 120 days, for 180 days when vaccinated at 2 g, and for 1 year if immunized at 4 g (Johnson et al. 1982b).

Joosten et al. (1995) showed that efficacy of oral vaccination with encapsulated vaccine preparations can also depend on age of fish when immunized. *V. anguillarum* vaccine encapsulated into *Artemia nauplii* was used to vaccinate common carp orally when 15, 29, and 58 days old and gilthead seabream when 57 and 69 days old. Carp immunized at 58 to 69 days developed an insignificantly higher immune response, but seabream immunized at the same age produced a significantly higher response.

Environmental conditions under which fish are vaccinated are important because they can influence stress factors that adversely affect immune response (Ellis 1981). The immune system's ability to respond to antigens is compromised by poor water quality (low oxygen, high ammonia, etc.) and during stressful conditions, fish produce corticosteroids that suppress immune response (Pickering and Pottinger 1985). Sublethal concentrations of toxicants, particularly phenols, will severely reduce immunity and render aquatic animals more susceptible to infectious disease (Ellis 1988). High population density will also reduce immune response. It was shown that channel catfish vaccinated against *E. ictaluri* and placed in high stocking densities had poorer survival than vaccinated fish stocked at lower densities (Plumb et al. 1993). It has also been reported in the United Kingdom that ERM trout vaccinations failed when environmental conditions were less than favorable or when fish were in poor condition (Rogers 1991).

Water temperature is the principal immunomodulator in fish (Avtalion 1981; Johnson et al. 1982b; Bly and Clem 1991a). Temperature does not determine whether an immune response occurs, but it regulates the rapidity and degree to which it develops. The optimum temperature for immunization corresponds to the optimum temperature in which the fish normally lives. Generally, salmonids respond better when vaccinated at 15 to 18°C, whereas warmwater fish respond better when vaccinated at temperatures above 20°C. Plumb et al. (1986) showed that channel catfish immunized and held at 12 to 15°C developed a slow but longer-lasting humoral antibody response than did fish immunized and maintained at 28°C. When warmwater fish are vaccinated and held at an optimum temperature for a few days to 2 weeks before temperature is reduced, the immune system is primed and a memory may occur that will later become activated (Avtalion 1981; Plumb et al. 1986). Although Bly and Clem (1991a) demonstrated temperature-dependent immunity, they proposed that channel catfish may be immunocompromised during winter. It appears that immunization at low temperatures may actually suppress immune response, but a secondary response may occur when fish are again exposed to that antigen when water temperatures warm (Bly and Clem 1991b). Low water temperatures suppress T-dependent lymphocyte antibody functions in channel catfish and possibly other fishes as well (Bly et al. 1986). Seasonal modulation of immune response may also occur in salmonids (Zeeman 1986). Rainbow trout vaccinated and held at 18°C before winter, when water temperatures would be declining, produced lower antibody titers than fish vaccinated at the same temperature just prior to spring, when water temperatures would be rising.

Except in some regions where trout and salmon are routinely vaccinated for specific diseases, vaccination does not yet play an intricate role in health maintenance of fish. When taking into consideration the current attitude toward drugs, namely: time, expense, and legal questions involved in new drug approval; low drug approval rate; and current effort being expended in vaccine research and development, one would assume that vaccination will eventually become a more routine and accepted part of aquaculture.

REFERENCES

Alderman, D. J., and C. Michel. 1992. Chemotherapy in aquaculture today. In: *Chemotherapy in Aquaculture*, edited by C. Michel and D. J. Alderman, 3–24. Paris: Office International des Epizooties.

Anderson, D. P. 1992. Immunostimulants, adjuvants, and vaccine carriers in fish: Applications to aquaculture. *Annual Reviews of Fish Diseases* 2:281–307.

Austin, B., J. Rayment, and D. J. Alderman. 1983. Control of furunculosis by oxolinic acid. *Aquaculture* 31:101–108.

Avtalion, R. R. 1981. Environmental control of the immune response in fish. *CRC Critical Reviews in Environmental Contamination* 11:163–188.

Bennett, A. J., P. W. Whitby, and G. Coleman. 1992. Retention of antigenicity by a fragment of *Aeromonas salmonicida* 70 kDa serine protease which includes the primary substrate binding site expressed as 8-galactosidase hybrid proteins. *Journal of Fish Diseases* 15:473–484.

Bly, J. E., and L. W. Clem. 1991a. Temperature-mediated processes in teleost immunity: in vitro immunosupression induced by in vivo low temperature in channel catfish. *Veterinary Immunology and Immunopathology* 28:365–377.

Bly, J. E., and L. W. Clem. 1991b. Temperature-mediated processes in teleost immunity: in vivo low temperature immunization does not induce tolerance in channel catfish. *Fish & Shellfish Immunology* 1991: 229–231.

Bly, J. E., T. M. Buttke, E. F. Meydrech, and L. W. Clem. 1986. The effects of in vivo acclimation temperature on the fatty acid composition of channel catfish (*Ictalurus punctatus*) peripheral blood cells. *Comparative Biochemistry and Physiology* 83:791–795.

Bowser, P. R., G. A. Wooster, and J-M. Hsu. 1994. Laboratory efficacy of enrofloxacin for the control of *Aeromonas salmonicida* infection in rainbow trout. *Journal of Aquatic Animal Health* 6:288–291.

Boyd, C. E. 1990. *Water Quality in Ponds for Aquaculture.* Auburn, AL: Alabama Agricultural Experiment Station, Auburn University.

Boyd, C. E., and S. Pippopinyo. 1994. Factors affecting respiration in dry pond bottom soils. *Aquaculture* 120:283–293.

Boyd, C. E., and B. J. Wattson. 1989. Aeration systems in Aquaculture. *CRC Critical Reviews in Aquatic Sciences* 1:425–472.

Boyd, C. E., R. P. Romaire, and E. Johnston. 1978. Predicting early morning dissolved oxygen concentrations in channel catfish ponds. *Transactions of the American Fisheries Society* 107:484–492.

Boyd, C. E., P. Munsiri, and B. J. Hajek. 1994. Composition of sediment from intensive shrimp ponds in Thailand. *World Aquaculture* 25:53–55.

Chaslus-Dancla, E. 1992. Les probleme d'antibioresistance chez les especes animales autres que les poissons. In: *Chemotherapy in Aquaculture*, edited by C. Michel and D. J. Alderman, 243--253. Paris: Office International des Epizooties.

Cooper, R. K., II, E. B. Shotts, Jr., and L. K. Nolan. 1996. Use of a mini-transposon to study chondroitinase activity associated with *Edwardsiella ictaluri*. *Journal of Aquatic Animal Health* 8:319–324.

Dickerson, H. W., J. Brown, D. L. Dawe, and J. B. Gratzek. 1984. *Tetrahymena pyriformes* as a protective antigen against *Ichthyophthirius multifiliis* infections: comparison between isolates and ciliary preparations. *Journal of Fish Biology* 24:423–528.

Dixon, B. A. 1994. Antibiotic resistance of bacterial fish pathogens. *Journal of the World Aquaculture Society* 25:60–63

Doulos, S. K., A. J. Garland, J. R. Marshall, and M. D. White. 1994. Comparison of two methods of oxygen supplementation for enhancing water quality in fish culture. *The Progressive Fish-Culturist* 56:130–134.

Duff, D. C. B. 1942. The oral immunization of trout against *Bacterium salmonicida*. *Journal of Immunology* 44:87–94.

Ellis, A. E. 1981. Stress and the modulation of defense mechanisms in fish. In: *Stress and Fish*. edited by A. D. Pickering, 147–169. London: Academic Press.

Ellis, A. E. 1988. *Fish Vaccination*. London: Academic Press.

Extralabel Drug Use in Animals. 1996. Department of Health and Human Services, Food and Drug Administration. *Federal Register*, 61(217):57731–57746

Goven, B. A., D. L. Dawe, and J. B. Gratzek. 1981. Protection of channel catfish (*Ictalurus punctatus*) against *Ichthyophthirius multifiliis* (Fouget) by immunization with varying doses of *Tetrahymena pyriformes* (Lwoff) cilia. Aquaculture 23:269–273.

Groman, D. B., and G. W. Klontz. 1983. Chemotherapy and prophylaxis of bacterial kidney disease with erythromycin. *Journal of the World Mariculture Society* 14:226–235.

Guide to drug, vaccine, and pesticide use in aquaculture. 1994. Prepared by the Federal Joint Subcommittee on Aquaculture, U.S. Department of Agriculture.

Herwig, N. 1979. *Handbook of Drugs and Chemicals Used in the Treatment of Fish Diseases*. Springfield, MO: Charles C. Thomas.

Johnson, K. A., J. K. Flynn, and D. F. Amend. 1982a. Onset of immunity in salmonid fry vaccinated by direct immersion in *Vibrio anguillarum* and *Yersinia ruckeri* bacterins. *Journal of Fish Diseases* 5:197–205.

Johnson, K. A., J. K. Flynn, and D. F. Amend. 1982b. Duration of immunity in salmonids vaccinated by direct immersion in *Yersinia ruckeri* and *Vibrio anguillarum*. *Journal of Fish Diseases* 5:207–213.

Joosten, P. H. M., M. Avilés-Trigueros, P. Sorgeloos, and J. H. W. M. Rombout. 1995. Oral vaccination of juvenile carp (*Cyprinus carpio*) and gilthead seabream (*Sparus aurata*) with bioencapsulated *Vibrio anguillarum* bacterin. *Fish & Shellfish Immunology* 5:289–299.

Klesius, P. H., and C. A. Shoemaker. 1999. The development and use of modified live *Edwardsiella ictaluri* vaccine against enteric septicemia of catfish. In: *Veterinary Vaccines and Diagnostics*, edited by Ronald D. Schultz. Vol. 41, *Advances in Veterinary Medicine* 523–537.

Lai-Fa, Z., and C. E. Boyd. 1988. Nightly aeration to increase the efficiency of channel catfish production. *The Progressive Fish-Culturist* 50:237–242.

Lawrence, M. L., R. K. Cooper, and R. L. Thune. 1997. Attenuation, persistence, and vaccine potential of an *Edwardsiella ictaluri purA* mutant. *Infection and immunity* 65:4642–4651.

Leong, J. C., and J. L. Fryer. 1993. Viral vaccines for aquaculture. *Annual Reviews of Fish Diseases* 3:225–240.

Leong, J. C., and ten others. 1992. Biotechnological approaches to the development of salmonid fish vaccines. In: *Proceedings of the OJI International Symposium on Salmonid Diseases*, edited by T. Kimura, 250–255. Sapporo, Japan: Hokkjaido University Press.

Lorio, W. J. 1994. Production of channel catfish in ponds with water recirculation. *The Progressive Fish-Culturist* 56:202–206.

Lovell, T. 1989. *Nutrition and Feeding of Fish*. New York: Van Nostrand Reinhold.

Minor use guidance document: A guide to the approval of animal drugs for minor uses and for minor species [draft]. 1996. Rockville, MD: FDA Center for Veterinary Medicine, Food and Drug Administration.

Moffitt, C. M. 1992. Survival of juvenile chinook salmon challenged with *Renibacterium salmoninarum* and administered oral doses of erythromycin thiocyanate for different durations. *Journal of Aquatic Animal Health* 4:119–125.

Newman, S. G. 1993. Bacterial vaccines for fish. *Annual Reviews of Fish Diseases* 3:145–185.

Novoa, B., S. Blake, B. L. Nicholson, and A. Figueras. 1995. Comparison of different procedures for serotyping aquatic birnavirus. *Applied and Environmental Microbiology* 61:1925–2929.

Pickering, A. D., and T. G. Pottinger. 1985. Cortisol can increase the susceptibility of brown trout *Salmo trutta* L., to disease without reducing the white blood cell count. *Journal of Fish Biology* 27:611–619.

Piedrahita, R. L. 1991. Simulation of short-term management actions to prevent oxygen depletion in ponds. *Journal of the World Aquaculture Society* 22(3): 157–166.

Plumb, J. A., and S. Vinitnantharat. 1990. Dose titration of sarafloxacin (A-56620) against *Edwardsiella ictaluri* infection in channel catfish. *Journal of Aquatic Animal Health* 2:194–197.

Plumb, J. A., M. L. Wise, and W. A. Rogers. 1986. Modulary effects of temperature on antibody response and specific resistance to challenge of channel catfish, *Ictalurus punctatus,* immunized against *Edwardsiella ictaluri. Veterinary Immunology and Immunopathology* 12:297–304.

Plumb, J. A., S. Vinitnantharat, V. Abe, and R. P. Phelps. 1993. Density-dependent effect on oral vaccination of channel catfish against *Edwardsiella ictaluri. Aquaculture* 122:91–96.

Plumb, J. A., S. Vinitnantharat, and W. D. Paterson. 1994. Optimum concentration of *Edwardsiella ictaluri* vaccine in feed for oral vaccination of channel catfish. *Journal of Aquatic Animal Health* 6:118–121.

Plumb, J. A., C. C. Sheifinger, T. R. Shryock, and T. Goldsby. 1995. Susceptibility of six bacterial pathogens of channel catfish to six antibiotics. *Journal of Aquatic Animal Health* 7:211–217.

Press, C. M., and A. Lillehaug. 1995. Vaccination in European salmonid aquaculture: A review of practices and prospects. *British Veterinary Journal* 151:45–69.

Rogers, C. J. 1991. The use of vaccination and antimicrobial agents for control of *Yersinia ruckeri. Journal of Fish Diseases* 14:291–301.

Rogers, W. A., and D. Xu. 1992. Protective immunity induced by a commercial *Vibrio* vaccine in hybrid striped bass. *Journal of Aquatic Animal Health* 4:303–305.

Schlotfeldt, H. J. 1992. Current practices of chemotherapy in fish culture. In: *Chemotherapy in Aquaculture: From Theory to Reality,* edited by C. Michel and D. J. Alderman, 25–38. Paris: Office International des Epizooties.

Schnick, R. A. 1992. An overview of the regulatory aspects of chemotherapy in aquaculture. In: *Chemotherapy in Aquaculture: From Theory to Reality,* edited by C. Michel and D. J. Alderman, 71–79. Paris: Office International des Epizooties.

Schnick, R. A., F. P. Meyer, and D. L. Gray. 1989. *A guide to approved chemicals in fish production and fishery resource management.* Washington, DC: University of Arkansas Cooperative Extension Service and U.S. Fish and Wildlife Service.

Shoemaker, C. A., and P. H. Klesius. 1997. Protective immunity against enteric septicaemia in channel catfish, *Ictalurus punctatus* (Rafinesque), following controlled exposure to *Edwardsiella ictaluri. Journal of Fish Diseases* 320:361-368.

Shoemaker, C. A., P. H. Klesius, and J. A. Plumb. 1997. Killing of *Edwardsiella ictaluri* by macrophages from channel catfish immune and susceptible to enteric septicemia of catfish. *Veterinary Immunology and Immunopathology* 58:181–190.

Stoffregen, D. A., P. R. Bowser, and J. G. Babish. 1996. Antibacterial chemotherapeutants for finfish aquaculture: a synopsis of laboratory and field efficacy and safety studies. *Journal of Aquatic Animal Health* 8:181–207.

Straus, D. L., and C. S. Tucker. 1993. Acute toxicity of copper sulfate and chelated copper to channel catfish *Ictalurus punctatus. Journal of the World Aquaculture Society* 24:390–395.

Thiery, M., F. Lecocq-Xhonneux, I. Dheur, A. Renard, and P. de Kinkelin. 1990. Molecular cloning of the mRNA coding for the G protein of the viral haemorrhagic septicaemia (VHS) of salmonids. *Veterinary Microbiology* 23:221–226.

Thune, R. L., R. Cooper, D. H. Fernandez, and M. L. Lawrence. 1997. Live attenuated mutants of *Edwardsiella ictaluri* are effective vaccines against enteric septicemia of catfish (ESC). *Catfish Farmers of American Research Symposium,* Nashville, TN. Abstract.

Tucker, C. S., and C. E. Boyd. 1977. Relationships between potassium permanganate treatment and water quality. *Transactions of the American Fisheries Society* 106:481–488.

Tucker, C. S., J. A. Steeby, J. E. Waldrop, and A. B. Garrard. 1992. Production characteristics and economic performance for four channel catfish, *Ictalurus punctatus,* pond stocking density-cropping system combinations. In: *Recent Developments in Catfish Aquaculture,* edited by D. Tave and C. S. Tucker, 333–352. Binghamton, NY: Food Products Press.

Uno, K., M. Kato, T. Aoki, S. S. Kubota, and R. Ueno. 1992. Pharmacokinetics of nalidixic acid in cultured rainbow trout and amago salmon. *Aquaculture* 102:297–307.

Vallejo, A. N., N. W. Miller, and L. W. Clem. 1992. Antigen processing and presentation in teleost immune response. *Annual Review of Fish Diseases* 2:73–90.

Vaughan, L. M., P. R. Smith, and T. J. Foster. 1993. An aromatic-dependent mutant of the fish pathogen *Aeromonas salmonicida* is attenuated in fish and is effective as a live vaccine against the salmonid disease furunculosis. *Infection and Immunity* 61:2172–2181.

Wedemeyer, J. W. 1996. *Physiology of Fish in Intensive Culture Systems.* New York: Chapman & Hall.

Wellborn, T. L., Jr. 1985. Control and therapy. In: *Principal Diseases of Farm Raised Catfish,* edited by J. A. Plumb, 50–67. Southeastern Cooperative Series Bulletin No. 225. Auburn, AL: Alabama Agricultural Experiment Station, Auburn University.

Xu, L., D. V. Mourich, H. M. Engelking, S. Ristow, J. Arnzen, and J. C. Leong. 1991. Epitope mapping and characterization of the infectious hematopoietic necrosis virus glycoprotein, using fusion proteins synthesized in *Escherichia coli. Journal of Virology* 65:1611–1615.

Zeeman, M. G. 1986. Modulation of the immune response in fish. *Veterinary Immunology and Immunopathology* 12:235–241.

II
Viral Diseases

Compared with the research history of bacterial and parasitic diseases of aquatic animals, the study of fish viral diseases is relatively new. Infectious pancreatic necrosis virus (IPNV) of salmonids was the first proved viral fish disease, even though some might argue that infectious hematopoietic necrosis virus (IHNV) was demonstrated earlier. Infectious pancreatic necrosis virus was described by Wood et al. (1955) and viral etiology demonstrated by Wolf et al. (1959 and 1960). The first major literary compilation on fish viral diseases was *Viral Diseases of Poikilothermic Vertebrates,* published by the New York Academy of Sciences (Whipple 1965) and included contributions by fisheries scientists from the United States, Europe, and Asia. The most extensive work published on fish viruses to date is *Fish Viruses and Fish Viral Diseases* (Wolf 1988), a detailed scholarly documentation of all fish viruses and viral diseases known at that time.

The study of fish virology has grown rapidly since its inception. McAllister (1979) discussed about 30 virus diseases that affect fish. Wolf (1988) listed 59 fish diseases that were in some way associated with viral agents, 35 of which had been isolated in tissue culture. In 1993, the number of fish virus diseases was updated to approximately 95 (Hetrick and Hedrick 1993). Ahne (1994) indicated, however, that only about 60 different viruses were known from finfish.

Increased fish virus detection may be attributed to more extensive surveillance, a greater number of fish health–oriented laboratories that have virus detection and research capabilities, the conduciveness of intensive aquaculture to expression of viral infections, the availability of more and better-trained fish disease diagnosticians, and development of improved and more sensitive virus-detection methods and reagents. Research of viral fish diseases remains a high priority because of their potential to cause catastrophic losses and the tumor-producing ability of some of these agents. Several new fish viruses are being discovered every year.

Comparatively few fish viruses cause severe disease in aquaculture, but when they do, results can be devastating. Most known fish viruses have been reported in freshwater cultured species of high economic value, some occur in marine fish only, and others are found in both environments. Until fish are exposed to environmental stressors, the devastating potential of some viruses may not be realized; therefore, stressors often stimulate discovery and identification of previously unknown viruses that affect natural freshwater and/or marine fish populations.

Although the emphasis of this book is on viral and bacterial diseases of cultured fishes, it is becoming increasingly clear that viruses are more widespread in natural fish populations than was once suspected. There is also evidence to indicate that in some instances, viruses are being transmitted directly from wild populations to cultured fishes and vice versa.

An important virus characteristic is that it must invade a living cell in order to replicate. Viruses cannot reproduce themselves but must rely on synthetic cell machinery to produce new virus. A virus may manifest itself in several ways, the most drastic being production of hemorrhagic inflammation and primary necrosis, which often results in high mortality. Tumorous growths on the skin and fins can be caused by a virus infection, or there may be total absence of pathology. Viruses vary in virulence; several are capable of killing a high percentage of infected fish (IPNV of trout), some have relatively low virulence (golden shiner virus), whereas others are avirulent (catfish reovirus). Also, virulent viruses can be carried for extended periods by healthy fish without the host showing any overt signs of infection.

Viral infections can be confirmed by isolation of the causative agent in cell culture, by producing an identical disease in susceptible animals by injection of cell-free filtrates from diseased animals, or by observation of the virus in tissues by electron microscopy. The first isolated fish virus in cell culture was reported by Wolf and Quimby (1962). This pioneering technology has led to major advancements in the field of fish virology during the last three decades. Other research tools such as serological, radiological, and other technological procedures are gradually being incorporated into fish virus detection but have not yet replaced tissue culture for this purpose.

Viruses often affect cultured cells in a particular way. Cell injury (cytopathic effect—CPE) can be detected with relatively low magnification light microscopy. The CPE produced by a virus in cell culture is commonly characteristic of a particular type or group of viruses but is seldom specific enough for definitive identification.

Multiple, interrelated factors determine to what extent a virus will affect a fish population; water temperature is among the most important. Typically, fish viruses have a temperature range in which they are most pathogenic. Channel catfish virus is a good example; usually severe mortality occurs in young-of-the-year fish when water temperatures are above 25°C, moderate mortality occurs at 21 to 24°C, and little, if any mortality occurs below 18°C. Age also affects a fish's susceptibility to virus infection, with younger fish generally suffering greatest mortality. Although severe mortality is rare among fish older than 1 year, they often become carriers and serve as virus reservoirs for as long as they remain alive. Stressful effects of transport, handling, poor water quality, high stocking densities, and inadequate nutrition can also affect severity of a virus infection.

Viral agents with the greatest potential to cause disease in cultured fish are included in this text and are organized according to the primary family or group of fish they infect. Only minor attention is given to nonvirulent viruses or those that have been detected only by electron microscopy. No attempt has been made to include all viruses that affect fish.

REFERENCES

Ahne, W. 1994. Viral infections of aquatic animals with special reference to Asian aquaculture. *Annual Review of Fish Diseases* 4:357–388.

Hetrick, F. M., and R. P. Hedrick. 1993. New viruses described in finfish from 1988–1992. *Annual Review of Fish Diseases* 3:187–207.

McAllister, P. E. 1979. Fish viruses and viral infections. In: *Comprehensive Virology,* edited by H. Fraenkel-Conrat and R. R. Wagnor, 401–470. New York: Plenum Press.

Whipple, H. E. (Editor) 1965. Viral diseases of poikilothermic vertebrates. *Annals of the New York Academy of Science* 126: 680.

Wolf, K. 1988. *Fish Viruses and Fish Viral Diseases.* Ithaca, NY: Cornell University Press.

Wolf, K., C. E. Dunbar, and S. F. Snieszko. 1960. Infectious pancreatic necrosis of trout, a tissue-culture study. *Progressive Fish Culturist* 22:64–68.

Wolf, K, and M. C. Quimby. 1962. Established erythermic line of fish cells in vitro. *Science* 135:1065.

Wolf, K., S. F. Snieszko, and C. E. Dunbar. 1959. Infectious pancreatic necrosis, a virus-caused disease of fish. *Excerptas Medica* 13:228. Abstract.

Wood, E. M., S. F. Snieszko, and W. T. Yasutake. 1955. Infectious pancreatic necrosis in trout. *American Medical Association Archives in Pathology* 60:26–28.

6 Catfish Viruses

The most serious virus disease of cultured catfishes is channel catfish virus disease (CCVD). There has also been a catfish reovirus (CRV) isolated from channel catfish, and iridoviruses have been found in black bullhead and European catfish (sheatfish or wels).

CHANNEL CATFISH VIRUS DISEASE

Channel catfish virus disease, an acute, highly communicable infection of cultured juvenile channel catfish was discovered in 1968 (Fijan et al. 1970). The etiological agent of CCVD is channel catfish virus (CCV), a member of the family Herpesviridae with *Herpesvirus ictaluri* suggested as the specific epithet (Wolf and Darlington 1971).

Geographical Range and Species Susceptibility

Channel catfish virus has been isolated from channel catfish in most southern states and several other areas of the United States. The virus was also isolated from a group of channel catfish fry shipped from the United States to Honduras in 1972. These fish were destroyed, and the virus apparently failed to become established in that country. There have also been reports of a disease resembling CCVD in other countries where channel catfish have been introduced.

Channel catfish appear to be the most susceptible catfish species to CCV. Experimental infections have been induced in fingerling blue catfish and channel catfish X blue catfish hybrids by virus injection, but no infection occurred when CCV was introduced orally or by cohabitation of naive fish with virus-infected channel catfish (Plumb and Chappel 1978). Contrary to experimental refractiveness to CCV, the virus was isolated from diseased fingerling blue catfish in Missouri in 1988 (Southeastern Cooperative Fish Disease Project 1988). Brown and yellow bullheads cannot be experimentally infected with the virus by injection or feeding. European catfish are also resistant to the virus (Plumb et al. 1985).

When CCV was fed to different strains of channel catfish fry, a variation in susceptibility among strains of the species was noted (Plumb et al. 1975). Young fish that were the product of crossbreeding between different channel catfish strains were more resistant to CCV than were inbred parental strains, which would suggest a strain hybrid vigor.

Clinical Signs

Clinical signs of CCV vary with some or all of the following being present (Figure 6.1): distension of the abdomen, exophthalmia, pale or hemorrhagic gills, and hemorrhage at the base of fins and throughout the skin on the ventral surface (Fijan et al. 1970). Infected fish swim erratically, sometimes rotating about their longitudinal axis. Moribund fish hang head up at the surface or sink to the bottom, become quiescent, and respire weakly but rapidly just prior to death.

Internal signs include presence of a clear straw-colored fluid in the peritoneal cavity and a general hyperemia throughout the visceral cavity, although the liver and kidneys can be pale. The spleen is generally dark red and enlarged, and the stomach and intestine are void of food but contain a mucoid secretion.

FIGURE 6.1. A 10-cm channel catfish naturally infected with channel catfish virus. Note the abdominal distension and exophthalmia. The upper lobe of the caudal fin (arrow) has a columnaris lesion.

Diagnosis

When a sudden increase in morbidity occurs among young channel catfish during the summer, CCV should be suspected and fish necropsied for virus. Virus is detected by inoculating filtrates made from whole fry or visceral tissue into appropriate cell cultures. The cell line of choice is channel catfish ovary (CCO) because of its high sensitivity to CCV and increased viral productivity (Bowser and Plumb 1980). Intranuclear CCV replication occurs in CCO cells at 15 to 35°C, with 30 to 35°C being optimum. Cytopathic effect may be visible in 12 to 24 hours at 30°C if an active CCV infection is in progress. Initial CPE is indicated by foci of pyknotic cells, which coalesce into a multinucleated syncytium that is connected to surrounding normal cells by strands of protoplasm resembling irregular wheel spokes (Figure 6.2). As CPE progresses, cells release from the culture vessel surface and form a loose network. Syncytia in susceptible cell cultures are presumptive identification of CCV; positive identification can be made by serum neutralization using CCV specific antiserum.

Brown bullhead (BB) cells and cell cultures derived from walking catfish kidney (K1K) are also susceptible to CCV, but they are not as sensitive as CCO cells (Bowser and Plumb 1980; Noga and Hartmann 1981). Cell lines developed from Japanese striped knife jaw and amberjack and the hybrid of kelp and spotted grouper showed CPE when infected with CCV. Other poikilothermic and mammalian cell culture systems are refractory to the virus (Fernandez et al. 1993).

Virus Characteristics

Although CCV has been recognized as a herpesvirus since its characterization, it has since been further categorized as an alphaherpesvirus (Roizman 1990) on the basis of its short life cycle and rapid destruction of tissue culture cells. Using immediate-early transcription from CCV genome, Silverstein et al. (1995) characterized two immediate-early transcripts that support CCV's close relationship to the alphaherpesvirus subfamily. Comparisons of the nucleotide sequence with other herpesviruses suggests that CCV is not closely related to any of the well-characterized ones (Davison 1992).

Channel catfish virus is an enveloped icosahedral virus with DNA genome and has a nucleocapsid diameter of 95 to 105 nm. Enveloped virions have a diameter of 175 to 200 nm. Infectivity is inactivated by either 20% ether or 5% chloroform. Channel catfish virus is heat labile at 60°C for 1 hour, unstable in seawater, and sensitive to ultraviolet light, requiring 20 to 40 minutes for inactivation (Robin and Rodrique 1980).

The virus survives for less than 24 hours on dried fish netting or glass cover slips; it retains infectivity in pond water for about 2 days at 25°C and for 28 days at 4°C. In dechlorinated tap water, infectivity is retained for 11 days at 25°C and for

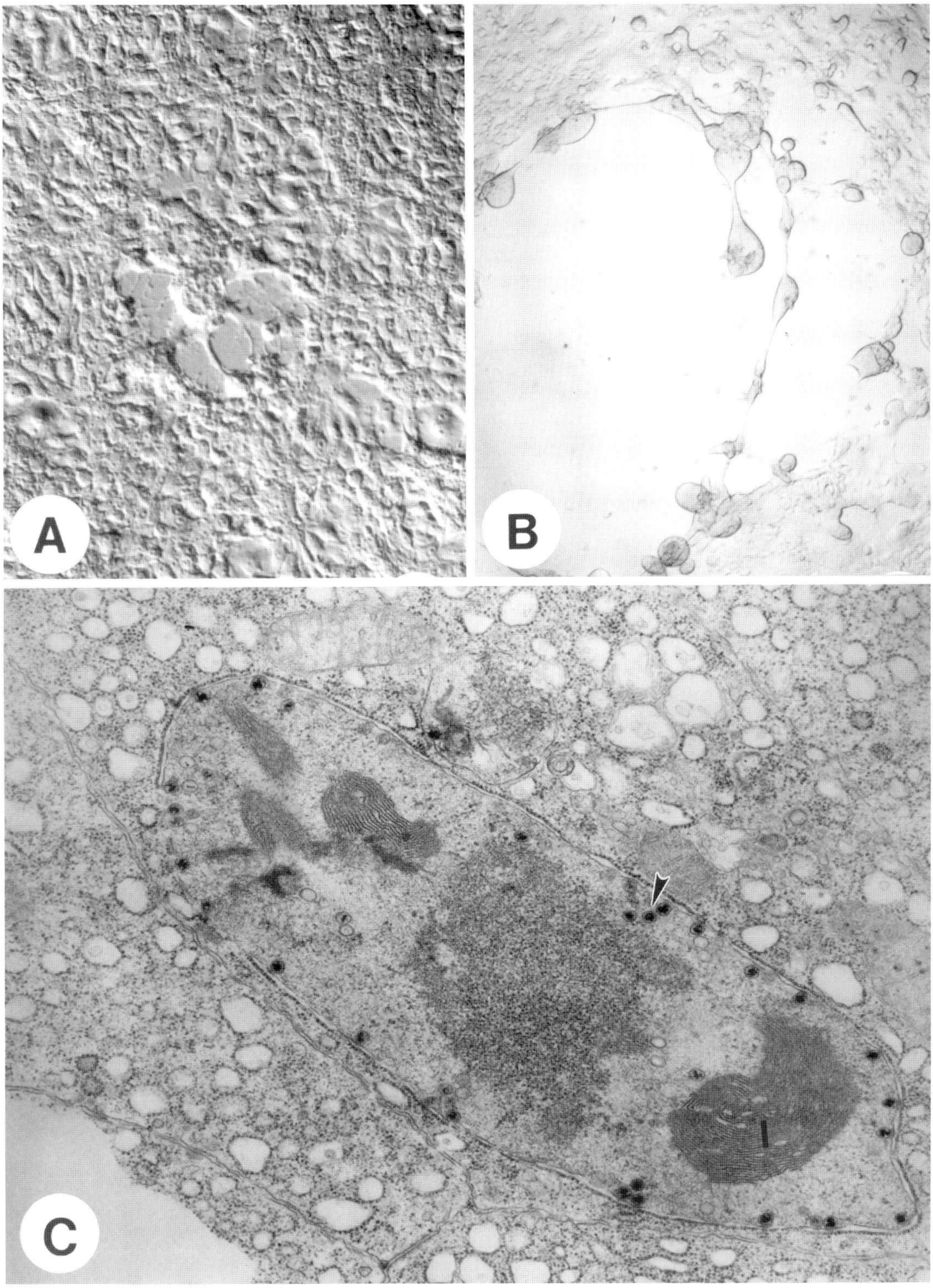

Figure 6.2. Channel catfish ovary cells infected with channel catfish virus and incubated at 30°C. (A) *Focal cytopathic effect (CPE) 19 hours after inoculation surrounded by normal cells.* (B) *Total CPE 24 hours after inoculation.* (C) *Channel catfish virus* (arrow) *in the nucleus of a spleen cell from experimentally infected channel catfish. Note the hyalinelike inclusion bodies* (I). *(From Plumb and Gaines [1974]; reprinted with permission of Blackwell Scientific Publications Ltd.)*

more than 2 months at 4°C. Under experimental conditions, CCV infectivity is almost immediately neutralized when introduced into pond mud. Infectious CCV cannot be isolated from decomposing infected fish at 22°C 48 hours after death. Virus is recoverable for up to 14 days from viscera of whole iced fish, for 162 days from fish frozen at –20°C, and for 210 days from fish frozen at –80°C (Plumb et al. 1973).

Epizootiology

Outbreaks of CCVD in channel catfish have been diagnosed during May through October, when water temperatures are above 25°C. In Alabama, 94% of 53 confirmed CCVD cases reported between 1991 and 1996 occurred during June through September, whereas none were diagnosed from November through April. Epizootics on catfish farms occur most frequently during years of high water temperature and in heavily stocked fingerling ponds. In many cases, outbreaks are preceded by handling and transport.

As long as active infection is present, fry and fingerling channel catfish transmit CCV horizontally by contact and through the water. Under experimental conditions, it is also possible to transmit virus from infected moribund or dead fish to healthy fish through water. Other CCV transmission modes are intramuscular or intraperitoneal injection, incorporation of virus into feed, or swabbing gills with a saline solution containing virus. Infectivity can be experimentally induced in fish weighing up to 10 g by waterborne exposure, but injection is required to infect larger fish. Occasionally, 1-year-old fish will suffer a CCVD outbreak, which is generally associated with low mortality and usually follows stressful transport or handling. Often CCVD occurs in conjunction with a secondary *Flavobacterium columnare* infection that prolongs mortality.

Incubation time between fish exposure to virus and appearance of clinical signs and morbidity is inversely related to water temperature (Plumb 1972a). Experimental infection at 30°C was followed by clinical signs in 32 to 42 hours, with death occurring several hours later (Plumb and Gaines 1975). At 20°C, incubation was 10 days. Under field conditions at 25 to 30°C, healthy channel catfish fingerlings develop disease within 72 to 78 hours postexposure, and up to 100% may die within 6 days. In a documented case of CCVD, a group of fry held at 28°C in troughs receiving well water developed clinical CCVD when 21 days old, and 72 hours later, 100% were dead (Plumb 1972b). These fish had been negative for CCV in cell culture when assayed 7 days prior to disease outbreak. Disease developed rapidly, even though infected fish were not exposed to a known source of virus between days 14 and 21; therefore, it must be assumed that the fry were infected by brood stock. Young-of-the-year fish less than 4 months old are most susceptible to CCV, and the younger and/or smaller the fish, the higher the mortality.

Channel catfish virus is communicable from the time clinical signs appear until soon after death. In experimentally infected juveniles, virus began to decline in surviving fish 120 hours postinfection (Plumb 1971). It is impossible to isolate virus by routine cell culture procedures once clinical signs disappear. Kancharla and Hanson (1996) found that CCV was shed by infected juveniles and could be detected in water for up to 12 days postinfection. Virus was also found in internal organs, skin, and gills. Fish infected with thymidine kinase (TK)–negative CCV shed less virus than do fish infected with unaltered virus. Recombinant TK–negative CCV will replicate in cell cultures; however, 100 times more virus is required to kill fish than to produce CPE in TK cells (Zhang and Hanson 1995).

Adult fish are considered to be possible sources of CCV infection by vertical transmission, but to date, this has not been conclusively demonstrated. Although active CCV infections are relatively easy to diagnose by virus isolation in cell culture, detection of covert CCV in a carrier population is difficult. Plumb (1973) detected CCV antibodies in channel catfish that had been exposed to virus and suggested that this may be a method for separating adult fish that had been exposed to CCV from those that had not. Amend and McDowell (1984) used serum neutralization tests on adult catfish to separate possible carrier from noncarrier populations. They later demonstrated that 7 of 17 major brood populations tested showed positive CCV serum neutralization. Virus was subsequently isolated from young-of-the-year fish at 2 of the 7 CCV antibody-positive farms. Channel catfish virus neutralizing antibodies were found in adult channel catfish for 2 years following last known exposure to virus (Hedrick et al. 1987). In this same study, adults not exposed to CCV as juveniles were experimentally infected, and one died apparently as a result of the virus. Surviving adults produced CCV antibody for 6 months, although titers were low. It was suggested that CCV does become latent in epizootic survivors and that expression of certain viral antigens, or periodic virus reactivation, may be a stimulus for continued anti-CCV antibody production.

Studies strongly support the idea that adult fish do carry CCV and can possibly develop active infections. Plumb et al. (1981) demonstrated that CCV nucleic acid could be detected in gonads of adult fish by immunofluorescence. Bowser et al. (1985) isolated CCV from adult channel catfish in winter when water temperatures were 6 to 8°C and later from individuals of the same population when fish were stressed by steroid injection. Wise et al. (1985) used a CCV-specific DNA probe to demonstrate clearly the presence of CCV nucleic acid in various tissues of adult channel catfish, including adults from which virus had been isolated. It was further shown by a CCV–nucleic acid probe that CCV genetic material was present in adults as well as their offspring, indicating vertical transmission (Wise et al 1988). Bird et al. (1988) used molecular cloning of CCV genome fragments to suggest that CCV can persist in channel catfish in a dormant or transcriptionally active state without causing clinical signs. Polymerase chain reaction (PCR) was used by Boyle and Blackwell (1991) to detect latent CCV, and Baek and Boyle (1996) refined the procedure with nested primers that enabled detection of CCV DNA in the blood of brood stock.

Pathological Manifestations

A generalized viremia is established within 24 hours after experimental infection. The kidneys, liver, spleen, and intestines become active sites of virus replication 24 to 48 hours after infection, and virus can sometimes be isolated from brain tissue after 48 hours (Plumb 1971). Peak virus titers occur in the kidney and intestine 72 hours after infection and in the spleen, brain, and liver at 96 hours. Virus titers in muscle tissue remains comparatively low.

Histopathological changes are similar in both natural and experimental CCV infections (Wolf et al. 1972; Plumb et al. 1974). Renal hematopoietic tissue is edematous and has extensive necrosis and cellular dissolution that is coupled with increased numbers of macrophages. The liver develops regional edema, necrosis, and hemorrhage, and hepatic cells have eosinophilic intracytoplasmic inclusions. Pancreatic acinar cells are necrotic. The submucosa of the digestive tract is edematous and has focal areas of macrophage concentrations and hemorrhage. The spleen shows extensive reduction of lymphoid tissue and becomes congested with erythrocytes. Virus particles have been seen in electron micrographs of the liver, kidney, and spleen of infected fish (Figure 6.2).

Significance

The effects of a CCVD outbreak on individual channel catfish farms can be great because of the potential for exceptionally high mortality among young-of-the-year fish. However, CCVD comprises only 1% to 2% of total disease cases recorded at fish disease diagnostic laboratories in geographical areas where channel catfish are cultured. Although individual outbreaks of CCVD can be very significant, its overall effect on the channel catfish industry has not been great.

OTHER VIRUSES OF CATFISH

Catfish reovirus has also been isolated from channel catfish but has not been a significant problem. On the other hand, an iridovirus isolated from black bullhead and sheatfish has been implicated in significant mortality in these species.

Catfish Reovirus

A reovirus was isolated from overtly healthy juvenile channel catfish during a CCV assay project in California (Amend et al. 1984). The virus, named catfish reovirus (CRV), is an aquareovirus that is nonpathogenic when injected into naive channel catfish. The virus replicates in BB and CHSE-214 (chinook salmon embryo) cells, but fathead minnow (FHM) cells are refractory. Optimum incubation temperature is 25 to 30°C. Characterization of CRV by Hedrick et al. (1984) shows that the icosahedral virions have a double capsid and a diameter of 75 nm. Nucleic acid staining suggests a double-stranded RNA with 11 segments. Serum neutralization tests made with CRV, chum salmon (CSV), and golden shiner virus (GSV), both of which are aquareoviruses, indicate that these viruses are distinctly different but yet some cross-reactivity occurs. Using morphological and biochemical properties of four aquareoviruses (CRV, GSV, CSV, and $13p_2$ from American oyster) Winton et al. (1987) showed that these viruses were unlike other genera of Reoviridae and formed the basis of a new genera, the aquareoviruses. Because CRV causes no pathology in injected channel catfish, it is considered to be insignificant.

Catfish Iridovirus

This virus was isolated from cultured black bullhead in France (Pozet et al. 1992). Diseased fish showed body and muscle edema and petechiae around pectoral and abdominal girdles and viscera. Gills were pale, and ascites was present in the body cavity. Catfish iridovirus (CIV) was isolated in epithelioma papillosum of carp (EPC) cells but also grew well in BF-2 and CCO cells at temperatures of 15 to 25°C. The virus forms very small round plaques under agarose overlays, with BF-2 cells being most adaptable to plaquing. The hexagonal virions measure 150 to 160 nm in diameter. Biochemical tests and inhibition tests with 5-iodo-2-deoxyuridine indicate a DNA genome; these results combined with virus morphology are compatible with iridovirus characteristics. The CIV was isolated from dead fish from a population in which 100% succumbed to infection during a 1-week period. Experimental infection trials via intraperitoneal injection and waterborne exposure of subadult and adult black bullheads held at 18°C resulted in high mortality. High levels of virus were found in internal organs (kidney and spleen) of infected fish, and necrosis occurred in the spleen and kidney. Subsequently, most virus-exposed fish died. The full significance of CIV is not yet known, but it could be important where black bullheads are cultured.

Sheatfish Iridovirus

The sheatfish iridovirus (SHV) was isolated in 1988 from cultured European catfish (sheatfish or wels) in a recirculating system in Germany (Ahne et al. 1990) and so far as is known the virus is confined to this species. Moribund European catfish fry infected with this virus show spiral swimming and hemorrhage in the skin and internal organs. The virus can be isolated in BF-2 cells when incubated at 25°C. Experimental infection of sheatfish fry by waterborne exposure resulted in 100% mortality within 8 days (Ogawa et al. 1990). Horizontal transmission from infected fish to naive fry resulted in 100% mortality in 11 days at 25°C.

Internal tissue pathology of infected fish included a generalized destruction of hematopoietic tissue of the spleen and kidney. The most common lesion of the gill was hyperplasia and edema of the epithelium in primary and secondary lamellae. Less severe pathology was seen in the skin, heart, eye, liver, pancreas, brain, and digestive tract. Indications are that SHV could be a serious problem in culture of European catfish.

MANAGEMENT OF CATFISH VIRUSES

The only practical measures for controlling catfish viruses are management, avoidance, quarantine of infected stocks, and proper sanitation. Because mortality of CCV-infected fish is closely correlated to water temperature, in certain circumstances a reduction in temperature from an optimum of greater than 25°C to less than 19°C will be beneficial. During periods when CCVD is most likely to occur (June through September), every effort should be made to eliminate any adverse environmental conditions in which susceptible channel catfish are being reared. High fingerling stocking rates should be avoided; water temperatures should be kept at moderate levels if possible; transportation of susceptible fish during summer should be minimized; and when bacterial infections develop, especially "columnaris," they should be treated.

Additional precautions to avoid CCV would include surveying brood stock for CCV-neutralizing antibodies or use of other sensitive virus-detection methods. As CCV-specific nucleic acid probes are perfected, detection of carrier fish may become more reliable. Application of PCR techniques may also provide an accurate and sensitive method for detecting CCV carrier fish. Baek and Boyle (1996) used a nested PCR procedure to detect fewer than 10 copies of CCV DNA (about 1 fg) in asymptomatic adult channel catfish. Although not yet available to the catfish industry because of technical and economic constraints, the PCR technique may be applicable to sensitive CCV detection and provide a means for avoiding this virus. Suhalim (1997) used a Falcon assay screening test (FAST)–enzyme linked immunosorbent assay (ELISA) procedure to measure CCV antibody in serum of juvenile and adult channel catfish that had been exposed to CCV or had produced CCV-infected offspring. There was no correlation between the FAST–ELISA optical density levels and serum neutralization titers. However, the FAST–ELISA distinctly separated sera from experimentally CCV-infected and noninfected channel catfish.

Vaccination is being considered as a means to combat CCVD. The virus becomes attenuated when passed frequently and rapidly in K1K cells (Walczak et al. 1981). When channel catfish are injected with, or bathed in, the attenuated virus they become immune to wild-type (virulent) CCV. The potential for developing a CCV subunit vaccine was explored by Awad et al. (1989), and Hanson and Thune (1993) used a genetic engineering approach to vaccine development. A thymidine kinase–negative CCV mutant used by Zhang and Hanson (1995) provided immunological protection of juvenile channel catfish when challenged with a lethal dose of wild-type CCV. The degree of protection provided by TK-negative CCV was related to mutant dose.

Ponds from which CCV diseased fish have been removed should be disinfected with 40 mg of chlorine per liter and then drained. Although evidence suggests that epizootic survivors may be stunted (McGlamery and Gratzek 1974), they can be grown to a marketable size, providing affected fish are segregated from ponds containing healthy, susceptible channel catfish. Under no circumstances should CCV epizootic survivors be stocked in noninfected waters, nor should they knowingly be used as brood stock. No management strategy for control of catfish reovirus or sheatfish iridoviruses has been established.

REFERENCES

Ahne, W., M. Ogawa, and J. J. Schlotfeldt. 1990. Fish viruses: transmission and pathogenicity of an icoshedral cytoplasmic deoxyribovirus isolated from sheatfish (*Silurus glanis*). *Journal of Veterinary Medicine* B37:187–190.

Amend, D. F., and T. McDowell. 1984. Comparison of various procedures to detect neutralizing antibody to the channel catfish virus in California brood channel catfish. *The Progressive Fish Culturist* 46:6–12.

Amend, D. F., T. McDowell, and R. P. Hedrick. 1984. Characteristics of a previously unidentified virus from channel catfish (*Ictalurus punctatus*). *Canadian Journal of Aquatic Sciences* 41:807–811.

Awad, M. A., K. E. Nusbaum, and Y. J. Brady. 1989. Preliminary studies of a newly developed subunit vaccine for channel catfish virus disease. *Journal of Aquatic Animal Health* 1:233–237.

Baek, Y.-S., and J. A. Boyle. 1996. Detection of channel catfish virus in adult channel catfish by use of a nested polymerase chain reaction. *Journal of Aquatic Animal Health* 8:97–103.

Bird, R. C., K. E. Nusbaum, A. E. Screws, R. R. Young-White, J. M. Grizzle, and M. Toivio-Kinnucan. 1988. Molecular cloning of fragments of the channel catfish virus (Herpesviridae) genome and expression of the encoded mRNA during infection. *American Journal of Veterinary Research* 49:1850–1855.

Bowser, P. R., and J. A. Plumb. 1980. Channel catfish virus: comparative replication and sensitivity of cell lines from catfish ovary and the brown bullhead. *Journal of Wildlife Diseases* 16:451–454.

Bowser, P. R., A. D. Munson, H. H. Jarboe, R. Francis-Floyd, and R. P. Waterstrat. 1985. Isolation of channel catfish virus from channel catfish *Ictalurus punctatus* (Rafinesque) broodstock. *Journal of Fish Diseases* 8:557–561.

Boyle, J., and J. Blackwell. 1991. Use of polymerase chain reaction to detect latent channel catfish virus. *American Journal of Veterinary Research* 52:1965–1968.

Davison, A. J. 1992. Channel catfish virus: A new type of herpesvirus. *Virology* 186:9–14.

Fernandez, R. D., M. Yoshimizu, T. Kimura, Y. Ezura, K. Inouye, and I. Takami. 1993. Characterization of three continuous cell lines from marine fish. *Journal of Aquatic Animal Health* 5:127–136.

Fijan, N. N., T. L. Wellborn, Jr., and J. P. Naftel. 1970. An acute viral disease of channel catfish. *U. S. Fish Wildlife Service Technical Paper* no. 43, 11 p.

Hanson, L., and R. L. Thune. 1993. Characterization of thymidine kinase encoded by channel catfish virus. *Journal of Aquatic Animal Health* 5:199–204.

Hedrick, R. P., R. Rosemark, D. Aronstein, J. R. Winton, T. McDowell, and D. F. Amend. 1984. Characteristics of a new reovirus from channel catfish (*Ictalurus punctatus*). *Journal of General Virology* 65:1527–1534.

Hedrick, R. P., J. M. Groff, and T. McDowell. 1987. Response of adult channel catfish to waterborne exposures of channel catfish virus. *The Progressive Fish-Culturist* 49:181–187.

Kancharla, S. R., and L. A. Hanson. 1996. Production and shedding of channel catfish virus (CCV) and thymidine kinase negative CCV in immersion exposed channel catfish fingerlings. *Diseases of Aquatic Organisms* 27:25–34.

McGlamery, M. H., Jr., and J. B. Gratzek. 1974. Stunting syndrome associated with young channel catfish that survived exposure to channel catfish virus. *The Progressive Fish-Culturist* 36:38–41.

Noga, E., and J. X. Hartmann. 1981. Establishment of walking catfish (*Clarias batrachus*) cell lines and development of a channel catfish (*Ictalurus punctatus*) virus vaccine. *Canadian Journal of Fisheries and Aquatic Sciences* 38:925–930.

Ogawa, M., W. Ahne, T. Fischer-Scherl, R. W. Hoffmann, and J. J. Schlotfeldt. 1990. Pathomorphological alterations in sheatfish fry *Silurus glanis* experimentally infected with an iridovirus-like agent. *Diseases of Aquatic Organisms* 9:187–191.

Plumb, J. A. 1971. Tissue distribution of channel catfish virus. *Journal of Wildlife Diseases* 7:213–216.

Plumb, J. A. 1972a. Effects of temperature on mortality of fingerling channel catfish (*Ictalurus punctatus*) experimentally infected with channel catfish virus. *Journal of the Fisheries Research Board of Canada* 30:568–570.

Plumb, J. A. 1972b. Some biological aspects of channel catfish virus disease. Auburn, AL. Auburn University. Doctoral Dissertation.

Plumb, J. A. 1973. Neutralization of channel catfish virus by serum of channel catfish. *Journal of Wildlife Diseases* 9:324–330.

Plumb, J. A., and J. Chappel. 1978. Susceptibility of blue catfish to channel catfish virus. *Proceedings of the Annual Conference Southeastern Association of Fish Wildlife Agencies* 32:680–685.

Plumb, J. A., and J. L. Gaines, Jr. 1975. Channel catfish virus disease. In: *The Pathology of Fishes,* edited by R. E. Ribelin and G. Migaki, 287–302. Madison, WI: University of Wisconsin Press.

Plumb, J. A., L. D. Wright, and V. L. Jones. 1973. Survival of channel catfish virus in chilled, frozen and decomposing channel catfish. *The Progressive Fish-Culturist* 35:170–172.

Plumb, J. A., J. L. Gaines, E. C. Mora, and G. G. Bradley. 1974. Histopathology and electron microscopy of channel catfish virus in infected channel catfish, *Ictalurus punctatus* (Rafinesque). *Journal of Fish Biology* 6:661–664.

Plumb, J. A., O. L. Green, R. O. Smitherman, and G. B. Pardue. 1975. Channel catfish virus experiments with different strains of channel catfish. *Transactions of the American Fisheries Society* 104:140–143.

Plumb, J. A., R. L. Thune, and P. H. Klesius. 1981. Detection of channel catfish virus in adult fish. *Developments in Biological Standardization* 49:29–34.

Plumb, J. A., V. Hilge, and E. F. Quinlan. 1985. Resistance of the European catfish (*Silurus glanis*) to channel catfish virus. *Journal of Applied Ichthyology* 1:87–89.

Pozet, F., M. Morand, A. Moussa, C. Torhy, and P. de Kinkelin. 1992. Isolation and preliminary characterization of a pathogenic icosahedral deoxyribovirus from the catfish (*Ictalurus melas*). *Diseases of Aquatic Organisms* 14:35–42.

Robin, J., and A. Rodrique. 1980. Resistance of herpes channel catfish virus (HCCV) to temperature, pH, salinity and ultraviolet irradiation. *Reviews in Canadian Biology* 39:153–156.

Roizman, B. 1990. Herpesviridae: a brief introduction. In: *Virology,* 2nd ed., edited by B. N. Fields et al., 841–847. New York: Raven Press, Ltd.

Silverstein, P. S., R. C. Bird, V. L. van Santen, and K. E. Nusbaum. 1995. Immediate-early transcription from the channel catfish virus genome: characterization of two immediate-early transcripts. *Journal of Virology* 69:3161–3166.

Southeastern Cooperative Fish Disease Project. 1988. 24th Annual Report. Auburn, AL: Auburn University.

Suhalim, R. 1997. Detection of channel catfish virus antibody using Falcon assay screening test-enzyme linked immunosorbent assay. Auburn, AL: Auburn University. M.S. thesis.

Walczak, E. M., E. J. Noga, and J. X. Hartmann. 1981. Properties of a vaccine for channel catfish virus disease and a method of administration. *Developments in Biological Standardization* 49:419–429.

Winton, J. R., C. N. Lannan, J. L. Fryer, R. P. Hedrick, T. R. Meyers, J. A. Plumb, and T. Yamamoto. 1987. Morphological and biochemical properties of four members of a novel group of reoviruses isolated from aquatic animals. *Journal of General Virology* 68:353–364.

Wise, D. A., P. R. Bowser, and J. A. Boyle. 1985. Detection of channel catfish virus in asymptomatic adult catfish, *Ictalurus punctatus* (Rafinesque). *Journal of Fish Diseases* 8:485–493.

Wise, D. A., S. F. Harrell, R. L. Busch, and J. A. Boyle. 1988. Vertical transmission of channel catfish virus. *American Journal of Veterinary Research* 49:1506–1509.

Wolf, K., and R. W. Darlington. 1971. Channel catfish virus: a new herpesvirus of ictalurid fish. *Journal of Virology* 8:525–533.

Wolf, K., R. L. Herman, and C. Carlson. 1972. Fish viruses: histopathologic changes associated with experimental channel catfish virus disease. *Journal of the Fisheries Research Board of Canada* 29:149–150.

Zhang, G. H., and L. A. Hanson. 1995. Deletion of thymidine kinase gene attenuates channel catfish herpesvirus while maintaining infectivity. *Virology* 209:658–663.

7 Carp and Minnow Viruses

Several viruses have been isolated from species of the minnow family. The most pathogenic of these are spring viremia of carp (SVC) and fish pox (herpesvirus of carp) in common carp and the golden shiner virus (GSV) of golden shiners.

SPRING VIREMIA OF CARP

Spring viremia of carp is a subacute to chronic disease of subadult cultured common carp. The virus was found in the former Yugoslavia and isolated in tissue culture from diseased fish with clinical signs of infectious dropsy (ID) (Fijan et al. 1971). Consequently, it was proposed that infectious dropsy syndrome be divided into two etiologically and clinically independent diseases: SVC, causative agent *Rhabdovirus carpio*, and carp erythrodermatitis (CE), a bacterial skin infection with necrotic lesions surrounded by hemorrhagic areas (Fijan 1972). Bootsma et al. (1977) demonstrated that CE was associated with the bacterium *Aeromonas salmonicida achromogenes* (atypical *A. salmonicida*), which was also later shown to be the causative agent of goldfish ulcer disease. *Aeromonas hydrophila* (*A. punctata*), historically thought to be the causative agent of ID, can actually complicate both SVC and CE infections by causing a secondary infection that is probably responsible for the dropsy associated with the disease.

Geographical Range and Species Susceptibility

Spring viremia of carp occurs primarily in Europe, having been confirmed in the former Yugoslavia, Hungary, Czechoslovakia, Austria, Bulgaria, France, Germany, Romania, Spain, Great Britain, and the former Soviet Union (Fijan et al. 1971; Wolf 1988). Curiously, SVC has not been reported outside this region in spite of carp being cultured in many parts of the world. A 1988 outbreak of SVC in Great Britain was thought to have resulted from importation of infected carp (Dixon et al. 1994). Although carp is a primary culture species in China, the disease has not yet been reported there.

Fish species naturally infected with SVC are common carp, crucian carp (goldfish), bighead carp, silver carp, and grass carp (Shchelkunov and Shchelkunova 1989). A population of sheatfish was naturally infected with the virus (Fijan et al. 1981), and guppies and northern pike fry have been experimentally infected (Ahne 1985a). Spring viremia of carp has also been experimentally transmitted to roach (Haenen and Davidse 1993).

Clinical Signs

At the onset of SVC, fish are attracted to the water inlet and seriously affected fish become moribund, respire slowly, and lie on their side (Fijan 1972). Fish experimentally infected by cranial injection express similar behavior. External clinical signs include dark pigmentation, a pronounced enlargement of the abdominal area, exophthalmia, a prolapsed and inflamed anus, and pale gills with distinct petechiae (Figure 7.1). Internally, a generalized hyperemia is apparent with peritonitis, enteritis, and hemorrhages in the kidney, liver, and air bladder. Also, the liver is edematous, and adhesions and hemorrhages are present in muscle tissue. In fry and small carp, the swim bladder becomes inflamed and shows focal hemorrhaging or petechiae.

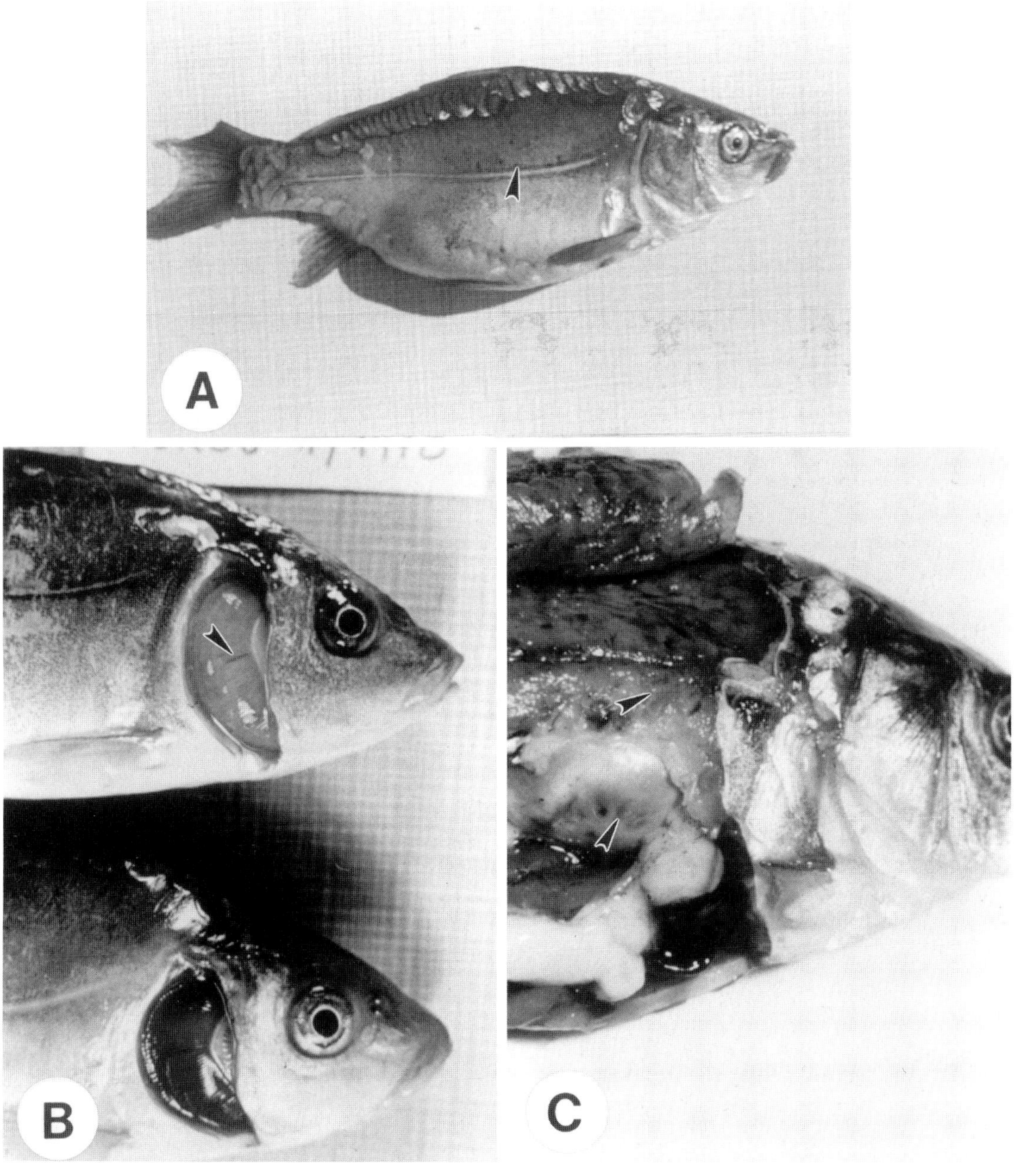

FIGURE 7.1. Common carp infected with spring viremia of carp virus. (A) *Petechiae in the skin* (arrow). (B) *Upper fish with pale but hemorrhaged* (arrow) *gills and lower fish with normal gills.* (C) *Petechiae in the muscle and swim bladder* (arrows) *and pale liver* (L). *(Photographs A and B courtesy of N. Fijan; photograph C by P. Ghittino.)*

Diagnosis

Diagnoses of SVC is by isolation of *R. carpio* from diseased fish in cell culture. Virus replicates between 10 and 30°C, with an optimum being 21°C. Under normal conditions, virus can be isolated only from fish during clinical infection. Fathead minnow (FHM) and epithelioma papillosum of

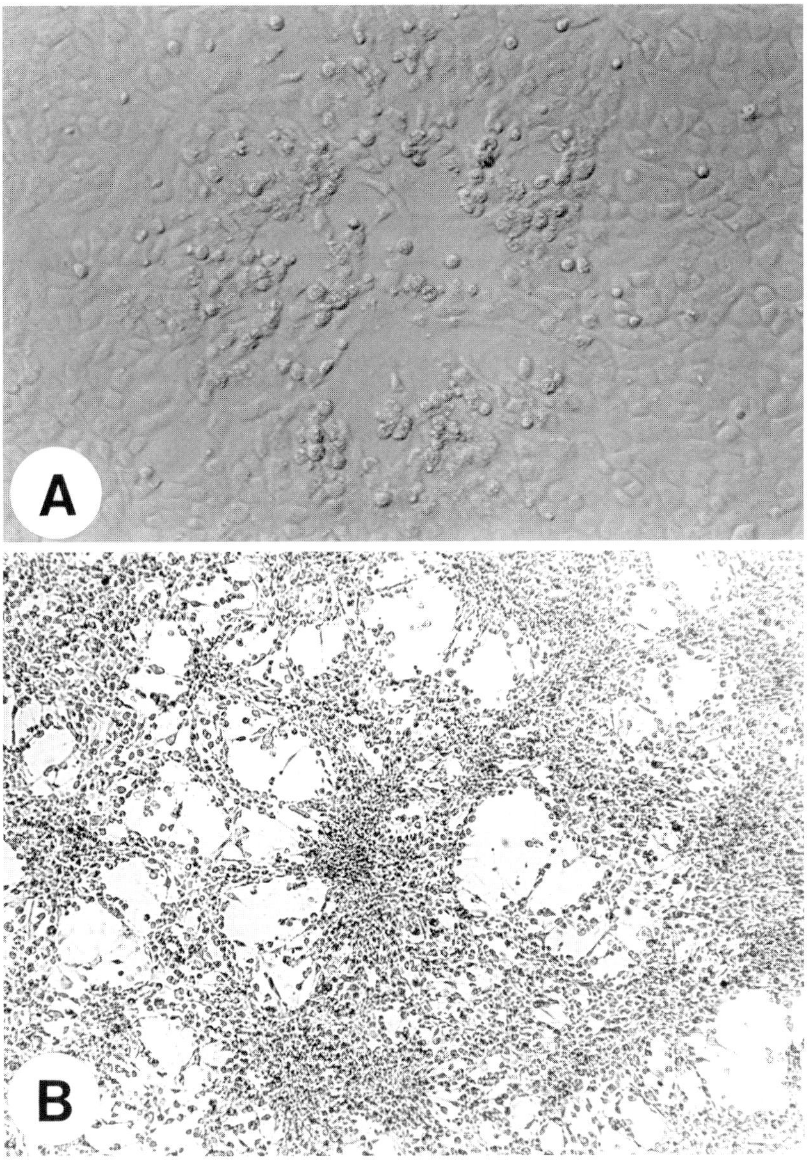

FIGURE 7.2. (A) *Focal cytopathic effect (CPE) of* Rhabdovirus carpio *(spring viremia of carp) in epithelioma papillosum of carp (EPC) cells. Infected cells are rounded and detached (×230). (Photograph by N. Fijan).* (B) *Massive* R. carpio *CPE in fathead minnow cells 72 hours after infection. (Photographs courtesy of W. Ahne.)*

carp (EPC) cell lines, as well as primary carp ovary cells, are susceptible to spring viremia of carp virus (SVCV) and yields of 108 plaque-forming units (PFU)/mL can be obtained. The BB, RTG-2 (rainbow trout gonad), and BF-2 (bluegill fry) cells are also susceptible to SVC, but cytopathic effect (CPE) develops more slowly and virus yield is lower than in other cell lines (Fijan et al. 1971; de Kinkelin and Le Berre 1974). Focal CPE includes granulation of chromatin in the nucleus with thickening or lysis of the nuclear membrane and cytoplasmic degeneration. Advanced CPE consists of degeneration and rounding of cells that then detach from the culture surface (Figure 7.2).

Spring viremia of carp virus can be positively identified by neutralization tests using specific SVCV antiserum, but some cross-reactive neutralization may occur with other fish rhabdoviruses (Table 7.1). The antigen can be detected in frozen liver, kidney, and spleen tissues by either immunoperoxidase (enzyme linked immunosorbent assay [ELISA]) or fluorescent antibody (Faisal and Ahne 1984). These serological techniques can also be used to identify SVCV in FHM cell cultures 12

TABLE 7.1. CROSS-NEUTRALIZATION TESTS WITH FIVE RHABDOVIRUSES FROM FISH AGAINST HOMOLOGOUS AND HETEROLOGOUS ANTISERA

	50% plaque reduction titer of virus[a]				
Antisera	SVC	SBI	PFV	IHN	VHS
SVC	6500	6200	20	20	10
SBI	980	980	68	50	10
PFV	33	28	200	27	10
IHN	25	25	27	1700	10
VHS	10	10	10	10	2300

Source: Hill et al. (1975).
[a]Reciprocal of antiserum dilution giving 50% plaque reduction against the specified virus.
IHN, infectious hematopoietic necrosis; PFV, pike fry rhabdovirus; SBI, swim bladder infection of carp; SVC, spring viremia of carp; VHS, viral hemorrhagic septicemia.

hours after inoculation as compared with the 48 hours required for CPE detection in inoculated cell cultures. More recently, Way (1991) used an ELISA system to detect SVCV antigen in inoculated cell cultures and in tissue extracts from clinically infected fish within 1 hour. This system also detected virus in subclinical fish but with less accuracy and some false positives were noted. The antibody in SVCV–ELISA reagents also reacted with pike fry rhabdovirus but at a very low level.

Virus Characteristics

Rhabdovirus carpio is a bullet-shaped RNA virus measuring approximately 70 × 180 nm (Ahne 1976). Virus activity is destroyed when exposed to 45°C, ether, or pH 3 (de Kinkelin and Le Berre 1974). In tissue culture medium containing 5% serum, the virus survives at a low level for 28 days at 23°C and for over 6 months at 4 and –20°C. Infectivity is reduced if serum is not included in the medium. Most disinfectants (formalin, sodium hydroxide, and chlorine) kill the virus in minutes, but it remains infective for up to 42 days in water and mud.

Virus isolates from carp with swim bladder inflammation (SBI), named SBI virus, appear to have the same characteristics as SVCV (Bachmann and Ahne 1973). Hill et al. (1975) compared SVC with rhabdoviruses that infect other fish species and showed that virus isolated from SBI-infected carp was serologically identical to SVCV. Biochemical properties of SVC and pike fry rhabdovirus (PFR) are similar, and there may be a serological relationship between these two rhabdoviruses of the vesiculovirus group (Enzmann et al. 1981). Immunochemical comparisons indicate that SVCV and PFR share common antigenic determinants on the G, N, and M protein, which can cause a problem in distinguishing them by indirect fluorescent antibody technique. They can, however, be differentiated by ELISA and serum neutralization tests. Spring viremia of carp virus is distinctly different from infectious hematopoietic necrosis virus (IHNV) and viral hemorrhagic septicemia virus (VHSV) of salmonids (lassovirus group) (Table 7.1).

Epizootiology

Spring viremia of carp occurs primarily in carp pond populations 1 year old or older. Epizootics of SVC were initially confirmed during the spring of 1969 and 1970 in the former Yugoslavia when water temperatures ranged from 12 to 22°C (Fijan et al. 1971). The optimum temperature for this disease is 16 to 17°C. Following waterborne virus exposure, 50-g carp helped clarify what role water temperature played in SVC development (Baudouy et al. 1980). Clinical signs occurred between 11 and 18°C. Some fish survived infection and became immune to the disease, whereas others died. Below 10°C, no immunity developed and disease was always fatal. The researchers suggested that elevated water temperatures in the spring are not necessarily a factor in the disease process.

Severity of SVC varies from farm to farm and from pond to pond and year to year on a given farm (Fijan 1972). After consecutive years with no SVC outbreaks, heavy losses can be experienced for several years. What causes these severity variations in SVC outbreaks is unknown but probably results from a combination of susceptible fish, a virulent virus, and conducive environmental conditions. Carp that survived an SVCV infection in the laboratory were resistant to repeated challenges with virulent virus but remained susceptible to carp erythrodermatitis. This further supports different etiologies for SVC and CE.

Spring viremia of carp is transmitted by waterborne exposure; cohabitation of infected and healthy fish; contaminated food; and intraperitoneal, intramuscular, or intracranial injection. Fish parasites (for example, fish louse, *Argulus foliaceus*, and *Pisciocola geometra*) can also transmit SVC to uninfected fish (Ahne 1985b). Incubation time for

SVC varies from 6 to 60 days depending on water temperature and exposure method. Survival of carp experimentally infected with virus by injection in the swim bladder was 8.9 days compared with 12.8 days for intraperitoneal injection and 12.9 days for intracranial injection (Varovic and Fijan 1973). Mortality of common carp infected with SVCV varies, but Haenen and Davidse (1993) reported a 97.7% mortality over 35 days following waterborne exposure to 105.8 $TCID_{50}$/mL of water.

Shedding of SVCV by adult carp during spawning was reported by Békési and Csontos (1985). They isolated SVCV from 3 of 491 ovarian fluid samples assayed, but none from 211 seminal fluid samples. Whether virus was actually inside the eggs was not determined. It does, however, indicate that some adult survivors of SVCV infections can be carriers, even though survivors generally do not shed infectious virus after clinical disease dissipates.

Pathological Manifestations

During waterborne exposure, SVCV enters the host through the gills, where virus replicates and then spreads via the bloodstream. Primary viremia becomes evident at 6 days postinfection (Ahne 1978). Target organs are kidney, liver, spleen, heart, and alimentary canal. Clinical disease is evident 7 days postexposure, and virus is shed by intestinal mucous casts and feces at this time.

Nagele (1977) described the histopathology of experimentally infected carp fingerlings. The liver has edema, necrotic blood vessels, and focal necrosis in the parenchyma. Vessels in the peritoneum are hyperemic, and intestines become inflamed. The spleen is hyperemic and hyperplastic. Renal tubules become clogged with casts, and tubule cells develop cytoplasmic inclusions. The swim bladder has extensive focal areas of hemorrhage.

Significance

Spring viremia of carp is an extremely important disease of cultured carp in Europe, where it is most prevalent in yearling fish, with up to 70% mortality of infected populations. Adult fish are also susceptible but to a lesser degree. The overall impact of SVC has been somewhat offset by the fact that in years following an epizootic, mortalities are greatly reduced in a given year's class of carp, a likely result of increased immunity among epizootic survivors.

FISH POX (HERPESVIRUS OF CARP)

Fish pox, also known as carp pox or epithelioma papillosum, is one of the oldest known fish diseases, being recorded as early as 1563 (Hedrick and Sano 1989). It takes the form of a benign hyperplastic, epidermal papilloma that occurs on common carp. On the basis of electron microscopy, Schubert (1966) proposed that fish pox was caused by a virus and Sonstegard and Sonstegard (1978) further suggested that it was caused by a herpesvirus. Sano et al. (1985a and 1985b) confirmed the previous findings when they isolated the virus from epithelioma growths on Japanese ornamental (Asagi) carp. Because the disease is not a true "pox," they proposed that it be called papillosum cyprini, which is caused by *Herpesvirus cyprini*, also referred to as carp herpesvirus (CHV).

Geographical Range and Species Susceptibility

Fish pox has been reported from most European countries, China, Japan, Russia, Israel, Korea, Malaysia, and the United States. Common carp and Asagi carp (koi carp) are its primary hosts. Other cyprinids—crucian carp, willow shiner, and grass carp—are not susceptible (Sano et al. 1985b; Sano et al. 1991). Fish pox was reported for the first time in North America in ide that had been imported from Europe 1 year prior to disease appearance (McAllister et al. 1985) and later in koi carp in California (Hedrick et al. 1990). According to Jiang (1995), fish pox is a major disease in cultured common carp in northern and eastern China.

Clinical Signs

Fish pox are benign, hyperplastic, papillomatous growths in the epithelium of carp (T. Sano et al. 1985b). The tumors are milky-white to gray and are raised about 1 to 3 mm above the skin on the head, fins, or any body surface (Figure 7.3). Lesions are generally small, smooth, raised growths that may increase in size and cover large areas of the body surface before they eventually regress and disappear. Infected fish show no distinct behavioral signs, and there is usually no morbidity among infected adult fish, but an exception was reported in China (Jiang 1995).

Juvenile carp, however, are adversely affected

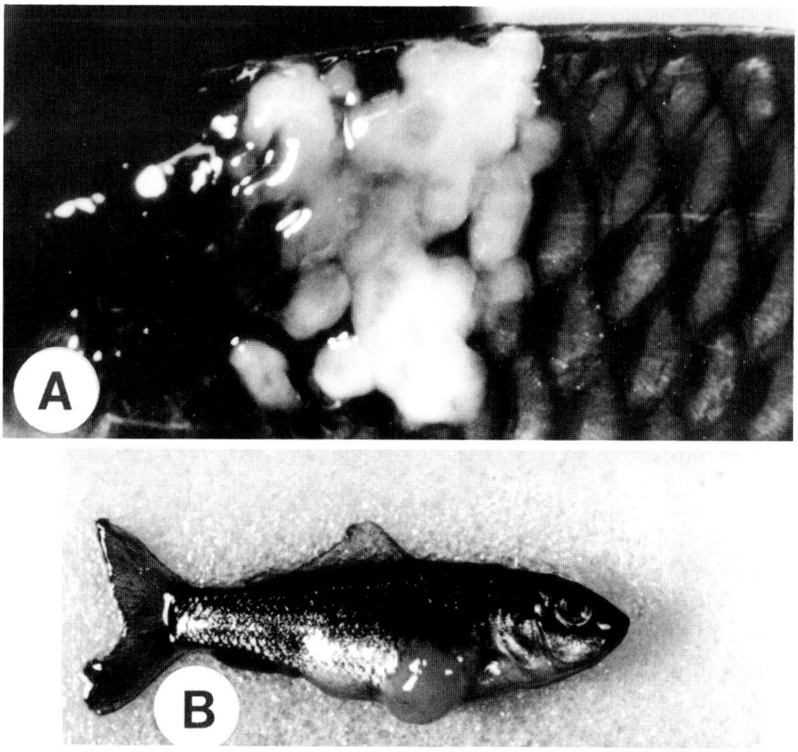

FIGURE 7.3. (A) Massive papilloma on common carp naturally infected with herpesvirus of carp (fish pox). (B) Papilloma on juvenile carp that survived experimental infection by immersion (From Sano et al. 1991b; reprinted with permission of Blackwell Scientific Publications Ltd.)

by CHV herpesvirus and can suffer high mortality (N. Sano et al. 1991). Two-week-old common carp fry infected with CHV showed loss of appetite and swam in rigid lines with intermittent immobility, had distended abdomens, exophthalmia, darkened pigmentation, and hemorrhages on the operculum and abdomen.

Diagnosis

When clinical signs of fish pox appear, virus isolation should be attempted in FHM or EPC cells (N. Sano et al. 1991). Virus titers in fish tissues range from $10^{5.8}TCID_{50}$/mL in FHM cells to $10^{2.5}TCID_{50}$/mL in EPC cells. Optimum cell culture incubation temperature is 20°C; replication will also occur at 15°C, but not at 10 or 25°C. Carp herpesvirus–infected FHM cells develop vacuolation, rounding, and slight pyknosis in focal areas 5 days after inoculation. Cellular disruption occurs slowly but will expand to 60% or 70% of the cell sheet with occasional cellular recovery. Syncytia formation, characteristic of cell lines infected with most herpesviruses, does not occur; however, intranuclear, Cowdry type A inclusion bodies develop in infected FHM cells (Figure 7.4). T. Sano et al. (1985b) found CHV antigen in neoplastic tissues at all stages of tumor development, especially in older and recurrent tumors. Identification of CHV is made by serum neutralization tests. N. Sano et al. (1992) also detected CHV in infected carp fry by in situ hybridization with biotinylated probes.

Virus Characteristics

Herpesvirus cyprini is an icosahedral DNA virus with a capsid that measures 110 to 113 nm in diameter (Figure 7.4) (T. Sano et al. 1985b) and with its envelope, measures 190 nm. *H. cyprini* is sensitive to ether, pH 3, 50°C for 1 hour, and iododeoxyuridine (IUdR). Basically, only one CHV strain is recognized, but T. Sano et al. (1991) noted slight differences in the viral genome of two different isolates by restriction endonuclease cleavage profiles.

Epizootiology

Most data concerning fish pox have been generated from cultured fish populations; its incidence

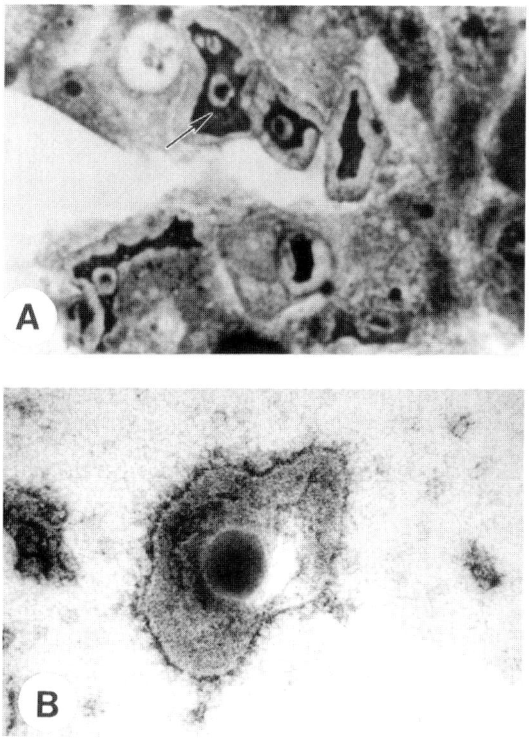

FIGURE 7.4. (A) *Fathead minnow cells infected with herpesvirus of carp showing Cowdry type A inclusion bodies* (arrow). (B) *Herpesvirus of carp with electron dense nucleocapsid and envelope (Photographs courtesy of T. Sano.)*

in wild populations is not known but is assumed to be low. Jiang (1995) reported death of 200 kg of CHV-infected caged common carp of unspecified fish size in a large reservoir in China. It has been suggested that young fish are infected by brood fish, but whether this occurs vertically and/or horizontally is unknown. Released virus from ruptured cells sloughed from the skin is probably instrumental in transmission. Evidence indicates that the disease can be transmitted by cohabitation or experimentally by intramuscular and intraperitoneal injection of media from infected cell cultures. Time between exposure and tumor development is 5 months to a year at 15°C (T. Sano et al. 1991; McAllister et al. 1985). T. Sano et al. (1985a) found that CHV can be transmitted to 30-day-old carp and will manifest itself in papillomas 6 months later. The first herpesvirus isolation from a fish papilloma (fish pox) was reported by T. Sano et al. (1985b).

Epizootiological confusion surrounding fish pox was clarified by T. Sano et al. (1991). They infected 2-week-old common carp fry by immersion at 20°C, which resulted in viremia and clinical disease. From 86% to 97% of virus-exposed fish died, compared with 3% mortality in controls. Carp herpesvirus was isolated from infected fish for 3 weeks postexposure. When 4- and 8-week-old common carp were infected with CHV, mortality was 20% and 0%, respectively; however, neoplasia developed in 55% of survivors 6 months postexposure to CHV and virus was isolated from tumors. A second set of neoplasms developed in 83% of the fish 7.5 months after initial tumors had sloughed, and CHV was again isolated from all necropsied moribund fish and survivors. Papillomas also developed in 13% of adult mirror carp and 10% of adult fancy carp 5 months after intraperitoneal injection with virus.

The foregoing studies suggest that CHV replicates in the epidermal tissue and can be either suppressed or generated by water temperature. Although CHV-induced papillomas occur at an optimum temperature of 15°C, lesion regression begins when water temperature is raised to 20°C and accelerates as temperature is increased to 30°C (N. Sano et al. 1993b). The regression mechanism was not understood until Morito and Sano (1990) showed that peripheral blood lymphocytes (PBLs) play an important role in this process. When carp with CHV papillomas were injected with anti–carp PBL serum, regression was delayed, suggesting that spontaneous cytotoxicity of carp PBL contributes to carp papilloma regression. This would explain the cycle of neoplasia, with an induction phase, desquamation, and recurrence being related to environmental and hormonal cycles and PBLs.

The CHV genome was traced in carp by in situ hybridization with a biotinylated probe following acute infection (N. Sano et al. 1993b). The viral genome was detected in the cranial nerve ganglia, subcutaneous tissue, and spinal nerves. Complete viral antigen was not detected, nor was virus isolated from these fish. When papillomas did appear, the viral genome was found in the same tissue as before and virus antigen and infective virus particles were present in the papillomas. The viral genome was still present in the spinal nerves, cranial nerve ganglia, and subcutaneous tissue after papilloma regression. These authors suggested that CHV becomes latently established in these tissues and is associated with lesion recurrence.

Pathological Manifestations

Histopathology of fish pox tumors was described by T. Sano et al. (1985a and b) and McAllister et al. (1985). Scaled areas of the skin show hyperplastic epidermal cells and similar lesions develop on fins and where no epidermal mucus or alarm substance cells are present. The compact dermis layer is thickened, and eosinophiles and lymphocytes in the loose connective tissue or hyperdermis suggests an inflammatory response. The oncogenic nature of CHV in carp was demonstrated by T. Sano et al. (1985a) when they experimentally produced tumors on fish skin and fins. Experimentally infected juvenile common carp developed a viremia and had necrosis in the liver parenchyma, kidneys, and lamina propria of the intestinal mucosa. Moribund fish developed severe hepatitis and extensive necrosis in the liver, limited necrosis in renal tissue, and epithelial erosion and edema of gills.

Significance

When very young carp are infected with CHV, high mortality can occur and survivors may develop tumors within a year. Fish pox is regarded as benign and has not been associated with mortality in the tumor state. From an aesthetic point of view, however, CHV-infected fish are unsightly and undesirable for food or display, particularly "fancy carp." Because mortality can be high in juveniles and tumors can lower market value of adults, one must conclude that CHV is of significance to the cultured carp producer.

GOLDEN SHINER VIRUS

An increasing number of viruses belonging to the family Reoviridae have been isolated from fish (Hetrick and Hedrick 1993). Recently, an aquareovirus genus was formed to include reoviruses from fish and shellfish (Francki et al. 1991). Golden shiner virus is the type species of the new genus. Generally, aquareoviruses exhibit low or no pathogenicity for fish; however, GSV and several others have the capacity to cause significant mortality under certain circumstances. Golden shiner virus was first isolated in 1977 from cultured golden shiners in which it produces a mild viremia (Plumb et al. 1979).

Geographical Range and Species Susceptibility

Golden shiner virus appears to be confined to the southeastern region of the United States and to California, where golden shiners are grown commercially (Plumb et al. 1979; Hedrick et al. 1988). The virus has also been isolated from grass carp, but other fish species are apparently refractive.

Clinical Signs

Shiners infected with GSV may swim lethargically at the surface, diving when disturbed; may hold fins close to the body; and may lie on the bottom of aquaria or tanks (Plumb et al. 1979). External clinical signs are not obvious, but the back and head of infected fish are reddish (erythema) as a result of hemorrhage in the skin and dorsal musculature. Petechiae are present in visceral fat and intestinal mucosa. Occasionally, open ulcerative lesions will develop in the skin; however, these lesions may be due to secondary bacterial infections.

Diagnosis

Golden shiner virus can most readily be isolated from the kidney, liver, and spleen by inoculation of organ filtrates onto FHM cells. Virus replication in FHM cells will occur at 20 to 30°C, with an optimum being 28 to 30°C. No replication occurs at 15 or 35°C (Brady and Plumb 1991). Some virus replication accompanied by CPE will also occur in chinook salmon embryo cells (CHSE-214) and BB cells, but virus yield is less than in FHM.

At approximately 24 hours postinoculation, focal CPE occurs in FHM monolayers incubated at 30°C. By 48 hours, CPE will envelop the entire cell sheet (Figure 7.5). Infected cells will become large and rounded, detach from the culture vessel surface, and float in the media as spherical, highly vacuolated debris. If very low levels of virus (less than 10^1 $TCID_{50}$/mL) are inoculated onto a cell sheet, initial foci may be overgrown by apparently healthy cells, leaving no evidence of infection other than a few large spherical cells floating in the media.

Virus Characteristics

The aquareoviruses are icosahedral and have a double capsid that measures 70 to 75 nm in diameter

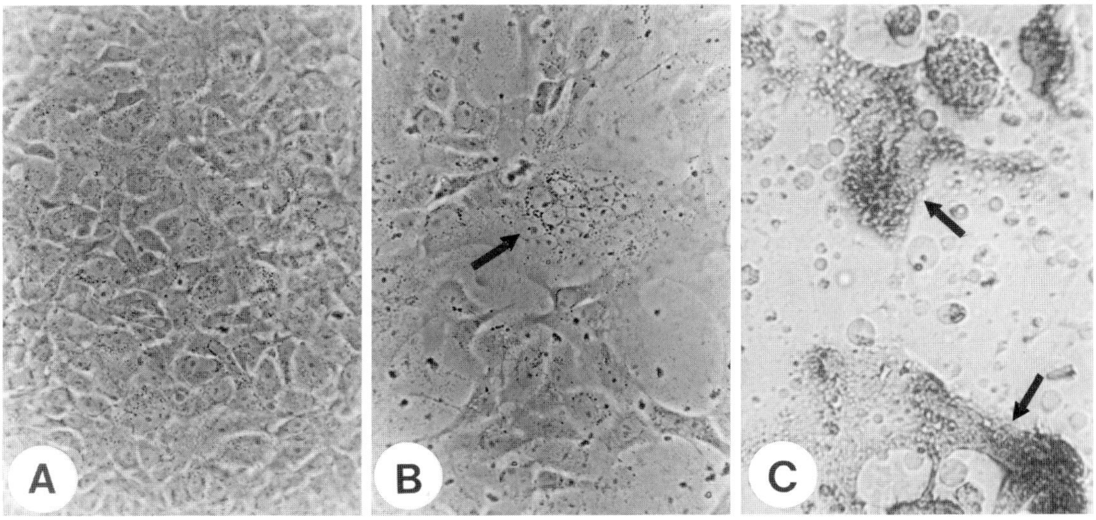

FIGURE 7.5. (A) Noninfected control fathead minnow (FHM) cell cultures. (B) Focal infection of golden shiner virus cytopathic effect (CPE) in FHM with multinucleated cells. (C) Massive CPE with rounded, vacuolated FHM cells that have released from the substrate. (From Plumb et al. 1979; reprinted with permission of the Canadian Journal of Fisheries and Aquatic Sciences.*)*

and a double-stranded RNA genome (Plumb et al. 1979; Schwedler and Plumb 1980; Winton et al. 1987). Several studies comparing GSV, catfish reovirus (CRV), chum salmon reovirus (CBSV), and $13p_2$ reovirus (of American oysters) indicate a serological relationship, but additional research is required before an accurate conclusion can be drawn (Winton et al. 1987; Brady and Plumb 1988).

Epizootiology

Although not a major disease, under certain conditions GSV can cause significant losses. Normally, GSV causes a chronic mortality of less than 10%; however, if fish are stressed in holding tanks, mortality can increase to 75% and disease may be compounded by secondary columnaris. Crowding will increase numbers of GSV-positive fish from very low to as high as 50% (Figure 7.6) (Schwedler and Plumb 1982). Intermediate size (7 to 10 cm) fish appear to be more susceptible than smaller fish. Limited experimental GSV transmission studies indicate that virus is not easily transmitted by cohabitation but is readily transmitted by injection. Possible vertical transmission has not been investigated.

Early documented cases of GSV disease occurred during summer when water temperatures exceeded 25°C; however, A. J. Mitchell (Stuttgart National Aquaculture Research Center, Stuttgart, Arkansas, personal communication) reported that high losses to GSV have been reported from March

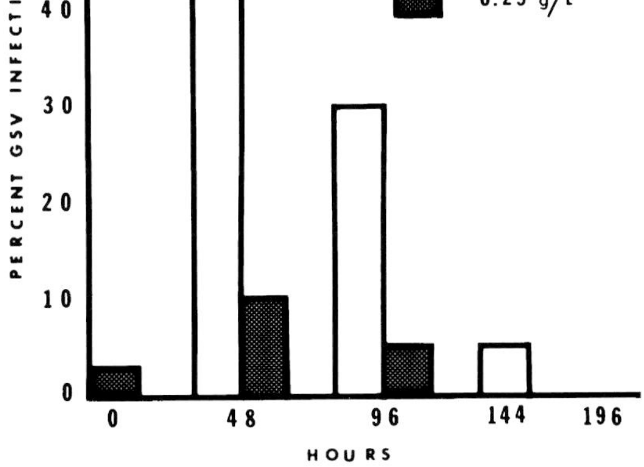

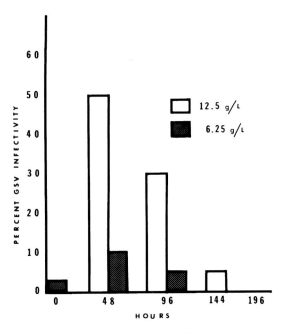

FIGURE 7.6. Effect of crowding on golden shiners infected with golden shiner virus and stocked in troughs at two different densities (From Schwedler and Plumb 1982; reprinted with permission of the Journal of Wildlife Diseases.*)*

to November. He also noted that the virus impact of GSV on the golden shiner industry is unclear because when clinical signs and rate of mortality indicate a possible GSV infection, virus can be isolated in only about 25% of the cases. Therefore, other contributing etiologies may also be involved where GSV is suspect.

Pathological Manifestations

In addition to presence of petechiae in the skin and occasionally the internal organs, the most notable histopathology of GSV disease is invasion of lymphocytes into the liver and presence of intestinal hemorrhage.

Significance

The disease has little impact on the aquaculture industry based on its modest pathogenicity and low frequency of occurrence. Generally, mortality is low in culture ponds, but potential for high losses does exist when golden shiners are stressed in holding tanks. This is especially true if a secondary infection of columnaris is present. Larger golden shiners have been in greater demand by the sport fishing industry as bait fish in recent year, and because the disease seems to target larger fish, its future impact may be more significant.

OTHER VIRUSES OF CARP AND MINNOWS

Several other viruses that have been reported in different species of cyprinids (goldfish, common carp, grass carp, and black carp) are also worth noting.

Goldfish Iridoviruses

Two viruses, tentatively identified iridoviruses, were isolated from swim bladders of healthy goldfish by Berry et al. (1983). Isolates were designated goldfish virus-1 (GFV-1) and goldfish virus-2 (GFV-2). These enveloped, DNA, icosahedral viruses measure 180 nm in diameter.

Goldfish Herpesvirus

A herpesvirus was isolated from cultured goldfish in Japan (Jung and Miyazaki 1995). The disease, named herpesviral hematopoietic necrosis (HVHN), caused a severe epizootic, but no external clinical signs were apparent. Internally, a softening and discoloration of the spleen and kidney were noted, and necrotic foci were seen in hematopoietic tissue. Virus was isolated in both FHM and EPC cells. The enveloped virion measures 170 to 220 nm and is sensitive to IUdR, pH 3, and ether.

Grass Carp Hemorrhagic Virus

The disease known as grass carp hemorrhagic virus disease (GCHV) was first reported in China in 1972 (Nie and Pan 1985); however, it was not until the virus was isolated by Chen and Jiang (1984) that its causative agent (grass carp hemorrhagic virus) was classified as an aquareovirus.

The disease occurs in China, where it primarily infects grass carp. Other susceptible species include black carp, gudgeon (topmouth), and rare minnow (Ding et al. 1991; Wang et al. 1994). The virus replicates in silver carp but causes no clinical signs.

Infected grass carp develop hemorrhage in the skin, base of fins, gill covers, and mouth cavity; become exophthalmic; and have pale gills (Shen 1978). The most dramatic sign of GCHV is a severely hemorrhagic intestinal tract with some hemorrhaging occurring in the liver, spleen, kidney, and musculature.

Grass carp hemorrhagic virus can be isolated in a cell line developed from gonadal tissue (Shen 1978) or cells derived from the snout of grass carp. Cytopathic effect appears in 3 to 4 days when incubated at 28°C.

The icosahedral virus measures 60 to 78 nm in diameter and contains an RNA genome composed of 11 segments, along with other reovirus characteristics (Ke et al. 1990). According to Jun et al. (1997), methods for detection and identification of GCHV include immunoenzyme staining, staphylococcal coagglutination test, immunofluorescence, and ELISAs. These authors also described a method for detecting GCHV based on a reverse transcription-polymerase chain reaction. The technique was able to detect GCHV in infected cell culture fluids and infected grass carp and rare minnow tissue, whether or not fish exhibited hemorrhagic clinical signs.

The virus infects primarily fry and yearling grass carp and can cause very high losses. Morbidity in these fish can be more than 65% after experimental infection (Shen 1978; Shao et al. 1990). Mortality is less in yearling fish than in fry, in which up to 80% mortality has been noted. Most epidemics occur during warm summer months as water temperatures rise to 25 to 30°C. Under these conditions, GCHV becomes one of the most serious diseases to affect grass carp in China.

Grass Carp Rhabdovirus

In Hungary, a rhabdovirus that is distinctly different from the reovirus of grass carp was isolated from 2-year-old apparently healthy grass carp (Ahne et al. 1987). The rhabdovirus was isolated in goldfish (CAR) and FHM cells, where it replicated at 15 to 25°C. The virus measures 170 to 220 X 50 to 55 nm and has an RNA genome. The role that grass carp rhabdovirus plays in the health of grass carp is unknown.

MANAGEMENT OF CYPRINID VIRUSES

Spring viremia of carp is best managed and controlled by quarantining infected fish from healthy populations. Disinfection of holding facilities will prevent transmission to naive fish that are later placed in those units. For years, it has been a goal of fish farmers to prevent the spread of SVCV by detecting virus carriers and eliminating them as brood stock. Because isolation of virus from carriers is not possible, serological detection to indicate prior exposure has been pursued. A competitive immunoassay to detect SVCV antibodies in fish was developed by Dixon et al. (1994). This technique detected SVCV-positive antibody in 88.5% of a carp population known to have been exposed to the virus, compared with 34.6% positive detection by serum neutralization. In testing sera from three other groups, 8% to 12% were positive by the competitive immunoassay method versus 0% positive by serum neutralization. The competitive assay has potential for use in large-scale screening of fish populations, but before that happens, additional validation with a wider range of sera is needed.

Vaccination for SVCV may be beneficial when practical. Fish exposed to inactivated SVCV produce antibodies that confer varying degrees of disease resistance during subsequent virus exposures. Carp were first experimentally vaccinated by immersion and hyperimmunization for SVCV by Fijan et al. (1977a); however, immune response was only measured by serum neutralization. Fijan et al. (1977b) reported that protective immunity in carp was obtained for 9 months after vaccination by intraperitoneal injection with a viable SVCV isolate that had been passed on human diploid and FHM cells.

Immunity against spring viremia of carp is temperature dependent. Fish vaccinated by intraperitoneal injection develop protective immunity at 20 to 22°C but not at 10°C. Successful mass carp vaccinations by injection have been reported; however, vaccination must be done when water temperatures are 18°C or above. Mass vaccination for SVC remains problematical.

There is no known control for carp herpesvirus, and the only management practice is avoidance. Carp shipped from CHV-infected areas into regions where it has not been reported should be carefully selected from stocks historically free of the disease because virus is likely transferred from one geographical area to another in covertly infected fish. Given that tumor development follows long incubation, it is advisable to quarantine newly transported carp for 1 year before they are released into communal waters. Proper management of GSV is based on elimination of overcrowding in conjunction with treatment of external bacterial infections.

Vaccination against grass carp reovirus by injection has been practiced in China for several years using two vaccine preparations; one is made by extracting virus antigen from visceral organs of moribund and dead fish, and the second is virus grown in tissue culture. Both are formalin inactivated, and juvenile fish are injected intraperitoneally with these preparations. In laboratory studies, Zhang et al. (1990) achieved an average of 68% survival in fish vaccinated by immersion with virus plus adjuvant (scopolamine) versus 0% survival of controls following challenge. They also acquired 85% survival versus 40% survival in vaccinated and control fish groups, respectively, under field conditions. Use of GCHV vaccine, along with improved fish culture environmental conditions, has been accredited with reduced disease severity in cultured grass carp in China.

REFERENCES

Ahne, W. 1976. Untersuchungen uber die stabilitat des karpfen pathogenen virusstammes. *Fish und Umwelt* 2:121–27.

Ahne, W. 1978. Uptake and multiplication of spring viremia of carp virus in carp, *Cyprinus carpio* L. *Journal of Fish Diseases* 1:265–268.

Ahne, W. 1985a. Viral infection cycles in pike (*Esox lucius* L.), *Journal of Applied Ichthyology* 1:90–91.

Ahne, W. 1985b. *Arqulus foliaceus* L. and *Philometra qeometra* L. as mechanical vectors of spring viremia of carp virus (SVCV). *Journal of Fish Diseases* 8:241–242.

Ahne, W., Y. Jiang, and I. Thomsen. 1987. A new virus isolated from cultured grass carp *Ctenopharyngodon idella. Diseases of Aquatic Organisms* 3:181–185.

Bachmann, P. A., and W. Ahne. 1973. Isolation and characterization of agent causing swim bladder inflammation in carp. *Nature* 244:235–237.

Baudouy, A. M., M. Danton, and G. Merle. 1980. Viremie printaniere de la carpe: etude experimental de l'infection evoluant a differentes temperatures. *Annales d'Virologie* 131E:479–488.

Békési, L., and L. Csontos. 1985. Isolation of spring viraemia of carp virus from asymptomatic broodstock carp, *Cyprinus carpio* L. *Journal of Fish Diseases* 8:471–472.

Berry, E. S., T. B. Shea, and J. Gabliks. 1983. Two eiridovirus isolates from *Carassius auratus* (L.). *Journal of Fish Diseases* 6:501–510.

Bootsma, R., N. Fijan, and J. Blommaert. 1977. Isolation and preliminary identification of the causative agent of carp erythrodermatitis. *Veterinarski Arhives* 47:291–302.

Brady, Y. L., and J. A. Plumb. 1988. Serological comparison of golden shiner virus, chum salmon virus, reovirus $13p_2$ and catfish reovirus. *Journal of Fish Diseases* 11:441–443.

Brady, Y. J., and J. A. Plumb. 1991. Replication of four aquatic reoviruses in experimentally infected golden shiners, (*Notemigonus chrysoleucas*). *Journal of Wildlife Diseases* 27:463–466.

Chen, Y., and Y. Jiang. 1984. Morphological and physicochemical characterization of the hemorrhagic virus of grass carp [Chinese]. *Kexue-Tongboa* 29:832–835.

de Kinkelin, P., and M. Le Berre. 1974. Rhabdovirus des poissons. II. Proprietes in vitro du virus de la viremie printaniere de la carpe. *Annales d'Microbiologie* (Paris) 125A:113–124.

Ding, Q., L. Yu, X. Wang, and L. Ke. 1991. Study on infecting other fishes with grass carp hemorrhagic virus [Chinese with English abstract]. *Chinese Journal of Virology* 6:371–373.

Dixon, P. F., A. M. Hattenberger-Baudouy, and K. Way. 1994. Detection of carp antibodies to spring viraemia of carp virus by competitive immunoassay. *Diseases of Aquatic Organisms* 19:181–186.

Enzmann, P. J., B. Maier, and K. Bigett. 1981. Biochemical data on the RNA of VHS-V and RVC indicating a possible serological relationship. *Bulletin of the European Association of Fish Pathologists* 1:37–39.

Faisal, M., and W. Ahne. 1984. Spring viraemia of carp virus (SVCV): comparison of immunoperoxidase, fluorescent antibody and cell culture isolation techniques for detection of antigen. *Journal of Fish Diseases* 7:57–64.

Fijan, N. 1972. Infectious dropsy in carp—a disease complex. In: *Diseases of Fish Symposium of the Zoological Society of London,* edited by L. E. Mawdesley-Thomas, 39–51. London: Academic Press.

Fijan, N., C. Petrinec, D. Sulimanovic, and L. O. Zwillenberg, 1971. Isolation of the viral causative agent from the acute form of infectious dropsy of carp. *Veterinarski Arhives* 41:125–138.

Fijan, N., Z. Petrinec, Z. Stancl, M. Dorson, and M. Le Berre. 1977a. Hyperimmunization of carp with *Rhabdovirus carpio. Bulletin Office International des Epizooties* 87:439–440.

Fijan, N., Z. Petrinec, Z. Stancl, N. Kezic, and E. Teskeredzie. 1977b. Vaccination of carp against spring viremia: comparison of intraperitoneal and peroral application of live virus to fish kept in ponds. *Bulletin Office International des Epizooties* 87:441–442.

Fijan, N., Z. Matasin, Z. Jeney, J. Olah, and L. O. Zwillenberg, 1981. Isolation of *Rhabdovirus carpio* from sheatfish (*Silurus glanis*). In: *Fish, Pathogens and Environment in European Polyculture,* edited by J. Oláh, K. Molnàr and Z. Jeney, 48–58. Szarvas, Hungary: Fisheries Research Institute.

Francki, R. I. B., C. M. Fauquet, D. L. Knudson, and F. Brown, Editors. 1991. Classification and nomenclature of viruses. *Archives in Virology Supplement 2.* New York: Springer-Verlag.

Haenen, O. L., and A. Davidse. 1993. Comparative pathogenicity of two strains of pike fry rhabdovirus and spring viremia of carp virus for young roach, common carp, grass carp and rainbow trout. *Diseases of Aquatic Organisms* 15:87–92

Hedrick, R., J. M. Groff, T. McDowell, and W. H. Wingfield. 1988. Characteristics of reoviruses isolated from cyprinid fishes in California, USA. In: *Viruses of Lower Vertebrates,* edited by W. Ahne and E. Kurstak, 241–249. Berlin: Springer-Verlag.

Hedrick, R. P., and T. Sano. 1989. Herpesviruses of fishes. In: *Viruses of Lower Vertebrates,* edited by W. Ahne and E. Kurstak, 161–170. Berlin: Springer-Verlag.

Hedrick, R. P., J. M. Groff, M. S. Okihiro, and T. S. McDowell. 1990. Herpesviruses detected in papillomatous skin growths of koi carp (*Cyprinus carpio*). *Journal of Wildlife Diseases* 26:578–581.

Hetrick, F. M., and R. P. Hedrick. 1993. New viruses described in finfish from 1988–1992. *Annual Review of Fish Diseases* 3:187–192.

Hill, B. J., B. O. Underwood, C. J. Smale, and F. Brown. 1975. Physico-chemical and serological characterization of five rhabdoviruses infecting fish. *Journal of General Virology* 27:369–378.

Jiang, Y. 1995. Advances in fish virology research in China. In: *Diseases in Asian Aquaculture II,* edited by M. Shariff, J. R. Arthur and R. P. Subasinghe, 211–225. Manila, Phillipines: Fish Health Section, Asian Fisheries Society.

Jun, L., W. Tiehui, Y, Yonglan, L. Hangin, L. Renhou, and C. Hongxi. 1997. A detection method for grass carp hemorrhagic virus (GCHV) based on a reverse transcription-polymerase chain reaction. *Diseases of Aquatic Organisms* 29:7–12.

Jung, S. G., and T. Miyazaki. 1995. Herpesviral haematopoietic necrosis of goldfish, *Carassius auratus* (L). *Journal of Fish Diseases* 18:211–220.

Ke, L., Q. Fang, and Y. Cai. 1990. Characteristics of a novel isolate of grass carp hemorrhagic virus [Chinese with English abstract]. *Acta Hydrobiologica Sinica* 14:155–159.

McAllister, P. E., B. C. Lidgerding, R. L. Herman, L. C. Hoyer, and J. Hankins. 1985. Viral disease of fish: first report of carp pox in golden ide (*Leuciscus idus*) in North America. *Journal of Wildlife Diseases* 21:199–204.

Morita, N., and T. Sano. 1990. Regression effect of carp, *Cyprinus carpio* L., peripheral blood lymphocytes on CHV-induced carp papilloma. *Journal of Fish Diseases* 13:505–511.

Nagele, R. D. 1977. Histopathological changes in some organs of experimentally infected carp fingerlings with *Rhabdovirus carpio. Bulletin Office International des Epizooties* 87:449–450.

Nie, D-S., and J-P. Pan. 1985. Diseases of grass carp (*Ctenopharyngdon idellus* Vallenciennes, 1844) in China, a review from 1953 to 1983. *Fish Pathology* 20:323–330.

Plumb, J. A., P. R. Bowser, J. M. Grizzle, and A. J. Mitchell. 1979. Fish viruses: a double-stranded RNA icosahedral virus from a North American cyprinid. *Journal of the Fisheries Research Board of Canada* 36:1390–1394.

Sano, N., R. Honda, H. Fukuda, and T. Sano. 1991. *Herpesvirus cyprini:* restriction endonuclease cleavage profiles of the viral DNA. *Fish Pathology* 26:207–212.

Sano, N., M. Sano, T. Sano, and R. Hondo. 1992. *Herpesvirus cyprini:* detection of the viral genome in situ hybridization, *Journal of Fish Diseases* 15:153–162.

Sano, N., M. Moriwake, R. Hondo, and T. Sano. 1993a. *Herpesvirus cyprini:* a search for viral genome in infected fish by in situ hybridization. *Journal of Fish Diseases* 16:495–499.

Sano, N., M. Moriwake, and T. Sano. 1993b. *Herpesvirus cyprini:* Thermal effects on pathogenicity and oncogonicity. *Gyobyo Kenkyu* 28:171–175.

Sano, T., H. Fukuda, and M. Furukawa. 1985a. *Herpesvirus cyprini:* Biological and oncogenic properties. *Fish Pathology* 20:381–388.

Sano, T., H. Fukuda, M. Furukawa, H. Hosoya, and Y. Moriya. 1985b. A herpesvirus isolated from carp papilloma in Japan. In: *Fish and Shellfish Pathology,* edited by A. E. Ellis, 307–311. Orlando, FL: Academic Press.

Sano, T., N. Morita, N. Shima, and M. Akimoto 1991. *Herpesvirus cyprini:* lethality and oncogenicity. *Journal of Fish Diseases* 14:533–543.

Schubert, G. 1966. The infective agent in carp pox. *Bulletin Office International des Epizooties* 65:1011–1022.

Schwedler, T. E., and J. A. Plumb. 1980. Fish viruses: serologic comparison of the golden shiner and infectious pancreatic necrosis viruses. *Journal of Wildlife Diseases* 16:597–599.

Schwedler, T. E., and J. A. Plumb. 1982. Golden shiner virus: effects of stocking density on incidence of viral infection. *The Progressive Fish-Culturist* 44:151–152.

Shao, J., S. Mao, and Y. Shen. 1990. Studies on the isolation and pathogenicity of two kinds of hemorrhagic virus of grass carp, *Ctenopharynogodon idellus* [Chinese with English abstract]. *Journal of Hangzhou University* 17:74–79.

Shchelkunov, I. S., and T. I. Shchelkunova. 1989. *Rhabdovirus carpio* in herbivorous fishes: isolation, pathology, and comparative susceptibility of fishes. In: *Viruses of Lower Vertebrates,* edited by W. Ahne and E. Kurstake, 333–348. Berlin: Springer-Verlag.

Shen, L. Z. 1978. Studies on the causative agent of hemorrhage of the grass carp (*Ctenopharyngdon idellus*) [Chinese with English summary]. Acta Hydrobiologica Sinica 6:322–329.

Sonstegard, R. A., and K. S. Sonstegard. 1978. Herpesvirus-associated epidermal hyperplasia in fish (carp). In: *Proceedings International Symposium on Oncogenic Herpesvirus,* edited by G. de The, W. Henle and F. Rapp, 863–868. Scientific publication no. 24. Lyon, France: International Agency Research for Cancer Science.

Varovic, K., and N. Fijan. 1973. Osjetljivost sarana prema rhabdovirus carpio pri raznim nacinima inokulacije [English summary]. *Veterinarski Arhives* 43:271–276.

Wang, T., H. Chen, H. Liu, Y. Yi, and W. Guo. 1994. Preliminary studies on the susceptibility of *Gobiocypris rarus* to hemorrhagic virus of grass carp [Chinese with English abstract]. *Acta Hydrobiologia Sinica* 18:144–149.

Way, K. 1991. Rapid detection of SVC virus antigen in infected cell cultures and clinically diseased carp by enzyme-linked immunosorbent assay (ELISA). *Journal of Applied Ichthyology* 7:95–107.

Winton, J. R., C. N. Lannan, J. L. Fryer, R. P. Hedrick, T. R. Meyers, J. A. Plumb, and T. Yamamoto. 1987. Morphological and biochemical properties of four members of a novel group of reoviruses isolated from aquatic animals. *Journal of General Virology* 68:353–364.

Wolf, K. 1988. *Fish Viruses and Fish Viral Diseases.* Ithaca, NY: Cornell University Press.

Zhang, N., G. Yang, W. Yin, J. Shen, and Z. Cao. 1990. Scopolamine enhances the immunity of carp using immersion vaccine [Chinese]. *Journal of Zhejiang College of Fisheries* 9:79–83.

8 Eel Viruses

Viruses have been reported in European eel, American eel, and Japanese eel. Isolates have come mainly from Europe and Japan and occasionally from North America and elsewhere. Virus impact on cultured eel populations is difficult to assess because many isolates have come from apparently normal fish. Also, numerous reported viruses from eels remain poorly classified (Ahne et al. 1987). Five viruses that affect eels will be discussed: eel virus European (EVE), eel rhabdovirus (multiple designations), anguillid herpesvirus, eel iridovirus (EV-102), and stomatopapilloma.

EEL VIRUS EUROPEAN

Eel virus European was isolated from European eels cultured in Japan (T. Sano 1976). Later, it was suggested the disease be called viral kidney disease because of how severely it affected that particular organ (T. Sano et al. 1981).

Geographical Range and Species Susceptibility

Although EVE was initially isolated from European eels in Japan, it was also found in Japanese eels and tilapia in Taiwan (T. Sano 1976; Chen et al. 1985). T. Sano et al. (1981) stated that rainbow trout are not susceptible to EVE. A virus similar to EVE, but characterized as infectious pancreatic necrosis virus (IPNV) strain Ab, was isolated from normal young eels in England (Hudson et al. 1981), and in 1988, the author isolated IPNV (Ab serotype) from moribund cultured American eels in Florida (serotyped by B. Hill, Disease Laboratory, Weymouth, England, personal communication).

Clinical Signs

Eels experimentally infected with EVE become rigid with muscle spasms and have congested anal fins (T. Sano 1976; Ueno et al. 1983). A microscopic gill examination shows marked hyperplasia of the lamellar epithelium and lamellae fusing and clubbing of filaments, giving them a swollen appearance. The underside of affected fish show slight petechiation. Ascitic fluid can be present in the abdominal cavity, the alimentary tract is void of food, and the kidney, hypertrophied.

Diagnosis

Eel virus European can be isolated from internal organs of affected fish in BF-2 (bluegill fry), CHSE-214 (chinook salmon embryo), or RTG-2 (rainbow trout gonad) cells incubated at 10 to 20°C; however, cell lines from eels should be equally satisfactory (Chen and Kou 1981). T. Sano (1976) noted that cytopathic effect (CPE) of EVE is characterized by nuclear pyknosis and that cells retain an elongated shape almost identical to CPE caused by IPN virus.

Virus Characteristics

The EVE agent has biophysical characteristics (size, morphology, and so forth) of IPNV-Ab. It is an RNA, icosahedral virus with double capsid and, more importantly, is closely related serologically to the IPNV-Ab strain (Hedrick et al. 1983). Okamoto et al. (1983) compared 10 strains of IPNV to EVE and found the eel virus to be more closely related to the IPNV-Ab serotype than to any other viral strain.

The similarity of EVE to other aquatic birnaviruses was demonstrated by Chi et al. (1991) using monoclonal antibody (MAB) to virus isolated from Japanese eels affected with branchionephritis. Three of six MABs identified epitopes possessed by IPN-Ab virus only; one recognized an epitope on both the Ab and Sp serotypes; and two MABs exhibited epitopes on Ab, Sp, and VR-299 IPNV serotypes. Clearly, the EVE virus is closely related to the Birnavirus group.

Epizootiology

Although cultured young European and Japanese eels appear to be most susceptible to EVE, it has low virulence for young rainbow trout in experimental infections. The virus has a propensity for producing disease at temperatures from 8 to 14°C, especially in small eels (T. Sano 1976). Eel virus European mortalities may be as high as 60% in 4- to 26-g eels. T. Sano et al. (1981) reported 55% to 75% mortality of experimentally infected 9- to 12-g Japanese eels. Ueno et al. (1983) determined that an EVE isolate from Japanese eels was not infectious to common carp at 20 to 25°C but that tilapia were severely affected at 10 to 16°C, temperatures that approach stressful or lethal limits for tilapia.

Eel virus European can be transmitted by waterborne exposure, and fish-to-fish transmission seems to be a likely possibility. Wolf (1988) postulated that virus was transmitted from Europe to Japan in elvers and then to Taiwan in eyed trout eggs.

Japanese eels are affected by a gill disease commonly called branchionephritis, from which EVE has been isolated; however, the virus does not cause this condition in experimentally infected fish (Ueno et al. 1983). The relationship of branchionephritis disease of Japanese eels to EVE is still inconclusive, although there appears to be a strong connection.

Pathological Manifestations

Fins and gills develop congestion with branchionephritis also present in gills. Occasionally diffused congestion is noted in the abdomen (T. Sano 1976). Histopathological findings show proliferative glomerulonephritis, and necrosis in the renal interstitium nephrons, hyaline droplet degeneration, and cloudy swelling in the tubular epithelial cells. Focal necrosis occurs in the liver, and extensive necrosis is present in the spleen of some eels (Egusa et al. 1971).

Significance

Significance of EVE in cultured eels is not clear; however, high mortality in eels in Japan and Taiwan have been associated with its presence.

EEL RHABDOVIRUSES

There are a large number of rhabdoviruses that have been isolated from cultured eels (Castric et al. 1984) but seldom cause significant disease.

Geographical Range and Species Susceptibility

A rhabdovirus was isolated from American eels (EVAs) imported into Japan from Cuba (T. Sano 1976). Rhabdoviruses are also found in eels in Europe (Castric et al. 1984).

Clinical Signs

Because most eel rhabdoviruses have been isolated from normal, apparently healthy elvers, there are few clinical signs associated with infection. T. Sano (1976) reported, however, that naturally EVA infected American eel elvers imported from Cuba displayed hemorrhages and congestion on the body's underside and on anal and pectoral fins. Fish also turned their heads downward and gills showed hemorrhage and degradation of bone or cartilage. A disease called rhabdoviral dermatitis, described by Kobayashi and Miyazaki (1996), produces cutaneous lesions and mild histopathology of internal organs.

Diagnosis

For isolation of eel rhabdoviruses, filtrates from homogenates of brain, kidney, liver, and spleen tissue are inoculated onto BF-2, eel kidney (EK-1), eel ovary (EO-2), fathead minnow (FHM), or RTG-2 cells and incubated at 15 or 20°C, depending on cell line (Wolf 1988). Cytopathic effect consists of pyknosis and cytoplasmic granulation followed by cell lysis. In a comparison of CPE of EVA and eel

TABLE 8.1. RHABDOVIRUSES ISOLATED FROM EELS (*ANGUILLA* SPP.)

Rhabdovirus group	Isolate designation	Eel species	Geographical origin
Lyssavirus	B_{12}	*A. anguilla*	Europe
	C_{26}	*A. anguilla*	Europe
Vesiculovirus	Eel virus American (EVA)	*A. rostrata*[a]	Japan
	Eel virus European–X (EVEX)	*A. anguilla*[b]	Japan
	C_{30}	*A. anguilla*	Europe
	B_{44}	*A. anguilla*	Europe
	D_{13}	*A. anguilla*	Europe

Source: Castrick and Chastil (1980) and Castric et al. (1984).
[a]From elvers from Cuba.
[b]From elvers from Europe.

virus European (unknown)-X, Nishimura et al. (1981) showed that effects were similar on RTG-2 cells and CPE was also similar to that of infectious hematopoietic necrosis virus (IHNV).

Virus Characteristics

At least seven different isolates of rhabdoviruses have been made from eels and are designated as: eel virus American (EVA), eel virus European-X (EVEX), B_{12}, B_{44}, C_{26}, C_{30}, and D_{13} (Table 8.1) (Wolf 1988). These viruses, several of which came from apparently healthy eels, were analyzed by Castric et al. (1984); two of the viruses belonged to the Lyssavirus genus (B_{12} and C_{26}) and five were members of the Vesiculovirus genus (EVA, EVEX, C_{30}, B_{44}, and D_{13}). The rhabdovirus reported by Kobayashi and Miyazaki (1996) was similar to EVA.

Epizootiology

Of all the eel viruses that have been isolated, the rhabdoviruses may be of greatest interest to fish virologists. These viruses seem to be very prevalent in some established wild populations of European eels but do not appear to cause any significant problem among cultured populations. Castric and Chastel (1980) showed that eel rhabdovirus strain B_6 was experimentally infectious to rainbow trout fry by bath exposure but was not pathogenic for elvers or young eels. Strain B_{12}, C_{30}, and EVEX produced no significant mortality in either fish species. The American eel from Cuba infected with EVA suffered high mortality at 20 to 27°C (T. Sano 1976).

Infectivity trials of EVA and EVEX in rainbow trout, Japanese eel, common carp, and ayu resulted in the death of only rainbow trout (Nishimura et al. 1981). Infectivity in trout was more severe at 15 and 20°C than at 10°C, which led investigators to be concerned with the possibility of introducing another trout pathogen into Japan.

Pathological Manifestations

Infected American eel from Cuba had hemorrhaging and necrosis of the branchial vessels, skeletal muscle, and kidneys (T. Sano 1976). European eel infected with rhabdovirus showed no pathology. Experimental EVA and EVEX infections in rainbow trout produced hemorrhagic skeletal muscle, intensive necrosis, hemorrhage of the kidneys' hematopietic tissue and focal necrosis of the liver, spleen, and pancreas (Nishimura et al. 1981). Histopathology was comparable to that associated with IHNV infections in salmonids.

Significance

Currently, these viruses do not appear to be a significant cause of eel disease; however, as eel culture intensifies they could become more important.

ANGUILLID HERPESVIRUS

A previously unidentified herpesvirus was isolated from eel in two locations by M. Sano et al. (1990). They designated the virus Anguillid herpesvirus 1, and *Herpesvirus anguillidae* (HVA) as the Latinized name; the agent is called "eel herpesvirus."

Geographical Range and Species Susceptibility

Anguillid herpesvirus has only been found in limited eel populations in Japan, but a probable herpesvirus found in skin lesions of European eels in Europe has been identified by electron microscopy (Békési et al. 1986). The relationship, if any, of the Japanese isolate to the agent found in Europe is unknown, however, the herpesvirus in Japan was isolated from both Japanese and European eels (M. Sano et al. 1990).

Clinical Signs

Affected eels had varying degrees of erythema on the skin and gills.

Diagnosis

Anguillid herpesvirus is diagnosed by isolation in cell culture. Although HVA produces CPE in BF-2, FHM, and RTG-2 cells, virus replication is best in eel kidney cells (EK-1) and eel ovary cells (EO-2). Cytopathic effect consisted of syncytium formation and rounded cells with Cowdry type A intranuclear inclusions. In cell culture, the optimum incubation temperature for HVA is 20 to 25°C but replication occurs from 10 to 37°C (M. Sano et al. 1990).

Virus Characteristics

Herpesvirus anguillidae clearly conforms to the herpesvirus criteria while being serologically distinct from channel catfish virus, salmonid herpesvirus type 2, and cyprinid herpesvirus type 1 (M. Sano et al. 1990). The DNA genome, icosahedral nucleocapsid, measures 110 nm, and the mean enveloped virion diameter is 200 nm. Twenty-five polypeptides are present in the virion, with molecular weights of 19.5 kD to 320 kD.

Epizootiology

Anguillid herpesvirus was initially isolated from Japanese eels that suffered 1% mortality at one facility and European eels that suffered 6.8% mortality at a second facility (M. Sano et al. 1990). These fish were being cultured at high density in recirculating systems at 30°C. It was speculated that unfavorable environmental conditions may have caused stress and onset of the herpesvirus infection. Virus could be isolated 2 weeks after initial detection but not 1 month later, suggesting progression into a dormant state.

Pathological Manifestations

Irregular-sized eosinophilic granulations appeared in the epidermal tissue, which became necrotic with partial sloughing. Blood capillaries increased in the dermis, but melanophores contracted; branchial lamellae of the gills became fused, followed by necrosis; and necrotic hepatic cells were scattered throughout the liver.

Significance

Significance of eel herpesvirus is unknown.

EEL IRIDOVIRUS

The Japanese eel iridovirus (EV-102) was isolated from cultured eels in Japan during a routine farm survey (Sorimachi 1984).

Geographical Range and Species Susceptibility

Japanese eel iridovirus is known only in Japan, where it is pathogenic to Japanese eel but not to European eel.

Clinical Signs

Experimental infection of Japanese eels with EV-102 produced skin depigmentation and fin congestion, and increased skin mucus appeared 3 to 5 days postinfection (Wolf 1988).

Diagnosis of Eel Iridovirus

Japanese eel iridovirus replicates in EPC and RTG-2 cells as well as EK-1 and EO-2 cells when incubated at 15 to 20°C (Wolf 1988).

Virus Characteristics

Biophysical and biochemical characteristics of eel iridovirus is typical of the group. It is an icosahedral particle that measures 100 to 200 nm in diameter and contains a DNA genome (Sorimachi and Egusa 1982).

Epizootiology

After experimentally infecting eels up to 120 g with EV-102 by immersion or injection, clinical signs appeared in 3 to 5 days at 15 to 23°C (Sorimachi 1984). Mortalities began shortly after onset of clinical signs at about 1 week postexposure and continued for 2 weeks. Deaths stopped and clinical signs abated when water temperature reached 23°C, but 70% mortality had already occurred at that point.

Pathological Manifestations

The pathological manifestations of eel iridovirus is unknown.

Significance

With only one case of eel iridovirus being reported, its significance would appear to be minimal.

STOMATOPAPILLOMA

Stomatopapilloma, also known as cauliflower disease because of lesion morphology, is a tumorous disease of eels that has been recognized since 1910 (Wolf 1988). Several viruses have been associated with these lesions, but their role in tumor production is unclear.

Geographical Range and Species Susceptibility

Stomatopapilloma affects only European eel and occurs mainly in tributaries and brackish waters of the Baltic and North Seas of northern Europe (Wolf 1988). It has also been reported in Great Britain (Delves-Broughton et al. 1980).

Clinical Signs

Chronic, benign epidermal neoplasia usually develop on the mouth, nares, and head of eels. Lesions may cover the entire head and occlude the oral cavity and occasionally occur on fins and other body locations (Peters 1975). The lesions are amorphous, irregular in shape, have a pebbly surface, and are usually light in color but can be pigmented.

Diagnosis

Diagnosis of stomatopapilloma is based on morphology of head lesions, histopathology, and electron microscopy. Virus has been isolated from actual tumors and visceral organ tissue in RTG-2 and FHM cells.

Virus Characteristics

Ahne et al. (1987) reported that nine virus isolates have been made from European eels with stomatopapilloma. The first virus isolation was from blood of affected fish and was named "eel virus (Berlin)" (Pfitzner and Schubert 1969). Characterization of the virus indicates an icosahedral virion with an RNA genome that is morphologically similar to IPNV (Schwanz-Pfitzner et al. 1984). It most likely is IPNV serotype Ab and, therefore, would be classified as a birnavirus. Wolf and Quimby (1970) isolated a virus from European eel with stomatopapilloma and designated it EV-1. This virus was plaque purified and then designated EV-2 for characterization and was classified as an orthomyxolike virus (Nagabayashi and Wolf 1979). The icosahedral virus is RNA, measures 80 to 140 nm in diameter, and replicates in FHM cells at 16 to 18°C. Cytopathic effect, which appears after 10 hours, consists of cell fusion, syncytia, and irregularly rounded cell masses.

Three viruses were isolated from a single European eel with stomatopapilloma by Ahne and Thomsen (1985). These viruses were identified as IPNV-Ab from the blood and as *Rhabdovirus anguilla*-subtype EVEX and an unclassified agent from the tumor. The viruses produced CPE in FHM cells in 3 days. Ahne et al. (1987) determined that of nine virus isolates taken from stomatopapilloma-infected eels, five taken from blood and identified by serum neutralization tests were IPNV-subtype Ab and four taken from tumors were identified as *Rhabdovirus anguilla*-subtype EVEX.

Epizootiology

Stomatopapilloma of European eels has been of interest to fishery scientists for years. The disease is still not fully understood, and its etiology remains unresolved. That at least nine different viruses were isolated from infected fish does not necessarily indicate that any were responsible for the tumors. As noted, viruses have been isolated from blood, visceral organs, and actual tumor tissue, but experiments to induce stomatopapilloma lesions with these viruses were unsuccessful.

Stomatopapilloma appears to be a disease primarily of wild eel populations, with no specificity for age or size. The number of affected eels doubled between 1957 and 1967 from 5.6% to 11.9% in the Elbe River in Germany (Koops et al. 1970). Tumors were found on 15- to 35-cm eels in the 3- to 4-year age group in fresh water in the Lower Elbe River (Peters 1975). Incidence of tumors was low (2–3%) in spring and autumn compared with summer months, when incidence increased to about 28%. A similar seasonal incidence was found in Scottish rivers, where the disease was not known until 1980 (Hussein and Mills 1982). Although not totally eliminating a viral etiology, Peters (1975) speculated that seasonal fluctuations in tumor growth were due to water temperature and oxygen content, but he conceded that chemical pollutants could also be a factor. No such water quality correlations were found by Hussein and Mills (1982) in comparing incidence of stomatopapilloma in four tributaries of the River Tweed in Scotland. Stomatopapilloma probably does not kill eels directly, but head tumors can be so massive that feeding is affected, causing as much as a 50% loss in fat content in affected fish (Koops et al. 1970).

Pathological Manifestations

Stomatopapilloma lesions are hyperplastic skin epithelium that might or might not be pigmented. The resultant tumor consists of fibroepithelium that most commonly is seen on the upper and lower jaw and occasionally on fins and lateral areas (Koops et al. 1970). Pathological changes begin at the basal layer of the epidermis, followed by transformation of cuboidal cells into elongate cells with spindle-shaped nuclei. Papillomas are smaller (up to 300 mm^3) in cooler weather and larger in August–September, when they may reach 900 mm^3 and show signs of invasiveness (Peters 1975).

Significance

Currently, stomatopapilloma is of little concern to the eel culture industry. Its presence does not cause large mortality but could have an adverse economic effect on marketability of cultured eels.

MANAGEMENT OF EEL VIRUSES

There are no known management procedures specifically for prevention and/or control of eel viruses; however, avoidance whenever possible is advisable. Also, for viruses that are more virulent at low temperatures increasing water temperature to more than 23°C may be beneficial.

REFERENCES

Ahne, W., and I. Thomsen. 1985. The existence of three different viral agents in a tumor bearing European eel (*Anguilla anguilla*). *Journal of Veterinary Medicine* 32:228–235.

Ahne, W., I. Schwanz-Pfitzner, and I. Thomsen. 1987. Serological identification of 9 viral isolates from European eels (*Anguilla anguilla*) with stomatopapilloma by means of neutralization tests. *Journal of Applied Ichthyology* 3:30–32.

Békési, L. I., I. Horvath, E. Kovacs-Gayer, and G. Csaba. 1986. Demonstration of herpesvirus-like particles in skin lesions of the European eel (*Anguilla anguilla*). *Journal of Applied Ichthyology* 4:190–192.

Castric, J., and C. Chastel. 1980. Isolation and characterization attempts of three viruses from European eel, *Anguilla anguilla;* preliminary results. *Annals in Virology* 13E:435–448.

Castric, J., D. Rasschaert, and J. Bernard. 1984. Evidence of Lyssaviruses among rhabdovirus isolates from the European eel *Anguilla anguilla. Annals in Virology* 135E:35–55.

Chen, S. N., and G. H. Kou. 1981. A cell line of Japanese eel (*Anguilla japonica*) ovary. *Fish Pathology* 16:129–137.

Chen, S. N., G. H. Kou, R. P. Hedrick, and J. L. Fryer. 1985. The occurrence of viral infections of fish in Taiwan. In: *Fish and Shellfish Pathology,* edited by A. Ellis, 313–319. London: Academic Press.

Chi, S. C., S. N. Chen, and G. H. Kou. 1991. Establishment, characterization and application of monoclonal antibodies against eel virus European (EVE). *Fish Pathology* 26:1–7.

Delves-Broughton, J., J. K. Fawell, and D. Woods. 1980. The first occurrence of "cauliflower disease" of eels *Anguilla anguilla* L. in the British Isles. *Journal of Fish Diseases* 3:255–256.

Egusa, S., H. Hirose, and H. Wakabayashi. 1971. A report of investigations on branchionephritis of cultured eel II. Conditions of the gills and serum ion concentrations. *Fish Pathology* 6:57–61.

Hedrick, R. P., J. L. Fryer, S. N. Chen, and G. H. Kou. 1983. Characteristics of four birnaviruses isolated from fish in Taiwan. *Fish Pathology* 18:91–97.

Hudson, E. B., D. Bucke, and A. Forrest. 1981. Isolation of infectious pancreatic necrosis virus from eels, *Anguilla anguilla* in the United Kingdom. *Journal of Fish Diseases* 4:429–431.

Hussein, S. A., and D. H. Mills. 1982. The prevalence of "cauliflower disease" of eel, *Anguilla anguilla* L., in tributaries of the River Tweed, Scotland. *Journal of Fish Diseases* 5:161–165.

Kobayashi, T., and T. Miyazaki. 1996. Rhabdoviral dermatitis in Japanese eel, *Anguilla japonica. Fish Pathology* 31:183–190.

Koops, H., H. Mann, I. Pfitzner, O. J. Schmid, and G. Schubert. 1970. The cauliflower disease of eels. In: *Diseases of Fishes and Shellfishes,* edited by S. F. Snieszko, 191–295. Special publication no. 5. Washington, DC: American Fisheries Society.

Nagabayashi, T., and K. Wolf. 1979. Characterization of EV-2, a virus isolated from European eels (*Anguilla anguilla*) with stomatopapilloma. *Journal of Virology* 30:358–364.

Nishimura, T., M. Toba, F. Ban, N. Okamoto, and T. Sano. 1981. Eel rhabdovirus, EVA, EVEX and their infectivity to fishes. *Fish Pathology* 15:173–184.

Okamoto, N., T. Sano, R. P. Hedrick, and J. L. Fryer. 1983. Antigenic relationships of selected strains of infectious pancreatic necrosis and European eel virus. *Journal of Fish Diseases* 6:19–25.

Peters, G. 1975. Seasonal fluctuations in the incidence of epidermal papillomas of the European eel *Anguilla anguilla* L. *Journal of Fish Biology* 7:415–422.

Pfitzner, I., and G. Schubert. 1969. Ein virus aus dem blut mit blumendohlkrankheit behalfteter ale. *Zeitschrift fuer Naturforschung* 24b:789–790.

Sano, M., H. Fukuda, and T. Sano. 1990. Isolation and characterization of a new herpesvirus from eel. In: *Pathology in Marine Science,* edited by F. O. Perkins, and C. T. Cheng, 15–31. New York: Academic Press.

Sano, T. 1976. Viral diseases of cultured fishes in Japan. *Fish Pathology* 19:221–226.

Sano, T., N. Okamoto, and T. Nishimura. 1981. A new viral epizootic of *Anguilla japonica* Temminck and Schlegel. *Journal of Fish Diseases* 4:127–139.

Schwanz-Pfitzner, I., M. Özel, G. Darai, and H. Gelderblom. 1984. Morphogenesis and fine structure of eel virus (Berlin), a member of the proposed birnavirus group. *Archives of Virology* 81:151–162.

Sorimachi, M. 1984. Pathogenicity of ICD virus isolated from Japanese eel [English abstract]. *Bulletin National Research Institute of Aquaculture* 6:71–75.

Sorimachi, M., and S. Egusa. 1982. Characteristics and distribution of viruses isolated from pond-cultured eels. *Bulletin National Research Institute of Aquaculture* 6:71–75.

Ueno, Y., S. Chen, G. Kou, R. P. Hedrick, and J. L. Fryer. 1983. Characterization of a virus isolated from Japanese eels (*Anguilla japonica*) with nephroblastoma. *Bulletin of the Institute Zoological Academy of Sinica* (Taipei) 23:47–55.

Wolf, K. 1988. *Fish Viruses and Fish Viral Diseases.* Ithaca, NY: Cornell University Press.

Wolf, K., and M. Quimby. 1970. Virology of eel stomatopapilloma. In: *Progress in Sport Fishery Research 1970,* 94–95. Resource publication no. 106. Washington, DC: U.S. Fish and Wildlife Service.

9 Pike Viruses

Northern pike is not a prototype aquaculture species, although it is cultured in North America to some degree and more extensively in northern Europe. Because of its strong piscivorous feeding habits, northern pike are seldom grown larger than a few centimeters in length before being stocked into larger bodies of water where live food is available. Juvenile northern pike in Europe are afflicted by a highly virulent virus, and in North America, northern pike and muskellunge are susceptible to at least two tumorous diseases.

PIKE FRY RHABDOVIRUS

Pike fry rhabdovirus disease (PFRD) is an acute, highly contagious disease, commonly called "head disease" in cultured fry and "red disease" in cultured fingerling northern pike (Bootsma 1971; de Kinkelin et al. 1973). The etiological agent, pike fry rhabdovirus (PFR), was characterized by de Kinkelin et al. (1973). It was noted that "head disease" had previously been reported in the Netherlands in 1959 and "red disease" in 1956; however, the disease has not been severe in recent years (Békési et al. 1984).

Geographical Range and Species Susceptibility

Pike fry rhabdovirus has been isolated from northern pike in hatcheries in The Netherlands and in Germany, and a serologically identical rhabdovirus was isolated from grass carp (Ahne 1975a), tench, white bream, European catfish (Jørgensen et al. 1989), brown trout (Adair and McLoughlin 1986), roach (Haenen and Davidse 1989), and gudgeon, an ornamental fish accidentally introduced into Europe with grass carp (Ahne 1975b; Ahne and Thomsen 1986). These isolates have subsequently been identified as PFR. Pike fry rhabdovirus disease has also been reported in Hungary but has not been reported in North America.

Clinical signs

Two forms of PFRD occur in northern pike, one that is commonly known as "head disease" affects swimming fry (de Kinkelin et al. 1973). Infected fry have a raised cranium (Figure 9.1), a result of hydrocephalus; are exophthalmic; and exhibit poor growth. The second form occurs in larger fish, which develop severe hemorrhages along the lateral trunk musculature, thus the name "red disease" (Figure 9.1). Hemorrhagic areas are swollen, gills are pale, and infected fish exhibit bilateral exophthalmia, abdominal distension, and ascites. In both forms, fish lose schooling behavior and individuals either swim lethargically at the surface or lie listlessly on the bottom. Although fish species other than northern pike can become infected with PFR, clinical signs of these infections are poorly defined.

Diagnosis

Pike fry rhabdovirus is detected by isolation of virus in fathead minnow (FHM) or epithelioma papillosum of carp (EPC) cells (de Kinkelin et al. 1973). Kidney, spleen, and intestines provide the best tissue sources for virus isolation. Infected cells incubated at 20°C develop focal areas of rounded cells in 2 to 3 days. Virus identification can be obtained by

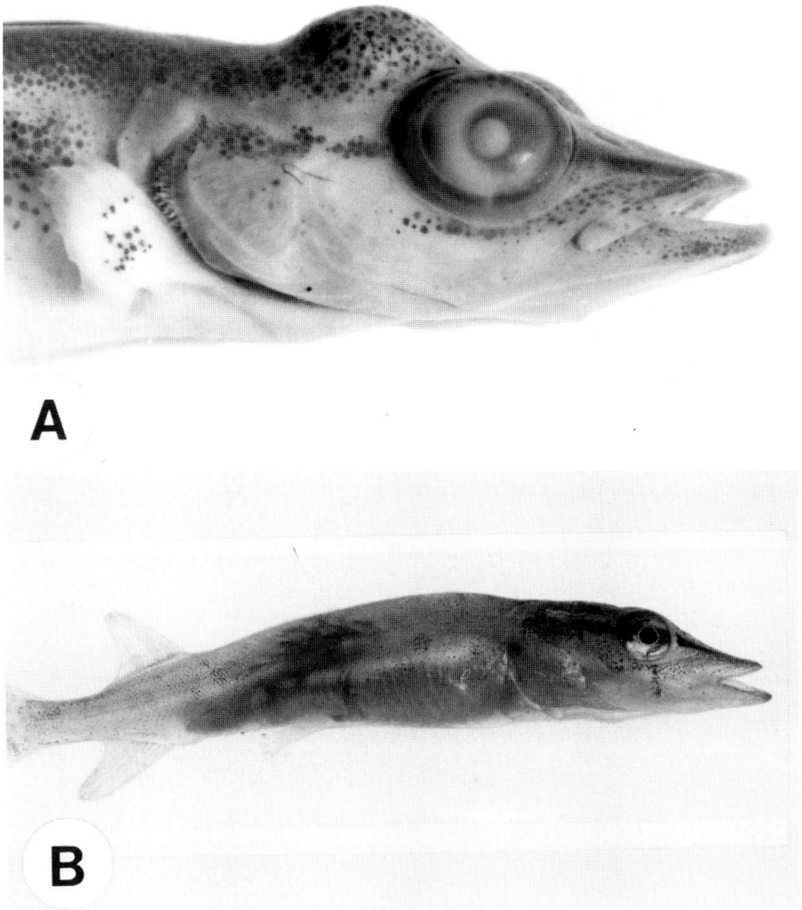

FIGURE 9.1. (A) *Pike infected with pike Rhabdovirus showing "head disease" with hydrocephalus.* (B) *Pike with "red disease" showing hemorrhaging in the lateral musculature. (Photographs courtesy of R. Bootsma.)*

serum neutralization, but when using homologous rabbit antiserum, PFR neutralization is complement dependent (Clerx et al. 1978). Also, other viruses (such as spring viremia of carp [SVC]) can be partially neutralized by PFR-specific antiserum (Ahne 1986). This partial cross-reactivity can be removed by reacting the antiserum with other rhabdoviruses of fish. Pike fry rhabdovirus is serologically distinct from viral hemorrhagic septicemia virus (VHSV) and infectious hematopoietic necrosis virus (IHNV) but some cross-reactivity occurs with spring viremia of carp virus (SVCV) (see Table 7.1).

Virus Characteristics

The causative agent of PFRD is a rhabdovirus that measures 125 to 155 nm in length and about 80 nm in diameter (Bootsma and van Vorstenbosch 1973). It is enveloped and has a single-strand, nonsegmented RNA genome (Ahne 1975b). Pike fry rhabdovirus is highly infectious to FHM cells at 10 to 28°C, with titers reaching greater than 10^9 plaque-forming units (PFUs) per mL (de Kinkelin et al. 1973). Epithelioma papillosum of carp, BB, and RTG-2 (rainbow trout gonad) cells are also susceptible, but virus yield is lower than in FHM cells. Infectivity persists in tissue culture medium with serum at 4°C for 5 months but decreases at higher temperatures. Seventy percent of infectivity is lost in 3 days in water at 14°C.

Epizootiology

Fry and small fingerling northern pike (up to 5 cm) are susceptible to PFR, and 100% mortality can occur (Bootsma 1971). Haenen and Davidse

(1993) showed that a PFR isolate from asymptomatic roach was highly pathogenic to young roach (85% mortality) after waterborne exposure, mildly pathogenic to carp (45% mortality), and avirulent to rainbow trout. These authors concluded that PFR could pose a health threat to cyprinid farms as well as cultured northern pike farms. The disease occurs in early spring when water temperatures are approximately 10°C.

The natural transmission mechanism for PFR is not known. Bootsma et al. (1975) were unable to isolate the virus from frozen kidney, liver, spleen, intestine, or gonads of brood pike even though progeny of these fish were infected. Bootsma and van Vorstenbosch (1973) also experimentally infected pike eggs with PFR, and within weeks of hatching, there was 100% fry mortality. Definitive proof of vertical transmission, however, is lacking. Horizontal transmission of PFR has been experimentally demonstrated after waterborne exposure and injection of juvenile northern pike (Ahne 1985). Fry (0.2 g) bathed in virus and then fed to larger pike (60 g) transmitted PFR to recipient fish, but deaths or overt disease did not occur.

Pike fry rhabdovirus disease can be confused with other viruses that affect pike. Ahne (1985) demonstrated susceptibility of pike fry (0.2 g) to other viruses by exposing them to water bath containing infectious pancreatic necrosis virus (IPNV), SVCV, or VHSV, with a resultant mortality of 42% to 90%. Grass carp reovirus caused 64% mortality under similar circumstances. Dorson et al. (1987) confirmed that 10-day- to 2-month-old northern pike fry are susceptible to IPNV, IHNV, VHSV, and perch rhabdovirus, all of which can cause hemorrhaging, exophthalmia, and up to 74% mortality. When virus is isolated from northern pike fry, a definitive virus identification should be made. It was postulated by Ahne (1986) that northern pike could play the role of vector for other pathogenic viruses normally not found in this species.

Pathological Manifestations

In "head disease," there is an accumulation of cerebral fluid in the ventricle of the mesencephalon (Bootsma 1971). Petechia can be found in the brain, spinal cord, and spleen; and kidney tubules show degeneration with necrosis of epithelial cells. "Red disease" is characterized by petechia throughout the spinal cord, pancreas, and hematopoietic tissue of the kidney, but severe muscle hemorrhage is the most dramatic histopathological lesion, with erythrocytes and inflammatory cells congregating between bundles of necrotic muscle. There is extensive necrosis of excretory kidney tubules, and the cerebral region becomes edematous. The liver, heart, and gastrointestinal tract appear normal in both forms of disease.

Significance

When PFRD occurs in juvenile northern pike, it can be devastating. In 1972, only 0.6% of 1.85 million cultured young northern pike in The Netherlands reached a length of 4 to 5 cm, and it was estimated that approximately 86% of the mortality was due to PFRD. Pike fry rhabdovirus disease has had a significant effect on northern pike fry culture in the past, but owing to egg disinfection with iodine at spawning, incidence of disease has been reduced (P. de Kinkelin, Laboratoire de Ichtyopathologie, Thiverval-Grignon, France, personal communication). Research by Haenen and Davidse (1993) indicates that seed carp farms could also be at risk by the virus.

OTHER VIRUSES OF ESOCIDS

As previously noted, northern pike are also experimentally susceptible to IPNV, VHSV, and SVCV (Ahne 1985). Several other viruses that are also known to infect northern pike and other members of the esocid family are *Esox* lymphosarcoma, and *Esox* sarcoma, which have been reported from the northern United States and Canada. They were originally considered to be separate diseases, but according to McAllister (1979), these conditions are the same. Incidence of lymphosarcoma in wild northern pike and muskellunge approaches 21% in some populations (Sonstegard 1976). The disease is characterized by macroscopically visible nodular lesions in subepidermal connective tissue from which tumors metastasize into the muscle and then into internal organs. It is caused by a putative retrovirus that measures about 100 nm in diameter (Sonstegard 1976). The disease, described as a stem cell lymphoreticular neoplasm, appears to be temperature related and usually causes death (Sonstegard 1976). Lymphosarcoma has been experimentally transmitted by injection of filtered tumor homogenate, implantation of neoplastic tissue, and cohabitation (Sonstegard 1976). The significance of *Esox* lymphosarcoma is not great owing to lack of distribution, but the disease has received attention because some fish tumor cells associated with *Esox* lymphosarcoma are morphologically similar to structures found in several types of human lymphoma and leukemia (Dawe et al. 1976).

Esox epidermal hyperplasia is a retrovirus-caused infection of northern pike and muskellunge in Canada (Sonstegard 1976) and pike in Sweden (Winqvist et al. 1968), where disease incidence has been as high as 6% in some populations. The epidermal lesions measure about 5 to 10 mm in diameter and are 1- to 3-mm thick with no evidence of metastasis or muscle invasion. According to McAllister (1979), the virus has no known adverse effect on fish and no transmission studies have been reported.

MANAGEMENT OF PIKE VIRUSES

In The Netherlands, for the last two decades, management practices initiated by Bootsma et al. (1975) have reduced PFR impact on northern pike. They demonstrated that a solution of 25 mg of iodine per liter, in the form of Wescodyne, deactivated 99.99% of PFR in 30 seconds. Routinely bathing pike eggs in this solution after fertilization interrupts the pike fry rhabdovirus infection cycle.

REFERENCES

Adair, B. M. and M. McLoughlin. 1986. Isolation of pike fry rhabdovirus from brown trout (*Salmo trutta*). *Bulletin of the European Association of Fish Pathologist* 6(3):85–86.

Ahne, W. 1975a. A rhabdovirus isolated from grass carp (*Ctenopharyngodon idella* Val.). *Archives of Virology* 48:181–185.

Ahne, W. 1975b. Biological properties of a virus isolated from grass carp (*Ctenopharyngodon idella* Val.). In: *Wildlife Diseases,* edited by L. A. Page, 135–140. New York: Plenum Press.

Ahne, W. 1985. Viral infection cycles in pike (*Esox lucius* L). *Journal of Applied Ichthyology* 1:90–91.

Ahne, W. 1986. Unterschiedliche biologische eigenschaften 4 *cyprinidenpathogener rhabdovirusisolate. Journal of Veterinary Medicine* 33:253–259.

Ahne, W., H. Mahnel, and P. Steinhagen. 1982. Isolation of pike fry rhabdovirus from tench, *Tinca tinca* L., and white bream, *Blicca bioerkna* (L.). *Journal of Fish Diseases* 5:535–537.

Ahne, W., and I. Thomsen. 1986. Isolation of pike fry rhabdovirus from *Pseudorasbora parva* (Temminck and Schlegel). *Journal of Fish Diseases* 9:555–556.

Békési, L., G. Majoros, and S. Edit. 1984. Csukaivadék (*Esox lucius* L.) elhllását okozó rhabdovirus megjelenése hasánkban. *Magyar Allatorvosok Lapja* 39:231–234.

Bootsma, R. 1971. Hydrocephalus and red-disease in pike fry *Esox lucius* L. *Journal of Fish Biology* 3:417–419.

Bootsma, R., and C. J. A. H. V. van Vorstenbosch. 1973. Detection of a bullet-shaped virus in kidney sections of pike fry (*Esox lucius* L.) with red-disease. *Netherlands Journal of Veterinary Science* 98:86–90.

Bootsma, R., P. de Kinkelin, and M. Le Berre. 1975. Transmission experiments with pike fry (*Esox lucius* L.) rhabdovirus. *Journal of Fish Biology* 7:269–276.

Clerx, J. P. M., M. C. Horzinek, and A. D. Osterhaus. 1978. Neutralization and enhancement of infectivity of non-salmonid fish rhabdoviruses by rabbit and pike immune sera. *Journal of General Virology* 40:297–308.

Dawe, C. J., W. G. Banfield, R. Sonstegard, C. W. Lee, and H. J. Michelitch. 1976. Cylindroid lamella-like particle complexes and nucleoid intracytoplasmic bodies in lymphoma cells of northern pike (*Esox lucius*). *Progress in Experimental Tumor Research* 20:166–180.

de Kinkelin, P., B. Galimard, and R. Bootsma. 1973. Isolation and identification of the causative agent of "red disease" of pike (*Esox lucius* L. 1766). *Nature* 241:465–467.

Dorson, M., P. de Kinkelin, C. Torchy, and D. Monge. 1987. Sensibilite du brochet (*Esox lucius*) a differents virus de salmonides (NPI, SHV, NHI) et au rhabdovirus de la perche. *Bulletin d' Francie Peche Piscicultura* 307:91–101.

Haenen, O. L. M., and A. Davidse. 1989. Isolation of pike fry rhabdovirus from roach (*Rutilus rutilus*). *Bulletin of the European Association of Fish Pathologist* 9(5):116.

Haenen, O. L. M., and A. Davidse. 1993. Comparative pathogenicity of two strains of pike fry rhabdovirus and spring viremia of carp virus for young roach, common carp, grass carp and rainbow trout. *Diseases of Aquatic Organisms* 15:87–92.

Jørgensen, P. E. V., N. J. Olesen, W. Ahne, and N. Lorenzen. 1989. Spring viremia of carp and PFR viruses: serological examination of 22 isolates indicates close relationship between the two fish rhabdoviruses. In: *Viruses of Lower Vertebrates,* edited by W. Ahne and E. Kurstak, 349–399. Berlin: Springer-Verlag.

McAllister, P. E. 1979. Fish viruses and viral infections. In: *Comprehensive Virology,* edited by H. Fraenkel-Conrat and R. R. Wagner, 401–470. New York: Plenum Press.

Sonstegard, R. A. 1976. Studies of the etiology and epizootiology of lymphosarcoma in Esox (*Esox lucius* and *Esox masquinongy*). *Progress in Experimental Tumor Research* 20:141–155.

Winqvist, G., O. Ljungberg, and B. Hellstroem. 1968. Skin tumors of northern pike (*Esox lucius* L.). II. Viral particles in epidermal proliferations. *Bulletin of the Office of International Epizooties* 69:1023–1031.

10 Trout and Salmon Viruses

Viral diseases of salmonids are among the most studied infectious fish diseases. They are of special interest because of their pathogenicity, wide range of occurrence, epizootiology, and adaptability to management. About 20 viruses that infect salmonids, including those that are highly or intermediately pathogenic and those that apparently cause few problems, were listed and discussed by Sano (1995). Some of these viruses are readily isolated in cell culture, making them easy to work with; some can be experimentally transmitted only from fish to fish; and others are only detectable by electron microscopy. This chapter concentrates on viruses that are known to cause, or are apparently associated with, serious disease in salmonids.

INFECTIOUS HEMATOPOIETIC NECROSIS VIRUS

Infectious hematopoietic necrosis (IHN) is a highly contagious disease primarily of rainbow trout and several species of Pacific salmon. The etiological agent of IHN is a rhabdovirus, infectious hematopoietic necrosis virus (IHNV) (Amend et al. 1969). Episodes of IHN date back to the 1940s and 1950s when unexplained juvenile salmon mortalities were reported at hatcheries from California to Washington state. Rucker et al. (1953) suggested a viral etiology for these deaths, and Watson et al. (1954) presented strong evidence that this was the case. Related disease outbreaks in the Pacific Northwest have also been known at one time or another as Columbia River sockeye salmon disease, Oregon sockeye virus disease, sockeye salmon virus disease, and Sacramento River chinook disease (Amend et al. 1969). Comparative studies of histopathology and viral biochemical, biophysical, and serological properties determined that the viruses that caused these diseases were actually the same virus–IHNV (Amend et al. 1969; Amend and Chambers 1970; McCain et al. 1971).

Geographical Range and Species Susceptibility

Originally, IHNV was confined to anadromous salmon in tributaries that flowed directly into the Pacific Ocean from northern California to Alaska. But by the late 1960s, IHNV epizootics were confirmed in Minnesota (Plumb 1972), South Dakota, Montana, and West Virginia. The virus was later reported in rainbow trout in a closed system in New York (Wolf 1988). In addition to being enzootic on the west coast of North America, IHNV has now become established in the Hagerman Valley of Idaho and has been confirmed in China, France, Germany, Italy, Japan, Korea, and Taiwan (Sano 1976; Bovo et al. 1987; Hattenberger-Baudouy et al. 1989; LaPatra et al. 1989a; Park et al. 1993).

Species most susceptible to IHNV are rainbow trout (steelhead), and chinook, sockeye (kokanee) (Amend et al. 1969; Rucker et al. 1953), and chum salmon (K. Amos, Washington Department of Fish and Wildlife, personal communication). Wingfield and Chan (1970) listed coho salmon as being refractive to IHNV; however, LaPatra and Fryer (1990) isolated the virus from adult coho but alevin coho were resistant. Sano (1976) reported that amago and masu (yamame) were also susceptible. Infectious hematopoietic necrosis virus will infect brook trout and brown trout but to a much lesser degree than rainbow trout (LaPatra and

Fryer 1990). According to Follett et al. (1997) Arctic char, Arctic grayling, and pink salmon are refractory to experimental infection with IHNV, but lake trout will support virus replication with concomitant clinical signs. Mulcahy (1984) reported Atlantic salmon mortalities from a natural IHNV infection. Experimentally IHNV infected larval white sturgeon can possibly become viral vectors but fingerlings and adults are probably refractive (LaPatra et al. 1995).

Clinical Signs and Findings

Clinical signs of IHN may vary somewhat from species to species (Amend 1969; Mulcahy 1984). Larger fish in an infected fingerling population usually exhibit clinical signs before smaller fish of the same age. Infectious hematopoietic necrosis virus–infected fry and fingerlings are lethargic, avoid currents, and move to the edge of ponds or raceways or lie on the bottom respiring weakly. In the final stages of disease, they swim in circles and hang vertically or flash in a frenzy, followed quickly by death.

Externally, affected fish are dark, exophthalmic, have swollen abdomens, and pale gills. The base of fins may be hemorrhaged, petechiae are present on mouth and body surfaces, and chevronlike hemorrhages occur along the lateral musculature (Figure 10.1). An opaque, mucoid fecal cast trails from the vent. An abnormally dark (red) area can develop behind the head, in the abdominal region, or on the caudal peduncle (Burke and Grischkowsky 1984).

Internally, organs appear anemic with petechiae in the mesenteries, peritoneum, air sac, liver, and kidney. The digestive tract is void of food but contains mucoid fluid, and the body cavity may contain pale, yellowish, clear fluid. Amend et al. (1969) and Rucker et al. (1953) reported several deformities associated with IHNV: malformed heads, scoliosis, or lordosis in 5% to 60% of survivors.

Diagnosis

Clinical IHNV is diagnosed by conventional isolation of virus in cell culture. Whole fry or viscera of fingerlings are suitable material for virus isolation. In rainbow trout infected by immersion, IHNV titers can reach 10^9 PFU/g of body tissue within 5 days (Yamamoto et al. 1990). Significant viral infectivity can remain in tissue samples for up to 2 weeks at 4°C, and some preparations of ovarian fluid, eggs, serum, and brain retain virus for 5 weeks (Burke and Mulcahy 1983). Frozen samples stored at –20°C will retain infectivity for 3 to 5 months.

The CHSE-214 (chinook salmon embryo) and epithelioma papillosum of carp (EPC) cell lines are systems of choice for isolation, but IHNV can also be isolated in RTG-2, fathead minnow (FHM), and others (Yoshimizu et al. 1988a). It was shown by Ristow and Avila (1994) that a newly developed cell line from rainbow trout (RBTE 45) was susceptible to IHNV as were three cell lines from coho salmon. All produced better viral replication than CHSE-214 but EPC cells remained most sensitive. Yoshimizu et al. (1988a) indicated that IHNV would replicate in all 17 cell lines derived from salmonids and in 12 of 15 nonsalmonid fish cell lines tested. Cultures should be incubated at 15°C (never above 20°C) for up to 14 days. If IHNV is present in necropsied fish, ctyopathic effect (CPE) may appear in 2 to 3 days at 15°C, but it usually appears in 4 to 5 days. The CPE is initially characterized by pyknotic cells, followed by rounding and sloughing of these cells from the surface (Figure 10.2). Treatment of monolayers of IHNV-susceptible cells with 7% polyethylene glycol before inoculation with virus-suspect materials enhances plaque forming unit (PFU) titers by 4- to 17-fold (Batts and Winton 1989). Blind passage is recommended for samples that are negative on primary isolation attempts. Once IHNV is isolated, positive identification is made by conventional serum neutralization, enzyme linked immunosorbent assay (ELISA), or other serological methods.

A method using alkaline phosphatase immunocytochemical procedures to detect and identify IHNV in dried tissue cultures was developed by Drolet et al. (1993), and with this method, they were able to detect IHNV in dried tissue cultures that were more than 1 year old. The system was sensitive for five known IHNV types and no cross-reactivity was noted with six other fish rhabdoviruses or with infectious pancreatic necrosis virus (IPNV). The test was specific, sensitive, and was not affected by age of fixed or dried cell culture plates.

LaPatra et al. (1989b) found that polyclonal and monoclonal antibody can be used to detect IHNV in blood and organ smears of juvenile

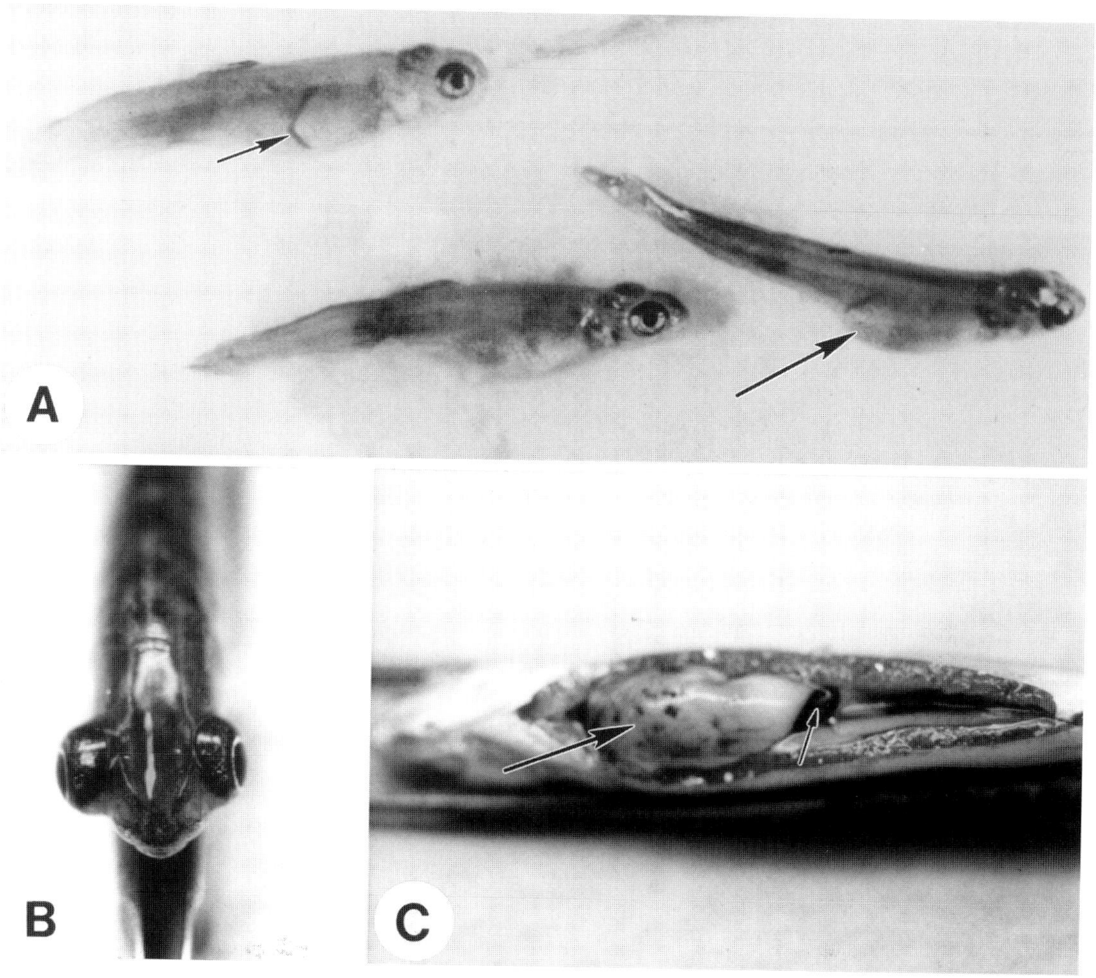

FIGURE 10.1. (A) Coho salmon infected with infectious hematopoietic necrosis virus. Note the chevron-shaped hemorrhages in the musculature (small arrow), *the hemorrhage behind the head and on the abdomen, and the swollen abdomen* (large arrow). *(Photograph A courtesy of J. Rohovec.)* (B) *Exophthalmia in hybrid rainbow trout with infectious hematopoietic necrosis virus (IHNV).* (C) *Petechial hemorrhage in liver* (large arrow) *and enlarged, dark spleen* (small arrow) *in hybrid rainbow trout with IHNV. (Photographs B and C courtesy of S. LaPatra, Clear Springs Food Inc.)*

salmonids and in ovarian fluid of adults. Virus was also detectable in cell cultures 48 hours after inoculation and was sensitive for all virus isolates. According to Arnzen et al. (1991), the fluorescent antibody technique (FAT) is as sensitive as plaque assay and requires less time to obtain a confirmed diagnosis; however, monoclonal antibody against IHNV nucleoprotein and glycoprotein detected stages of virus replication in cell cultures as early as 6 to 8 hours postinoculation. Dixon and Hill (1984) developed an ELISA technique for rapid IHNV identification in infected tissues, with results being obtained in 2 hours. An ELISA assay was developed by Medina et al. (1992) that detected as few as 70 PFU of virus in 100 µg of homogenized Atlantic salmon tissue and also detected virus in cell culture media. The system gave results in 1 day and also cross-reacted with other types of IHNV, enabling detection of all tested types. The immunodot blot method may also be used in detecting IHNV protein (McAllister and Schill 1986).

Molecular technology has recently been used

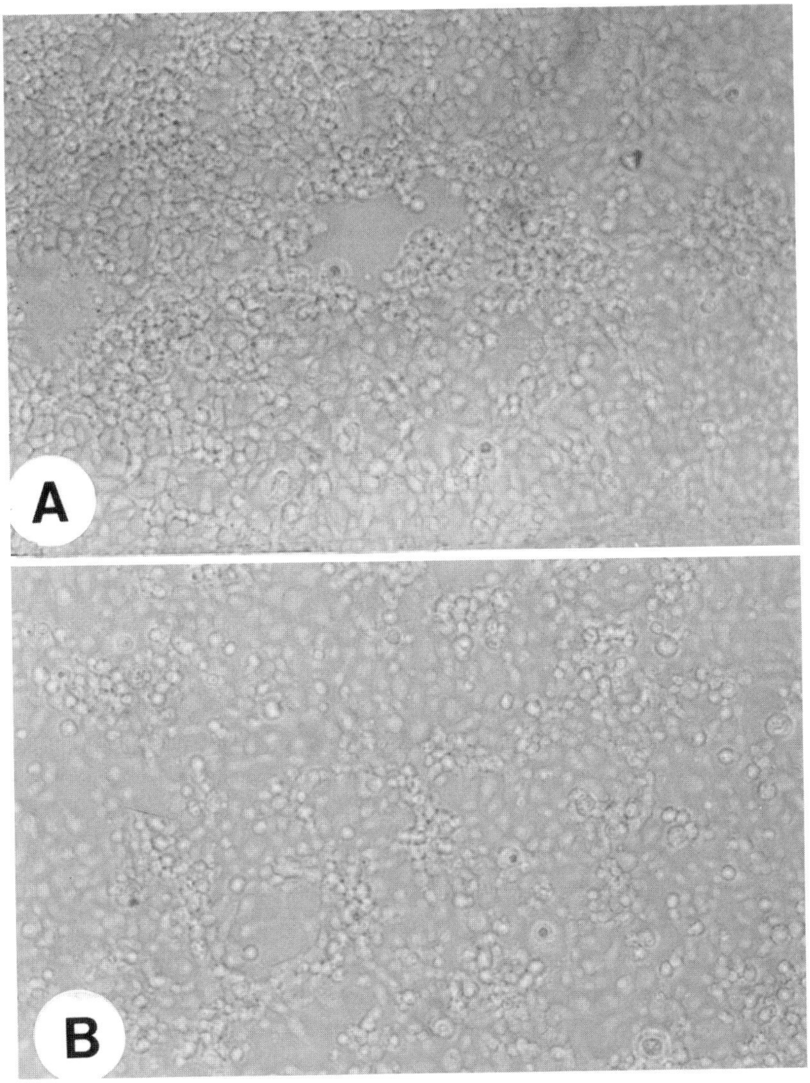

FIGURE 10.2. Cytopathic effect (CPE) of infectious hematopoietic necrosis virus in CHSE-214 cells incubated at 15°C. (A) Focal CPE at 48 hours and (B) total CPE at 60 hours after inoculation (×500).

extensively in detecting IHNV in fish tissue and cell cultures. Arakawa et al. (1990) used a nucleic acid probe in combination with polymerase chain reaction (PCR) to develop a rapid, highly specific, and sensitive IHNV detection method. A nonradioactive DNA probe developed by Deering et al. (1991) detects and identifies IHNV using a dot blot format with virus-infected tissue cultures. This technique proved to be sensitive in detecting very small concentrations of target viral mRNA. Any of these detection and identification procedures can be used for IHNV but are technically intensive and require specialized equipment. All are more rapid than standard tissue culture methods, and some are just as sensitive.

Virus Characteristics

Infectious hematopoietic necrosis virus is a rhabdovirus with a mean diameter of about 70 nm and a length of 170 nm (Amend and Chambers 1970). The enveloped RNA virus is heat, acid, and ether labile. Stability studies of IHNV are contradictory. Pietsch et al. (1977) reported that virus survived for 5 days or less in seawater and Hanks' balanced salt solution (HBSS), but with addition of 10% fetal calf serum, viability was extended to 12 days. Barja et al. (1983) demonstrated that IHNV survived for about 30 days in fresh water at 15°C and 17 days at 20°C. Survival was 17 days at 20°C and 22 days at 15°C in seawater. Infectivity of IHNV is reduced in distilled water at pH 5 and 9, but at pH 6 to 8, it remains infective for more than 10 days. Infectious hematopoietic necrosis virus does not survive well under dry conditions at any temperature, but freezing has little adverse effect.

McCain et al. (1971) demonstrated by plaque reduction that there was only one IHNV serotype, and this was confirmed by Engelking et al. (1991). On the basis of molecular weights of nucleocapsid proteins and glycoproteins, Hsu et al. (1986) classified the virus into five different electropherotypes: types 1, 2, 3, 4, and 5. Using monoclonal antibodies, Winton et al. (1988) separated IHNV isolates from Alaska, California, Idaho, Oregon, Washington, and Japan into four separate groups. Although only one IHNV serological strain has been identified, Arkush et al. (1989) showed that the U.S. isolate has a different polypeptide profile and is antigenically different from European isolates. Isolates of IHNV from France were typed by Danton et al. (1994), and only 5 of 27 isolates were recognized by monoclonal antisera Mab1NDW14D, previously considered to be a universal immunofluorescent reagent for IHNV. All 27 isolates did share the nucleocapsid epitope identified by Mab2NH105B, thought to be specific for IHNV electropherotype 2; however, these isolates did not belong to this electropherotype. It was suggested that IHNV had been in Europe for a sufficient amount of time to evolve separately from North American strains. In view of these different results, it appears that the issue of one or more IHNV strains remains in question.

Epizootiology

Infectious hematopoietic necrosis is generally considered to cause high mortality among young cultured salmonids. Actual percentage of deaths can vary from low to nearly 100% depending on fish species, age, and size; environmental conditions, particularly water temperature; and virus strain. The mortality pattern of IHNV is typical of highly virulent viruses that affect young fish. Under optimum conditions, death rate accelerates rapidly and often 70% or greater mortality will accrue within 10 days. Amend and Nelson (1977) demonstrated that sockeye salmon families, which had undergone selective breeding, suffered 52% to 98% mortality after artificial infection with IHNV. They concluded that these variances were due to genetic differences among the host. A 30% heritability of resistance to IHNV resulted from selective breeding, suggesting that more IHNV-resistant fish strains could be developed (McIntyre and Amend 1978). Using rainbow trout as the host, Idaho, Oregon, and California IHNV strains were compared. At 10°C, 62% mortality occurred in groups infected with the Idaho strain, 4% in groups infected with the Oregon strain, and 67% in groups infected with the California strain.

Water temperature and age of fish also influence mortality pattern. Most hatchery epizootics associated with high mortality involve fry or small fingerlings (Amend et al. 1969; Plumb 1972; LaPatra et al. 1990a). Pilcher and Fryer (1980) reported that fry up to 2 months of age suffered mortality exceeding 90% at 10°C. In 2- to 6-month-old fish, mortality is often more than 50% but can be as low as 10% in yearlings. LaPatra et al. (1990a) reported that sockeye salmon were susceptible to IHNV for up to 210 days of age (7.2 g) and rainbow trout up to 170 days (13.1 g). Outbreaks have occurred in 2-year-old sockeye salmon, but mortality has usually been comparatively low.

Infectious hematopoietic necrosis virus disease normally does not occur at temperatures above 15°C (Amend 1970; LaPatra et al. 1989a). Hetrick et al. (1979) showed that IHNV could cause 72% to 88% mortality in 0.2- to 0.3-g rainbow trout at 18°C and 72% to 94% at 3°C (Table 10.1). The mean days to death was 15 to 18 days at 3°C and 7 to 10 at 18°C. Infectious hematopoietic necrosis virus was isolated from infected dead fish at 21°C, but mortality among groups of infected fish and noninfected controls at the higher temperature was similar.

Survivors of IHNV infections are thought to become virus reservoirs, but it is not clear whether these fish actually continue to carry and shed virus after surviving an epizootic as juveniles or if they become reinfected horizontally at a later time.

TABLE 10.1. AVERAGE PERCENT MORTALITY OF RAINBOW TROUT (0.2 TO 0.3 G) NOT INFECTED AND INFECTED WITH INFECTIOUS HEMATOPOIETIC NECROSIS VIRUS BY WATERBORNE EXPOSURE AND HELD AT SEVEN TEMPERATURES

	Water temperature (°C)						
	3	6	9	12	15	18	21
Control	0	0	0	1	6	30	67
Infected	83	81	76	68	66	80	63

Source: Hetrick et al. (1979).

Amos et al. (1989) found that adult sockeye salmon captured as they left salt water during spawning migration were free of IHNV, and if allowed to become sexually mature in IHNV-free fresh water, they remained uninfected until spawning. Adults allowed to migrate naturally to sea and return to spawning grounds developed an IHNV-positive prevalence of 90% to 100%. These data strongly suggest that returning adults become infected from other sources rather than by reappearance of latent virus. In a study by Wingfield and Chan (1970), it was found that 34% of females and 5% of males assayed were IHNV carriers. Mulcahy et al. (1983a) sampled females from seven adult salmon populations and found that active infections ranged from 39% to 100% (average, 85%). The samples were taken from ovarian fluid or post-spawning tissue, and many virus titers exceeded 10^5 $TCID_{50}$/mL.

Prevalence of IHNV-positive adult female Alaskan sockeye salmon varied from 7% to 94% (average, 44%), and positive adult males varied from 0% to 48% (average, 13%) (Grischkowsky and Amend 1976). Mulcahy et al. (1984) studied the IHNV carrier rate of female sockeye salmon from August (prespawning) through October (peak spawning). In August, no assayed fish were positive for IHNV, but in October, 100% were positive (Table 10.2). As females progress from a prespawning to spawning condition, they shed greater numbers of virus, and potential for vertical transmission increases. Sperm could also be a vehicle for virus to enter the egg during fertilization because experimentally IHNV did adsorb sperm (Mulcahy and Pascho 1984). These researchers subsequently reported two instances of vertical IHNV transmission (Mulcahy and Pascho 1985).

To clarify the role of eggs in the transmission process, Yoshimizu et al. (1989) showed that IHNV does not survive in fertilized egg yolks prior to the eyed stage and that no virus was present on the egg surface. When eyed eggs were injected with IHNV, however, virus concentrations increased and 90% of the eggs died. Also, 90% of hatched masu salmon and 20% to 30% of hatched chum salmon from injected IHNV eyed eggs died. These authors doubt that vertical IHNV transmission occurs because of the virus's inability to survive in eggs before they become eyed.

TABLE 10.2. INCIDENCE OF INFECTIOUS HEMATOPOIETIC NECROSIS VIRUS IN PRESPAWNING AND SPAWNING FEMALE SOCKEYE SALMON IN 1980

Date	Number assayed	Percent positive	Condition
August 17–30	18	0	Prespawning
September 11–25	31	0	Prespawning
October 6–7	87	21.5	Spawning
October 15	69	96.5	Spawning
October 20–29	65	100	Spawning

Source: Mulcahy et al. (1984).

The first IHNV isolation from adult sockeye salmon while at sea was reported by Traxler et al. (1997). They investigated the possibility of vertical transmission by these virus-positive fish but found no evidence that IHNV was transmitted vertically during spawning, nor could they infect ova by hardening eggs in water containing large concentrations of IHNV. These data would tend to contradict earlier circumstantial evidence indicating that vertical transmission does occur and leaves this mode of transmission open to further study. The virus can be transmitted horizontally via waterborne exposure, feeding, or injection (Mulcahy et al. 1983b).

Infectious hematopoietic necrosis virus can be isolated from coelomic fluid. Mulcahy and Batts (1987) showed, however, that if nonerythrocytic cells are separated from coelomic fluid and incubated at 15°C for 7 days and then assayed for IHNV, the number of IHNV-positive fish will be higher than if only the coelomic fluid is assayed. Detection of specific antibody in fish serum may be used to determine prior exposure. Jørgensen et al. (1991) collected serum from rainbow trout and screened it for IHNV antibody by plaque neutralization test (PNT), ELISA, and FAT. The three methods detected antibody in 9 to 18 of 20 serum samples; ELISA proved to be the most sensitive and PNT the least. Under some circumstances, these assays could be useful in epizootiological IHNV studies, but because some sera cross-reacts with VHSV, the procedure probably could not be used when specific pathogen-free determinations are required, especially given that Eaton et al. (1991a) isolated IHNV and VHSV from the same fish.

A nonsalmonid IHNV reservoir is uncertain, but Mulcahy et al. (1990) did isolate high concentrations of virus from a leech (*Piscicola salmositica*) and copepod (*Salmincola* sp.) found on sockeye salmon. High IHNV concentrations were also found in detached leeches recovered from bottom gravel of a spawning area. Although the role, if any, these parasites play in IHNV transmission is not known, the possibility that skin parasites facilitate infection rate during spawning must be considered.

Infectious hematopoietic necrosis virus can also cause significant losses in wild juvenile salmon. Williams and Amend (1976) confirmed an IHNV epizootic in a naturally spawned sockeye salmon fry population in British Columbia and speculated that abnormally low egg survival in Chilko Lake was due to IHNV. The virus also caused significant mortality in the Weaver Creek spawning channel, where an estimated 50% of 16.8 million migrating sockeye salmon fry succumbed to IHNV (Traxler and Rankin 1989).

Pathological Manifestations

The hematopoietic tissue of the kidney and spleen is the most extensively involved tissue where degeneration and necrosis takes place, especially in the anterior kidney (Yasutake 1975). Pleomorphic intracytoplasmic inclusions frequently occur in pancreatic cells, and focal necrosis is present in the liver. Necrosis is also found in granular cells of the lamina propria, the stratum compactum, and stratum granulosum of the intestine. A sloughing of mucous membrane (possibly catarrhal inflammation) in the intestine may be the origin of fecal cast observed in moribund fish (Amend et al. 1969).

The pathogenesis of three IHNV strains were compared by LaPatra et al. (1990b). They found that the most virulent strain produced severe tissue injury in 3 to 4 days, whereas other strains required 8 to 10 days to cause similar pathology. The pathology of IHNV was also compared in selected tissues of rainbow trout and coho salmon by Helmick et al. (1995a). They found morphological differences in histopathology of the gills, esophagus/cardiac stomach region, anterior intestine, and pyloric caeca in the two species. Gills exhibited differences in distribution of mucous and chloride cells and size of pillar cells. The esophageal/cardiac regions differed with respect to epithelial cell architecture of mucus-secreting glands. In a companion study, Helmick et al. (1995b) used streptavidin gold labeled monoclonal antibody (1NDW14D) to follow pathogenesis and replication of IHNV from esophageal mucosa to the mucus-secreting serous cardiac glands in rainbow trout and coho salmon. Glandular cystic degeneration was seen as early as 1 hour postinfection and progressed into severe cystic degeneration in 24 hours. This process was more severe in rainbow trout than in coho salmon, leading to the conclusion that IHNV replicated more efficiently in mucus-secreting serous cardiac glands of rainbow trout than in coho salmon.

Pathogenesis studies by Yasutake (1975) indicate that IHNV target tissue is the hematopoietic tissue of the kidney, where a close correlation between histopathological progression and virus concentration was noted. About 3 days postinfection, focal concentrations of macrophages were observed in the kidney followed by involvement of the pancreas, liver, and presence of granular cells in the intestinal wall. Amend and Smith (1975) concluded that death probably resulted from a severe electrolyte and fluid imbalance caused by renal failure. The role of leukocytes in IHNV replication was shown in vitro by Chilmonczyk and Winton (1994).

LaPatra et al. (1989a) suggested that infection from waterborne exposure resulted from virus invading the integument and gills because an increase in detectable virus was noted in these tissues for 3 days postexposure. They also postulated that detectable virus in the mucus resulted from replication in the integument. Subsequent studies by Cain et al. (1996) found IHNV in cutaneous and intestinal mucus following injection or waterborne virus exposure. Some antiviral activity was confirmed in intestinal tract mucus, and the authors suggested that mucus from these areas could be an innate mechanism of viral resistance that may be important as a first line of defense against IHNV.

Yamamoto and Clermont (1990) described the sequential spread of IHNV to tissues of rainbow trout infected by immersion. Gill tissues had virus titers 16 to 20 hours postexposure, and virus spread rapidly to the kidney and spleen. In fact, Zhang and Congleton (1994) showed that IHNV could be isolated from water that contained infected juvenile steelhead trout several days before onset of clinical signs and mortality. Using whole-body assay and immunohistochemistry, Yamamoto et al. (1990) determined that initial lesions following infection occurred in the epidermis of pectoral fins, opercula, and on the ventral body surface. LaPatra et al. (1989c) also found virus in mucus of juvenile and adult fish infected with IHNV and postulated that the epithelium is a portal of entry for virus and a site of replication. Lesions did not become apparent in the kidneys until the third day. These findings were essentially confirmed by Drolet et al. (1994), who followed IHNV progression to various organs and tissues of steelhead trout fry after immersion and oral infection. The virus passed directly from the gills into the circulatory system, where virus was detected 2 days later. Virus was found in the brain, pancreas, and heart 5 days after exposure, with detection in other organs being intermediate. After gastrointestinal exposure, virus passed through the epithelium into the circulatory system and then to all organs and tissues. These experiments indicate that the epidermis and gills are important initial sites of IHNV replication.

Significance

Infectious hematopoietic necrosis virus is the most serious viral disease that affects trout and salmon in the Pacific Northwest. It has also become an important disease in Europe and Asia, particularly Japan and China (Sano et al. 1977; Jiang 1993). It was reported that in 1975, 70 million young salmon succumbed to IHNV infections in Japan. Clearly, the virus has been responsible for the death of millions of juvenile hatchery-reared Pacific salmon and rainbow trout and likely large numbers of wild salmonids as well.

INFECTIOUS PANCREATIC NECROSIS VIRUS

Infectious pancreatic necrosis (IPN) is generally characterized as a subacute disease of fry and fingerling salmonids, but IPN or IPN-like viruses have also been isolated from a variety of marine and freshwater fish and invertebrates. These isolates are combined into the aquatic birnavirus group (Hill and Way 1995). Infectious pancreatic necrosis virus was likely the etiological agent of "acute catarrhal enteritis" described by M'Gonigle (1941) in Canada, but Wood et al. (1955) were the first to describe the disease as "infectious pancreatic necrosis" and to propose its possible viral etiology. Wolf et al. (1960) were able to demonstrate the viral etiology by isolating and propagating the virus in cell culture, thus making IPNV the first proved viral disease of fish.

Geographical Range and Species Susceptibility

Infectious pancreatic necrosis virus, and IPNV-like agents, are some of the most widely distributed aquatic viruses, essentially occurring worldwide. Infectious pancreatic necrosis virus, or aquatic birnaviruses, have been found in 22 countries including North and South America, Europe, and Asia (Hill and Way 1995) (Table 10.3).

Infectious pancreatic necrosis virus has been isolated from 16 different species of salmonids (Wolf 1988). Brook and rainbow trout are the most susceptible, but virtually all salmonids are susceptible to some degree including Pacific and Atlantic salmon (McAllister et al. 1984; Christie and Haverstein 1989). Hill and Way (1995) listed 31 families of nonsalmonid fish, ranging from primitive Petromyzonidae (lampreys) to phylogenetically advanced Cichlidae (cichlids), 9 species of mollusks, 4 species of crustaceans, and 2 gastropods from which aquatic birnaviruses have been recovered. These animals come from fresh water and marine habitats in subarctic to tropical cli-

TABLE 10.3. COUNTRY AND YEAR REPORTED IN WHICH INFECTIOUS PANCREATIC NECROSIS VIRUS AND OTHER AQUATIC BIRNAVIRUSES HAVE BEEN ISOLATED

Country	Year reported
Canada	1982
Chile	1984
China	1989
Denmark	1969
France	1965
Germany	1975
Greece	1975
Finland	?
Italy	1972
Japan	1971
Korea	1984
New Zealand	?
Norway	1976
Scotland	1971
South Africa	?
Spain	1988
Sweden	1973
Taiwan	1983
Thailand	?
United Kingdom	?
United States	1955
Yugoslavia	1974

Source: McAllister and Reyes (1984); Wolf (1988); McAllister (1993); Hill and Way (1995).

mates. Although all aquatic birnaviruses are not overtly detrimental to the health of affected animals, clearly they are not host specific and appear to affect salmonids more severely than nonsalmonids. In a survey of fish farms in England and Wales, Buck et al. (1979) found that of 29 salmonid farms surveyed, 17% were rearing fish infected with IPNV, but there was no evidence that nonsalmonid fishes living in the effluent from these farms were infected with virus.

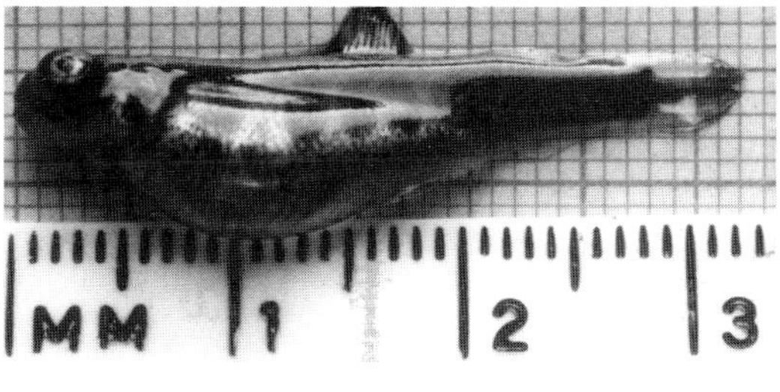

FIGURE 10.3. Rainbow trout infected with infectious pancreatic necrosis virus. Note the enlarged abdomen, exophthalmia, and very dark skin pigmentation. (Reprinted with permission of CRC Press.)

Clinical Signs and Findings

Larger, more robust fry and small fingerlings of susceptible salmonids are usually the first to develop clinical signs of IPNV-associated disease. Externally infected trout have an overall darker pigmentation, abdominal distension, exophthalmia, hemorrhages on the ventral surface and fins, and pale gills (Figure 10.3) (Snieszko et al. 1959; Wolf et al. 1960). Rapid whirling on the long axis followed by quiescence is a typical behavior of IPNV-infected trout. Internally, a general hemorrhagic appearance is obvious with petechiae throughout the viscera, especially in the pyloric caeca and adipose tissue. The spleen, heart, liver, and kidneys are unusually pale in advanced cases. The body cavity is filled with clear yellow fluid, the digestive tract is void of food, and the posterior stomach contains a gelatinous, mucoid (clear or milky) plug that is pathognomonic for IPNV. The gelatinous plug remains intact in 10% formalin; therefore, it is of diagnostic value in examining preserved specimens.

The clinical signs just described are present primarily in IPNV-infected salmonids and are not always seen in other birnavirus infected fish. Sano et al. (1981) reported that Japanese eels infected with IPNV had muscle spasms, retracted abdomens, and congestion of the anal fin, abdomen, and gills. These fish also had ascites and mild hypertrophy of the kidneys. Infectious pancreatic necrosis virus–infected larval seabass displayed spiral swimming, swim bladder distension, fecal casts, exophthalmia, and sloughing of gut epithelium (Bonami et al. 1983). Infected striped bass showed no overt clinical signs of disease other than darker than normal pigmentation (Schutz et al. 1984). Atlantic menhaden from the Chesapeake Bay demonstrated a spinning behavior (Stephens et al. 1980), and gizzard shad from which an IPN-like virus was isolated showed only dark pigmentation (Southeastern Cooperative Fish Disease Laboratory, Auburn University, Alabama, USA, unpublished). Infectious pancreatic necrosis virus–infected juvenile Atlantic halibut (0.1 to 3.5 g) had distended abdomens, exhibited uncoordinated swimming, the livers, kidneys, and intestines became necrotic but the pancreatic tissue was unaffected (Biering et al. 1994).

Diagnosis

In trout, presumptive clinical diagnosis of IPNV can be made based on mortality patterns and clinical signs combined with isolation in tissue culture. In nonsalmonids, however, IPNV or aquatic birnavirus infections may not be expressed in clinical disease or mortality and IPN diagnosis is confirmed by virus isolation and serological identification. Histological examination can be used to support a clinical diagnosis by detecting typical IPNV-associated histopathology (Buck et al. 1979; Roberts and McKnight 1976).

Virus isolation is obtained by homogenizing whole fish (fry) or visceral organs from IPNV-suspect fish. Homogenates are decontaminated by filtration or antibiotics and diluted to at least 1:100 and inoculated onto susceptible cells (RTG-2, CHSE-214, BF-2, and so on). Yoshimizu et al. (1988a) tested 32 salmonid and nonsalmonid cell lines and found 26 to be susceptible to IPN virus. Inoculated cells incubated at about 20°C will develop CPE consisting of cellular pyknosis, elongation, and lysis in 18 to 36 hours (Figure 10.4) (Wolf et al. 1960). Temperature range for IPNV replication is 4 to 27.5°C, but incubation temperature should be appropriate to cell line used, that is, neither CHSE-214 nor RTG-2 should be incubated at above 20°C. A blind passage at dilutions greater than 1:10 should be made in the absence of CPE af-

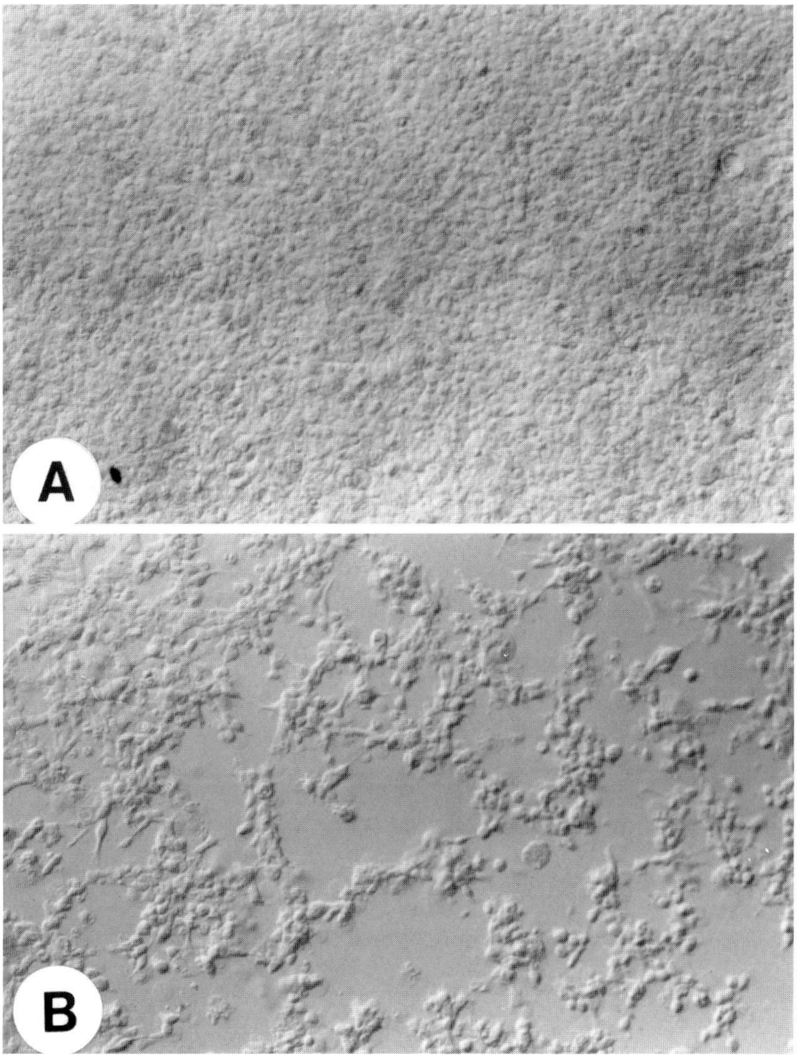

FIGURE 10.4. CHSE-214 cells infected with infectious pancreatic necrosis virus and incubated at 15°C. (A) Noninfected controls. (B) Extensive viral CPE 48 hours after inoculation (×500). (Photograph by J. Grizzle.)

ter 7 to 14 days of incubation. Clinically, IPNV-infected fish will have titers of 10^6 to 10^9 $TCID_{50}$/g of tissue but, in the carrier state titers, can be as low as 10^1 $TCID_{50}$/g of tissue.

Serum neutralization is usually used to identify positively IPNV and other aquatic birnaviruses. Other types of identification techniques used in identifying IPNV in cell cultures, cell culture fluids, and infected tissues are fluorescent antibody, ELISA, and immunoperoxidase systems (Hattore et al. 1984; Shankar and Yamamoto 1994; Hill and Way 1995). Nucleic acid probes and monoclonal antibodies have made rapid and accurate detection and identification of IPNV possible (Dominquez et al. 1990). Babin et al. (1990) described a capture immunodot procedure that uses monoclonal antibodies to detect IPNV in field identification studies. Using a purified reference IPNV strain in conjunction with immunogold and electron microscopy, Novoa (1996) distinguished between the Sp, Ab, and West Buxton (WB) strains of IPNV. A dot-blot hybridization test was developed to detect IPNV–RNA in fish cells and cell cultures, and the procedure was 100% effective in detecting IPN–RNA in inoculated cell cultures when applied 12 to 24 hours after inoculation, but it was less sensitive in fish tissue (Dopazo et al. 1994). These researchers concluded that the hybridization test is appropriate to identify IPNV when combined with virological diagnostic cell culture inoculation procedures.

A method for detecting IPNV by reverse transcription (RT)–PCR was described by Lopez-Lastra et al. (1994). Recently, Wang et al. (1997) reported development of a single-tube, noninterrupted RT–PCR procedure to detect IPNV in infected CHSE-214 cells. Amplified products were detected in cells infected with isolates of WB, Ja, Sp, Ab, VR299, 3372 MFK (isolated from eel in Taiwan), and CV-HB-1 (from hard clams in Taiwan) but not in uninfected cells nor cells infected with IHNV. Tests were run on infected cultures 5 to 6 days postinfection.

Detection tests for IPNV are crucial for identifying possible virus carrier populations and properly managing the disease. When individual rainbow trout were checked for IPNV antibodies, about 30% were virus positive by isolation compared with approximately twice that number being positive by ELISA (Dixon and de Groot 1996). These authors did note that not all virus-positive fish by isolation were identified as having been exposed by the ELISA method; however, the procedure can aid in detecting rainbow trout populations that have had previous IPNV exposure.

Virus detection in carrier fish using a nondestructive procedure involves collecting ovarian and seminal fluids or fecal material (Wolf et al. 1967). McAllister et al. (1987) found that ovarian fluid yields greater amounts of virus when the fluid is centrifuged into a pellet and sonicated before being inoculated into cell cultures. A more accurate and reliable method of detecting IPNV carrier fish is to process internal organ tissue from a population sample, but this is not always practical because it requires killing assayed animals.

Virus Characteristics

Infectious pancreatic necrosis virus is a member of the family Birnaviridae and is included in the large group of viruses called "aquatic birnaviruses" (Dobos et al. 1977; Hill and Way 1995). The virus is a nonenveloped icosahedral particle that averages 60 nm in diameter (range, 55–75 nm) (Dominquez et al. 1991). It consists of a single capsid layer and a genome of bisegmented double-stranded RNA. The virion is stable in acid, ether, and glycerol and is relatively heat stable. The virus is also stable at 4°C for 4 months in cell culture media containing serum, but a temperature of –20°C or lower is recommended for long-term storage. At –70°C, about 1 $\log_{10}$ of infectivity is lost per year. Most IPNV infectivity is lost in fresh water within 10 weeks at 4°C, with residual virus remaining for 24 weeks. Infectious pancreatic necrosis virus remains infective for up to 20 days in freshwater at 15°C and 15 days at 20°C compared with a 20-day survival time at both temperatures in seawater (Barja 1983). The virus will survive drying for more than 8 weeks (Desautels and MacKelvie 1975) and for up to 1 year after lyophilization in lactalbumin hydrolysate, lactose, powdered milk, or saline with 10% fetal calf serum (Wolf et al. 1969).

Most aquatic birnaviruses are antigenically related, regardless of geographic origin or animal species host. These isolates were placed in two serogroups, a major serogroup I, and a minor serogroup II (Hill and Way 1988; Caswell-Reno et al. 1989). Serogroup I contains nine cross-reactive but distinct serotypes including WB (VR-299), Sp, Ab, He, Te, Canada 1 (C1), Canada 2 (C2), Canada 3 (C3), and Jasper (Ja) (Table 10.4). A 10th serotype (N1) was proposed by Christie and Haverstein (1989). Serogroup II contains serotype TV-1 because it is serologically reactive to VR-299, Sp, Ab, and several other aquatic birnaviruses.

Using 196 aquatic birnavirus isolates from 56 animal species in 17 countries, Hill and Way (1995) applied serum neutralizaton and monoclonal antibodies to reclassify this group of viruses (Table 10.4). They proposed serogroup A (formerly serogroup I) and serogroup B (formerly serogroup II). Serogroup A contains most of the isolates from a variety of locations and are designated serotype A_1 through A_9, whereas serogroup B contains serotype B_1 from the United Kingdom. Serotype N1 was included in Serotype A_2. As suggested by Hill and Way (1995), for the sake of clarity, it is best to use both the old and new proposed serotype designations for a period until the new ones are accepted.

Nearly all aquatic birnaviruses isolated from salmonids in the United States belong to serotype A_1 (WB); however, examples of serotype A_3 (Ab) have also been isolated from other fish species, mollusks, and crustaceans (McAllister and Owens 1995). Serotypes A_6 (C1), A_7 (C2), A_8 (C3), and A_9 (Ja) have been found in Canadian fish populations (Lecomte et al. 1992), and serotypes A_2 (Sp), A_3 (Ab), A_4 (He), and A_5 (Te) have been found in Europe. Representatives of serotype A_1 (WB), A_2 (Sp), and A_3 (Ab) have also been detected in Asia.

TABLE 10.4. MAJOR SEROLOGICAL GROUPS OF INFECTIOUS PANCREATIC NECROSIS VIRUS (AQUABIRNAVIRUSES)

	Proposed Serotype[a]	Previous Serotype[b]	Animal group isolated from
Serogroup A (formerly I)	A_1	WB—West Buxton, USA	Rainbow trout
	A_2	Sp—Spjarup, Denmark	Rainbow trout (VR-299)
	A_3	Ab—Abild, Denmark	Rainbow trout
	A_4	He—W. Germany	Northern pike
	A_5	Te—United Kingdom	Telapia
	A_6	Can. 1—Canada	Atlantic salmon
	A_7	Can. 2—Canada	Atlantic salmon
	A_8	Can. 3—Canada	Arctic char
	A_9	Ja—Jasper, Alberta	Rainbow trout
Serogroup B (formerly II)	B_1	TV-1—United Kingdom	Telina

Source: Hill and Way (1988); Lecomte et al. (1992); Hill and Way (1995).
[a]Hill and Way (1995).
[b]Previous serotype and location of initial isolation. Hill and Way (1988).

Biering et al. (1997) serotyped 12 birnavirus isolates from Norway by PCR with monoclonal antibody and immunodot assay and found slight serological variations between them but concluded that the N1 and Sp/A_1 (A_2) strain were identical.

Many nonsalmonid isolates are of the A_3 (Ab) serotype and tend to be of lower virulence for trout (McAllister and Owens 1995). In an exhaustive IPNV study, Rodriguez et al. (1994) found that all IPNV strains in Spain were of the A_2 (Sp) or A_3 (Ab) serotype. Novoa and Fugueras (1996) detected the A_1 (WB) strain in turbot in Europe, a strain previously thought to be confined to North America. They also isolated six aquatic birnaviruses from turbot that were heterogenous and could not be serotyped, so these isolates were placed in the IPNV group based on genomic segment mobility.

Epizootiology

Infectious pancreatic necrosis virus is one of the more fascinating fish viruses because of its diverse host susceptibility and broad geographical range. Severity of IPNV is influenced by species, age, strain, and physiological condition of affected fish, virus strain, and environmental conditions. Manifestation of an IPNV infection varies from subclinical to acute, with nearly 100% mortality possible in some affected trout populations (Wolf 1988). In Spain, mortality related to naturally occurring infections of IPNV A_2 (Sp), and A_3 (Ab) strains in rainbow trout farms ranged from 20% to 60% (Rodriguez et al. 1994).

Experimental transmission of IPNV via waterborne exposure was demonstrated in 13-week-old Atlantic salmon and 15-week-old brook trout by Taksdal et al. (1997). Cumulative mortality in two tanks of Atlantic salmon was 76% and 82%, compared with 6% and 15% for control fish. Infected brook trout mortality was 42% and 43%, compared with 8% to 10% for control fish. Onset of clinical infection occurred first in brook trout and virus was isolated from all moribund and dead fish assayed. The risk factor of contracting IPNV by post-smolt Atlantic salmon after being moved from fresh water to marine culture sites was determined by Jarp et al. (1995). They found that on 124 sea sites, a 39.5% mortality could be attributed to IPNV and that risk factor was related to age, geographical location of the culture site, and method of fish transport.

The fact that water effluence from an IPNV-contaminated hatchery can cause problems downstream was emphasized by McAllister and Bebak (1997). They found no IPNV in streams or spring water above hatcheries, but virus was found in effluent streams 19.3 km below outfall of infected hatcheries. They also isolated IPNV from 3 of 11 brook trout but failed to isolate virus from 30 brown trout, 4 rainbow trout, or 61 nonsalmonid fish. These data emphasize the infective potential of water for susceptible salmonids below virus-contaminated facilities.

Optimum water temperature for IPNV development is 10 to 15°C. When trout fry are naturally exposed to IPNV, clinical signs can occur from 1 week to 6 months of age. Under experimental conditions, appearance of clinical signs can be as short as 3 to 5 days at 15°C. Brook trout reach peak

IPNV susceptibility between 30 and 50 days of age, and mortality becomes less severe as fish become older. At 10°C, 1-month-old brook trout suffered 85% cumulative mortality, but by 6 months of age, mortalities were less than 5% (McAllister and Owens 1986).

McAllister and McAllister (1988) demonstrated that IPNV can be transmitted from carrier striped bass to brook trout. Mortality of IPNV-infected nonsalmonids can also vary from none to high. Bonami et al. (1983) reported up to 95% mortality in cultured seabass fry, and Schutz et al. (1984) reported that 2 million, 4-week-old IPNV-infected striped bass died. Both of these reports carefully avoided claiming that IPNV actually killed the fish. An attempt to infect experimentally four strains of 1- to 120-day-old striped bass with IPNV did not produce clinical disease, and virus was subsequently isolated from only virus infected fish (Wechsler et al. 1986). Although it is questionable whether striped bass are killed by IPNV, McAllister and McAllister (1988) successfully transmitted virus from asymptomatic IPNV-carrier striped bass to juvenile brook trout, and 8% of the surviving brook trout became long-term virus carriers. Because striped bass may carry IPNV and serve as reservoirs for more susceptible species, all striped bass or their hybrids stocked into IPNV-free waters with access to trout populations should be virus free. Numerous reports describe isolation of IPNV or IPN-like virus from asymptomatic fish during routine virus assays, and any of these carriers can be a virus source for more susceptible species. The IPNV can also be transported by fish-eating birds because it survives in their digestive tracts and is released through excreta (Peters and Neukirch 1986).

Biering et al. (1994) isolated IPNV from Atlantic halibut that was experimentally infectious to 0.1- to 3.5-g halibut fry at 12 and 15°C with significant mortality. An experimental IPNV infection was later established in Atlantic halibut yolk-sac larvae and resulted in 100% mortality following waterborne exposure to 10^7 $TCID_{50}$/mL (Biering and Bergh 1996).

Virulence of isolates from salmonids, nonsalmonids, and molluscans was compared in brook trout by McAllister and Owens (1995). The trout and salmon isolates were more pathogenic to brook trout than were those from nonsalmonids or molluscans. Virulence was found in 5 of 6 salmonid, 10 of 16 nonsalmonid, and 2 of 5 molluscan IPNV isolates. These authors concluded that IPNV is widespread and may not always result from introduction of carrier salmonids into an area.

When an IPNV outbreak occurs in trout, some survivors probably become life-long carriers, but prevalence of carrier fish in a population can vary with the season. Mangunwiryo and Agius (1988) showed that percentage of fish shedding IPNV increased between March and June and declined during the latter part of the year. The heightened carrier state of these fish was inversely related to IPNV antibody levels, implying that high serum antibody titers reduce active virus production. Billi and Wolf (1969) found that when an IPNV-carrier population was sampled over time, more than 90% were carriers, but not all fish or even the same fish shed virus at each sampling.

A model for an IPNV carrier state in trout and cell cultures by Hedrick and Fryer (1982) showed that virus was produced by less than 1% of the cell population of either persistently infected cell lines or kidney cells of carrier trout. In cell culture, neither antibody nor interferon affected replication; however, defective virus did control or interfere with virus replication. A similar mechanism may function in IPNV-carrier fish. Replication of IPNV in leukocytes (in vivo) from the head kidney of Atlantic salmon suggests that these cells, especially the adherent macrophages, could play a role in maintaining a virus carrier state (Johansen and Sommer 1995).

Although adverse environmental conditions are not necessarily associated with IPNV epizootics, Roberts and McKnight (1976) linked temperature and transport stress to recurrent disease outbreaks in yearling rainbow trout. Frantsi and Savan (1971) described a similar pattern in yearling IPNV carrier fish after exposure to low oxygen concentrations. These observations suggest that stressful environmental conditions may cause a reoccurrence of disease, especially in older fish populations. Using naturally infected juvenile brook trout, McAllister et al. (1994) demonstrated that when fish were fed immunosuppressant triamcinolone acetonide (Kenolog) and water temperatures were raised from 12 to 18°C, IPNV prevalence and virus titers increased significantly, thus showing that "stress" can affect viral infections. This procedure could also be used to increase the probability of isolating IPNV from asymptomatic fish. Postsmolt Atlantic salmon appear to be more susceptible to IPNV when transferred to seawater, which could be related to stress (Stangeland et al. 1996). Following an experimental injection of IPNV, 50% mortality was experienced under seawater conditions, and extended clinical disease and higher mortality were noted in stressed fish that were possibly induced by the injection process.

Multiple infections involving two or more pathogens in the same fish is not uncommon. Yamamoto (1975a) demonstrated that IPNV and *Renibacterium salmoninarum* (bacterial kidney disease) occurred in the same population of rainbow and brook trout, and electron microscopy indicated that both pathogens occupied the same cell. It was noted by Evensen and Lorenzen (1997) that IPNV and *Flavobacterium psychrophilum*, the causative agent of bacterial coldwater disease, can coinfect juvenile rainbow trout. These disease agents can occur in the same cells, but generally, the virus appears in the exocrine pancreatic cells and the bacteria in the interstitial tissue surrounding pancreatic islets. These pathogens were detected by immunohistochemistry in asymptomatic fish as well as those with clinical signs of disease. It has also been shown that IHNV and IPNV occurred in the same rainbow trout fry, but concentrations of IHNV (2.8×10^5 PFU/g) were greater than IPNV (1.2×10^4 PFU/g) (Mulcahy and Fryer 1976). In Norway, both IPNV and erythrocytic inclusion body syndrome virus were present in Atlantic salmon (Jarp et al. 1996). Fish that had experienced an IPNV epizootic as fry in fresh water could have a recrudescence of disease when moved into salt water even though IPNV antibody may have been present. However, fish that had IPNV antibodies but had not experienced clinical IPNV in fresh water were protected against IPNV in seawater. This immunity could be important in controlling IPNV in sea-farmed salmon. There was no relationship between hypo-osmoregulatory capacity, infection with erythrocytic inclusion body syndrome (EIBS), or risk of clinical IPN after transfer to seawater.

Transmission of IPNV occurs horizontally from infected fish through water to susceptible fish, in which it is absorbed by the gills and gut. During an epizootic, virus titers in water can be as high as 10^5 $TCID_{50}$/mL (Barja et al. 1983). Carrier trout shed IPNV in reproductive products and feces. Vertical transmission was strongly suggested by the early work of Wolf et al. (1963), and evidence suggests the potential for it to occur via the gametes (Bootland et al. 1991). Circumstantial evidence, and some experimental data, suggest that vertical transmission is a possible but unpredictable event. Infectious pancreatic necrosis virus can be found in ovarian fluid and has been shown to be adsorbed onto sperm (Mulcahy and Pascho 1984). The presence of IPNV on fresh sperm of rainbow trout was confirmed by FAT and flow cytometer analysis in Spain (Saint-Jean et al. 1993). This would further substantiate a likely mode of transmission to progeny during spawning. An unusual means of IPNV dissemination could also occur when pituitary extracts from infected trout are injected into other fish species to stimulate ovulation (Ahne 1983).

Immunization of trout against IPNV is becoming a management reality but may not prevent vaccinated fish from becoming carriers and active virus shedders (Bootland et al. 1995). Brook trout vaccinated with inactivated IPNV developed a strong humoral immune response, but after injection with virulent IPNV, immunized fish and nonvaccinated controls shed virus in their feces and both male and female reproductive products contained virus. The IPNV was carried at least 15 weeks, leading to the conclusion that even though trout develop an immune response, they can still transmit virus during spawning activities.

The importance of IPNV as a pathogen of cultured trout, salmon, and occasionally other species is unquestioned, but its significance in wild populations is uncertain. Survivors of IPNV epizootics are often stocked into streams or lakes for sport fishing, and at times, these populations become reproductively self-sustaining. Yamamoto (1975b) found that a hatchery population of brook trout had an IPNV-carrier prevalence of 90%, and 2.5 years after stocking, the carrier prevalence was 69%. Shankar and Yamamoto (1994) determined that 44.4% of adult lake trout in Cornwall Lake, Alberta, Canada, were infected with IPNV and that when virus was transmitted to juvenile brook trout, a 74% mortality resulted. The same virus produced 10% mortality in experimentally infected young rainbow trout but was avirulent to young lake trout, implying that lake trout IPNV was not pathogenic for that species. Even though these fish were not affected by the virus, they could become reservoirs of infection for other species.

Pathological Manifestations

Pathological changes resulting from IPNV are typical of fish virus infections (Yasutake 1975). The most profound histopathological change, especially with serotype A_1 (VR-299) in trout, is the obliteration of most acinar tissue of the pancreas and its replacement by necrotic detritus containing fragmented acinar cells, pyknotic nuclei, and zymogen granules. In advanced cases, polymorphonuclear lymphocytes and monocytes increase, indicating presence of some inflammation. Initially, foci of necrosis are scattered throughout the acinar tissue and then rapidly spread to surrounding tissue. At the periphery of necrotic areas, many acinar cells will contain intracytoplasmic inclusions. Fatty tissue surrounding pancreatic tissue often becomes necrotic and necrosis frequently occurs in the islets of Langerhans. Smail et al. (1992) also noted extensive pathology in pancreatic acinar cells of IPNV serotype Sp (A_2)–infected Atlantic salmon.

Scattered to extensive necrosis occurs in voluntary muscles accompanied by mild inflammation. Excretory and hematopoietic kidney tissue also becomes necrotic to varying degrees. Congestion, hemorrhage, and edema occur in the epithelium of renal tubules. Sano (1971) found congestion and necrosis in liver tissue of rainbow trout.

Using rainbow trout as an experimental model, McKnight and Roberts (1976) described the sequence of histopathological changes that occurred in the pancreas, renal tissue, and intestinal mucosa. Changes in the pancreatic tissue were characterized by destruction of acinar tissue that degenerated into an amorphous eosinophilic mass. Contrary to earlier descriptions, the islets of Langerhans and fatty tissue were normal. Hematopoietic tissue in the kidney showed focal degeneration. The intestine was more severely affected than previously reported and exhibited necrosis of the intestine, sloughing of the epithelium, and accumulation of a pinkish or whitish catarrhal exudate in the lumen, all of which persisted to a diminishing degree for more than a year. These authors postulated that intestinal lesions contributed more to death of infected fish than did pancreatic lesions.

Significance

Because mortalities in susceptible fish populations are usually moderate to high, IPNV is a significant disease of cultured fishes. Despite extensive fish health inspections, IPNV has spread to most regions of the world where trout are cultured, and its geographic and host range continues to expand. All new IPNV incidences may not be due to movement of infected fish but the result of the ubiquitous nature of the virus. Infectious pancreatic necrosis virus is one of the most significant diseases in Norwegian Atlantic salmon culture, where it may cost the industry $US 4.5 to 5 million per year (Distribution and Economical Impact of the Disease IPN in Norway 1994).

INFECTIOUS SALMON ANEMIA VIRUS

Infectious salmon anemia (ISA) disease is a condition found in Atlantic salmon caused by infectious salmon anemia virus (ISAV). It was initially described by Thorud and Djupvik (1988).

Species Susceptibility and Geographical Range

Infectious salmon anemia occurs only in Atlantic salmon in Norway after smolts are transferred from fresh water to seawater cages. Sea run brown trout (sea trout) are likely asymptomatic virus carriers.

Clinical Signs and Findings

Atlantic salmon infected with ISAV have anemia (pale gills and internal organs) and high mortality. They are exophthalmic, develop ascites, and liver and spleen are congested and enlarged. The intestine is congested, and petechiae are present in visceral fat (Nylund et al. 1994). Hematocrits of infected fish are reduced from a normal of 35% to 48% to 12% to 25% (Dannevig et al. 1993). Experimentally infected brown trout developed no mortality or clinical signs other than significantly reduced hematocrits (Nylund and Jakobsen 1995).

Diagnosis

The virus has not been isolated in tissue culture; diagnosis is based on presence of clinical signs and electron microscopy. Sommer and Menner (1996) have, however, successfully cultivated the virus in primary in vitro cultures of Atlantic salmon head kidney leukocytes. Virus presence was also demonstrated by injecting ascitic fluid from the body cavity or prepared organ filtrates from suspect infected fish into suitable size, naive Atlantic salmon (Nylund et al. 1995a).

Virus Characteristics

Because ISAV has not been isolated in tissue culture, virus characterization is incomplete. Electron microscopy of tissues from Atlantic salmon with clinical ISA revealed enveloped virus budding from endothelial cells of the luminal side of the plasma membrane and some intracellular membranes (Koren and Nylund 1997). The 100-nm diameter virion has a regularly arranged filamentous nucleocapsid and a matrix proteinlike structure. Morphology and morphogenesis of ISAV resembles the orthomyxoviruses.

Epizootiology

When Atlantic salmon smolts are transferred from fresh water to marine culture cages, they become infected with ISAV from natural sources. It appears that brown trout serve as natural virus reservoirs in the marine environment, even though they show no clinical signs of disease (Nylund and Jakobsen 1995). Infectious salmon anemia virus is easily transmitted from infected brown trout or Atlantic salmon to healthy Atlantic salmon by injection of visceral tissue homogenates, peritoneal ascitic fluid, blood, or cohabitation (Olsen et al. 1992; Nylund et al. 1994 and 1995b). Mortality generally occurs as early as 18 days postinjection. Nylund et al. (1994) also suggested that sea lice (*Caligus elongatus* and *Lepeophtheirus salmonis*) may be instrumental in ISAV transmission in epizootic and enzootic phases and that passive transmission in seawater may be of lesser importance to the spread of virus than was originally thought. However, ISAV remained infectious after 20 hours in seawater and after 4 days in blood and kidney tissue stored at 6°C.

Mortalities due to ISAV can be high, with up to 50% occurring in naturally infected culture populations. Nylund et al. (1994) reported 6% to 82% cumulative mortality by day 30 following experimental infection and 45% to 100% mortality in these same populations 70 days postinfection. In other experimental infections using infected leukocytes from Atlantic salmon, up to 50% mortality occurred in 22 days (Dannevig et al. 1993). During cohabitation experiments, 57% to 60% of Atlantic salmon exposed to infected brown trout died in 48 days (Nylund and Jakobsen 1995).

Several risk factors influenced the incidence of ISA along the Norwegian coast (Jarp and Karlsen 1997). Disease incidence was significantly higher when farms were within 5 km of a salmonid slaughterhouse, and risk of infection increased by 8:1 if the culture site was closer than 5 km to an ISAV-positive site. Generally, in areas where density of sea-caught fish markets was high, there was an increased incidence of ISAV. Jarp and Karlsen (1997) concluded that ISAV is mainly transmitted from infected salmonid sources to clean sites through seawater; therefore, one could judge that the disease would lend itself to control by appropriate location of culture operations and slaughterhouses.

Pathological Manifestation

Infectious salmon anemia virus creates a viremia during infection. By electron microscopy, Nylund et al. (1995b) found virus in a variety of organs and tissues, including the liver, spleen, anterior and trunk kidney, urinary bladder, gut, somatic muscle, and several glands including thyroid and gonadal tissue. The most dramatic pathological manifestation of ISAV is a reduction in hematocrit (Olsen et al. 1992). The plasma cortisol level also increases. Atlantic salmon infected with ISAV show a significant overall increase in plasma glutathione, which may be due to intracellular release of this substance; however, there is a significant decrease in its level in the liver, which may affect the organ's ability to transform and excrete xenobiotics from the body (Hjeltnes et al. 1992). Dannevig et al. (1993) showed that ISA virus impairs its host's response to mitogens, which suggests a compromised immune system; however, they also demonstrated that this event was independent of the development of anemia.

Significance

Infectious salmon anemia virus is a serious disease in sea-pen–reared Atlantic salmon in Norway.

VIRAL HEMORRHAGIC SEPTICEMIA

Viral hemorrhagic septicemia (VHS) is an acute, highly infectious viral disease of cultured salmo-

TABLE 10.5. CHARACTERISTICS OF THE THREE DIFFERENT DISEASE FORMS OF VIRAL HEMORRHAGIC SEPTICEMIA IN SALMONIDS

Form	Behavior	External signs	Internal lesions
Acute	Poor feeding, erratic swimming, high mortality	Dark pigmentation, anemia, pale gills, exophthalmia	Multiple hemorrhages in skeletal muscle, eye, viscera; pale, gray liver; hyperemic kidney; swollen intestine, void of food
Chronic	Lower mortality, reduced activity	Dark pigmentation, exophthalmia, some hemorrhage	Liver pale to grayish; kidney not smooth
Nervous	Negligible mortality, erratic swimming		

Source: Ghittino (1965).

nids in Europe. It has a long history, with references being made to the condition in the mid-1930s (Wolf 1988). The disease was originally named Egtved disease after the village in Denmark where it was first found (Jensen 1965). There have been many common names associated with the disease, most of which reflect its viral nature or its kidney and liver tropism. The etiological agent of VHS is viral hemorrhagic septicemia virus (VHSV).

Geographical Range and Species Susceptibility

Since establishment of its viral etiology in the early 1960s (Jensen 1965), clinical VHS had been confined to the European continent until the virus was isolated from adult Pacific salmon returning to hatcheries in Washington state in 1988 (Hooper 1989). Wolf (1988) listed 13 European countries where VHSV had been found, and others including Spain and the United States have since been added.

Although rainbow trout is the most susceptible species to VHSV, the brown trout is also considered a natural host (de Kinkelin and Le Berre 1977). Dorson et al. (1991) demonstrated that hybrid rainbow and brook trout and Arctic char are resistant to VHSV but that lake trout are susceptible and exhibit clinical disease and significant mortality. It was reported by Meier (1985) that under experimental conditions, juvenile northern pike were more susceptible to VHSV than were rainbow trout. At least 13 fish species, including some hybrid salmonids, are experimentally or naturally susceptible to the virus (Wolf 1988). The list of susceptible species includes eight salmonid and five nonsalmonid species, including the marine seabass and turbot (Castric and de Kinkelin 1984). In 1990 and 1991, VHSV was isolated from Pacific cod with ulcerative lesions in Prince William Sound, Alaska, and infected Pacific herring showed similar lesions (Meyers et al. 1994). Wolf (1988) also listed nine fish species that are refractive to overt VHSV disease. Ironically, this list included chinook (Ord et al. 1976) and coho salmon (de Kinkelin et al. 1974), the two species in Washington from which VHSV was isolated in 1988.

Clinical Signs and Findings

Viral hemorrhagic septicemia of salmonids is categorized as an acute, chronic, or nervous form (Table 10.5). The description of each reflects behavior, overt pathological changes, and mortality pattern that typifies each particular form of the disease. Typically, infected salmonids do not feed, and behavior ranges from lethargic to hyperactive.

Acute VHS is the most serious disease form because of the high mortality associated with it. In this form, fish exhibit erratic and spiral swimming, dark skin, exophthalmia, pale gills indicative of anemia, and gills may be flecked with petechiae (Figure 10.5). Internally, there are multiple hemorrhages in the skeletal muscle and swim bladder, the kidney is swollen and hyperemic with liquefactive necrosis, and the liver is either hyperemic, gray, or yellowish. Subacute to chronic VHS is characterized by lower mortality. Affected fish have dark pigmentation (melanosis), exophthalmia, and swollen abdomens. Internally, the liver is pale. In the nervous stage (neurovegetative, or N-form), there is very low mortality but fish do exhibit poor balance, swim in a circle, and are somewhat anemic. Either environmental stress or handling can cause a less severe form of VHS to develop into a more severe one.

Nonsalmonids also show clinical signs when infected with VHSV. Naturally infected whitefish de-

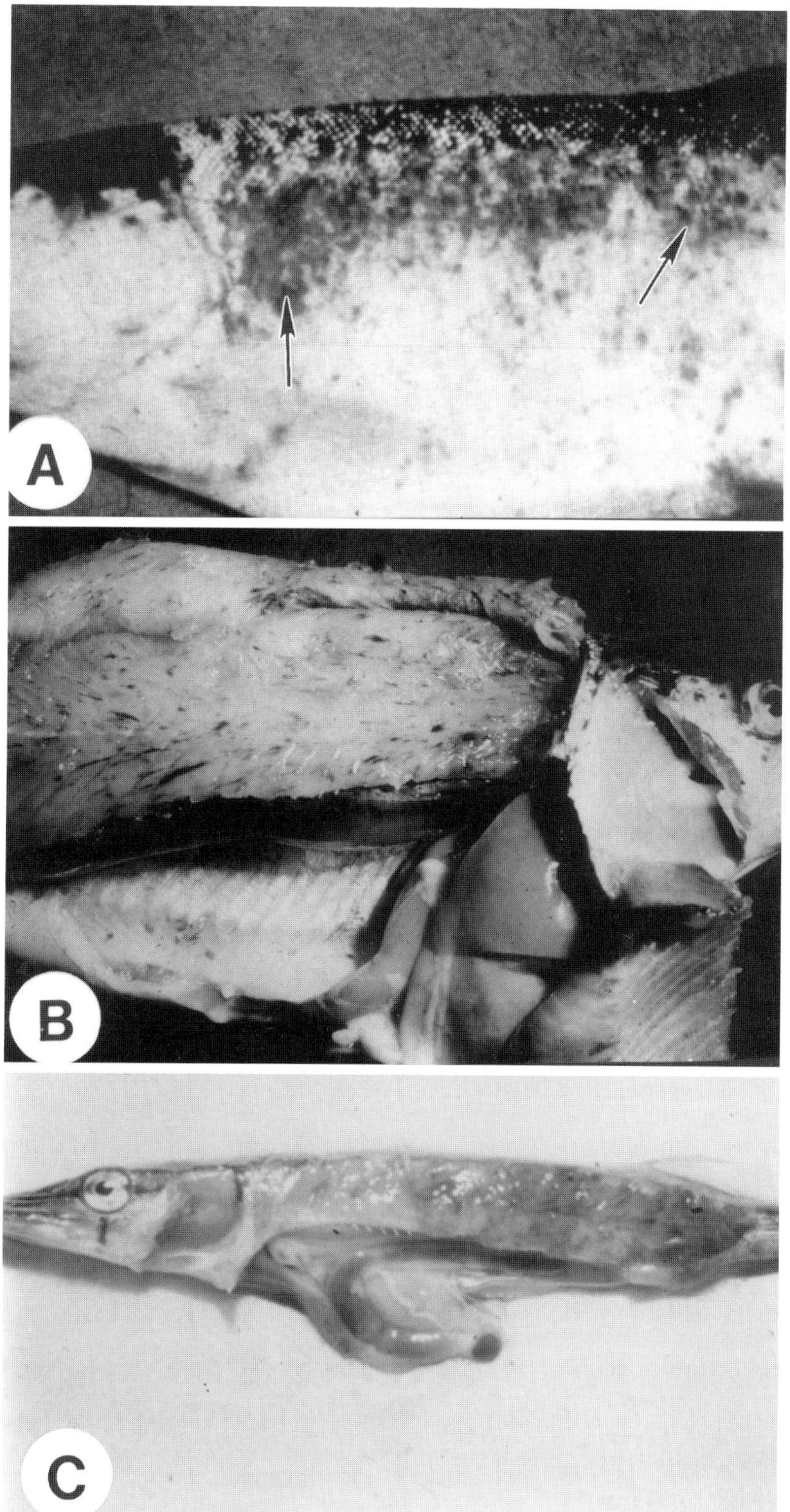

FIGURE **10.5.** *Fish infected with viral hemorrhagic septicemia infection.* (A) *Hemorrhage* (arrows) *in skin of rainbow trout.* (B) *Petechia in muscle, peritoneum, and pale liver and gills in rainbow trout. (Photographs A and B by P. Ghittino; photograph B reprinted with permission of CRC Press.)* (C) *Hemorrhaging in the muscle and pale gills of pike infected with viral hemorrhagic septicemia virus (VHSV). (Photograph courtesy of W. Ahne.)*

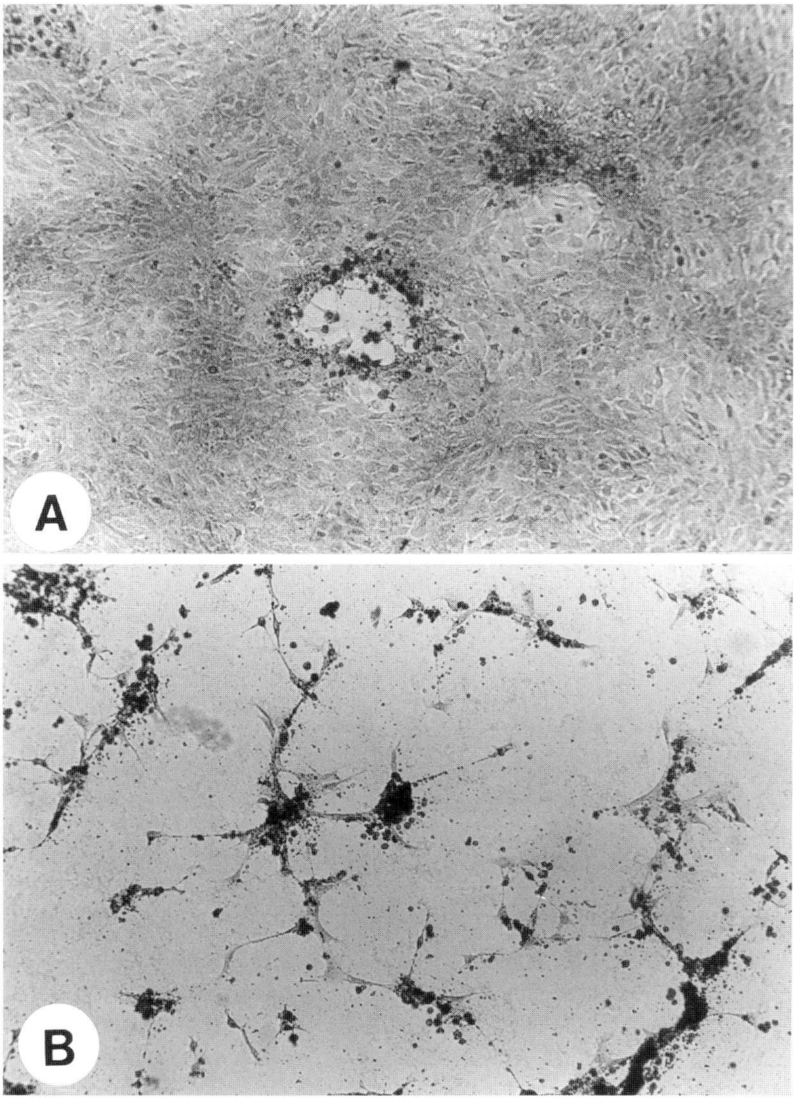

FIGURE 10.6. Viral hemorrhagic septicemia virus growth in fathead minnow (FHM) cells. (A) *Focal cytopathic effect (CPE) consisting of rounding and pyknotic cells.* (B) *Advanced CPE involving entire cell sheet (iodine stain). (Photographs courtesy of P. de Kinkelin.)*

veloped hemorrhages in the skeletal muscle and swim bladder (Ahne and Thomsen 1985). Naturally infected pike had exophthalmia, extensive hemorrhages in the skin, muscle, and kidney (Meier and Jørgensen 1980), and the body cavity contained ascitic fluid. Meier and Wahli (1988) found that in Switzerland, VHSV-infected grayling exhibited anemia, enteritis, and subcutaneous and intermuscular hemorrhage.

Detection

During an epizootic, VHSV can be readily isolated from clinically sick fish, but in later stages of infection, isolation becomes more difficult. Shortly after clinical disease disappears, virus can no longer be isolated from survivors. Kidney and spleen are the organs of choice when attempting VHSV isolation (Meier and Jørgensen 1980; Meier and Wahli 1988). The BF-2 cells are most susceptible to VHSV, but CHSE-214, EPC, RTG-2, or STE-137 cells may also be used. To distinguish VHSV presumptively from other fish viruses, it is important to maintain media at pH 7.6 to 7.7 and an incubation temperature of 15°C or less (Jensen 1965). Focal CPE consists of rounded cells followed by lysis and involvement of the entire cell sheet (Figure 10.6).

Positive VHSV identification is obtained by

plaque reduction, virus serum neutralization, ELISA, or FAT in conjunction with virus specific reagents. Fluorescent antibody techniques may be used to identify VHSV in tissue culture or frozen tissue sections from infected animals, but Ahne et al. (1986) reported a VHSV strain that could not be neutralized by VHSV antiserum but was positive by FAT.

In a given population, VHSV may coexist with IPNV or IHNV in the same fish; therefore, identification for all three viruses should be used when assaying trout and salmon populations, particularly when fish originate from an area where all three viruses are endemic. Infectious hematopoietic necrosis virus and VHSV were also isolated from coho salmon in the same watershed in the Pacific Northwest of North America by Eaton et al. (1991b). Monoclonal antibodies that do not bind with any other fish rhabdovirus and can identify a conserved epitope on VHSV or IHNV can be used to easily distinguish the two, regardless of origin (North America or Europe), when used in an immunodot assay (Ristow et al. 1991). Olesen and Jørgensen (1991) emphasized that the ELISA test for VHSV detection can be a valuable addition to traditional cell culture isolation when a rapid diagnosis is essential, but it should be used only in conjunction with isolation procedures for virus confirmation. These researchers found that 24 of 40 tested trout were VHSV positive by culture but that only 80% of the 24 culture-positive fish were positive by the ELISA. Detection of VHSV in tissues with avidin-biotin alkaline phosphatase was compared with virus isolation in BF-2 cells by Evensen et al. (1994). They found that about one-third fewer positive fish were detected by immunohistochemistry than by virus isolation in mostly asymptomatic experimentally infected rainbow trout. Some have proposed that immunohistochemistry and other rapid diagnostic procedures be used to replace more labor-intensive virus isolation in tissue culture, but this has not been universally adopted.

Detection of VHSV antibody by virus neutralization, FAT, or ELISA are methods used for determining possible virus carrier trout populations in Europe. Olesen et al. (1991a) noted that the ELISA technique was more accurate than serum neutralization or FAT but less specific than serum neutralization. Like IHNV, VHSV neutralization is complement dependent. After IHNV was found in France, Hattenberger-Baudouy et al. (1989) serologically surveyed an IHNV-infected rainbow trout population, which also carried VHSV, and found that some fish had neutralizing antibodies to both viruses. Jørgensen et al. (1991) cautioned against using antibody detection in fish sera to indicate presence of either VHSV or IHNV in populations in which both viruses may coexist because of potential cross-reactive antibody in the same fish.

Virus Characteristics

Viral hemorrhagic septicemia virus is a member of the family Rhabdoviridae. The virus is an enveloped, bullet-shaped particle that measures about 65 nm in diameter and 180 nm in length (Zwellenberg et al. 1965). The virus has a single-stranded RNA genome. Viral hemorrhagic septicemia virus is glycerin, heat, ether, and acid labile but is very stable at pH 5 to 10 (Ahne 1982). The virus is completely inactivated by 3% formalin, 2% NaOH, and 0.01% iodine in 5 minutes and in 2 minutes by 500 mg of chlorine per liter. Viral hemorrhagic septicemia virus is 90% inactivated after 14 days in stream or tap water at 10°C, 99% in mud at 4°C after 10 days, and drying at 15°C in 14 days. Gamma and ultraviolet (UV) irradiation completely inactivates the virus (de Kinkelin and Scherrer 1970; Ahne 1982).

Two or three serological VHSV strains were identified by Le Berre et al. (1977); however, Olesen et al. (1991b) identified four strains using monoclonal antibodies in neutralization tests. There are three recognized serotypes of VHSV (F1, F2, and 23.75) plus the North American isolate (Le Berre et al. 1977). Serological cross-reactivity occurs among the three serotypes in indirect fluorescent antibody assays and in some neutralization tests (Meier and Jørgensen 1980). McAllister and Owens (1987) identified all three serotypes by immunoblot ELISA with antiserum to F1 serotype.

Epizootiology

Originally, VHS was thought to be a nutritional disease; this was proved to be incorrect when Jensen (1965) isolated the VHSV agent and a more detailed description of its etiology was made by Zwellenberg et al. (1965).

Winton et al. (1991) showed that VHSV isolates from chinook and coho salmon in Washington were similar serologically and biochemically to European isolates. It was later determined, however, that this strain was less virulent than those from Europe. The Washington VHSV strain did not cause significant mortality in any of eight different salmonid species (including rainbow trout) exposed to it by waterborne exposure or injection. In comparative susceptibility studies, Follett et al. (1997) found that coho, chinook, pink, and sockeye salmon were refractory to the North American VHSV strain following waterborne virus exposure but that rainbow trout developed clinical signs and disease with 12% virus specific mortality. These data indicate that the North American strain was different pathogenically from typical European strains. An additional difference in the two strains was noted by Bernard et al. (1991), who compared the sequenced nucleoprotein genes of the North American strain and a European strain. This difference seems to suggest that the North American strain is not of European origin. Attempts were made to eliminate VHSV from the continental United States in 1988. Infected fish populations were destroyed and facilities disinfected, but it was later determined that the virus was enzootic and no more eradication efforts were attempted. According to Meyers et al. (1994), VHSV has become widely indigenous to Pacific herring populations in the Pacific Northwest of the United States.

There is no conclusive evidence that VHSV is transmitted vertically. However, VHSV was isolated from both milt and ovarian fluid of spawning salmonids in North America as well as from reproductive products and kidneys of coho salmon in the same water shed in Washington (Eaton et al. 1991b). Attempts to infect eggs experimentally with the virus were unsuccessful. In a hatchery environment, VHSV can be transferred on surfaces of animate and inanimate objects and can persist in water for several days (de Kinkelin and Scherrer 1970). The virus can be transmitted by injection, brushing the gills with infected material, cohabitation of naive fish with infected fish, and waterborne exposure (de Kinkelin et al. 1979). Oral infection is difficult, although pike fry were successfully infected when fed virus-contaminated trout flesh (Ahne 1985). Mortality in infected fish varies with species, age, and size of fish; method of infection; and water temperature.

Incubation time between exposure and overt appearance of VHSV is temperature dependent and can vary from 5 days to 4 weeks (Jørgensen 1973). Viral hemorrhagic septicemia occurred 5 days postinjection at 9°C and 8 to 10 days after exposure by immersion at 12°C. According to de Kinkelin (1983), incubation time was 3 to 10 days at 7 to 14°C and a subclinical infection eventually occurred at 2°C. Rasmussen (1965) found that VHSV outbreaks in rainbow trout were associated with low water temperatures of 5 to 10°C and that water temperature directly affected duration and severity of disease. Optimum water temperature for VHSV is between 8 and 12°C, and temperature fluctuations of a few degrees seem to trigger death. Clinical disease in hatchery populations normally does not occur at temperatures above 14°C.

The younger the fish, the more susceptible they are to VHSV. In trout populations, fish that weigh 0.5 to 4 g are most susceptible (de Kinkelin et al. 1979). As fish become larger, they gain greater resistance, but fish up to 200 g have been reported to develop VHSV (Castric and de Kinkelin 1980). Typical of several viral fish diseases, the healthier-appearing individuals in a population are the first to show clinical disease. It was also noted that VHSV-associated mortality was 80% in rainbow trout 1 month after they were moved from fresh water to full-strength seawater. Also, in Danish mariculture facilities, VHSV-related mortalities of 15% to 50% occurred in naturally infected rainbow trout 18 months after stocking (Horlyck et al. 1984). In a study in Switzerland, Meier and Wahli (1988) noted that mortality in both rainbow trout and grayling was 56% to 100%, respectively, depending on the virus, fish strain, and the environment. Georgetti (1980) postulated that up to 30% of cultured trout in Italy were killed by VHSV.

When seabass and turbot were injected with VHSV, greater than 90% mortality resulted (Castric and de Kinkelin 1984). Meier et al. (1986) reported lower mortalities in whitefish (28% to 50%) than in rainbow trout (40% to 80%) at temperatures of 11 to 14°C. Although mortalities from VHSV in nonsalmonid fish varies, Ahne (1985) induced 90% mortality in 0.2-g northern pike infected by waterborne exposure, and when these virus-infected pike were fed to 1-year-old, 60-g pike, clinical disease and 30% mortality occurred. Smaller pike were susceptible to IPNV, pike fry rhabdovirus (PFRV), and spring viremia of carp virus (SVCV) but older fish were refractive. Viral hemorrhagic septicemia virus was isolated from epidermal ulcerative lesions on Pacific herring (Meyers et al. 1994), but its role in these lesions was unknown. The herring population was diminished by one-third during the time disease was occurring. It was speculated that VHSV is simply an opportunistic pathogen that is indigenous to Pacific herring.

The kidney, spleen, brain, and digestive tract are sites where VHSV is most abundant, and clinically infected fish shed virus in the feces, urine, and reproductive fluids. Although survivor fish generally do not shed virus, a pathogen source during absence of clinical disease can likely be attributed to an asymptomatic infected fish species (Jørgensen 1982). Wolf (1988) noted that more VHSV-resistant salmonids (brown trout) may be virus reservoirs, but as shown with isolates from turbot and cod, these isolates are often avirulent.

Jørgensen (1982) reported that VHSV-infected rainbow trout (45% of those tested) were found in a stream below the outfall of a hatchery in Denmark, but it could not be determined whether infected fish were hatchery escapees or had become infected via contaminated water. Fish above the hatchery were not infected. Opinions differ as to whether birds are a virus source. Eskildsen and Jørgensen (1973) reported that VHSV was not present in the feces of 41 seagulls, but other reports claim that herons can be a virus vector.

Pathological Manifestations

Viral hemorrhagic septicemia virus produces hemorrhages, usually petechiae, throughout the musculature, visceral organs, and gills of affected fish (Yasutake 1975; Meier and Jørgensen 1980). These hemorrhages occur because the endothelial cells are major targets for VHSV replication (de Kinkelin et al. 1979). Hematopoietic tissue of the kidney is severely affected and is characterized by hyperplasia and necrosis. Melanomacrophage centers are destroyed and leukopenia is present along with congestion. The liver exhibits focal necrosis and degeneration with vacuolated pyknotic and karyolytic hepatic cells. Although small hemorrhages are present in the musculature, muscle cells are essentially normal.

In naturally occurring infections of VHSV, gill epithelium is a possible route of virus entry where it might or might not replicate (Neukirch 1984). Virus does replicate near its portal of entry in blood capillary walls of the skin and underlying muscle and is then transported to the kidney and spleen, where more active replication occurs. According to de Kinkelin et al. (1979), the time between waterborne exposure to VHSV and recovery of the replicated agent from internal organs is about 24 hours.

Significance

Historically, VHSV has been the most serious viral disease to affect European trout culture, and it still continues to be a major health concern. With its detection along the northwest coast of North America in 1988, new emphasis has been placed on its potential to affect salmon culture throughout the world. The establishment of VHSV in non-salmonid marine species emphasizes its importance. It is now recognized that where VHSV occurs, there is always the possibility of a highly communicable disease outbreak. Currently, however, VHSV is not a problem in salmon culture in the Pacific Northwest because the indigenous strain is avirulent. Viral hemorrhagic septicemia is also important as a specific pathogen free (SPF) certifiable disease and occurs on the United States Title 50 list (Salmonid Importation Regulations: Fish Health 1993). This virus has caused great economic impact on various regulatory bodies in their effort to prevent its introduction.

ERYTHROCYTIC INCLUSION BODY SYNDROME

Erythrocytic inclusion body syndrome, a viral infection that causes anemia in cultured coho and chinook salmon, was first recognized in 1982 in the northwestern United States (Holt and Rohovec 1984; Leek 1987). The disease is caused by the erythrocytic inclusion body syndrome virus (EIBSV) (Holt and Rohovec 1984).

Geographical Range and Species Susceptibility

Erythrocytic inclusion body syndrome occurs in cultured populations of coho and chinook salmon along the Columbia River and its tributaries in Oregon and Washington (Leek 1987). The disease has also been reported in Canada, Chile, Ireland, Japan, and Norway (Lunder et al. 1990; Rodger et

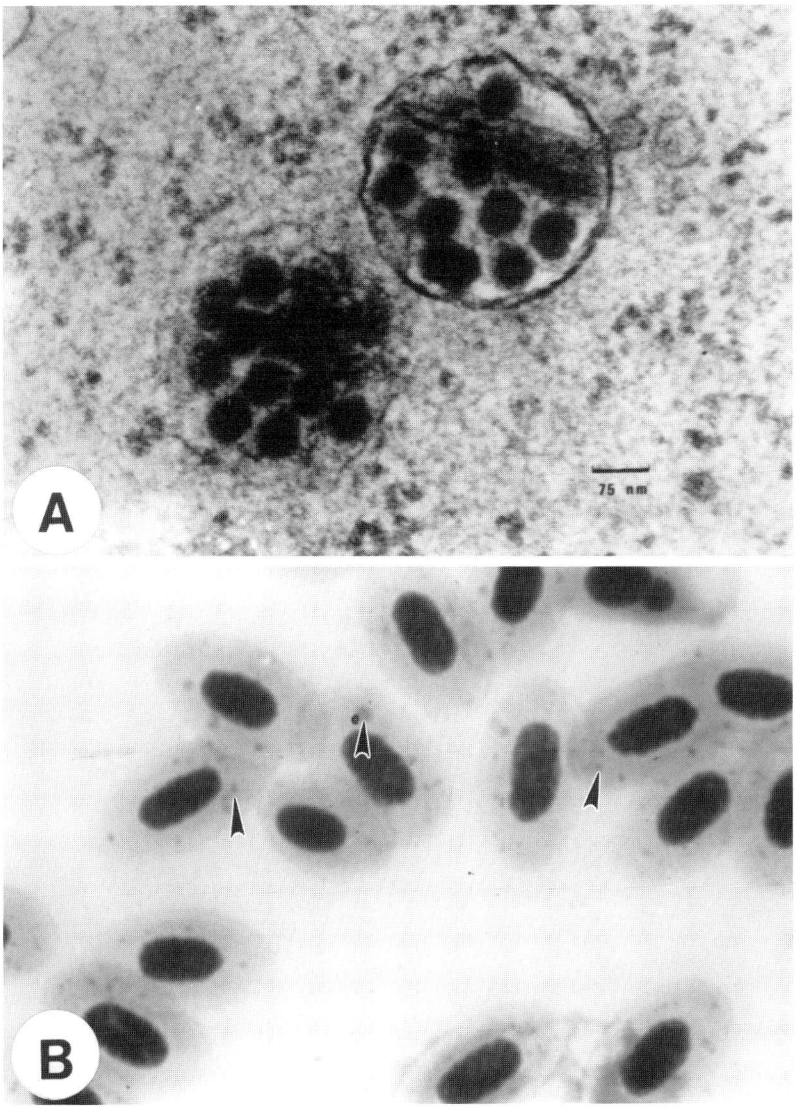

FIGURE 10.7. Erythrocytic inclusion body syndrome (EIBS) of salmon. (A) *Electron micrograph of EIBS virus within membrane-bound organelle. (Arakawa et al. 1989; reprinted with permission of Springer-Verlag Publishing Co.)* (B) *Erythrocytes with multiple cytoplasmic inclusion bodies* (arrows) *typical of EIBS-infected fish. (Photograph courtesy of S. Leek.)*

al. 1991; Takahashi et al. 1992a). Okamoto et al. (1992) found that chum and masu salmon are as susceptible to EIBS virus as are coho salmon. Rainbow and cutthroat trout can be experimentally infected, but disease severity in trout is mild compared with that in salmon (Piacentine et al. 1989).

Clinical Signs and Findings

Other than severe erythrocytic anemia and pale gills and internal organs, no behavioral or external clinical signs have been described (Holt and Rohovec 1984). Takahashi et al. (1992a) found that EIBS-infected fish had hyperemia of the intestine, dilatation of the spleen, hyperemia and hemorrhage of the atrium, the stomach contained mucus, and the liver appeared yellowish. Infected erythrocytes develop cytoplasmic inclusions that stain more lightly than the cell nucleus (Figure 10.7). Hematocrits of normal salmon are 40% to 45%, whereas those of EIBS-infected fish are often in the mid-30% range and can plummet to 15% and even 1% to 4% late in an infection.

Diagnosis

Erythrocytic inclusion body syndrome is presumptively diagnosed by staining erythrocytes with

TABLE 10.6. THE PROGRESSIVE STAGES OF EPIZOOTIC ERYTHROCYTIC INCLUSION BODY SYNDROME IN COHO SALMON HELD AT 12°C

Stage	Days postinfection	Hematocrit (% packed RBC)	Cytoplasmic inclusions	Fish condition
I	10	45	None	Normal
II	10–13	35–40	Low	Normal
III	18–25	25±	High	RBC lysis
IV	30–40	35–40	Low	Recovery
V	45	45	None	Normal

Source: Piacentini et al. (1989).
RBC, red blood cell.

pinacyanol chloride or Leishman-Giemsa to detect basophilic cytoplasmic inclusions that measure 1 to 8 µm in diameter (Figure 10.7) (Leek 1987; Arakawa et al. 1989). These inclusions stain pale purple, lavender, or pink. Each erythrocyte contains one round to ovoid inclusion that measures 0.8 to 3 µm in diameter, and their presence by electron microscopy is confirmatory of EIBS. The etiological agent has not yet been isolated in tissue culture. Nine different piscine cell lines inoculated with filtrates from EIBS-infected fish failed to develop CPE during a 30-day incubation period (Leek 1987; Piacentini et al. 1989).

Virus Characteristics

The EIBS virus is distinctly different from the iridovirus of viral erythrocytic necrosis. Because of its size, possession of a lipid envelope, and its presence in organelles, the EIBS virus most resembles members of the family Togaviridae (Arakawa et al. 1989). Membrane-bound inclusions in the erythrocytes contain viral particles that are 75 to 100 nm in diameter (Figure 10.7). The viruses are either associated with viroplasmlike material or are contained within membrane-bound organelles.

Epizootiology

An EIBS infection is not necessarily lethal, but mortality can occur during severe anemia. Piacentini et al. (1989) described five (I through V) disease stages of EIBS when fish were held at 12°C (Table 10.6). Each EIBS stage is characterized by specific changes in hematocrit values and presence of erythrocytic cytoplasmic inclusions. The entire cycle of infection from exposure to recovery required about 45 days, after which surviving fish appeared normal and were resistant to reinfection. Takahashi et al. (1992b) described similar stages of EIBS development in Japanese sea-cage cultured coho salmon at 10°C.

Although Takahashi et al. (1992b) attributed a mortality of about 23% directly to EIBS, affected fish were severely infected with opportunistic pathogens (such as external fungi) and, as a result, suffered much higher than normal mortality. The majority of deaths attributed to EIBS are actually due to opportunistic bacterial and fungal infections that complicate the disease process. Fish infected with EIBS generally survive if they do not succumb to other pathogens. Takahashi et al. (1992a) believe, however, that EIBS is a major contributor to high mortality of saltwater net-cage–reared coho salmon in Japan and that the virus is responsible for great economic losses to aquaculturists in that country. They also reported that water temperature affects the progression and duration of EIBS; the disease cycle begins more quickly at 15 to 18°C (5 days) and is more prolonged at 6 to 9°C (32 days).

Atlantic salmon that suffer from EIBS in fresh water appear to be less susceptible to recrudescence when transferred to seawater (Jarp et al. 1996). Fish classified as EIBS-positive before seawater transfer suffered 0.5% cumulative mortality, but fish groups that were EIBS-negative in fresh water had 4.9% mortality in seawater. Exposure to EIBS virus in fresh water appears to provide some protection after smoltification.

Experimental transmission of EIBS can be achieved by injecting naive fish with filtrates (0.22 µm) of homogenized kidney, spleen, or blood from afflicted fish (Piacentini et al. 1989). Typical inclusions appear in erythrocytes in about 14 days. The disease was also successfully transmitted to coho salmon by cohabitation with infected fish in 14 to 28 days. A wide size range of fish are susceptible to the EIBS virus. Okamoto et al. (1992) found that 1.2 to 220 g coho salmon are equally sensitive. Takahashi et al. (1992a) speculated that the EIBS virus could have been introduced into Japan via infected eggs.

Pathological Manifestations

Pathology of EIBS has not been fully described, but the major histopathological change is lysis of infected erythrocytes. Approximately 35% of erythrocytes and hemoblasts of infected fish will have cytoplasmic inclusions (Takahashi 1992a). The number of erythrocytes in the anterior kidney is reduced, and the parenchyma and hematopoietic areas show necrosis and congestion. Phagocytes, many containing eosinophilic vacuoles infiltrate the kidney, and epithelial cells of the renal tubules show hyaline droplet degeneration. Similar pathology occurs in the spleen with cellular vacuolation, and limited necrosis of hepatocytes occurs in the liver sinusoids.

Significance

By itself, the EIBS virus may not be a significant pathogen, but its presence reduces the host's natural resistance to other infectious agents. Secondary infections become invasive and often cause affected fish to succumb. In some aquaculture situations, EIBS virus is thought to cause significant economic losses (Takahashi et al. 1992b).

SALMONID HERPESVIRUS TYPE 1

As many as 10 herpesviruses have been isolated from trout and salmon (Tanaka et al. 1992); however, the two United States isolates are considered the same (salmonid herpesvirus type 1 [SHV-1]), and three from Japan are the same (salmonid herpesvirus type 2 [SHV-2]). An additional five isolates from coho and masu salmon in Japan are unresolved. Because SHV-1 and SHV-2 are distinctly different viruses and cause distinct clinical disease, they are discussed separately.

Salmonid herpesvirus type 1 causes a mild viral disease of rainbow trout and was first described by Wolf and Taylor (1975), who named the virus *Herpesvirus salmonis*. Isolation of a very similar herpesvirus from steelhead trout was reported and designated salmonid herpesvirus type 1 by Hedrick et al. (1986).

Geographical Range and Species Susceptibility

The original SHV-1 isolate was made in Washington state (Wolf and Taylor 1975), and later, the virus was isolated from fish in seven hatcheries and two lakes in California (Eaton et al. 1989). Although steelhead and rainbow trout are natural hosts of SHV-1, chum salmon fry and chinook salmon are experimentally susceptible, whereas coho and Atlantic salmon and brook and brown trout are resistant (Wolf and Taylor 1975; Eaton et al. 1989).

Clinical Signs and Findings

Adult fish from which the original SHV-1 isolate was obtained showed distinctly darkened pigmentation, which was the only clinical sign noted. Approximately 2 weeks after experimental exposure, juvenile rainbow trout stopped eating, became lethargic, laid on the tank bottom, and responded to physical stimulation with erratic swimming (Wolf and Smith 1981). Some fish also became dark and most developed exophthalmia accompanied by hemorrhage. Gills were pale and abdomens distended, hemorrhages developed on the fins of some fish, and mucoid fecal casts were evident. Internally, the peritoneal cavity had abundant ascites, which occasionally was bloody and gelatinous. The intestines were flaccid, the liver hemorrhagic, mottled or friable, and the kidneys pale.

Diagnosis

Salmonid herpesvirus type 1 disease is diagnosed by isolation of virus in susceptible cell cultures. The virus replicates in CHSE-214, rainbow trout fin (RTF-l) and RTG-2 cells, but not in three nonsalmonid cell lines (BB, BF-2, or FHM) (Wolf and Taylor 1975). Virus replication occurs at 5 to 10°C. Cytopathic effect, consisting principally of syncytia, was depressed or variable at 15°C, with complete inhibition at higher temperatures. Identification of SHV-1 is obtained by serum neutralization tests.

Virus Characteristics

Salmonid herpesvirus type 1 is a typical herpesvirus that measures 90 to 95 nm. The virion is enveloped and contains a DNA genome (Wolf et al. 1978; Hedrick et al. 1987). The herpesvirus isolate made from steelhead trout in California proved to be similar, but distinguishable, from the initial isolate; therefore, two strains of SHV-1 are recognized and no further distinction will be made between them. Hedrick et al. (1987) serologically compared North American isolates of SHV-1 to Japanese isolates of SHV-2 (*Oncorhynchus masou* virus [OMV], Yamame tumor virus [YTV], and nerca virus Tawata Lake [NeVTA]) and found that the United States isolates were distinctly different from those of Japan. Based on these differences, isolates from North America were named salmonid herpesvirus type 1 and those from Japan, salmonid herpesvirus type 2 (Hedrick and Sano 1989). Eaton et al. (1991a) also compared the five herpesviruses from salmonids by DNA homology and confirmed that SHV-1 was distinctly different from SHV-2.

Epizootiology

The first isolation of SHV-1 was made from ovarian fluid of postspawning adult rainbow trout that experienced up to 50% mortality at a federal hatchery in Washington state (Wolf and Taylor 1975). All brood stock infected with the initial SHV-1 were destroyed, and the hatchery, thoroughly disinfected. The virus was not isolated again until more than 10 years later in California (Hedrick et al. 1987). During the California outbreaks in 1985–1986, virus (SHV-1) was isolated from 5.8% of ovarian fluid samples and 4.5% of pooled tissue samples collected from steelhead trout at one hatchery (Eaton et al. 1989). It has been theorized that vertical transmission of SHV-1 occurs because of its presence in ovarian fluids.

Salmonid herpesvirus type 1 occurs from January through April when water temperatures are 10°C or less. The incubation period in experimental infections is 25 days or more. Wolf and Smith (1981) reported that mortality continued for about 50 days after onset of disease, but Eaton et al. (1989) were unable to kill fish with their SHV-1 isolates.

Pathological Manifestations

Wolf and Smith (1981) described histopathologic changes in fish that had been experimentally infected with SHV-1. Heart tissue was edematous and necrotic with loss of striations in muscle fibers; kidneys, livers, and spleens were edematous, hyperplastic, congested, and necrotic with syncytia in spleens. Eaton et al. (1989) described hypertrophy, edema, and hemorrhaged gill epithelium that separated from connective tissue. Also, fused liver cells had enlarged nuclei containing Feulgen-positive, eosinophilic Cowdry type A inclusion bodies; skeletal muscle of some infected fish was hemorrhaged and necrotic; and syncytia were present in liver hepatocytes.

Significance

Because SHV-1 has very low pathogenicity and is infrequently encountered, its importance to trout culture is believed to be minimal.

SALMONID HERPESVIRUS TYPE 2

Salmonid herpesvirus type 2 was isolated from adult masu in Japan (Kimura et al. 1981a). The virus is also known as *Oncorhynchus masou* virus, or OMV. Two other fish viruses isolated in Japan are similar to OMV, namely, nerka virus Towada Lake, Akita Prefecture (NeVTA); and yamame tumor virus (YTV) (Sano et al. 1983; Hedrick and Sano 1989).

Geographical Range and Species Susceptibility

Salmonid herpesvirus 2 (OMV) is known only in Japan, where it causes mortality in juvenile masu and coho salmon and tumors on adults (Kimura et al. 1981b). Experimentally, sockeye (kokanee) and chum salmon and rainbow trout are also susceptible (Kimura et al. 1981c and 1983).

Clinical Signs and Findings

Clinically, SHV-2 manifests itself in two different ways depending on age of fish. In juveniles, OMV produces an acute disease that kills 30– to 150-day-old fish, whereas neoplasia develops in some

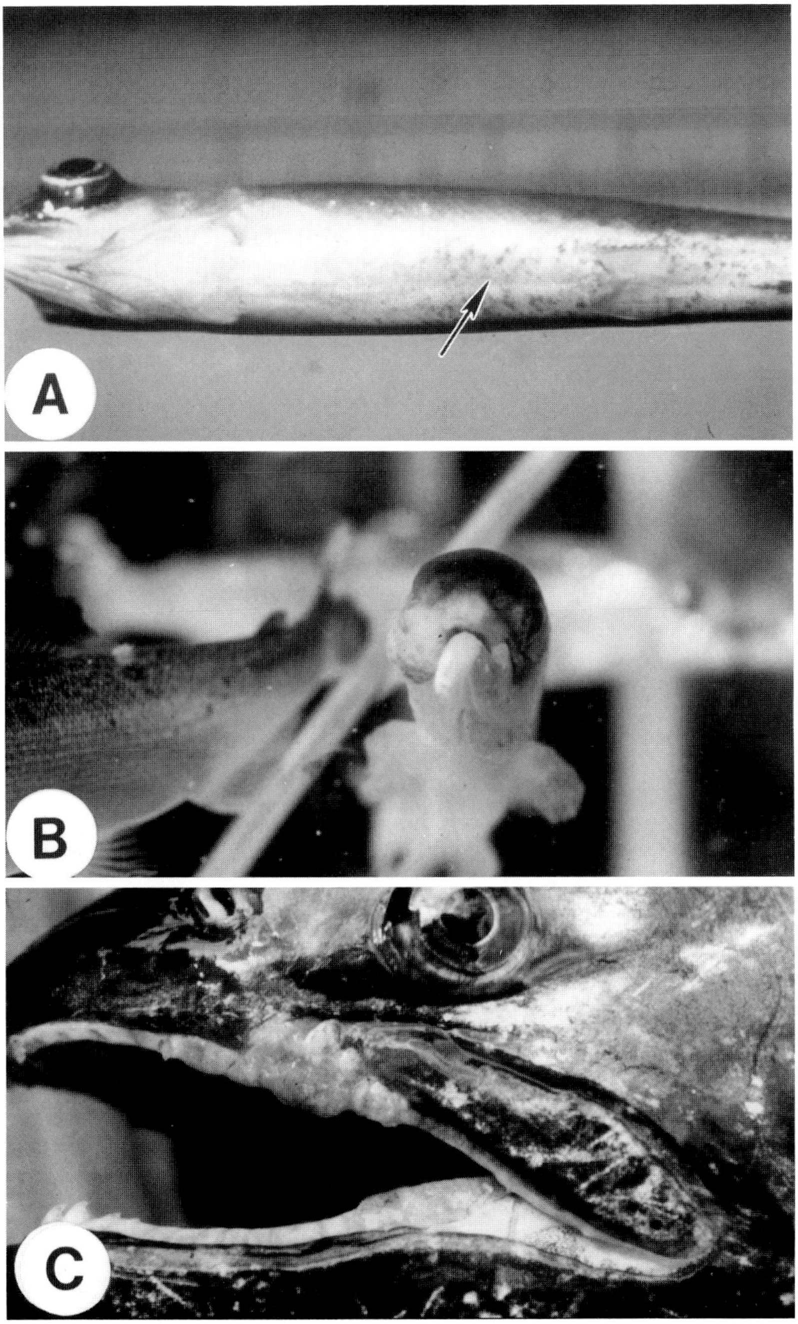

FIGURE 10.8. *Salmonid herpes virus type 2 (*Oncorhynchus masou *virus) in chum salmon.* (A) *Experimentally infected young fish with petechial hemorrhage on underside* (arrow) *and exophthalmia.* (B) OMV *tumor on mandible of young fish. (Yoshimizu et al. 1987; reprinted with permission of Springer-Verlag Publishing Co.)* (C) *Tumor on upper jaw of adult fish. (Photograph by M. Yoshimizu; reprinted with permission of CRC Press.)*

older fish. Infected juvenile fish become inappetent and exophthalmic, and petechiae appear on body surfaces especially under the jaw and belly (Figure 10.8). Internally, livers of infected salmon fry are mottled with white areas, and in advanced cases, the liver is totally white. The spleen may be swollen and macroscopically the kidney appears normal (Kimura et al. 1981b and c).

Beginning about 4 months after virus exposure and persisting for at least 1 year, neoplastic tumors will develop on surviving kokanee, chum, coho, and masu salmon and rainbow trout. These tumors occur mainly around the mouth in the maxillary and mandibular (Figure 10.8) regions but may also be found on gill covers, body surface, and cornea, but rarely in the kidney (Kimura et al. 1981b; Yoshimizu et al. 1987).

Diagnosis

During active infections, SHV-2 is isolated in cell cultures from clinically ill salmon less than 6 months old, normal adult salmon, or fish with tumors. The virus will replicate in all cell lines of salmonid origin including CHSE-214 and RTG-2, where yield is about 10^6 $TCID_{50}$/mL (Kimura et al. 1981a). Optimum cell culture incubation temperature is 10 to 15°C, but viral replication will occur at 5 to 16°C. Cytopathic effect develops in 5 to 7 days at 10 to 15°C and is characterized by cells becoming rounded, syncytium formation, and eventually lysis. Cowdry type A inclusion bodies are present in syncytial cells. Inoculated cultures that do not develop CPE in 10 to 14 days should be blind passed.

Isolates of OMV can be distinguished from other fish herpesviruses by serum neutralization tests (Hedrick et al. 1987). Gou et al. (1991) prepared recombinant plasmids as a DNA probe and were able to detect SHV-2 DNA in salmonid tissues 2 weeks before virus could be isolated from asymptomatic fish. A monoclonal antibody preparation against a recent isolate was cross-reactive with OMV and SHV-1 but not with IPNV or IHNV (Hayashi et al. 1993). Tumors may be confirmed by histology. The SHV-2 can be detected in fish tissue and infected cell cultures by indirect FAT using antisera to the nucleocapsid (Kumagai et al. 1995a). This simple, rapid, and reliable procedure was used to detect the virus in liver smears that were more than 6 months old.

Virus Characteristics

Salmonid herpesvirus type 2 is a typical herpesvirus. Without the envelope, the icosahedral nucleocapsid measures 100 to 115 nm in diameter and the enveloped particle measures 220 to 240 nm (Tanaka et al. 1987). The SHV-2 agent is serologically and morphologically similar, and possibly identical, to YTV and NeVTA agents (Sano et al. 1983). With the use of DNA, Eaton et al. (1991b) further demonstrated that OMV and YTV are similar but distinctly different from NeVTA; however, they still placed the three strains in SHV-2.

Epizootiology

Salmonid herpesvirus type 2 (OMV, YTV, and NeVTA) is widespread in Japan (Sano 1988); however, owing to management practices, the disease has disappeared from anadromous runs in some regions (Tanaka et al. 1992).

Salmonid herpesvirus type 2 is probably transmitted vertically during spawning via contaminated ovarian fluid and horizontally via water. Studies of SHV-2 by Kimura et al. (1981b) indicated that 35% to 60% of 3- to 5-month-old chum salmon died during a 60-day period following infection by immersion in water containing 100 $TCID_{50}$/mL. Kimura et al. (1981c) compared the susceptibility of 0- to 7-month-old chum, masu, kokanee, and coho salmon and rainbow trout at 10 to 15°C. Cumulative mortalities in chum salmon ranged from 3% for 0 age to as high as 98% for 3-month-old fish, but only 7% and 2% mortality occurred in 6- and 7-month-old chum salmon, respectively. Similar cumulative mortalities were recorded for sockeye and masu salmon; however, 1-month-old sockeye suffered 100% mortality. Coho salmon and rainbow trout were less susceptible to SHV-2 (OMV), with the highest mortality reaching 39% and 29%, respectively. Sano et al. (1983) reported a 65% mortality of 5-month-old yamame (masu) when injected intraperitoneally with 10^3 $TCID_{50}$ of SHV-2 (YTV) but only 15% mortality of those injected with 10^2 $TCID_{50}$. Tanaka et al. (1984) noted an 87% mortality of masu salmon during a 4-month period when infected at 1 month of age and 63% mortality when infected at 3 months.

The oncogenic nature of SHV-2 (OMV) provides a study model for herpesvirus-induced tumors. Sixty percent of masu that survived SHV-2 (OMV) epizootics exhibited tumors, 45% of which occurred around the mouth (Kimura 1981a; 1981b). Sano et al. (1983) provided the first information to authenticate the oncogenicity of a herpesvirus isolated directly from fish tumor and for which River's postulates were fulfilled. Working with the SHV-2 (YTV) agent, they found that tumors developed primarily on mandibles or premaxillae of masu and chum salmon 10 to 13 months after exposure. A comparison of tumor incidence in four different salmonid species exposed to SHV-2 between 1 and 7 months of age was conducted by Yoshimizu et al. (1987) (Figure 10.9). Tumors took 120 to 360 days to develop but devel-

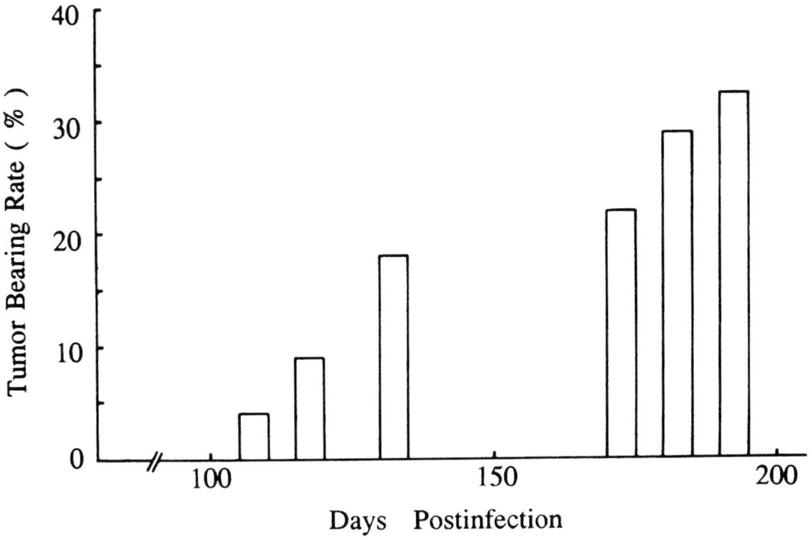

FIGURE 10.9. Increasing incidence of SHV-2 (OMV) tumor induction in chum salmon. (From Yoshimizu et al. 1987.)

oped in 100% of 3-month-old challenged masu, 12% of rainbow trout, 35% of coho salmon, and 40% to 71% of chum salmon.

Pathological Manifestations

Histopathologically, OMV disease varies with species and age of infected fish (Kimura et al. 1983; Tanaka et al. 1984). In the viremia stage of disease, the kidneys of masu, chum, and coho salmon were usually the principal target organ as judged by severity of histopathological changes. Epithelium of the mouth, jaw, operculum, and head in 1-month-old masu was necrotic, with focal necrosis of hepatic cells in the liver, spleen, and pancreas developing later. In 3-month-old masu, necrosis of hematopoietic tissue was seen simultaneously with hyperplasia, and syncytial cells were present in the kidney. As infection progressed, the kidney returned to normal, but the liver, which was initially marked only with focal necrosis, became more extensively necrotic. Also, liver hepatocytes displayed margination of chromatin. Cell degeneration occurred in the spleen, pancreas, cardiac muscle, and brain.

Tumors of differentiated epithelium developed 4 months to 1 year after infection in 12% to 100% of surviving chum, coho, and masu salmon and rainbow trout. Kumagai et al. (1995b) found that coho salmon survivors that had been injected with SHV-2 developed basal cell tumors on the gills 18 months after inoculation. Generally, tumors developed earlier in chum and coho salmon than in rainbow trout and masu salmon (Yoshimizu et al. 1989). Most tumors formed around the mouth and head, but tumors were also present in decreasing frequency on/in caudal fin, gill cover, body surface, cornea, and kidney. Tumors can include abundantly proliferative, well-differentiated squamous epithelial cells in which the papilloma is supported by a stroma of fine connective tissue, and several layers of squamous epithelial cells are present in a papillomatous array (Yoshimizu et al. 1988b).

Significance

Salmonid herpesvirus-2 is a major problem among land-locked salmon populations and coastal cage–reared coho salmon in northern Japan (Yoshimizu 1989). The virus was isolated from all but four locations sampled in northern Japan between 1976 and 1987. A significant number of 1- to 5-month-old salmon may be killed during a SHV-2 outbreak, and epizootic survivors tend to develop tumors and become virus reservoirs. Because SHV-2 has been isolated from ovarian fluids of adult fish, transmission during spawning is most likely; however, treatment of eggs with iodophores significantly reduces this possibility.

EPIZOOTIC EPITHELIOTROPIC DISEASE

Epizootic epitheliotropic disease (EED) is an acute viral disease of lake trout (McAllister and Herman

1989; Bradley et al. 1989), and since 1983, acute mortalities have occurred in juvenile lake trout populations in several state and federal hatcheries in the Great Lakes basin. Bradley et al. (1989) estimated that in that region, 15 million cultured lake trout died from undetermined disease between 1985 and 1989. Much speculation was made as to the cause of these mortalities, but no definitive etiology was established until transmission studies showed that the disease was infectious. McAllister and Herman (1989) demonstrated that a transmittable, filterable agent was involved and presented electron micrographs that showed putative herpesvirus particles in tissues from naturally and experimentally infected fish. Similarly, Bradley et al. (1989) demonstrated that a filterable agent was involved and virus capsids could be recovered from affected fish by isopycnic centrifugation. It is now accepted that the etiological agent of EED is a virus that has been named "epizootic epitheliotropic disease virus" (EEDV).

Geographical Range and Species Susceptibility

Epizootic epitheliotropic disease is known to occur only in the Great Lakes region of North America. The only known hosts are lake trout and lake trout X brook trout hybrids. The virus has not been observed in wild or feral stocks, although they are considered as possible virus reservoirs. Brown, brook, and rainbow trout as well as chinook and Atlantic salmon are refractive to EEDV.

Clinical Signs and Findings

Clinical signs of EED include lethargy, riding high in the water, sporadic swimming, and corkscrew diving (McAllister and Herman 1989; Bradley et al. 1989). Hemorrhages occur in the eyes and occasionally the mouth. Epithelial hyperplastic lesions, which are gray to white mucoid blotches, develop on the jaws, inside the mouth, and on the body and fins of infected fish. Secondary fungal infections (presumably *Saprolegnia*) develop on the eyes, fins, and body surface. The only visible internal lesion is a swollen spleen. Hematocrits are slightly elevated, averaging about 45% compared with a normal 40%.

Diagnosis

Epizootic epitheliotropic disease is presumptively diagnosed in lake trout by recognition of clinical signs in fish that are experiencing acute mortality. Electron micrographs of epithelial tissue reveal EEDV particles that can be recovered from skin tissue by ultracentrifugation and visualization by electron microscopy. Attempts to isolate the virus in CHSE-214, EPC, FHM, and RTG-2 cell cultures failed; therefore, no in vitro virological techniques are currently available for proper identification. No EEDV antisera are available for diagnostic and virus identification purposes.

Virus Characteristics

The etiological agent of EED is a virus that has been transmitted by filtrates from infected fish. The virus has tentatively been placed in the family Herpesviridae, even though it has not been characterized. Electron microscopy of harvested material from infected tissues by ultracentrifugation revealed hexagonal, unenveloped viral particles that measured 100 to 105 nm in diameter and enveloped particles that measured 220 to 235 nm (Bradley et al. 1989). McAllister and Herman (1989) found two types of ellipsoid to spherical viral particles, one of which measured 130 to 175 nm in diameter with a diffused electron-dense nucleoid with a diameter of 75 to 100 nm. The other particle type measured 150 to 200 nm in diameter with a diffuse nucleoid measuring 75 to 100 nm.

Epizootiology

Epizootics of EEDV can occur in spring or fall when water temperatures are 6 to 15°C, but they tend to occur primarily in spring when water temperatures are 8 to 10°C. Epizootic epitheliotropic disease has been diagnosed only in cultured lake trout in which mortality can approach 100% in fish up to 14 months of age. Survivors of an epizootic shed the infectious agent. Experimentally, EEDV was transmitted to naive lake trout and lake trout X brook trout hybrids by cohabitation and via waterborne exposure in a contaminated water bath containing filtered (0.22 μm) material from homogenates of diseased fish (McAllister and Herman 1989; Bradley et al. 1989).

Epizootic epitheliotropic disease is stress mediated, and almost any stressful condition will precipitate its onset. The process of anesthetizing and clipping fins for identification purposes prior to stocking appears to be a major predisposing factor in mortality of fingerling and yearling lake trout (Bradley et al. 1989).

According to R. W. Horner (Illinois Department of Natural Resources, personal communication), no new cases of EEDV have occurred since infected hatcheries were depopulated and sterilized in the early 1990s. Although millions of virus-infected lake trout were stocked into the Great Lakes before the disease was known to be a virus, no cases of EEDV have been detected in these wild populations.

Pathological Manifestations

Histopathology is the same in EEDV-infected lake trout whether disease occurs naturally or is experimentally induced (Bradley et al. 1989). The epithelium contains hypertrophied cells and areas of hyperplasia with lymphocytic infiltration, cell degeneration, and necrosis. Macrophages laden with cellular debris are common in the kidneys and sinusoids of the liver. In advanced cases, gills develop inconsistent edema. Intranuclear inclusion bodies are sometimes seen in epidermal cells.

Significance

A filterable, infectious agent has been implicated in at least some EED cases, and strong evidence points to a herpesvirus etiology. The virus is capable of killing large numbers of juvenile lake trout.

OTHER VIRUSES OF SALMONIDS

Other virus diseases have been detected in trout and salmon, but reported incidences are low, and, consequently, they have received little attention. However, some of these viruses have serious disease potential. In his review of fish viruses and viral diseases of salmonids, Sano (1995) listed a total of 24 conditions in which a virus was the known or suspect etiological agent, many of which have been discussed.

Aquareovirus Diseases

New aquareoviruses are being isolated more frequently and from a variety of aquatic animals. LaPatra and Hedrick (1995) reported that at least 25 new aquareovirus isolates were made in the state of Washington alone between 1990 and 1995. Many of these aquareoviruses are not pathogenic but were isolated from apparently healthy fish during routine virus assays. Chum salmon reovirus (CSV) is pathogenic, results in mortality, and causes moderate to severe multifocal liver necrosis (LaPatra and Hedrick 1995). Landlocked salmon virus (LSV) is a reovirus-like agent that was isolated in Japan and apparently causes little injury (Hsu et al. 1989). Several cell lines, including AS, BF-2, BB, CCO, and CHSE-214 cells, are sensitive to this virus.

• Cutthroat Trout Virus

A small hexagonal RNA-containing virus was isolated from ovarian fluid of adult cutthroat trout by Hedrick et al. (1991). Although there was no virus-induced mortality, the virus was experimentally infective to brown and brook trout. Following waterborne exposure, virus was reisolated from the kidney and spleen in CHSE-214 cells at 15°C for up to 5 weeks postinfection.

Pacific Salmon Anemia Virus

The Pacific salmon anemia virus (PSAV) isolated from anemic coho salmon was reported by Holt and Rohovec (1984). It is apparently different from the EIBS virus. The virus is considered a togavirus and was transmitted to fall chinook salmon via stomach gavage during force feeding. The virus was found in the cytoplasm of erythrocytes of juvenile coho and spring chinook salmon and caused anemia and high mortality. Affected fish had a high incidence of fungus on the dorsal and caudal fins and snout.

Salmon Pancreas Disease Virus

A pathological condition in the pancreas of cultured Atlantic salmon has been recognized in Ireland, Scotland, France, Norway, Spain, and the United States (Munro et al. 1984). The etiology was unknown until Nelson et al. (1995) isolated a

Togaviridae-like agent from diseased Atlantic salmon in CHSE-214 cells where cytopathic effect developed in 28 days at 15°C. When filtrates from infected cells were injected into naive Atlantic salmon, clinical signs and pathology were consistent with those described for pancreas disease of Atlantic salmon.

MANAGEMENT OF SALMONID VIRUSES

Fish virus diseases are not amenable to treatment with drugs or chemotherapeutants, but some may be controlled entirely through management: avoidance, prevention, and vaccination.

Avoidance

Avoidance is the most effective and economical way of controlling some salmonid viruses. Specific pathogen free (SPF) certification is used to ensure that eggs of a specified brood stock population are free of specific pathogens. This certification is based on a sample assay of broodfish from an egg production facility before using those fish as an egg source (Thoeson 1994). These procedures have been used widely in trout and salmon aquaculture since the mid-1960s and have undoubtedly thwarted the spread of IPNV as well as other fish viral agents. For example, according to Rodriguez et al. (1994), a large number of embryonated trout eggs are imported into Spain from the United States each year. To date, North American VR-299 (serotype A1) does not exist in Spain, and this can be attributed to effective IPNV-free certification prior to egg shipment. Ideally, brood stock should be certified free of a target virus for 3 consecutive years, with two sampling periods being made each year. When eggs are brought into a production facility from virus-free sources, they should be hatched and reared through the susceptible fingerling stage in a closed, fish-free water supply. Because nonsalmonid fish can be carriers of IPNV and other viruses, fish that are to come into contact with cultured salmonids should also be virus free. The SPF technique is used to help prevent the transmission of IHNV, VHSV, and possibly other virus diseases and works well providing brood stock from which eggs are taken are held in a closed water system and have no contact with waters containing possible carrier fish. Although this approach is a credible procedure, it is by no means a perfect system, because there are examples in which the spread of virus has not been totally prevented.

Avoidance of salmonid herpesvirus type 2 is enhanced by screening brood stock for virus and disinfecting egg by immersion in iodophores at 50 mg/L for 20 minutes (Kimura and Yoshimizu 1989). Egg disinfection can also reduce the impact of other salmonid viruses by eliminating pathogens that might be attached to egg surfaces. Eggs should be incubated in water from a fish-free source, or if eggs are incubated in open water supplies, water should be irradiated with ultraviolet light at $3.0 \times 103\ \mu W/s \cdot cm^2$ (Yoshimizu et al. 1986). Fingerlings should also be grown in UV-treated water. The incidence of SHV-2 has declined in Japan since initiation of egg disinfection with iodophores and UV irradiation of incubation and culture waters.

Avoidance of infectious salmon anemia disease can be enhanced by locating an Atlantic salmon culture site no closer than 5 km to another site and by decontamination of virus-positive sites (Jarp and Karlsen 1997). Implementation of good sanitation practices by fish farmers can also reduce risk of ISA. Currently, no management or control procedures, other than avoidance, are known to prevent dissemination of the EIBS virus or salmonid herpesvirus-1; however, Wolf (1988) recommended that sound hatchery sanitation practices be followed as a control for SHV-1.

Management and control of EEDV rely on avoidance. Hatcheries in which EEDV had been reported were disinfected with chlorine, and stringent regulations were imposed on reintroduction of fish. As of 1997, there had been no reoccurrences of the disease (R. Horner, personal communication). Experimentally, treating water with high doses of UV irradiation prevents waterborne transmission of EEDV.

Prevention of Virus Diseases

If potential IHNV-positive brood stock cannot be avoided (for example, sea-run Pacific salmon), the management approach outlined for Alaskan salmon operations by Meyers et al. (1990) should be applied. According to K. Amos (personal communication), these same procedures are used in the state of Washington to reduce potential for IHNV transmission during manual spawning of steelhead and all salmon, particularly sockeye. The following brood stock/egg management procedures can reduce hatched fry mortality attributable to IHNV from the 90% range to less than 10%.

1. IHNV-free water is used for egg incubation and rearing. If fish-free water is not available, depuration is accomplished by ultraviolet treatment.
2. A strict disinfection policy is used for utensils, facilities, field clothing, personnel, and external surfaces of brood fish during egg and milt collection.
3. Eggs from each female are fertilized separately with milt from 1 or 2 males. Each spawn of eggs from a specific mating is water-hardened separately in 100 mg/L solution of iodophor for 60 minutes.
4. Eggs from 80 to 100 females are pooled into stacked incubators with upwelling water.
5. Eggs in each hatching unit are kept physically isolated from eggs in other hatching units.
6. Alevins are transported and released when yolk is depleted or after rearing fry for 3 to 6 weeks.

For management of IHNV after clinical disease has already occurred, Amend (1970) recommended increasing water temperature to 18°C to help arrest IHNV epizootics. Hetrick et al. (1979) pointed out that increasing water temperature does not stop mortality of already infected fish and that epizootic survivors may still become virus carriers. Reducing water temperature to 4 or 5°C during peak susceptibility periods may have a beneficial effect on reducing effects of IPNV virus (Frantsi and Savan 1971).

To control VHSV, it is essential to avoid wide fluctuations in water temperature and to manage other environmental stressors carefully. Programs to manage VHSV in Europe advocate controlled movement of fish, destruction of infected populations when feasible, and disinfection of facilities before restocking with VHSV SPF fish or eggs (Daelman 1996).

In areas where IPNV, IHNV, or VHSV are indigenous, the culture of more resistant fish strains or species should be considered. Using male and female rainbow trout from a resistant stock, Kaastrup et al. (1991) showed that inherited resistance to VHSV was carried by the male rather than the female. In view of this, a breeding program to enhance resistance would appear to be feasible, and already such a program has begun on Danish rainbow trout farms. Also, Dorson et al. (1991) showed that triploid rainbow and brook trout hybrids were resistant to VHSV and IHNV but remained susceptible to IPNV.

Several antiviral agents have been used experimentally in an attempt to retard virus development in fish and cell cultures, but to date, none have proved to be practical. Addition of 0.14 mg of free iodine per liter to hatchery waters will help prevent rainbow trout from becoming infected by waterborne exposure to IHNV (Batts et al. 1991). Some antiviral agents have been noted to reduce SHV-2 mortality. Kimura et al. (1983) showed that Acyclovir quinine reduced tumor development when fish were immersed for 30 minutes each day for 2 months in 25 µg/mL. Iododeoxyuridine gave even better results with an oral dosage of 15 µg/fish per day for 2 months; however, neither drug is approved for use in food fish, and this approach is probably too expensive and labor intensive to be used on a large scale.

Vaccination

Considerable research has been expended to develop a vaccine for IPNV, IHNV, or VHSV, using several different kinds of antigen preparations. Attenuated and killed virus and subunit preparations have been developed and tested experimentally. When fish were exposed to an attenuated IHNV and then challenged, a significant degree of protection was detectable (Fryer et al. 1976). A formalin-killed preparation of IHNV was shown to be protective when administered by intraperitoneal injection or by hyperosmotic infiltration (Nishimura et al. 1985). Engelking and Leong (1989) protected kokanee salmon from five strains of IHNV by injection with a glycoprotein from a single strain of the virus. Gillmore et al. (1988) reported successful preliminary results from a genetically engineered IHNV subunit vaccine that was composed of viral glycoprotein produced through recombinant DNA and introduced in *Escherichia coli*. Oberg et al. (1991) expressed the ribonucleoprotein gene of IHNV in *E. coli* through plasmid induction, but this protein alone does not induce immunity or antibody. If, however, it is administered with the bacterial glycoprotein lysate, resistance to IHNV occurs.

In an attempt to show humoral antibody and protective response of rainbow trout to multiple

IHNV vaccine exposures, Ristow et al. (1993) found that when results of ELISA and plaque neutralization (PNT) tests were correlated, antibody levels in 5X vaccinated fish were significantly higher at 20 days postvaccination, but at 80 days, there was no difference in antibody levels. These data would indicate a short-term benefit to multiple vaccination but no long-term benefits. Exposure of rainbow trout to an avirulent cutthroat trout virus (CTV) provided significant protection when they were exposed to virulent IHNV (Hedrick et al. 1994). Although the protective mechanism has not been identified, it is thought to be associated with interferonlike activity in the anterior kidney cells.

Passive immunization also occurred in rainbow trout following injection of sera from survivors of an IHNV epizootic, with neutralization titers of 640 to 1280 being recorded (LaPatra et al. 1993). Relative protection ranged from 63% to 100% in these fish, clearly demonstrating that the presence of antibodies does provide fish immunological protection.

With the use of monoclonal and polyclonal antibodies, LaPatra et al. (1994a) identified 10 antigenic IHNV groups, 91% of which fell into three groups. When susceptible fish were exposed to seven of these antigenic groups, they experienced mortality of 14% to 92%, but more importantly, relative protection of infected fish following vaccination with immune serum ranged from 91% to 100%. After passive immunotherapy, rainbow trout were protected against IHNV by injection even though their virus neutralization antibody titers were low (LaPatra et al. 1994b). These data indicate that vaccination against one strain of IHNV may protect against other strains as well. Emmenegger et al. (1997) synthesized three putative antigenic peptide determinants on the IHNV glycoprotein and injected them into rainbow trout and analyzed the fish's antibody response by ELISA and virus neutralization. They found these antibody responses to be generally low and inconsistent and concluded that a better understanding of how a fish's immune system works is needed to facilitate development of an efficacious vaccine against IHNV.

Currently, there is much research being carried out to study the potential for using immunization as a method for prevention and control of IPNV. In fact, a commercially available subunit IPNV vaccine is now being used in Europe (see Table 5.5). Bootland et al. (1990) showed that immersion of 2–, 3–, and 6-week-old brook trout in a formalin-inactivated IPNV preparation enhanced relative percent survival, but as fish became older protection decreased. Vaccination did not prevent IPNV infection in any age group of fry. In an effort to develop a viral subunit vaccine for IPNV, Manning and Leong (1990) inserted the cDNA of IPNV into *E. coli.* Rainbow trout were vaccinated by immersion in the bacterial lysate for 20 minutes and challenged 20 days later; protection to IPNV was demonstrated.

In spite of encouraging early vaccination results, it may be difficult to vaccinate against IHNV or IPNV successfully for several reasons: (1) very young fish are most susceptible, (2) it takes time to develop a protective immunity, and (3) temperatures at which salmonids hatch and grow are not very conducive to rapid development of a strong immune response.

The possibility of immunizing trout and salmon against VHSV was proposed in the mid-1970s by de Kinkelin and Le Berre (1977), but to date it has not evolved into an effective practice. Experimental vaccination programs for VHSV in Europe that use both killed and live preparations were reviewed by de Kinkelin (1988). Killed VHSV vaccines require injection but are effective in eliciting protection. Live, attenuated, vaccines can be applied via immersion and have provided a moderate to high degree of protection in laboratory trials. The potential of subunit vaccines that use monoclonal antibodies against the VHS viral glycoprotein have also been investigated (Lorenzen et al. 1988). Two of these monoclonal antibodies provided strong protection in fish after passive immunization, but obviously much work still needs to be done in this area.

REFERENCES

Ahne, W. 1982. Vergleichende untersuchungen uber die stabilitat von vier fischpathgenen viren (VHSV, PFR, SVCV, IPNV). *Zeitschrift Veternarski (B)* 29:457–476.

Ahne, W. 1983. Presence of infectious pancreatic necrosis virus in the seminal fluid of rainbow trout, *Salmo gairdneri* Richardson. *Journal of Fish Diseases* 6:377.

Ahne, W. 1985. Viral infection cycles in pike (*Esox lucius* L.). *Journal of Applied Ichthyology* 2:90–91.

Ahne, W., and I. Thomsen. 1985. Occurrence of VHS in wild white fish (*Coregonus* sp.). *Journal of Veterinary Medicine* 32:73–75.

Ahne, W., P. E.V. Jørgensen, N. J. Olesen, W. Schafer, and P. Steinhagen. 1986. Egtved virus: occurrence of strains not clearly identifiable by means of virus neutralization tests. *Journal of Applied Ichthyology* 4:187–189.

Amend, D. F. 1970. Control of infectious hematopoietic necrosis virus disease by elevating the water temperature. *Journal of the Fisheries Research Board of Canada* 27:265–270.

Amend, D. F., and V. C. Chambers. 1970. Morphology of certain viruses of salmonid fishes. I. In vitro studies of some viruses causing hematopoietic necrosis. *Journal of the Fisheries Research Board of Canada* 27: 1285–1293.

Amend, D. F., and J. R. Nelson. 1977. Variation in the susceptibility of sockeye salmon *Oncorhynchus nerka* to infectious hematopoietic necrosis virus. *Journal of Fish Biology* 11:567–573.

Amend, D. F., and L. Smith. 1975. Pathophysiology of infectious hematopoietic necrosis virus disease in rainbow trout: hematological and blood chemical changes in moribund fish. *Inflammation and Immunity* 11:171–179.

Amend, D. F., W. T. Yasutake, and R. W. Mead. 1969. A hematopoietic virus disease of rainbow trout and sockeye salmon. *Transactions of the American Fisheries Society* 98:796–804.

Amos, K. H., K. A. Hooper, and L. LeVander. 1989. Absence of infectious hematopoietic necrosis virus in adult sockeye salmon. *Journal of Aquatic Animal Health* 1:281–283.

Arakawa, C. K., D. A. Hursh, C. N. Lannan, J. S. Rohovec, and J. R. Winton. 1989. Preliminary characterization of a virus causing infectious anemia among stocks of salmonid fishes in the western United States. In: *Viruses of Lower Vertebrates,* edited by W. Ahne, and E. Kurstak, 442–450. Berlin: Springer-Verlag.

Arakawa, C. K., R. E. Deering, K. H. Higman, R. J. Oshima, P. J. O'Hara, and J. R. Winton. 1990. Polymerase chain reaction (PCR) amplification of a nucleoprotein gene sequence of infectious hematopoietic necrosis virus. *Diseases of Aquatic Organisms* 8:165–170.

Arkush, K. D., G. Bovo, P. de Kinkelin, J. R. Winton, W. H. Wingfield, and R. P. Hedrick. 1989. Biochemical and antigenic properties of the first isolates of infectious hematopoietic necrosis virus from salmonid fish in Europe. *Journal of Aquatic Animal Health* 1:148–53.

Arnzen, J. M., S. S. Ristow, C. P. Hesson, and J. Lientz. 1991. Rapid fluorescent antibody tests for infectious hematopoietic necrosis virus (IHNV) utilizing monoclonal antibodies to the nucleoprotein and glycoprotein. *Journal of Aquatic Animal Health* 3:109–113.

Babin, M., C. Hernandez, C. Sanchez, and C. Dominquez. 1990. Detection rapida del virus de la necrosis pancreatica infecciosa por enzimo-imuno-adsorcion de captura. *Medical Veterinarian* 7:557.

Barja, J. L., A. E. Toranzo, M. L. Lemos, and F. M. Hetrick. 1983. Influence of water temperature and salinity on the survival of IPN and IHN viruses. *Bulletin of the European Association of Fish Pathologist* 3(4):47–50.

Batts, W. N., and J. P. Winton. 1989. Enhanced detection of infectious hematopoietic necrosis virus and other fish viruses by pretreatment of cell monolayers with polyethylene glycol. *Journal of Aquatic Animal Health* 1:284–291.

Batts, W. N., M. L. Landolt, and J. R. Winton. 1991. Inactivation of infectious hematopoietic necrosis virus by low levels of iodine. *Applied Environmental Microbiology* 57: 1379–1385.

Bernard, J., M. Bremont, and J. Winton. 1991. Sequence homologies between the N genes of the 07-71 and Makah isolates of viral hemorrhagic septicemia virus. *Proceedings of the Second International Symposium on Viruses of Lower Vertebrates* 190. Abstract.

Biering, E., and Ø. Bergh. 1996. Experimental infection of Atlantic halibut, *Hippoglossus hippoglossus* L., yolk-sac larvae with infectious pancreatic necrosis virus: detection of virus by immunohistochemistry and in situ hybridization. *Journal of Fish Diseases* 19:405–413. Abstract.

Biering, E., F. O. M. Nilsen, F. Rødseth, and J. Glette. 1994. Susceptibility of Atlantic halibut *Hippoglossus hippoglossus* to infectious pancreatic necrosis virus. *Diseases of Aquatic Organisms* 20:183–190.

Biering, E., H. P. Melby, and S. H. Mortensen. 1997. Sero- and genotyping of some marine aquatic birnavirus isolates from Norway. *Diseases of Aquatic Organisms* 28:169–174.

Billi, J. L., and K. Wolf. 1969. Quantitative comparison of peritoneal washes and feces for detecting infectious pancreatic necrosis (IPN) virus carrier brook trout. *Journal of the Fisheries Research Board of Canada* 26:1459–1465.

Bonami, J. R., F. Cousserans, M. Weppe, and B. J. Hill. 1983. Mortalities in hatchery-reared sea bass fry associated with Birnavirus. *Bulletin of the European Association of Fish Pathologist* 3:41.

Bootland, L. M., P. Dobos, and R. M. W. Stevenson. 1990. Fry age and size effects on immersion immunization of brook trout, *Salvelinus fontinalis* Mitchell, against infectious pancreatic necrosis virus. *Journal of Fish Diseases* 13:113–125.

Bootland, L. M., P. Dobos, and R. M. W. Stevenson. 1991. The IPNV carrier state and demonstration of vertical transmission in experimentally infected brook trout. *Diseases of Aquatic Organisms* 10:13–21.

Bootland, L. M., P. Dobos, and R. M. W. Stevenson.

1995. Immunization of adult brook trout, *Salvelinus fontinalis* (Mitchell), fails to prevent infectious pancreatic necrosis virus (IPNV) carrier state. *Journal of Fish Diseases* 18:449–458.

Bovo, G., G. Giorgett, P. E. V. Jørgensen, and N. J. Olesen. 1987. Infectious hematopoietic necrosis: first detection in Italy. *Bulletin of the European Association of Fish Pathologist* 7:24.

Bradley, T. M., D. J. Medina, P. W. Chang, and J. McClain. 1989. Epizootic epitheliotropic disease of lake trout (*Salvelinus namaycush*): history and viral etiology. *Diseases of Aquatic Organisms* 7:195–201.

Buck, D., J. Finlay, D. McGregor, and C. Seagrave. 1979. Infectious pancreatic necrosis (IPN) virus: its occurrence in captive and wild fish in England and Wales. *Journal of Fish Diseases* 2:549–553.

Burke, F., and D. Mulcahy. 1983. Retention of infectious haematopoietic necrosis virus infectivity in fish tissue homogenates and fluids stored at three temperatures. *Journal of Fish Diseases* 6:543–547.

Burke, L., and R. Grischkowsky. 1984. An epizootic caused by infectious hematopoietic necrosis virus in an enhanced population of sockeye salmon, *Oncorhynchus nerka* (Walbaum), smolts at Hidden Creek, Alaska. *Journal of Fish Diseases* 7: 421–429.

Cain, K. D., S. E. Lapatra, T. J. Baldwin, B. Shewmaker, J. Jones, and S. S. Ristow. 1996. Characterization of mucosal immunity in rainbow trout *Oncorhynchus mykiss* challenged with infectious hematopoietic necrosis virus: identification of antiviral activity. *Diseases of Aquatic Organisms* 27:161–172.

Castric, J., and P. de Kinkelin. 1980. Occurrence of viral hemorrhagic septicemia in rainbow trout *Salmo gairdneri* Richardson reared in sea water. *Journal of Fish Diseases* 3:21–27.

Castric, J., and P. de Kinkelin. 1984. Experimental study of the susceptibility of two marine fish species, sea bass (*Dicentrarchus labrax*) and turbot (*Scophthalmus maximus*), to viral haemorrhagic septicemia. *Aquaculture* 41:203–212.

Caswell-Reno, P., V. Lipipun, P. W. Reno, and V. L. Nicholson. 1989. Use of a group-reactive and other monoclonal antibodies in an enzyme immunodot assay for identification and presumptive serotyping of aquatic birnaviruses. *Journal of Clinical Microbiology* 27:1924–1929.

Chilmonczyk, S., and J. R. Winton. 1994. Involvement of rainbow trout leukocytes in the pathogenesis of infectious hematopoietic necrosis. *Diseases of Aquatic Organisms* 19:89–94.

Christie, K. E., and L. L. Haverstein. 1989. A new serotype of infectious pancreatic necrosis virus (IPN N 1). In: *Viruses of Lower Vertebrates,* edited by W. Ahne, and E. Kurstak, 279–283. Berlin: Springer-Verlag.

Daelman, W. 1996. Animal health and the trade in aquatic animals within and to the European Union. *Revue Scientifique et Technique* 15:711–722.

Dannevig, B., K. Falk, and J. Krogsurd. 1993. Leukocytes from Atlantic salmon, *Salmo salar* L., experimentally infected with infectious salmon anemia (ISA) exhibit an impaired response to mitogens. *Journal of Fish Diseases* 16:351–359.

Danton, M., S. S. Ristow, A. M. Hattenberger-Baudouy, and P. de Kinkelin. 1994. Typing of French isolates of infectious haematopoietic necrosis virus (IHNV) with monoclonal antibodies using indirect immunofluorescence. *Diseases of Aquatic Organisms* 18:223–226.

Deering, R. E., C. K. Arakawa, K. H. Oshima, P. J. O'Hara, M. L. Landolt, and J. R. Winton. 1991. Development of a biotinylated DNA probe for detection and identification of infectious hematopoietic necrosis virus. *Diseases of Aquatic Organisms* 11:57–65.

de Kinkelin, P. 1983. Viral haemorrhagic septicemia. In: *Antigens of Fish Pathogens: Development and Production for Vaccines and Serodiagnostics,* edited by D. P. Anderson, M. Dorson, and Ph. Dubourget, 51–62. Lyon, France: Collection Fondation Marcel Merieux.

de Kinkelin, P. 1988. Vaccination against viral haemorrhagic septicaemia. In: *Fish Vaccination,* edited by A. E. Ellis, 172–191. London: Academic Press.

de Kinkelin, P., and M. Le Berre. 1977. Demonstration de la protection de la truite arc-enciel contre la SHV, par l'administration d'un virus inactive. *Bulletin of the Office International Epizooties* 87(5–6):401–402.

de Kinkelin, P., and R. Scherrer. 1970. Le virus D'Egtved. *Annales de Recherches Veterinaires* 1:17–30.

de Kinkelin, P., M. Le Berre, and A. Meurillon. 1974. Septicemie hemorrhagique virale: Demonstraton de l'etat refractaire du saumon coho *(Oncorhynchus kisutch). Bulletin France Pisicicultura* 253:166-176.

de Kinkelin, P., S. Chilmonczyk, M. Dorson, M. Le Berre, and A. M. Baudouy. 1979. Some pathogenic facets of rhabdoviral infection of salmonid fish. In: *Mechanisms of Viral Pathogenesis and Virulence,* edited by P. Bachmann. *Proceedings of the Munich Symposium on Microbiology,* 357–375. Munich, June 6–7.

Desautels, D., and R. M. MacKelvie. 1975. Practical aspects of survival and destruction of infectious pancreatic necrosis virus. *Journal of the Fisheries Research Board of Canada* 32:523–531.

Distribution and economical impact of the disease IPN in Norway. 1994. Trondheim, Norway: SVL-Trondheim, Task Division.

Dixon, P. F., and J. de Groot. 1996. Detection of rainbow trout antibodies to infectious pancreatic necrosis virus by an immunoassay. *Diseases of Aquatic Organisms* 26:125–132.

Dixon, P. F., and B. J. Hill. 1984. Rapid detection of fish

rhabdoviruses by the enzyme-linked immunosorbent assay (ELISA). *Aquaculture* 42:1–12.

Dobos, P., B. J. Hill, R. Hallett, D. T. C. Kells, H. Becht, and D. Teninges. 1977. Biophysical and biochemical characterization of five animal viruses with bisegmented double-stranded RNA genomes. *Journal of Virology* 32:593–605.

Domínguez, J., R. P. Hedrick, and J. M. Sánchez-Vizcaine. 1990. Use of monoclonal antibodies for detection of infectious pancreatic necrosis virus by the enzyme-linked immunosorbent assay (ELISA). *Diseases of Aquatic Organisms* 8:157–163.

Dopazo, C. P., F. M. Hetrick, and S. K. Samal. 1994. Use of cloned cDNA probes for diagnosis of infectious pancreatic necrosis virus infections. *Journal of Fish Diseases* 17:1–16.

Dorson, M., B. Chevassus, and C. Torchy. 1991. Comparative susceptibility of three species of char and of rainbow trout X char triploid hybrids to several pathogenic salmonid viruses. *Diseases of Aquatic Organisms* 11:217–224.

Drolet, B. S., J. S. Rohovec, and J. C. Leong. 1993. Serological identification of infectious hematopoietic necrosis virus in fixed tissue culture cells by alkaline phosphatase immunocytochemistry. *Journal of Aquatic Animal Health* 5:265–269.

Drolet, B. S., J. S. Rohovec, and J. C. Leong. 1994. The route of entry and progression of infectious haematopoietic necrosis virus in *Oncorhynchus mykiss* (Walbaum): a sequential immunohistochemical study. *Journal of Fish Diseases* 17:337–347.

Eaton, W. D., W. H. Wingfield, and R. P. Hedrick. 1989. Prevalence and experimental transmission of the steelhead herpesvirus in salmonid fishes. *Disease of Aquatic Organisms* 7:23–30.

Eaton, W. D., J. Hulett, R. Brunson, and K. True. 1991a. The first isolation in North America of infectious hematopoietic necrosis virus (IHNV) and viral hemorrhagic septicemia virus (VHSV) in coho salmon from the same watershed. *Journal of Aquatic Animal Health* 3:114–117.

Eaton, W. D., W. H. Wingfield, and R. P. Hedrick. 1991b. Comparison of the DNA homologies of five salmonid herpesviruses. *Fish Pathology* 26:183–187.

Emmenegger, E., M. Landolt, S. LaPatra, and J. Winton. 1997. Immunogenicity of synthetic peptides representing antigenic determinants on the infectious hematopoietic necrosis virus glycoprotein. *Diseases of Aquatic Organisms* 28:175–184.

Engelking, H. M., and J. C. Leong. 1989. Glycoprotein from infectious hematopoietic necrosis virus (IHNV) induces protective immunity against five IHNV types. *Journal of Aquatic Animal Health* 1:291–300.

Engelking, H. M., J. B. Harry, and J. C. Leong. 1991. Comparison of representative strains of infectious hematopoietic necrosis virus by serological neutralization and cross-protection assays. *Applied Environmental Microbiology* 57:1372–1378.

Eskildsen, U. K., and P. E. V. Jørgensen. 1973. On the possible transfer of trout pathogenic viruses by gulls. *Rivista Italiana di Piscicoltura Ittiopatologia* 8(4):104–105.

Evensen, Ø., and E. Lorenzen. 1997. Simultaneous demonstration of infectious pancreatic necrosis virus (IPNV) and *Flavobacterium psychrophilum* in paraffin-embedded specimens of rainbow trout *Oncorhynchus mykiss* fry by use of paired immunohistochemistry. *Diseases of Aquatic Organisms* 29:227–232.

Evensen, Ø., W. Meier, T. Wahli, N. J. Olesen, P. E. V. Jørgensen, and T. Håstein. 1994. Comparison of immunohistochemistry and virus cultivation for detection of viral haemorrhagic septicaemia virus in experimentally infected rainbow trout *Oncorhynchus mykiss*. *Diseases of Aquatic Organisms* 20:101–109.

Follett, J. E., T. R. Meyers, T. O. Burton, and J. L. Geesin. 1997. Comparative susceptibilities of salmonid species in Alaska to infectious hematopoietic necrosis virus (IHNV) and North American viral hemorrhagic septicemia virus (VHSV). *Journal of Aquatic Animal Health* 9:34–40.

Frantsi, C., and M. Savan. 1971. Infectious pancreatic necrosis virus—temperature and age factors in mortality. *Journal of Wildlife Disease* 7:249–255.

Fryer, J. L., J. S. Rohovec, G. L. Tebbit, J. S. McMichael, and K. S. Pilcher. 1976. Vaccination for control of infectious diseases in Pacific salmon. *Fish Pathology* 10:155–164.

Georgetti, G. 1980. Rhabdoviruses in salmonids, infectious pancreatic necrosis in salmonids, spring viremia of carp. *Bulletin of the International Office of Epizoties* 93(9–10):1017–1024.

Ghittino, P. 1965. Viral hemorrhagic septicemia (VHS) in rainbow trout in Italy. *Annals of the New York Academy of Science* 126:468–478.

Gillmore, R. D., Jr., H. M. Engelking, D. S. Manning, and J. C. Leong. 1988. Expression in *Escherichia coli* of an epitope of the glycoprotein of infectious hematopoietic necrosis virus protects against viral challenge. *Bio/Technology* 6:295.

Gou, D. F., H. Kubota, M. Onuma, and H. Kodoma. 1991. Detection of salmonid herpesvirus (*Oncorhynchus masou* virus) in fish by southern-blot technique. *Medical Science* 53:43.

Grischkowsky, R. S., and D. F. Amend. 1976. Infectious hematopoietic necrosis virus: prevalence in certain Alaskan sockeye salmon, *Oncorhynchus nerka*. *Journal of Fisheries Research Board of Canada* 33:186–188.

Hattenberger-Baudouy, A. M., M. Danton, G. Merle, C. Torchy, and P. de Kinkelin. 1989. Serological evidence of infectious hematopoietic necrosis in rainbow trout from a French outbreak of disease. *Journal of Aquatic Animal Health* 1:126–134.

Hattore, M., H. Kodama, S. Ishiguro, A. Honda, T. Mikami, and H. Izawa. 1984. In vitro and in vivo detection of infectious pancreatic necrosis virus in fish by enzyme-linked immunosorbent assay. *American Journal of Veterinary Research* 45:1876–1879.

Hayashi, Y., H. Izawa, T. Mikami and H. Kodoma. 1993. A monoclonal antibody cross-reactive with three salmonid herpesviruses. *Journal of Fish Diseases* 16:479–486.

Hedrick, R. P., and J. L. Fryer. 1982. Persistent infections of salmonid cell lines with infectious pancreatic necrosis virus (IPNV): A model for the carrier state in trout. *Fish Pathology* 16:163–172.

Hedrick, R. P., and T. Sano. 1989. Herpesvirus of fishes. In: *Viruses of Lower Vertebrates*, edited by W. Ahne and E. Kurstak, 161–170. Berlin: Springer-Verlag.

Hedrick, R. P., W. D. Eaton, L. Chan, and W. Wingfield. 1986. Herpesvirus salmonis (HPV): first occurrence in anadromous salmonids. *Bulletin of the European Association of Fish Pathologist* 6:66–68.

Hedrick, R. P., T. McDowell, W. D. Eaton, T. Kimura, and T. Sano. 1987. Serological relationships of five herpesviruses isolated from salmonid fishes. *Journal of Applied Ichthyology* 3:87–92.

Hedrick, R. P., S. Yun, and W. H. Wingfield. 1991. A small RNA virus isolated from salmonid fishes in California, USA. *The Canadian Journal of Fisheries and Aquatic Sciences* 48:99–104.

Hedrick, R. P., S. E. LaPatra, S. Yun, K. A. Lauda, G. R. Jones, J. L. Congleton, and P. de Kinkelin. 1994. Induction of protection from infectious hematopoietic necrosis virus in rainbow trout *Oncorhynchus mykiss* by pre-exposure to the avirulent cutthroat trout virus (CTV). *Diseases of Aquatic Organisms* 20:111–118.

Helmick, C. M., J. F. Bailey, S. LaPatra, and S. Ristow. 1995a. Histological comparison of infectious hematopoietic necrosis virus challenged juvenile rainbow trout *Oncorhynchus mykiss* and coho salmon *O. kisutch* gill, esophagus/cardiac stomach region, small intestine and pyloric caeca. *Diseases of Aquatic Organisms* 23:175–187.

Helmick, C. M., J. F. Bailey, S. LaPatra, and S. Ristow. 1995b. The esophagus/cardiac stomach region: site of attachment and internalization of infectious hematopoietic necrosis virus in challenged juvenile rainbow trout *Oncorhynchus mykiss* and coho salmon *O. kisutch*. *Diseases of Aquatic Organisms* 23:189–199.

Hetrick, F. M., J. L. Fryer, and M. D. Knittel. 1979. Effect of water temperature on the infection of rainbow trout *Salmo gairdneri* Richardson with infectious haematopoietic necrosis virus. *Journal of Fish Diseases* 2:253–257.

Hill, B. J., and K. Way. 1988. Proposed standardization of the serological classification of aquatic Birnaviruses [abstract]. *International Fish Health Conference,* 151. Vancouver, B.C., Canada: Fish Health Section/American Fisheries Society.

Hill, B. J., and K. Way. 1995. Serological classification of infectious pancreatic necrosis (IPN) virus and other aquatic birnaviruses. *Annual Review of Fish Diseases* 5:55–77.

Hjeltnes, B., O. B. Samuelsen, and A. M. Svardal. 1992. Changes in plasma and liver glutathione levels in Atlantic salmon *Salmo salar* suffering from infectious salmon anemia (ISA). *Diseases of Aquatic Organisms* 14:31–33.

Holt, R., and J. Rohovec. 1984. Anemia of coho salmon in Oregon. *American Fish Society/Fish Health Section News Letter* 12:4.

Hooper, K. 1989. The isolation of VHSV from chinook salmon at Glenwood Springs, Orcas Island, Washington. *Fish Health Section/American Fisheries Society News Letter* 17:1.

Horlyck, V., S. Mellergard, E. Dalsgaard, and P. E. V. Jørgensen. 1984. Occurrence of VHS in Danish maricultured rainbow trout, *Bulletin of the European Association of Fish Pathologist* 4:11–13.

Hsu, Y-L., H. M. Engelking, and J. C. Leong. 1986. Occurrence of different types of infectious hematopoietic necrosis virus in fish. *Applied Environmental Microbiology* 52:1353–1361.

Hsu, Y-L., B-S. Chen and J. L. Wu. 1989. Characteristics of a new reolike virus isolated from landlocked salmon (*Oncorhynchus masou*). *Fish Pathology* 24:37–45.

Jarp, J., and E. Karlsen. 1997. Infectious salmon anaemia (ISA) risk factors in sea-cultured Atlantic salmon *Salmo salar. Diseases of Aquatic Organisms* 28:79–86.

Jarp, J., A. G. Gjevre, A. B. Olsen, and T. Bruheim. 1995. Risk factors for furunculosis, infectious pancreatic necrosis and mortality in post-smolt of Atlantic salmon, *Salmo salar* L. *Journal of Fish Diseases* 18:67–78.

Jarp, J., T. Taksdal, and B. Tørud. 1996. Infectious pancreatic necrosis in Atlantic salmon *Salmo salar* in relation to specific antibodies, smoltification, and infection with erythrocytic inclusion body syndrome (EIBS). *Diseases of Aquatic Organisms* 27:81–88.

Jensen, M. H. 1965. Research on the virus of Egtved disease. *Annals of the New York Academy of Science* 126:422–426.

Jiang, L. 1993. Advances in fish virology research in China. In: *Diseases in Asian Aquaculture II,* edited by M. Sharief, R. Subashinghe and T. Flegel, 211–225, Manila, Philippines: Asian Fisheries Society.

Johansen, L.-H., and A.-I. Sommer. 1995. Multiplication of infectious pancreatic necrosis virus (IPNV) in head kidney and blood leukocytes isolated from Atlantic salmon, *Salmo salar* L. *Journal of Fish Diseases* 18:147–156.

Jørgensen, P. E. V. 1973. Artificial transmission of viral haemorrhagic septicemia (VHS) of rainbow trout. *Rivista Italiana di Piscicoltura Ittiopatologia* 8:101–102.

Jørgensen, P. E. V. 1982. Egtved virus: occurrence in inapparent infections with virulent virus in free-living rainbow trout, *Salmo gairdneri* Richardson, at low temperature. *Journal of Fish Diseases* 5:251–255.

Jørgensen, P. E. V., N. J. Olesen, N. Lorenzen, J. R. Winton, and S. S. Ristow. 1991. Infectious hematopoietic necrosis (IHN) and viral hemorrhagic septicemia (VHS): detection of trout antibodies to the causative viruses by means of plaque neutralization, immunofluorescence and enzyme-linked immunosorbent assay. *Journal of Aquatic Animal Health* 3:100–108.

Kaastrup, P., V. Horlyuck, N. J. Olesen, N. Lorenzen, P. E. V. Jørgensen, and P. Berg. 1991. Paternal association of increased susceptibility to viral haemorrhagic septicemia (VHS) in rainbow trout (*Oncorhynchus mykiss*). *Canadian Journal of Fisheries and Aquatic Sciences* 48:1188–1192.

Kimura, T., and M. Yoshimizu. 1989. Salmon herpesvirus: OMV, *Oncorhynchus masou* virus. In: *Viruses of Lower Vertebrates*, edited by W. Ahne and E. Kurstak, 171–183. Berlin: Springer Verlag.

Kimura, T., M. Yoshimizu, and M. Tanaka. 1981a. Studies on a new virus (OMV) from *Oncorhynchus masou*–II. Oncogenic nature. *Fish Pathology* 15:149–153.

Kimura, T., M. Yoshimizu, and M. Tanaka. 1981b. Fish viruses: tumor induction in *Oncorhynchus keta* by the herpesvirus. In: *Phyletic Approaches to Cancer*, edited by C. Dawe, 59–68. Tokyo: Japanese Scientific Society Press.

Kimura, T., M. Yoshimizu, M. Tanaka, and H. Sannohe. 1981c. Studies on a new virus (OMV) from *Oncorhynchus masou*–I. Characteristics and pathogenicity. *Fish Pathology* 15:143–147.

Kimura, T., S. Suzuki, and M. Yoshimizu. 1983. In vitro antiviral effect of 9-(2-hydroxyethoxymethyl) guanine on the fish herpesvirus, *Oncorhynchus masou* virus (OMV). *Antiviral Research* 3:93–101.

Koren, C. W. R., and A. Nylund. 1997. Morphology and morphogenesis of infectious salmon anaemia virus replicating in the endothelium of Atlantic salmon *Salmo salar*. *Diseases of Aquatic Organisms* 29:99–109.

Kumagai, A., H. Noriyuki, Y. Sato, K. Takahashi, T. Sano, and H. Fukuda. 1995a. Application of fluorescent antibody technique on diagnosis of salmonid herpesvirus 2 infection in maricultured coho salmon. *Fish Pathology* 30:59–65.

Kumagai, A., K. Takahashi, and H. Fukuda. 1995b. Pathogenicity of salmonid herpesvirus 2 isolated from maricultured coho salmon to salmonids. *Fish Pathology* 30:215–220.

LaPatra, S. E., and J. L. Fryer. 1990. Susceptibility of brown trout (*Salmo trutta*) to infectious hematopoietic necrosis virus. *Bulletin of the European Association of Fish Pathologist* 10:125–127.

LaPatra, S. E., and R. P. Hedrick. 1995. Pathogenesis of selected aquareovirus isolates in rainbow trout. *American Fisheries Society/Fish Health Section News Letter* 23(2):3–4.

LaPatra, S. E., J. L. Fryer, W. H. Wingfield, and R. P. Hedrick. 1989a. Infectious hematopoietic necrosis virus (IHNV) in coho salmon. *Journal of Aquatic Animal Health* 1:277–280.

LaPatra, S. E., K. A. Robert, J. S. Rohovec, and J. L. Fryer. 1989b. Fluorescent antibody test for the rapid diagnosis of infectious hematopoietic necrosis. *Journal of Aquatic Animal Health* 1:29–36.

LaPatra, S. E., J. S. Rohovec, and J. L. Fryer. 1989c. Detection of infectious hematopoietic necrosis virus in fish mucus. *Fish Pathology* 24:197–202.

LaPatra, S. E., W. J. Groberg, J. S. Rohovec, and J. L. Fryer. 1990a. Size-related susceptibility of salmonids to two strains of infectious hematopoietic necrosis virus. *Transactions of the American Fisheries Society* 119:25–30.

LaPatra, S. E., J. M. Groff, J. L. Fryer, and R. P. Hedrick. 1990b. Comparative pathogenesis of three strains of infectious hematopoietic necrosis virus in rainbow trout *Oncorhynchus mykiss*. *Diseases of Aquatic Organisms* 8:105–112.

LaPatra, S. E., T. Turner, K. A. Lauda, G. R. Jones, and S. Walker. 1993. Characterization of the humoral response of rainbow trout to infectious hematopoietic necrosis virus. *Journal of Aquatic Animal Health* 5:165–171.

LaPatra, S. E., K. A. Lauda, and G. R. Jones. 1994a. Antigenic variants of infectious hematopoietic necrosis virus and implications for vaccine development. *Diseases of Aquatic Organisms* 20:119–126.

LaPatra, S. E., K. A. Lauda, G. R. Jones, S. C. Walker, and W. D. Shewmaker. 1994b. Development of passive immunotherapy for control of infectious hematopoietic necrosis. *Diseases of Aquatic Organisms* 20:1–6.

LaPatra, S. E., G. R. Jones, K. A. Lauda, T. S. McDowell, R. Schneider, and R. P. Hedrick. 1995. White sturgeon as a potential vector of infectious hematopoietic necrosis virus. *Journal of Aquatic Animal Health* 7:225–230.

Le Berre, M., P. de Kinkelin, and A. Metzger. 1977. Identification serologique des rhabdovirus de salmonides. *Bulletin of the Office of International Epizooties* 87(5–6):391–393.

Lecomte, L., M. Arella, and L. Berthiaume. 1992. Comparison of polyclonal and monoclonal antibodies for serotyping infectious pancreatic necrosis virus (IPN) strains isolated in eastern Canada. *Journal of Fish Diseases* 15:431–436.

Leek, S. L. 1987. Viral erythrocytic inclusion body syndrome (EIBS) occurring in juvenile spring chinook salmon (*Oncorhynchus tshawytscha*) reared in fresh water. *Canadian Journal of Fish and Aquatic Sciences* 44:685–688.

Lopez-Lastra, M., M. Gonzalez, M. Jashes, and A. M. Sandino. 1994. A detection method for infectious pancreatic necrosis virus (IPNV) based on reverse transcription (RT)-polymerase chain reaction (PCR). *Journal of Fish Diseases* 17:269–282.

Lorenzen, N., N. J. Olesen, and P. E. V. Jørgensen. 1988. Production and characterization of monoclonal antibodies to four Egtved virus structural proteins. *Diseases of Aquatic Organisms* 4:35–42.

Lunder, T., K. Thorud, T. T. Pappe, R. A. Holt, and J. S. Rohovec. 1990. Particles similar to the virus of erythrocytic inclusion body syndrome, EIBS, detected in Atlantic salmon (*Salmo salar*) in Norway. *Bulletin of the European Association of Fish Pathologists* 10:21–23.

Mangunwiryo, H., and C. Agius. 1988. Studies on the carrier state of infectious pancreatic necrosis virus infectious in rainbow trout, *Salmo gairdneri* Richardson. *Journal of Fish Diseases* 11:125–132.

Manning, S. D., and J. C. Leong. 1990. Expression in *Escherichia coli* of the large genomic segment of infectious pancreatic necrosis virus. *Virology* 179:16–25.

McAllister, K. W., and P. E. McAllister. 1988. Transmission of infectious pancreatic necrosis virus from carrier striped bass to brook trout. *Diseases of Aquatic Organisms* 4:101–104.

McAllister, P. E. 1993. Salmonid fish viruses. In: *Fish Medicine,* edited by M. K. Stoskopf, 380–408. Philadelphia: W. B. Saunders Co.

McAllister, P. E., and J. Bebak. 1997. Infectious pancreatic necrosis virus in the environment: relationship to effluent from aquaculture facilities. *Journal of Fish Diseases* 20:201–207.

McAllister, P. E., and R. L. Herman. 1989. Epizootic mortality in hatchery-reared lake trout *Salvelinus namaycush* caused by a putative virus possibly of the herpesvirus group. *Diseases of Aquatic Organisms* 6:113–119.

McAllister, P. E., and W. J. Owens. 1986. Infectious pancreatic necrosis virus: protocol for a standard challenge in brook trout. *Transactions of the American Fisheries Society* 115:466–470.

McAllister, P. E., and W. J. Owens. 1987. Identification of the three serotypes of viral hemorrhagic septicemia viruses by immunoblot assay using antiserum to serotype F1. *Bulletin of the European Association of Fish Pathologists* 7:90–92.

McAllister, P. E., and W. J. Owens. 1995. Assessment of the virulence of fish and molluscan isolates of infectious pancreatic necrosis virus for salmonid fish by challenge of brook trout, *Salvelinus fontinalis* (Mitchell). *Journal of Fish Diseases* 18:97–103.

McAllister, P. E., and X. Reyes. 1984. Infectious pancreatic necrosis virus: isolation from rainbow trout, *Salmo gairdneri* Richardson, imported into Chile. *Journal of Fish Diseases* 7:319–322.

McAllister, P. E., and W. B. Schill. 1986. Immunoblot assay: A rapid and sensitive method for identification of salmonid fish viruses. *Journal of Wildlife Diseases* 22:468–474.

McAllister, P. E., M. W. Newman, J. H. Sauber, and W. J. Owens. 1984. Isolation of infectious pancreatic necrosis virus (serotype Ab) from diverse species of estuarine fish. *Helgolander Wissenschaftliche Meeresuntersuchungen* 37:317–328.

McAllister, P. E., W. J. Owens, and T. M. Ruppenthal. 1987. Detection of infectious pancreatic necrosis virus in pelleted cell and particulate components from ovarian fluid of brook trout *Salvelinus fontinalis. Diseases of Aquatic Organisms* 2:235–237.

McAllister, P. E., W. J. Owens, and W. B. Schill. 1994. Effects of water temperature and immunosuppressant treatment on prevalence and titer of infectious pancreatic necrosis virus in naturally infected brook trout. *Journal of Aquatic Animal Health* 6:133–137.

McCain, B. B., J. L. Fryer, and K. S. Pilcher. 1971. Antigenic relationship in a group of three viruses of salmonid fish by cross neutralization. *Proceedings of the Society of Experimental Biology and Medicine* 317:1042–1046.

McIntyre, J. D., and D. J. Amend. 1978. Heritability of tolerance for infectious hematopoietic necrosis in sockeye salmon (*Oncorhynchus nerka*). *Transactions of the American Fisheries Society* 107:305–308.

McKnight, I. J., and R. J. Roberts. 1976. The pathology of infectious pancreatic necrosis. 1. The sequential histopathology of the naturally occurring condition. *British Veterinary Journal* 132:76–85.

Medina, D. J., P. W. Chang, T. M. Fradley, M. T. Yeh, and E. C. Sadasiv. 1992. Diagnosis of infectious hematopoietic necrosis virus in Atlantic salmon *Salmo salar* by enzyme-linked immunosorbent assay. *Diseases of Aquatic Organisms* 13:147–150.

Meier, Von W. 1985. Virale haemorrhagic septikaemie: empfanglichkeit un epizootiologische rolle des hechts (*Esox lucius* L.). *Journal of Applied Ichthyology* 4:171–177.

Meier, W., and P. E. V. Jørgensen. 1980. Isolation of VHS virus from pike fry (*Esox lucius*) with hemorrhagic symptoms. In: *Fish Diseases,* edited by W. Ahne.

Third COPRAQ-session, 8–17. Berlin: Springer-Verlag.

Meier, W., and T. Wahli. 1988. Viral haemorrhagic septicemia (VHS) in grayling (*Thymallus thymallus*) L. *Journal of Fish Diseases* 11:481–487.

Meier, W., W. Ahne, and P. E. V. Jørgensen. 1986. Fish viruses: viral haemorrhagic septicemia in white fish (*Coregonus* sp.). *Journal of Applied Ichthyology* 4:181–186.

Meyers, T. R., S. Short, K. Lipson, W. N. Batts, J. R. Winton, J. Wilcock, and E. Brown. 1994. Association of viral hemorrhagic septicemia virus with epizootic hemorrhages of the skin in Pacific herring *Clupea harengus* from Prince William Sound and Kodiak Island, Alaska, USA. *Diseases of Aquatic Organisms* 19:27–37.

Meyers, T. R., J. B. Thomas, J. E. Follett, and R. R. Saft. 1990. Infectious hematopoietic necrosis virus: trends in prevalence and the risk management approach in Alaskan sockeye salmon culture. *Journal of Aquatic Animal Health* 2:85–98.

M'Gonigle, R. H. 1941. Acute catarrhal enteritis of salmonid fingerlings. *Transactions of the American Fisheries Society* 70:297–303.

Mulcahy, D. 1984. Atlantic Salmon (*Salmo salar*) are naturally susceptible to infectious hematopoietic necrosis (IHN) virus. *U.S. Fish Wildlife Service Research Information Bulletin* no. 84-84, 2p.

Mulcahy, D., and W. N. Batts. 1987. Infectious hematopoietic necrosis virus detected by separation and incubation of cells from salmonid cavity fluid. *Canadian Journal of Fisheries and Aquatic Sciences* 44:1071–1075.

Mulcahy, D., and R. J. Pascho. 1984. Adsorption of fish sperm of vertically transmitted viruses. *Science* 225:333–335.

Mulcahy, D., and R. J. Pascho. 1985. Vertical transmission of infectious haematopoietic necrosis virus in sockeye salmon, *Oncorhynchus nerka* (Walbaum): isolation of virus from dead eggs and fry. *Journal of Fish Diseases* 8:393–396.

Mulcahy, D., C. K. Jenes, and R. Pascho. 1984. Appearance and quantification of infectious hematopoietic necrosis virus in female sockeye salmon (*Oncorhynchus nerka*) during their spawning migration. *Archives in Virology* 80:171–181.

Mulcahy, D., D. Klaybor, and W. N. Batts. 1990. Isolation of infectious hematopoietic necrosis virus from a leech (*Piscicola salmositica*) and a copepod (*Salmincola* sp.), ectoparasites of sockeye salmon *Oncorhynchus nerka*. *Diseases of Aquatic Organisms* 8:29–34.

Mulcahy, D., R. J. Pascho, and C. K. Jenes. 1983a. Titer distribution patterns of infectious haematopoietic necrosis virus in ovarian fluids of hatchery and feral salmon populations. *Journal of Fish Diseases* 6:183–188.

Mulcahy, D., R. J. Pascho, and C. K. Jenes. 1983b. Detection of infectious haematopoietic necrosis virus in river water and demonstration of waterborne transmission. *Journal of Fish Diseases* 6:321–330.

Mulcahy, D. M., and J. L. Fryer. 1976. Double infection of rainbow trout fry with IHN and IPN virus. *Fish Health News* 5:5–6.

Munro, A. L. S., A. E. Ellis, A. H. McVicar, and H. A. McLay. 1984. An exocrine pancreas disease of farmed Atlantic salmon in Scotland. *Helgoländer Meeresunters* 37:571–586.

Nelson, R. T., M. F. McLoughlin, H. M. Rwoley, M. A. Platten, and J. I. McCormick. 1995. Isolation of a toga-like virus from farmed Atlantic salmon *Salmo salar* with pancreas disease. *Diseases of Aquatic Organisms* 22:25–32.

Neukirch, M. 1984. An experimental study of the entry and multiplication of viral haemorrhagic septicemia virus in rainbow trout, *Salmo gairdneri* Richardson, after water-borne infection. *Journal of Fish Diseases* 7:231–234.

Nishimura, T., and 11 others. 1985. A trial vaccination against rainbow trout fry with formalin killed IHN virus. *Fish Pathology* 20:435–443.

Novoa, B. 1996. Immunogold technique applied to electron microscopy of infectious pancreatic necrosis virus (IPNV). *Fish Pathology* 31:141–143.

Novoa, B., and A. Fugueras. 1996. Heterogeneity of marine birnaviruses isolated from turbot (*Scophthalmus maximus*). *Fish Pathology* 31:145–150.

Nylund, A., and P. Jakobsen. 1995. Sea trout as a carrier of infectious salmon anaemia virus (ISAV). *Journal of Fish Biology* 47:174–176.

Nylund, A., T. Hovland, K. Hodneland, F. Nilsen, and P. Lrvik. 1994. Mechanisms for transmission of infectious salmon anaemia (ISA). *Diseases of Aquatic Organisms* 19:95–100.

Nylund, A., S. Alexandersen, and J. B. Rolland. 1995a. Infectious salmon anemia virus (ISAV) in brown trout. *Journal of Aquatic Animal Health* 7:236–240.

Nylund, A, T. Hovland, K. Watanabe, and C. Endresen. 1995b. Presence of infectious salmon anaemia virus (ISAV) in different organs of *Salmo salar* L. collected from three fish farms. *Journal of Fish Diseases* 18:135–145.

Oberg, L. A., J. Wirkula, D. Mourich, and J. C. Leong. 1991. Bacterially expressed nucleoprotein of infectious hematopoietic necrosis virus augments protective immunity induced by the glycoprotein vaccine in fish. *Journal of Virology* 65:4486–4489.

Okamoto, N., K. Takahashi, M. Maita, J. S. Rohovec, and Y. Ikeda. 1992. Erythrocytic inclusion body syndrome: susceptibility of selected sizes of salmonid fish. *Fish Pathology* 27:153–156.

Olesen, N. J., N. Lorenzen, and P. E.V. Jørgensen. 1991a.

Detection of rainbow trout antibody to Egtved virus by enzyme-linked immunosorbent assay (ELISA), immuno fluorescence (IF), and plaque neutralization tests (50 PNT). *Diseases of Aquatic Organisms* 10:31–38.

Olesen, N. J., N. Lorenzen, and P. E. V. Jørgensen. 1991b. Serological differentiation of Egtved virus (VHSV) using monoclonal and polyclonal antibodies [abstract]. *V. International EAFP Conference*. Budapest, Hungary: 79.

Olesen, N. J., and P. E. V. Jørgensen. 1991. Rapid detection of viral haemorrhagic septicemia virus in fish by ELISA. *Journal of Applied Ichthyology* 7:183–189.

Olsen, Y. A., K. Falk, and O. B. Reite. 1992. Cortisol and lactate levels in Atlantic salmon *Salmo salar* developing infectious anaemia (ISA). *Diseases of Aquatic Organisms* 14:99–104.

Ord, W., M. Le Berre, and P. de Kinkelin. 1976. Viral hemorrhagic septicemia; comparative susceptibility of rainbow trout (*Salmo gairdneri*) and hybrids (*S. gairdneri x Oncorhynchus kisutch*) to experimental infection. *Journal of the Fisheries. Research Board of Canada* 33:1205–1208.

Park, M. A., S. G. Sohn, S. D. Lee, S. K. Chun, J. W. Park, J. L. Fryer, and Y. C. Hah. 1993. Infectious haematopoietic necrosis virus from salmonids cultured in Korea. *Journal of Fish Diseases* 16:471–478.

Peters, F., and M. Neukirch. 1986. Transmission of some fish pathogenic viruses by the heron, *Ardea cinerea. Journal of Fish Diseases* 9:539–544.

Piacentini, J. S., J. S. Rohovec, and J. L. Fryer. 1989. Epizootiology of erythrocytic inclusion body syndrome. *Journal of Aquatic Animal Health* 1:173–179.

Pietsch, J. P., D. F. Amend, and C. M. Miller. 1977. Survival of infectious hematopoietic necrosis virus held under various environmental conditions. *Journal of the Fisheries Research Board of Canada* 34:1360–1364.

Pilcher, K. S., and J. L. Fryer 1980. The viral diseases of fish: a review through 1978. Part I: diseases of proven etiology. *Critical Reviews in Microbiology* 7:289–364.

Plumb, J. A. 1972. A virus-caused epizootic of rainbow trout (*Salmo gairdneri*) in Minnesota. *Transactions of the American Fisheries Society* 101:121–123.

Rasmussen, C. J. 1965. A biological study of the Egtved disease (IHNV). *Annals New York Academy of Science* 126:427–460.

Ristow, S. S., and J. de Avila. 1994. Susceptibility of four new salmonid cell lines to infectious hematopoietic necrosis virus. *Journal of Aquatic Animal Health* 6:260–265.

Ristow, S. S., N. Lorenzen, and P. E. V. Jørgensen. 1991. Monoclonal-antibody-based immunodot assay distinguishes between viral hemorrhagic septicemia virus (VHSV) and infectious hematopoietic necrosis virus (IHNV). *Journal of Aquatic Animal Health.* 3:176–180.

Ristow, S. S., J. de Avila, S. E. LaPatra, and K. Lauda. 1993. Detection and characterization of rainbow trout antibody against infectious hematopoietic necrosis virus. *Diseases of Aquatic Organisms* 15:109–114.

Roberts, R. J., and I. J. McKnight. 1976. The pathology of infectious pancreatic necrosis. 2. Stress-mediated recurrence. *British Veterinary Journal* 132:209–214.

Rodger, H. D., E. M. Dinan, T. M. Murphy, and T. Lunder. 1991. Observation of erythrocytic inclusion body in Ireland. *Bulletin of the European Association of Fish Pathologist* 11:108–111.

Rodriguez, S., M. Pilar Vilas, and S. I. Pérez. 1994. Prevalence of infectious pancreatic necrosis virus on salmonid fish farms in Spain. *Journal of Aquatic Animal Health* 6:138–143.

Rucker, R. R., W. J. Whipple, J. R. Parvin, and C. A. Evans. 1953. A contagious disease of salmon possibly of virus origin. *U.S. Fish Wildlife Service Fisheries Bulletin* no. 76, 54:35–46.

Saint-Jean, S. R., S. E. P. Prieto, and M. P. V. Minondo. 1993. Flow cytometric analysis of infectious pancreatic necrosis virus attachment to fish sperm. *Diseases of Aquatic Organisms* 15:153–156.

Salmonid importation regulations: fish health, U. S. Fish & Wildlife Service Division of Fish Hatcheries. 1993. *Federal Register* 58, no. 213.

Sano, T. 1971. Studies on viral diseases of Japanese fishes–II. Infectious pancreatic necrosis of rainbow trout: pathogenicity of the isolates. *Bulletin of the Japanese Society of Scientific Fisheries* 37:499–503.

Sano, T. 1976. Viral diseases of cultured fishes in Japan. *Fish Pathology* 10:221–226.

Sano, T. 1988. Characterization, pathogenicity, and oncogenicity of herpesvirus in fish [abstract]. In: *American Fisheries Society, Fish Health Section International Fish Health Conference,* Vancouver, B.C., Canada: 157.

Sano, T. 1995. Viruses and viral diseases of salmonids. *Aquaculture* 132:43–52.

Sano, T., T. Nishimura, N. Okamoto, T. Yamazaki, J. Hanada, and Y. Watanabe. 1977. Studies on viral diseases of Japanese fishes: VI. Infectious hematopoietic necrosis (IHN) of salmonids in the mainland Japan. *Journal of Tokyo University of Fisheries* 63:81–85.

Sano, T., N. Okamoto, and T. Nishimura. 1981. A new viral epizootic of *Anguilla japonica* Temminck and Schlegel. *Journal of Fish Diseases* 4:127–139.

Sano, T., H. Fukuda, N. Okamoto, and F. Kaneko. 1983. Yamame tumor virus: lethality and oncogenicity. *Bulletin of the Japanese Society of Scientific Fisheries* 49:1159–1163.

Schutz, M., E. B. May, J. N. Kraeuter, and F. M. Hetrick. 1984. Isolation of infectious pancreatic necrosis virus from an epizootic occurring in cultured striped bass, *Morone saxatilis* (Walbaum). *Journal of Fish Diseases* 7:505–507.

Shankar, K. M., and T. Yamamoto. 1994. Prevalence and pathogenicity of infectious pancreatic necrosis virus (IPNV) associated with feral lake trout, *Salvelinus namaycush* (Walbaum). *Journal of Fish Diseases* 17:461–470.

Smail, D. A., D. W. Bruno, G. Dear, L. A. McFarlane, and K. Ross. 1992. Infectious pancreatic necrosis (IPN) virus Sp. Serotype in farmed Atlantic salmon *Salmo salar* L., post-smolts associated with mortality and clinical disease. *Journal of Fish Diseases* 15:77–83.

Snieszko, S. F., K. Wolf, J. E. Camper, and L. J. Pettijohn. 1959. Infectious nature of pancreatic necrosis. *Transactions of the American Fisheries Society* 88:289–293.

Sommer, A. I., and S. Menner. 1996. Propagation of infectious salmon anaemia virus in Atlantic salmon, *Salmo salar* L., head kidney macrophages. *Journal of Fish Diseases* 19:179–183.

Stangeland, K., S. Hoie, and T. Taksdal. 1996. Experimental induction of infectious pancreatic necrosis in Atlantic salmon, *Salmo salar* L., post-smolts. *Journal of Fish Diseases* 19:323–327.

Stephens, E. B., M. W. Newman, A. L. Zachary, and F. M. Hetrick. 1980. A viral aetiology for the annual spring epizootics of Atlantic menhaden *Brevoortia tyrannus* (Latrobe) in Chesapeake Bay. *Journal of Fish Diseases* 3:387–398.

Takahashi, R., N. Okamoto, M. Maita, J. S. Rohovec, and Y. Ikeda, 1992a. Progression of erythrocytic inclusion body syndrome in artificially infected coho salmon. *Fish Pathology* 27:89–95.

Takahashi, K., N. Okamoto, A. Kumagai, Y. Maita, and J. S. Rohovec. 1992b. Epizootics of erythrocytic inclusion body syndrome in coho salmon in seawater in Japan. *Journal of Aquatic Animal Health* 4:174–181.

Taksdal, T. K. Stangeland, and B. H. Dannevig. 1997. Induction of infectious pancreatic necrosis (IPN) in Atlantic salmon Salmo salar and brook trout *Salvelinus fontinalis* by bath challenge of fry with infectious pancreatic necrosis virus (IPNV) serotype sp. *Diseases of Aquatic Organisms* 28:39–44.

Tanaka, M., M. Yoshimizu, and T. Kimura. 1984. *Oncorhynchus masou* virus: Pathological changes in masu salmon (*Oncorhynchus masou*), chum salmon (*O. keta*) and coho salmon (*O. kisutch*) fry infected with OMV by immersion method. *Bulletin of the Japanese Society of Scientific Fisheries* 50:431–437.

Tanaka, M., M. Yoshimizu, and T. Kimura. 1987. Ultrastructures of OMV infected RTG-2 cells and hepatocytes of chum salmon *Oncorhynchus keta*. *Nippon Suisan Gakkaishi* 53:47–55.

Tanaka, M., M. Yoshimizu, and T. Kimura. 1992. Herpesvirus infection of salmonid fish. In: *Salmonid Diseases,* edited by T. Kimura, 111–117. Sapporo, Japan: Hokkaido University Press.

Thoeson, J. C. 1994. *Bluebook; Suggested procedures for the detection of certain finfish and shellfish pathogens, Fourth Edition.* Bethesda, MD: Fish Health Section/American Fisheries Society.

Thorud, K., and H. O. Djupvik. 1988. Infectious anaemia in Atlantic salmon (*Salmo salar* L.). *Bulletin of the European Association of Fish Pathologist.* 8:109–111.

Traxler, G. S., and J. B. Rankin. 1989. An infectious hematopoietic necrosis epizootic in sockeye salmon *Oncorhynchus nerka* in Weaver Creek Spawning channel, Fraser River system, B. C., Canada. *Diseases of Aquatic Organisms* 6:221–226.

Traxler, G. S., J. Roome, K. A. Lauda, and S. LaPatra. 1997. Appearance of infectious hematopoietic necrosis virus (IHNV) and neutralizing antibodies in sockeye salmon *Oncorhynchus nerka* during their migration and maturation period. *Diseases of Aquatic Organisms* 28:31–38.

Wang, W-S., Y-L. Wi, and J-S. Lee. 1997. Single-tube, non-interrupted reverse transcription PCR for detection of infectious pancreatic necrosis virus. *Diseases of Aquatic Organisms* 28:229–233.

Watson, S. W., R. W. Guenther, and R. P. Rucker. 1954. A virus disease of sockeye salmon: interim report. *U. S. Fish Wildlife Service Special Scientific Report* no. 138:1–36.

Wechsler, S. J., C. L. Schuts, P. E. McAllister, E. B. May, and F. M. Hetrick. 1986. Infectious pancreatic necrosis virus in striped bass *Morone saxatilis*: experimental infection of fry and fingerlings. *Diseases of Aquatic Organisms* 1:203.

Williams, I. V., and D. F. Amend. 1976. A natural epizootic of infectious hematopoietic necrosis in fry of sockeye salmon (*Oncorhynchus nerka*) at Chilko Lake, British Columbia. *Journal of the Fisheries Research Board of Canada* 33:1564–1567.

Wingfield, W. H., and L. D. Chan. 1970. Studies on the Sacramento River chinook disease and its causative agent. In: A *Symposium on Diseases of Fishes and Shellfishes,* edited by S. F. Snieszko, 307–318. American Fisheries Society special publication no. 5.Washington, DC: American Fisheries Society.

Winton, J. R., C. K. Arakawa, C. N. Lannan, and J. L. Fryer. 1988. Neutralizing monoclonal antibodies recognize antigenic variants among isolates of infectious hematopoietic necrosis virus. *Diseases of Aquatic Organisms* 4:199–204.

Winton, J. R., W. N. Batts, R. E. Deering, M. Brunson, K. Hooper, T. Nislizarea, and C. Stehr. 1991. Characteristics of the first North American isolates of viral hemorrhagic septicemia virus [abstract]. *Proceedings of the Second International Symposium on Viruses of Lower Vertebrates* 43. Corvallis, OR: Oregon State University.

Wolf, K. 1988. *Fish Viruses and Fish Viral Diseases.* Ithaca, NY: Cornell University Press.

Wolf, K., and C. E. Smith. 1981. Herpesvirus salmonis: pathological changes in parenterally infected rainbow trout, *Salmo qairdneri* Richardson, fry. *Journal of Fish Diseases* 4:445–457.

Wolf, K., and W. G. Taylor. 1975. Salmonid viruses: a syncytium-forming agent from rainbow trout. *Fish Health News* 4:3–4.

Wolf, K., S. F. Snieszko, C. E. Dunbar, and E. Pyle. 1960. Virus nature of infectious pancreatic necrosis in trout. *Society of Experimental and Biology and Medicine* 104:105–108.

Wolf K., M. C. Quimby, and A. D. Bradford. 1963. Egg-associated transmission of IPN virus of trouts. *Virology* 21:317–321.

Wolf, K., M. C. Quimby, C. P. Carlson, and G. L. Bullock. 1967. Infectious pancreatic necrosis: selection of virus-free stock from a population of carrier trout. *Journal of Fisheries Research Board of Canada* 25:383–391.

Wolf, K., M. C. Quimby, and C. P. Carlson. 1969. Infectious pancreatic necrosis virus: lyophilization and subsequent stability in storage at 4C. *Applied Microbiology* 17:623–624.

Wolf, R., R. W. Darlington, W. G. Taylor, M. C. Quimby, and Y. Wakabayashi. 1978. Herpesvirus salmonis: characterization of a new pathogen of rainbow trout. *Journal of Virology* 27:659–666.

Wood, E. M., S. F. Snieszko, and W. T. Yasutake. 1955. Infectious pancreatic necrosis in brook trout. *American Medical Association Archives of Pathology* 60:26–28.

Yamamoto, T. 1975a. Infectious pancreatic necrosis virus and bacterial kidney disease appearing concurrently in populations of *Salmo gairdneri* and *Salvelinus fontinalis. Journal of the Fisheries Research Board of Canada* 32:92–95.

Yamamoto, T. 1975b. Frequency of detection and survival of infectious pancreatic necrosis virus in a carrier population of brook trout (*Salvelinus fontinalis*) in a lake. *Journal of the Fisheries Research Board of Canada* 32:568–570.

Yamamoto, T., and T. J. Clermont. 1990. Multiplication of infectious hematopoietic necrosis virus in rainbow trout following immersion infection: organ assay and electron microscopy. *Journal of Aquatic Animal Health* 2:261–270.

Yamamoto, T., W. N. Batts, C. K. Arakawa, and J. R. Winton. 1990. Multiplication of infectious hematopoietic necrosis virus in rainbow trout following immersion infection: whole-body assay and immunohistochemistry. *Journal of Aquatic Animal Health* 2: 271–280.

Yasutake, W. T. 1975. Fish viral diseases: clinical, histopathological, and comparative aspects. In: *The Pathology of Fishes,* edited by W. E. Ribelin and G. Migaki, 247–271. Madison, WI: University of Wisconsin Press.

Yoshimizu, M., H. Takizawa, and T. Kimura. 1986. U. V. susceptibility of some fish pathogenic viruses. *Fish Pathology* 21:47–52.

Yoshimizu, M., M. Tanaka, and T. Kimura. 1987. *Oncorhynchus masou* virus (OMV): incidence of tumor development among experimentally infected representative salmonid species. *Fish Pathology* 22:7–10.

Yoshimizu, M., M. Ramei, S. Dirakbusarakum, and T. Kimura. 1988a. Fish cell lines: susceptibility to salmonid viruses. In: *Invertebrate and Fish Tissue Culture,* edited by Y. Kuroda, E. Kurstak, and K. Maramorosch, 207–210. Tokyo: Japan Scientific Societies Press.

Yoshimizu, M., M. Tanaka, and T. Kimura. 1988b. Histopathological study of tumors induced by *Oncorhynchus masou* virus (OMV) infection. *Fish Pathology* 23:133–138.

Yoshimizu, M., M. Sami, and T. Kimura. 1989. Survivability of infectious hematopoietic necrosis virus in fertilized eggs of masu and chum salmon. *Journal of Aquatic Animal Health* 1:13–20.

Zhang, Y., and J. L. Congleton. 1994. Detection of infectious hematopoietic necrosis (IHN) virus in rearing units for steelhead before and during IHN epizootics. *Journal of Aquatic Animal Health* 6:281–287.

Zwellenberg, L. O., M. H. Jensen, and H. L. Zwellenberg. 1965. Electron microscopy of the virus of viral haemorrhagic septicemia of rainbow trout (Egtved virus). *Archives Virusforsch* 17:1–19.

11 Sturgeon Viruses

Intensive culture of white sturgeon is steadily increasing in northern California and possibly other locations in the United States as well as northern Italy and eastern Europe. As a result of more-intensive culture, infectious diseases that affect the species are beginning to emerge. To date, four viruses have been associated with mortality in cultured, and to some extent, wild, white sturgeon populations: white sturgeon adenovirus (WSAV), white sturgeon iridovirus (WSIV), and two white sturgeon herpesviruses (WSHV-l and WSHV-2).

WHITE STURGEON ADENOVIRUS

White sturgeon adenovirus was identified in diseased juvenile white sturgeon between 1984 and 1986 (Hedrick et al. 1985), but since that time, no serious disease problems involving the virus have been reported.

Geographical Range and Species Susceptibility

The virus was initially observed in 0.5-g sturgeon at a farm in northern California.

Clinical Signs and Findings

Affected fish appear lethargic, are anorexic and emaciated, have pale livers, and their intestines are void of food. No epidermal lesions have been described.

Diagnosis

The virus is known only from electron microscopy of infected fish tissue that has large numbers of electron dense hexagonal virions in the cell nuclei.

Virus Characteristics

The virions average 74 nm in diameter. Methyl green pyronine stained intestinal tissue from infected fish showed nuclear inclusions that contained DNA. Hedrick et al. (1985) tentatively placed the virus in the family Adenoviridae on the basis of DNA genome, virus morphology, and absence of an envelope. Attempts to isolate the virus in white sturgeon spleen (WSS-l) and white sturgeon heart (WSH-l) cell cultures were unsuccessful.

Epizootiology

The initial WSAV epizootic was not explosive, but accumulative mortality approached 50% during 4 months. Transmission to noninfected sturgeon was only partially successful. Intraperitoneal injections of filtrates into naive sturgeon resulted in enlarged cell nuclei in the gut epithelium, but reaction was not as severe as that observed in naturally infected fish. Experimentally infected sturgeon were larger than those with naturally occurring infection, which may have affected results.

Pathological Manifestations

Histologically, epithelial cells lining the straight intestine and spiral valve exhibited nuclear hypertrophy. Nuclei of some infected cells (up to one third of the cells in some fish) were 5 times larger than those in noninfected cells. Infected cells continued to enlarge until they ruptured and cell content was released into the gut lumen.

Significance

When WSAV was first isolated and diagnosed, the disease caused significant mortality in that particular population, but whether it will adversely affect future sturgeon seed production is unknown.

WHITE STURGEON IRIDOVIRUS

White sturgeon iridovirus was first isolated in 1988 from cultured juvenile white sturgeon at several fish farms in northern California (Hedrick et al. 1990).

Geographical Range and Species Susceptibility

The disease has been observed in Oregon, Idaho, and all California white sturgeon farms where adequate numbers of fish were assayed for virus (R. P. Hedrick and W. H. Wingfield, 1992, unpublished). The virus has been isolated from juvenile sturgeon that ranged in length from 7 to 46 cm.

Clinical Signs and Findings

Affected fish appear weak, experience weight loss, cease swimming, and drop to the tank bottom, where they die. Gills are pale and display hyperplasia and necrosis of the pillar cells lining the lamellar vascular channels, and some petechia are present. Internally, there is no body fat, livers are pale, and the gut is void of food.

Diagnosis

White sturgeon iridovirus has been diagnosed by electron microscopy and by isolation in cell culture. Hedrick et al. (1992) reported isolation of WSIV in white sturgeon spleen (WSS-2) cells, where it induced cell enlargement and slow but progressive cell degeneration. Virus replication occurred at 10, 15, and 20°C (optimum temperature) but not at 5 or 25°C.

Virus Characteristics

Electron microscopy of infected degenerating gill tissue shows abundant numbers of icosahedral virions in the cell cytoplasm that measure about 262 nm. White sturgeon iridovirus has been placed in the iridoviruslike group because of its DNA genome and large size.

Epizootiology

White sturgeon iridovirus is highly virulent. In the original epizootic, one farm lost 95% of 200,000 juvenile sturgeon in 4 months (Hedrick et al. 1990). Initial clinical signs of disease began 10 days after experimental exposure to the virus and produced 80% mortality in 50 days at 15°C (Hedrick et al. 1992). No deaths occurred in experimentally infected channel catfish, striped bass, or chinook salmon; however, lake sturgeon did suffer low mortality.

Pathological Manifestations

Histologically, gills and skin have numerous hypertrophic and occasionally basophilic cells with swollen nuclei in the epithelium and epidermis.

Significance

Because of its pathogenic potential, WSIV could be a serious threat to cultured white sturgeon. There is also the possibility that it may appear in other areas where white sturgeon have been introduced.

WHITE STURGEON HERPESVIRUS

Two herpesviruses have been isolated from white sturgeon; white sturgeon herpesvirus-l (WSHV-l) in 1989 (Hedrick et al. 1991a) and white sturgeon herpesvirus-2 (WSHV-2) a year later (Hedrick and Wingfield unpublished).

Geographical Range and Species Susceptibility

White sturgeon herpesvirus-1 was first reported in California in sturgeon less than 10 cm in length

(Hedrick et al. 1991a). White sturgeon herpesvirus-2 was found on commercial farms in California and in wild sturgeon in Oregon that were trapped in the Columbia River below Bonneville Dam (Engelking and Kaufman 1996).

Clinical Signs and Findings

There are no specific external clinical signs associated with either herpesvirus disease. Fish continue to feed until death. Internally, the stomach and intestines are filled with fluid, but other tissues appear normal. Affected wild white sturgeon became listless and appeared to have stopped eating.

Diagnosis

White sturgeon herpesvirus-1 and WSHV-2 were isolated from epithelial tissue only in white sturgeon skin (WSSK-1) cells where syncytia were detected 3 days after inoculation (Hedrick et al. 1991a). Cytopathic effect of WSHV-2 consists of grapelike clusters at the foci of plaques with heteromorphic and fused cells that eventually float off the culture substrate. Total CPE occurred 5 to 7 days after inoculation at 15°C. Other susceptible cell lines are WSS and white sturgeon liver (WSL) cells. Virus identical to that seen in cells of infected fish was observed in inoculated WSSK-1 cells, where it grew at 10, 15 and 20°C but not at 5 or 25°C. Antiserum to WSHV-1 can be used to distinguish WSHV-l from WSHV-2 by serum neutralization.

Virus Characteristics

Mature particles were hexagonal, enveloped in cytoplasmic vacuoles, and possessed all morphological characteristics of a herpesvirus. The WSHV-1 virion nucleocapsids, 110 nm in diameter, were found within the nucleus and cytoplasm of infected cells.

Epizootiology

The original sturgeon population from which WSHV-1 was isolated suffered 97% mortality after being moved into a wet laboratory. However, the disease process was complicated by the presence of columnaris. Experimental virus transmission from cell cultures resulted in 35% mortality after sturgeon were exposed by immersion to $10^{5.3}$ $TCID_{50}$ per mL of water for 30 minutes at 15°C. Virus was recovered for 2 weeks postinfection but was not isolated at 4 weeks.

White sturgeon herpesvirus-2 (Hedrick and Wingfield 1991b) was again isolated from internal organs of captive adult brood stock and later from juvenile fish on commercial farms. In the fall of 1994, an apparent WSHV-2 was isolated from approximately 2-year-old wild white sturgeon (Engelking and Kaufman 1996). The virus isolated from these fish was completely neutralized by WSHV-2 antisera, confirming a second isolation of that virus. White sturgeon herpesvirus-2 appears to be more pathogenic under experimental conditions than does WSHV-l and can be found in both skin and internal organs. It is more lytic in cell culture and replicates in several white sturgeon cell lines.

Pathological Manifestations

During histopathological examination, H & E stained tissue exhibited focal or diffused dermatitis, and epidermal lesions were characterized by intercellular edema. Hydropic degeneration and hypertrophy of malphigian cells with loss of intercellular junctions were common. Chromatin margination was associated with flocculent nonmembrane-bound intranuclear inclusion.

Significance

Virus diseases that affect white sturgeon have to be considered significant. The fish has a relatively short history of culture, and already, four viruses that appear specific for the species have surfaced.

MANAGEMENT OF STURGEON VIRUSES

To date, no management procedures have been established to control virus diseases in cultured white sturgeon, although avoidance of infected populations is always a prudent consideration. It was noted by LaPatra et al. (1995) that white sturgeon could also be vectors for infectious hematopoietic necrosis virus (IHNV) and should be handled accordingly.

REFERENCES

Engelking, H. M., and J. Kaufman. 1996. White sturgeon viruses isolated from Columbia River white sturgeon.

American Fisheries Society/Fish Health Section News Letter 24(3):4–5.

Hedrick, R. P., J. Spears, M. L. Kent, and T. McDowell. 1985. Adenovirus-like particles associated with a disease of cultured white sturgeon, *Acipenser transmontanus. Canadian Journal of Fisheries and Aquatic Sciences* 42:1321–1325.

Hedrick, R. P., J. M. Groff, T. McDowell, and W. H. Wingfield. 1990. An iridovirus infection of the integument of the white sturgeon *Acipenser transmontanus. Diseases of Aquatic Organisms* 8:39–44.

Hedrick, R. P., T. S. McDowell, J. M. Groff, S. Yun, and W. H. Wingfield. 1991a. Isolation of an epitheliotropic herpesvirus from white sturgeon *Acipenser transmontanus. Diseases of Aquatic Organisms* 11:49–56.

Hedrick, R. P., T. S. McDowell, J. M. Groff, and S. Yun. 1991b. Characteristics of two viruses isolated from white sturgeon *Acipenser transmontanus.* In: *Proceedings Second International Symposium on Viruses of Lower Vertebrates,* 165–174. Corvallis, OR: Oregon State University.

Hedrick, R. P., T. S. McDowell, J. M. Groff, S. Yun, and W. H. Wingfield. 1992. Isolation and some properties of an iridovirus-like agent from white sturgeon *Acipenser transmontanus. Diseases of Aquatic Organisms* 12:75–81.

LaPatra, S. E., G. R. Jones, K. A. Lauda, T. S. McDowell, R. Schneider, and R. P. Hedrick. 1995. White sturgeon as a potential vector of infectious hematopoietic necrosis virus. *Journal of Aquatic Animal Health* 7:225–230.

12 Walleye Viruses

Walleye are not widely cultured; however, their commercialization in the northern United States merits discussion of diseases that affect them. Known viral diseases of walleye are epidermal hyperplasia, diffuse epidermal hyperplasia (walleye herpesvirus), walleye dermal sarcoma, and lymphocystis (see Figure 12.1 and Table 12.1) (Yamamoto et al. 1985). Lymphocystis is discussed elsewhere (Chapter 13) and is referred to in this section only for comparative purposes.

Epidermal growths of fish and their relationship to environmental carcinogens and virus-caused tumors have received a great deal of attention in recent years. There is no evidence, however, that fish tumors, or viruses associated with them, are harmful to humans. Viruses and associated lesions that affect walleye and other fishes have generally received little attention from traditional fish pathologists because of a lack of mortality involved; however, they are of interest to scientists who are concerned with environmental influences on tumors in general.

EPIDERMAL HYPERPLASIA

Epidermal hyperplasia is synonymous with walleye epidermal hyperplasia and discrete epidermal hyperplasia (Wolf 1988). Walker (1969) was the first to suspect that the disease was caused by a virus, and 15 years later, Yamamoto et al. (1985) described C-type retrovirus particles in electron micrographs of the lesions.

Geographical Range and Species Susceptibility

Epidermal hyperplasia has been found only in adult walleye in Saskatchewan and Manitoba, Canada, and Lake Oneida, New York. Walleye is the only known susceptible species.

Clinical Signs and Findings

Epidermal hyperplasia lesions are gently raised, clear, translucent, mucoidlike patches and are more discrete and less granular than lymphocystis or dermal sarcoma lesions (Yamamoto et al. 1985) (Figure 12.1). The more or less circular lesions appear singularly or in groups and can cover large areas of the body and/or fins, particularly fin margins. Lesion size varies in diameter from a few millimeters to several centimeters and are 1–2 mm in height. Demarcation between normal epidermis and tumor is sharply defined as opposed to the more diffused, flatter nature of herpesvirus hyperplasia.

Diagnosis

Epidermal hyperplasia is confined to the epidermis, where growths are distinct from those lesions associated with diffuse epidermal hyperplasia, dermal sarcoma, and lymphocystis (Table 12.1) (Yamamoto et al. 1985). It is possible to make a presumptive field diagnosis; however, confirmation should be made by histological procedures and/or electron microscopy.

Virus Characteristics

The virus, which has not yet been isolated in tissue culture, is a C-type, irregular-shaped retrovirus and measures approximately 120 nm in diameter.

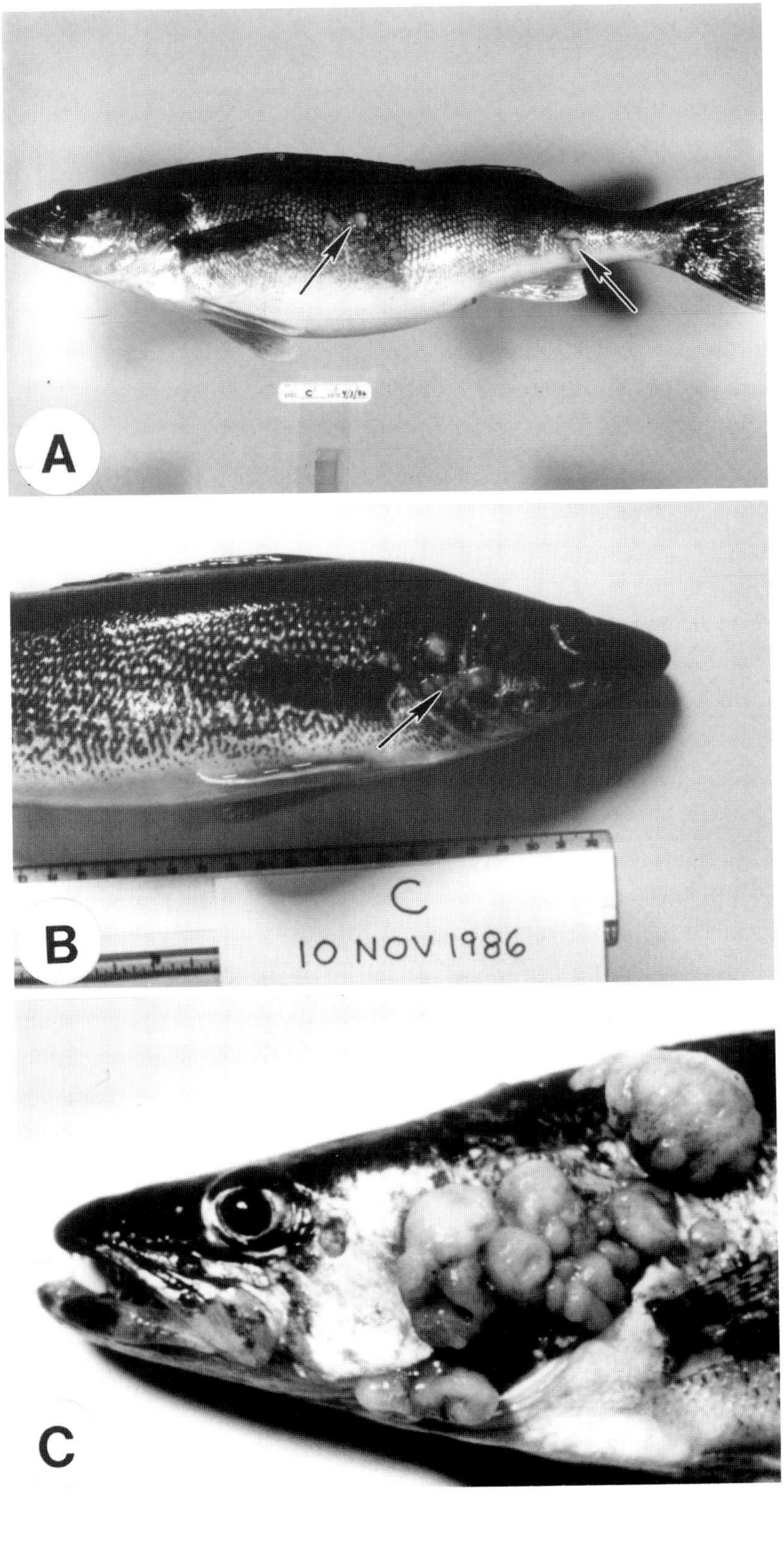

FIGURE 12.1. *Skin virus diseases of walleyes.* (A) *Walleye dermal sarcoma* (arrows). (B) *Diffuse epidermal hyperplasia* (arrow). (C) *Lymphocystis disease. (Photographs courtesy of P. R. Bowser.)*

Epizootiology

Epidermal hyperplasia appears to have little deleterious effect on infected walleye. In New York, the incidence of disease was reported to be about 5% in walleye populations, but infection rates of 0% to 20% were noted in a survey of nine Canadian walleye populations (Yamamoto et al. 1985). When the same Canadian populations were surveyed in spring (spawning season) and autumn, epidermal hyperplasia was found only in spring. Some fish were infected with lymphocystis or dermal sarcoma as well as epidermal hyperplasia. Yamamoto et al. (1985) found two types of epidermal hyperplasia on Canadian walleye, one involving the superficial epidermal hyperplasia lesion containing retrovirus and a more diffuse growth (diffuse epidermal hyperplasia).

Pathological Manifestations

Histopathologically, discrete epidermal hyperplasia lesions contain predominantly cuboidal cells with mucous cells at the surface (Figure 12.2). Lesions do not extend below the basal lamina (Yamamoto et al. 1985). Viral particles are randomly distributed in microvillilike extensions of the cell. Virions are not found in the cytoplasmic matrix but occur at the cell periphery.

Significance

Owing to a lack of mortality associated with epidermal hyperplasia, its significance is considered minimal.

DIFFUSE EPIDERMAL HYPERPLASIA

In central Canada, a herpesvirus was isolated from diffuse epidermal hyperplasia lesions on the skin of spawning walleye (Kelly et al. 1980). After further characterization, the virus was named *Herpesvirus vitreum* or walleye herpesvirus (Kelly et al. 1983). The herpesvirus infection was named "diffuse epidermal hyperplasia" by Yamamoto et al. (1985) because of the flat, diffused appearance of the skin lesion.

Geographical Range and Species Susceptibility

Diffuse epidermal hyperplasia has been found only in walleye in western Canada.

Clinical Signs and Findings

Walleye herpesvirus causes epidermal lesions that resemble thick areas of slime (Figure 12.1). The growths are flat and translucent with soft, swollen underlying tissue (Yamamoto et al. 1985). These lesions can measure several centimeters in diameter, spread laterally on the surface of the fish, and are not as pronounced as lymphocystis or dermal sarcoma. The transient lesions appear during spawning but have no apparent ill effects on the host.

Diagnosis

Diffuse epidermal hyperplasia (walleye herpesvirus) is diagnosed by lesion recognition, isolation of virus from lesions in walleye cell lines, and histological sectioning of diseased tissue.

Virus Characteristics

Walleye herpesvirus is icosahedral, measures 200 nm in diameter, presumably with an envelope, and has a DNA genome. It replicates in walleye ovary (WO), and walleye embryo (We-2) cells but not in CHSE-214 (chinook salmon embryo cells), BB, fathead minnow (FHM), or RTG-2 (rainbow trout gonad) cells. The primary cytopathic effect (CPE) consists of syncytium formation followed by cell lysis. Walleye herpesvirus replication occurs at 4 to 15°C in cell cultures; maximum replication occurs at 15°C; and no replication takes place at 20°C. At 15°C, the maximum titer in cell cultures is 10^5 PFU/0.1 mL after 10 to 13 days of incubation.

Epizootiology

In western Canada, diffuse epidermal hyperplasia is found in spawning walleye during early spring, but lesions disappear by late spring and early summer. Little is known about the effects of the virus or its associated lesions.

Pathological Manifestations

Cells in diffuse epidermal hyperplasia lesions are disorganized and have slightly enlarged nuclei that contain granular inclusions (Figure 12.2). Electron microscopy shows typical herpesvirus particles within the nucleus and mature enveloped virions near the cell periphery.

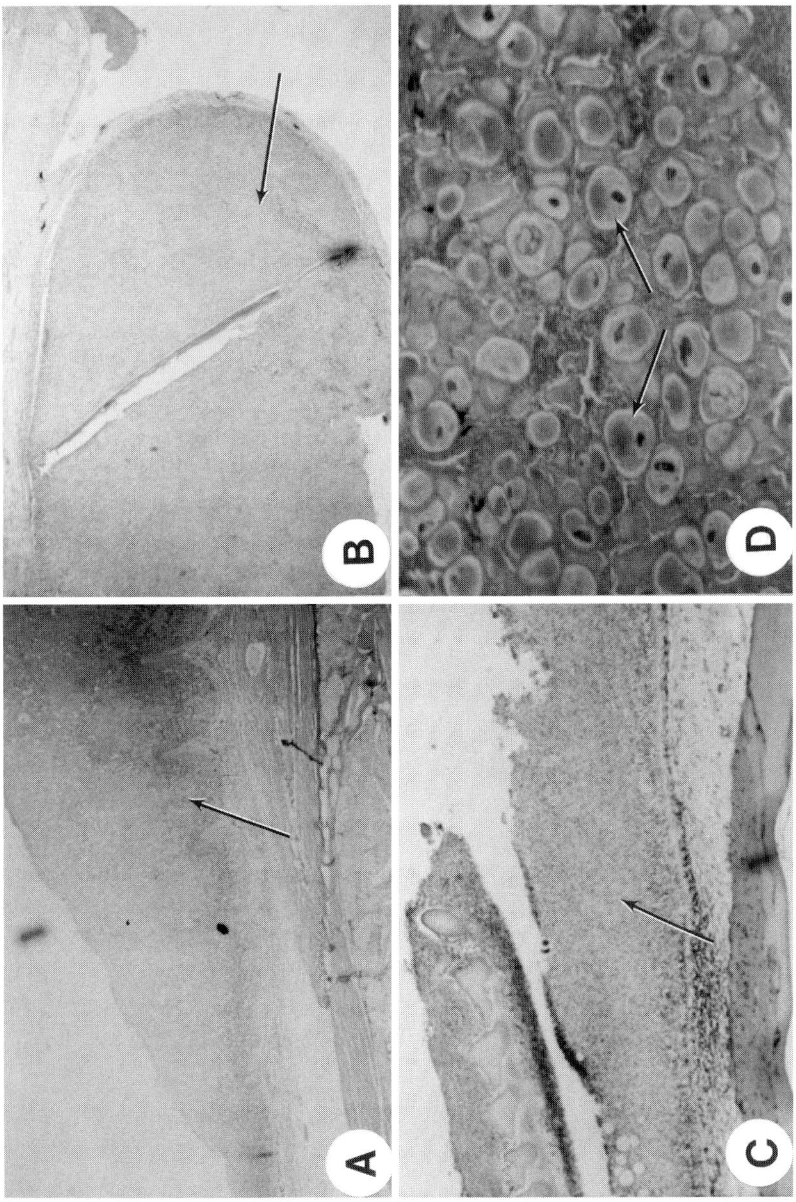

FIGURE 12.2. Histopathology of the four types of skin tumors (arrows) on walleye. (A) Discrete epidermal hyperplasia (arrow). (B) Walleye dermal sarcoma. (C) Diffuse epidermal hyperplasia (arrow). (D) Lymphocystis virus infected cells (arrow). (Photographs A, B, and C courtesy of P. R. Bowser.)

Significance

Other than its aesthetic effect on infected fish, the significance of diffuse epidermal hyperplasia is not fully understood.

WALLEYE DERMAL SARCOMA

According to Wolf (1988), walleye dermal sarcoma (WDS), caused by walleye dermal sarcoma virus (WDSV), was first described by Walker (1947), who recognized that the lesions were different from classical lymphocystis.

Geographical Range and Species Susceptibility

Dermal sarcoma lesions have been reported from adult walleye in Lakes Oneida and Champlain and a number of other lakes and rivers in New York, the Great Lakes region, and in central Canada. These tumors have not been reported from cultured walleye.

TABLE 12.1. COMPARISON OF VIRUS-CAUSED EPIDERMAL LESIONS ON WALLEYE AND THE PREVALENCE OF EACH TYPE OF DISEASE ON 25 FISH EXAMINED FROM A SINGLE POPULATION IN CANADA

Disease	Number positive	Lesion characteristic	Virus family	Size	Isolated
Epidermal hyperplasia	10	Clear, raised, distinct margin	Retroviridae	120 nm	No
Dermal sarcoma	7	Smooth, rounded	Retroviridae	135 nm	No
Diffuse epidermal hyperplasia	4	Mucoid, indistinct flat, translucent	Herpesviridae	100 nm	Yes
Lymphocystis	12	White, pink, large spherical cells, grapelike clusters	Iridoviridae	260 nm	Yes

Source: Yamamoto et al. (1985).

Clinical Signs and Findings

Walleye dermal sarcomas are spherical lesions that occur anywhere on the body and fins of adult walleye (Figure 12.1). The sarcomas are fine textured, variably firm, vascularized, and of varied color ranging from pink to white. Tumor size ranges from 1 to more than 10 mm, with a mean diameter of approximately 5 mm (Yamamoto et al. 1976).

Diagnosis

Although field observations may indicate the presence of dermal sarcoma because of morphological differences from lymphocystis, the only definitive way to identify the virus is to examine tumor tissue histologically.

Virus Characteristics

Martineau et al. (1991) homogenized 20 tumors, and after centrifugation, a C-type retrovirus was found in the sediment pellets by electron microscopy. Further evidence that the virus is a member of Retroviridae was presented by Bowser and Wooster (1994) when they demonstrated the presence of a lipid envelope. Viral RNA, presumably from the WDSV, was electrophoresed, and a resultant cDNA synthesized from viral RNA was hybridized with viral DNA in walleye tumors. This suggests that WDS results from an infection caused by a unique exogenous retrovirus of which the large genome is predominantly unintegrated in the tumor cells. The virus has not been isolated in cell cultures, but Martineau et al. (1992) have characterized its molecular structure. There is some disagreement about diameter of the retrovirus: Walker (1969) reported a diameter of 100 nm, Yamamoto et al. (1976) gave a slightly larger size of 135 nm, and Martineau et al. (1991) described a 90-nm retrovirus. These variances could be due to different methods of tissue fixation.

Epizootiology

Experimental transmission of walleye dermal sarcoma has rarely been documented, but Martineau et al. (1990) did successfully transmit the disease to healthy 4-month-old walleye by intramuscular injection with homogenized tumors from adult fish. Tumors developed in recipient fish 4 months later. In a subsequent study, transmission of walleye tumor-producing retrovirus was greater at 15°C than at 10 or 20°C (Bowser et al. 1990). Tumors developed in 8 weeks at 15°C. Bowser and Wooster (1991) showed that during an 18-week study period, tumors regressed either totally or partially when water temperatures rose from 7 to 29°C, but total regression was observed only in females. More recently, Bowser et al. (1996) successfully transmitted WDSV to 82% of young-of-the-year walleye by injecting them with infective material from tumors that had developed in the spring. Conversely, tumor material taken during autumn failed to elicit tumors in injected fish. In support of these findings, the authors identified WDSV RNA in fish injected with spring infectious material, but RNA was absent in fish that had been injected with extracts from fall tumors. These data support the theory that WDS develops in the spring and that when tumor regression occurs in the fall, infectious virus is no longer present.

In some walleye populations, frequency of neoplasia can be high (Table 12.1), and an individual fish may have more than one type of tumor. Yamamoto et al. (1985) examined 25 walleye and found that 12 had lymphocystis, 7 had dermal sarcoma, 4 had diffuse epidermal hyperplasia (walleye herpesvirus) and 10 had epidermal hyperplasia. Bowser et al. (1988) noted that on a seasonal basis, up to 27% of adult walleye in North America are affected with dermal sarcoma. They also noted that the number of infected fish in Lake Oneida was higher in spring and fall than during the summer.

Pathological Manifestations

Histopathological examination can be used to identify different virus-induced walleye growths definitively (Figure 12.2). Gross dermal sarcoma lesions may be confused with lymphocystis but probably not with epidermal hyperplasia (retrovirus) or diffuse epidermal hyperplasia (walleye herpesvirus) (Table 12.1, Figure 12.1). However, lymphocystis and dermal sarcoma tumor cells may also occur within the same lesion. Dermal sarcoma lesions consist of irregularly shaped tumors containing normal size cells that are often arranged in whorls (Figure 12.2), whereas lymphocystis cells are hypertrophied. Until recently, these tumors were thought to be noninvasive, but Ernest-Koons et al. (1996) discovered that after experimental exposure of 9-week-old walleye, some of the neoplasms, especially those collected at 84 days and later, did not remain strictly cutaneous. They were locally invasive and replaced normal muscle tissue. In one case, a head tumor had invaded the skull and caused the brain to be deformed.

With the use of in situ hybridization and immunohistochemical procedures, Poulet et al. (1995) showed that dermal sarcomas were associated with elevated transcriptionally active WDSV in neoplastic cells and that virus tropism extended beyond the mesenchymal fibroblastlike neoplastic cells to include mononuclear inflammatory and epidermal cells.

Significance

The appearance of dermal sarcoma–infected fish makes them unacceptable to those who engage in commercial fishing as well as to consumers. If walleye culture were to become a viable commercial industry, these virus-induced tumors could attract more attention.

MANAGEMENT OF WALLEYE VIRUSES

Because viral-induced hyperplasia and associated tumors have not been detected in cultured walleye, no management procedures have been needed. If, however, wild stocks are captured for broodstock, it would be prudent to ascertain their viral history if possible.

REFERENCES

Bowser, P. R., and G. Wooster. 1991. Regression of dermal sarcoma in adult walleyes (*Stizostedion vitreum*). *Journal of Aquatic Animal Health* 3:147–150.

Bowser, P. R., and G. A. Wooster. 1994. Ether sensitivity of the walleye dermal sarcoma virus. *Journal of Aquatic Animal Health* 6:178–179.

Bowser, P. R., M. J. Wolfe, J. L. Forney, and G. A. Wooster. 1988. Seasonal prevalence of skin tumors from walleye (*Stizostedion vitreum*) from Oneida Lake, New York. *Journal of Wildlife Diseases* 24:292–298.

Bowser, P. R., D. Martineau, and G. A. Wooster. 1990. Effects of water temperature on experimental transmission of dermal sarcoma in fingerling walleyes. *Journal of Aquatic Animal Health* 2:157–161.

Bowser, P. R., G. A. Wooster, S. L. Quackenbush, R. N. Casey, and J. W. Casey. 1996. Comparison of fall and spring tumors as inocula for experimental transmission of walleye dermal sarcoma. *Journal of Aquatic Animal Health* 8:78–81.

Earnest-Koons, K., G. A. Wooster, and P. R. Bowser. 1996. Invasive walleye dermal sarcoma in laboratory-maintained walleyes *Stizostedion vitreum. Diseases of Aquatic Organisms* 24:227–232.

Kelly, R. K., O. Nielsen, S. C. Mitchell, and T. Yamamoto. 1983. Characterization of *Herpesvirus vitreum* isolated from hyperplastic epidermal tissue of walleye, *Stizostedion vitreum vitreum* (Mitchell). *Journal of Fish Diseases* 6:249–260.

Kelly, R. K., O. Nielsen, and S. C. Yamamoto. 1980. A new herpes-like virus (HLV) of fish (*Stizostedion vitreum vitreum*). *In Vitro* 16:255.

Martineau, D., P. R. Bowser, and G. A. Wooster. 1990. Experimental transmission of dermal sarcoma in fingerling walleyes (*Stizostedion vitreum*). *Veterinary Pathology* 27:230–234.

Martineau, D., R. Renshaw, J. R. Williams, J. W. Casey, and P. R. Bowser. 1991. A large unintegrated retrovirus DNA species present in a dermal tumor of walleye (*Stizostedion vitreum*). *Disease of Aquatic Organisms* 10:153–158.

Martineau, D., P. R. Bowser, R. R. Renshaw, and J. W. Casey. 1992. Molecular characterization of a unique retrovirus associated with a fish tumor. *Journal of Virology* 66:596–599.

Poulet, F. M., V. M. Vogt, P. R. Bowser, and J. W. Casey. 1995. In situ hybridization and immunohistochemical study of walleye dermal sarcoma virus (WDSV) nucleic acids and proteins in spontaneous sarcomas of adult walleyes (*Stizostedion vitreum*). *Veterinary Pathology* 32:162–172.

Walker, R. 1947. Lymphocystis disease and neoplasia in fish. *Analytical Record* 99:559–560.

Walker, R. 1969. Virus associated with epidermal hyperplasia in fish. *National Cancer Institute Monographs* 31:195–207.

Wolf, K. 1988. *Fish Viruses and Fish Viral Diseases.* Ithaca, NY: Cornell University Press.

Yamamoto, T., R. D. MacDonald, D. C. Gillespie, and R. K. Kelly. 1976. Viruses associated with lymphocystis disease and dermal sarcoma of walleye (*Stizostedion vitreum vitreum*). *Journal of the Fisheries Research Board of Canada* 33:2408–2419.

Yamamoto, T., R. K. Kelly, and O. Nielsen. 1985. Morphological differentiation of virus associated skin tumors of walleye (*Stizostedion vitreum vitreum*). *Fish Pathology* 20:361–372.

13 Other Viral Diseases of Fish

Some viral fish diseases are less species or group specific than those previously discussed and/or the fish group they infect has not been represented. These viruses include epizootic hematopoietic necrosis (EHN), lymphocystis virus disease, pilchard herpesvirus disease, viral erythrocytic necrosis virus (VENV), viral nervous necrosis (VNN), and several others that are discussed in this chapter. Fish tumor-producing viruses, some of which have already been discussed, are also included.

EPIZOOTIC HEMATOPOIETIC NECROSIS

Epizootic hematopoietic necrosis disease is caused by epizootic hematopoietic necrosis virus (EHNV), an iridovirus found in Australia (Langdon et al. 1986). This virus is thought to be the first fish virus found in Australia.

Geographical Range and Species Susceptibility

Epizootic hematopoietic necrosis virus was initially isolated from wild redfin perch (yellow perch in North America) in New South Wales, Australia (Langdon et al. 1986) and, a short time later, from farmed rainbow trout (Langdon et al. 1988). The virus has also been found in Victoria and South Australia (Whittington et al. 1996). According to Langdon (1989), Macquarie perch, silver perch, mosquito fish, and mountain galaxias are also susceptible to EHNV by bath exposure, and Atlantic salmon are susceptible by intraperitoneal injection.

Clinical Signs and Findings

Epizootic hematopoietic necrosis in redfin perch and rainbow trout is characterized by slow or rapid spiraling to the surface and listless, ataxic swimming. The fish may even school and "headstand" above submerged objects or swim erratically. Redfin perch infected with EHNV become dark and anorexic. Petechial hemorrhages appear in skin, at base of fins, and on several internal organs (Langdon et al. 1986 and 1988). Infected fish develop a reddening of the head and nostrils and have focal or irregular, widespread muscular pallor, particularly in juvenile fish. Based on studies by Reddacliff and Whittington (1996), rainbow trout infected with EHNV have dark skin pigmentation and are inappetent and sometimes ataxic. Gross lesions include abdominal distension, swelling of spleen and kidney, and occasional pale foci in the liver.

Diagnosis

Epizootic hematopoietic necrosis virus can be isolated from infected fish in RTG-2 (rainbow trout gonad) or BF-2 (bluegill fry) cells with cytopathic effect (CPE) developing in 24 to 36 hours at 25°C (Langdon and Humphrey 1987). The virus is also infective in seven other fish cell lines, but CPE severity varies. Cytopathic effect begins as focal rounding and retraction of individual cells, followed by progressive lysis or detachment to form cell-free plaques.

Identification and detection methods were summarized by Whittington and Hyatt (1997). A direct antigen capture enzyme linked immunosorbent assay (ELISA) method for EHNV detection was developed by Steiner et al. (1991) using rabbit anti-EHNV primary immunoglobulin and sheep antirabbit secondary antibody. Serological methods for identifying EHNV include polyclonal antisera in serum neutralization tests, immunoperoxidase staining of histological sections, and immunogold electron microscopy. A polymerase chain reaction (PCR) procedure has also been developed for EHNV detection (Gould et al. 1995).

Virus Characteristics

Epizootic hematopoietic necrosis virus has a DNA-containing icosahedran that is unenveloped and measures 150 to 170 nm in diameter. The virus is classified in the genus *Ranavirus* of the family Iridoviridae (Hengstberger et al. 1993) and is similar to several other iridoviruses that have been isolated from fish and frogs (Ahne et al. 1997). Studies to determine the virus's sensitivity to desiccation, disinfectants, high and low pH, and temperature have indicated that EHNV can survive for extended periods in nature and on fomites.

Epizootiology

Epizootic hematopoietic necrosis virus is the first known fish iridovirus to cause severe necrosis of internal organ tissue. It also appears to be among the most virulent iridovirus isolated from fish to date. Outbreaks of EHNV in wild redfin perch populations have resulted in high mortality, ranging from about 40 to several thousand dead juvenile fish per 100 meters of shoreline per day (Langdon and Humphrey 1987). An estimated 95% decline in a wild redfin perch population was anecdotally described in one outbreak in New South Wales (Whittington et al. 1996). In another epizootic, which lasted 2 to 3 weeks, mortality may have reached 100% in adult fish.

The virus is reported to spread easily by fomites (nets, boots, and so forth) and can be transferred via live or frozen fish or in intestines of piscivorous birds (Whittington et al. 1996). The virus is resistant to drying and physical and chemical agents and can survive in frozen fish carcasses for at least a year. The virus has been found in healthy adult redfin perch, which are thought to be virus reservoirs in the ecosystem. Whittington et al. (1994) found, however, that redfin perch that had been exposed to EHNV showed no sign of infection 2 and 4 months later.

Initial experimental EHNV transmission was by injection of infected cell culture media into three adult redfin perch. Two died at 4 days and virus was isolated from both; the third yielded no virus when killed 21 days later (Langdon 1989). Whittington and Reddacliff (1995) discovered that redfin perch were highly susceptible to EHNV by bath inoculation with virus doses as low as 0.08 $TCID_{50}$/mL of water. At 19 to 21°C, incubation was 11 days, longer at cooler temperatures, and no infection occurred at temperatures below 12°C (Whittington and Reddacliff 1995). Rainbow trout were susceptible by intraperitoneal injection at 8 to 21°C and were persistently infected after 63 days but were not susceptible via experimental bath exposure. Rainbow trout of less than 12.5 cm are most susceptible, with mortality ranging from 1% to 8%. Common recrudescence was thought to result from contact with infected wild redfin perch. Most EHNV outbreaks at rainbow trout hatcheries can be correlated with water quality and poor management practices; when these problems are corrected, disease often disappears. Most likely, the redfin perch and rainbow trout epizootics were caused by the same virus, given the similar clinical signs and the nature of the virus isolates, and in all probability, the trout contracted the disease from redfin perch residing in the watershed (Langdon 1988).

Pathological Manifestations

In redfin perch, EHNV produces focal necrosis of the renal hematopoietic tissue, hepatocellular tissue, spleen, and gastrointestinal tract epithelium, and it is believed that death results from injuries to the kidney and spleen (Langdon 1989). Infected fish also develop encephalopathy in various degrees of severity. On the other hand, clinically diseased rainbow trout develop focal to extensive acute renal hematopoietic necrosis, small multiple focal hepatocellular necrosis, focal to acute splenic necrosis, mild branchial hyperplasia with occasional focal necrosis, congestion, edema in the wall of the swim bladder, and focal acute necrotizing myocarditis with necrotic cells and debris in the cardiac lumen.

Significance

The significance of EHNV is not clear, but when it becomes established in redfin perch or rainbow trout populations, significant losses can occur. The virus also has the capability to infect a wider range of fish hosts.

LYMPHOCYSTIS

Lymphocystis, a cellular hypertrophic disease primarily found on the body and fins, is the oldest and perhaps best known of all fish virus diseases. Although lymphocystis has been recognized since 1874, and theorized to be of a viral nature in 1920 (Weissenberg 1965), its viral etiology was not actually proved until the early 1960s (Wolf 1962). Its etiological agent is lymphocystis virus, for which River's postulates were fulfilled in 1964 (Wolf et al. 1966).

Geographical Range and Species Susceptibility

Lymphocystis is the most widely distributed of all fish viruses. It affects freshwater fish in most areas of the world including North and South America, Europe, Africa, Australia, and Asia (Nigrelli and Ruggieri 1965; Wolf 1988; Anders 1989). Lymphocystis is also found in fish from nearly all marine waters including the Atlantic and Pacific Oceans, Gulf of Mexico, Red Sea, Bering Sea, and the Mediterranean Sea to name a few (Anders 1989; Faisal 1989). Lymphocystis had not been reported in the Caribbean Sea until it was found in Caitipa mojarras and the exotic Indian glassfish (Bunkley-Williams et al. 1996; Williams et al. 1996).

Anders (1989) named 11 orders, 45 families, and 141 fish species that are susceptible to lymphocystis virus, and the list grows annually. The most prominent orders affected are Perciformes and Pleuronectiformes, which contribute 85% of susceptible species. The sunfishes (bass and bluegills), perch, seatrout, drum, butterfly fish, angelfish, cichlids, and flat fishes are most frequently infected. Susceptible fishes include cultured, wild, and traditional aquarium or ornamental species (Durham and Anderson 1978; Dukes and Lawler 1975). Because the international ornamental fish trade is vast and essentially unregulated in terms of disease, spread of lymphocystis has likely been enhanced by this industry (Williams et al. (1996). Salmonids, cyprinids, and ictalurids and members of other low phylogenetic orders do not appear to be susceptible.

Clinical Signs and Findings

Lymphocystis manifests itself as a series of greatly hypertrophied cells that occur in connective tissue below the epidermis. The cells are large, gray or whitish, and may measure up to 2 mm in diameter. These cells, which are easily seen with the unaided eye, may occur singularly or grouped together in grapelike clusters giving a tumorous appearance (Figure 13.1). Lesions can appear at any location but are more prevalent on fins, head, and lateral body surfaces. They have also been noted on eyes and gills (Dukes and Lawler 1975). Rarely, lymphocystis lesions have been found in the spleen, liver, and mesenteries of juvenile marine fish (Lawler et al. 1974); however, Colorni and Diamant (1995) described a significant lymphocystis infection in the spleen of red drum. There are no behavioral abnormalities associated with lymphocystis virus infections unless lesions interfere with respiration or feeding.

Diagnosis

Lymphocystis virus infections can be identified by gross clinical signs coupled with presence of hypertrophied cells with enlarged nuclei. These cells may contain Feulgen-positive inclusions in the cytoplasm that can only be seen in stained sections (Figure 13.2). Generally, virus isolation in tissue culture is not easily accomplished because of a long incubation period (Wolf et al. 1966). The virus has been isolated in BF-2 cells and largemouth bass cells. Also Bowden et al. (1995) used red drum dorsal fin cells (RDDF-1) incubated at 25°C to isolate the virus. Initial CPE became apparent in RDDF-1 cells 8 to 10 days after exposure and after 2 weeks developed enlarged cells, ranging from 40 to 90 μm, that floated in the media. The CPE that develops in infected cell cultures is similar to lymphocystis lesions on infected fish (Wolf and Carlson 1965). At 23 to 25°C, in vitro infected cells will increase in size and become basophilic. Six days postinfection, Feulgen-positive inclusions can be seen in the cytoplasm, followed by formation of a hyaline capsule at day 10. The cells reach maturity in 3 to 4 weeks and are identical in appearance to

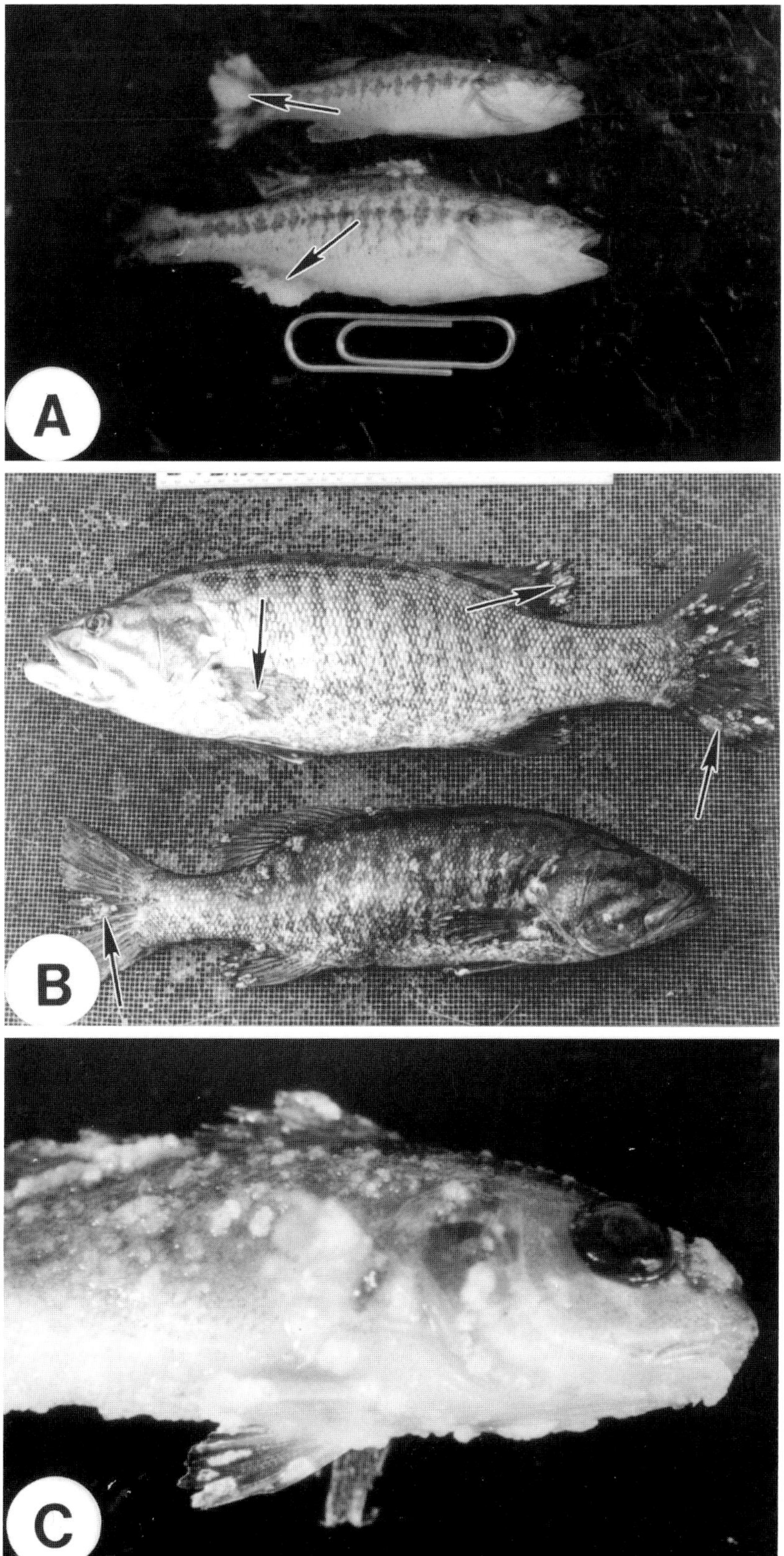

FIGURE 13.1. *Lymphocystis virus.* (A) *Large lymphocystis virus–induced lesions* (arrows) *on fins of a cultured 2-month-old spotted bass.* (B) *Lymphocystis lesions* (arrows) *on a 40-cm small-mouth bass from a wild population.* (C) *Lymphocystis lesion on an 8-cm croaker taken from a wild population and held in captivity where lesions developed (formalin preserved).*

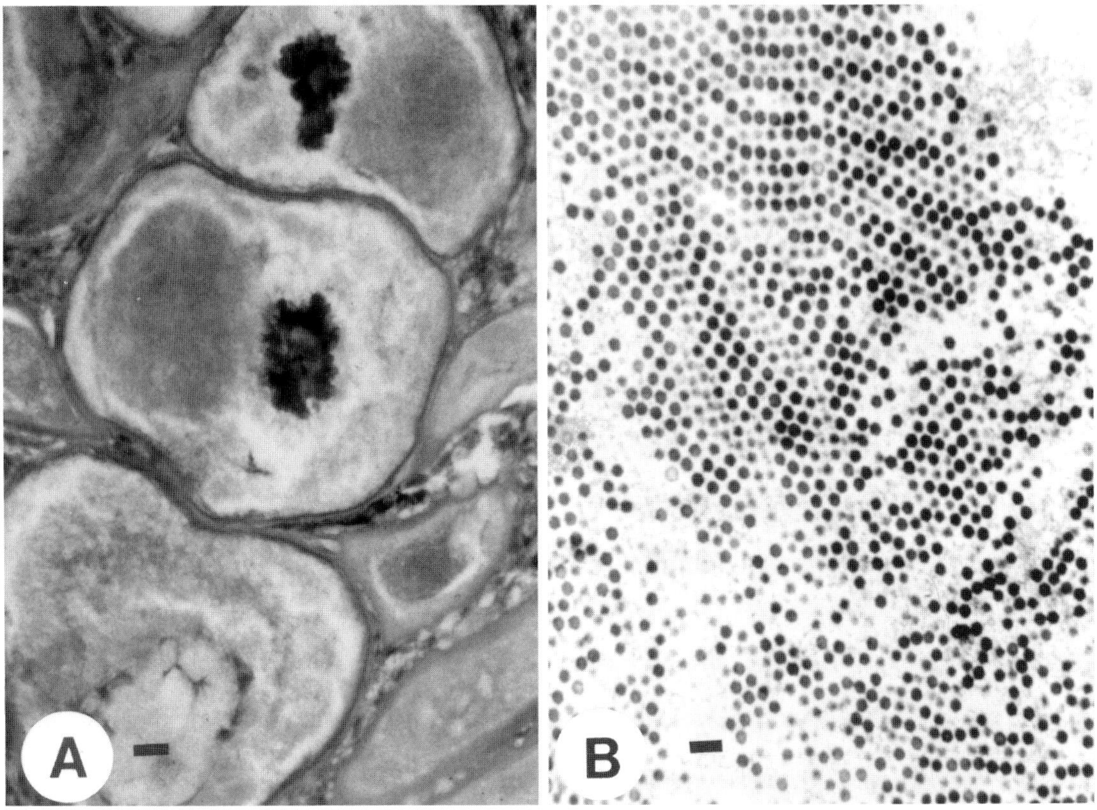

FIGURE 13.2. Lymphocystis virus. (A) *Histological section of a lymphocystis lesion showing the hypertrophied cell with an enlarged nucleus (*bar = *100 fm).* (B) *Electron micrograph of lymphocystis virus (*bar = *400 nm).*

those seen in infected fish, including an enlarged nucleus; however, cells are smaller.

Virus Characteristics

Lymphocystis virus is in the genus *Lymphocystivirus,* family Iridoviridae. It is icosahedral with a diameter of 145 to 330 nm (Robin and Bertholimue 1981) (Figure 13.2) and a double-stranded DNA genome that replicates in the nucleus but is assembled in the cytoplasm of infected cells. It has been noted that there may be different specific strains of lymphocystis virus based on species susceptibility and the fact that virus taken from one species is not always pathogenic to other species (Overstreet and Howse 1977). Peters and Schmidt (1995) found that lymphocystis virus in plaice developed anomalies in mature lymphocystis cells and underwent disintegration within the cells inclusion body, possibly caused by nucleases. It is speculated that the capsid is denatured by proteolytic enzymes. The conclusion was drawn that lymphocystis that occurs in plaice is not totally adapted to that species and would support the theory of more than one lymphocystis virus strain.

Epizootiology

Lymphocystis is a chronic, benign condition that seldom results in morbidity or death of fish, with cultured red drum being a possible exception (Colorni and Diamant 1995). The primary problem caused by the disease is that infected fish are unattractive and less marketable. Studies in California indicated that lesions can become so large that the buccal cavity is occluded and fish die of starvation (McKosker 1969). Less severely affected fish may experience weight loss; infected walleye from Lake Michigan weighed less than noninfected fish (Petty and Magnuson 1974). The presence of large body or fin lesions may also render fish more vulnerable to gill nets.

Transmission of lymphocystis virus is influenced by several factors; host species, fish density, injury, parasite load, environmental conditions, and pollution levels. The virus can be transmitted by cohabitation or spraying infected material on scarified skin (Wolf et al. 1966). Lymphocystis was also experimentally transmitted in material from cultured red drum to naive drum by intramuscular injection, but attempts at oral transmission were unsuccessful (Bowden et al. 1995). In natural waters, lymphocystis virus is presumably transmitted to healthy fish when infected cells rupture and virus is released into the water. Ryder (1961) found that injuries sustained during spawning activities led to a higher incidence of lymphocystis. Handling, netting, tagging, and so forth, may also increase disease incidence because lesions tend to develop where scales are disturbed or fins damaged (Clifford and Applegate 1970).

Some external parasites (copepods or isopods) that disrupt the protective mucous layer may increase transmission efficiency (Nigrelli 1950). Lawler et al. (1974) postulated that parasitic isopods (*Lironeca ovalis*) may have been instrumental in a lymphocystis virus infection of silver perch by causing skin irritation, thus allowing virus to invade the tissue. Although lymphocystis is usually self limiting, Bowden et al. (1995) noted that "lymphocystis in intensive red drum culture systems is frequently a disease of high morbidity" and can result in considerable economic loss because of unsightly lesions. They also pointed out that lymphocystis predisposes red drum to more pathogenic diseases such as amyloodinium and vibriosis.

It has been suggested that an increased occurrence of lymphocystis is influenced by water pollution (Alpers et al. 1977). Anders and Yoshimizu (1994) proposed that pollutants adversely affect the immune and endocrine systems of fish, allowing activation of latent skin tumor–producing viruses. In Norway, Reiersen and Fugelli (1984) described a higher lymphocystis incidence in flounder in polluted waters than in those in less polluted waters but cautioned against drawing any definitive conclusions because of a complex disease–environmental relationship. On the basis of limited sampling, Bunkley-Williams et al. (1996) concluded that virus prevalence was higher in areas where lymphocystis infected fish were cleaned and offal returned to the sea enabling other fish to come into contact with the virus.

Lymphocystis occurs in all sizes and ages of fish, but Yamamoto et al. (1985) did report a higher virus incidence in small walleye (250 to 600 g) than in medium (600 to 1200 g) or larger fish. They also noted that the female infection rate was three times greater than that of males. In a population dynamics study of flounder, Lorenzen et al. (1991) determined that a higher incidence of lymphocystis occurred in younger fish because they were held in high-density conditions and had not yet acquired disease immunity as had older fish. In Israel, lymphocystis lesions were seen on the body and fins of cultured red drum that had been shipped from Texas. Lesions first appeared on only a few fish when they weighed about 20 g, but 2 months later, disease had spread to several hundred fish in a population of 10,000. Lesions identical to those on the fins and body were subsequently seen in the spleen and heart but none were found in the liver or kidney. The infected cells measured about 430 µm in diameter.

Under experimental conditions, lymphocystis incubation time depends on temperature. At 10 to 15°C, incubation may take up to 6 weeks for lesions to develop, compared with 5 to 12 days at 20 to 25°C (Cook 1972; Roberts 1975).

Incidence of lymphocystis infections may reach 100% in some crowded fish populations (Durham and Anderson 1978). In contrast, in wild marine populations infection incidence may range from 4% to 57%, with the highest incidence usually occurring in spring and summer (Amin 1979). However, Reiersen and Fugelli (1984) reported a disease increase in winter. Lymphocystis lesions usually do not cover large areas of the body or fins; however, Overstreet (1988) reported that naturally infected croakers had lesions on 60% to 80% of their body surface, which was an unusual level of infection. Experimental infections or those that develop under confined conditions tend to produce lesions that cover a larger portion of the body surface.

Pathological Manifestations

Lymphocystis virus infects fibroblastic connective tissue cells of the skin and occasionally the internal organs. The hypertrophied cells may be 20 to 1000 times larger than the normal 10 to 100 µm cell (Figure 13.2) (Anders 1989). These cells are generally round to oval and easily visible. Histologically, they contain enlarged, centrally located nucleus and Feulgen-positive, basophilic, ribbon-shaped cytoplasmic inclusions containing viral DNA (Dunbar and Wolf 1966). Morphology of the inclusions can vary with fish species but are considered sites of viral maturation (Wolf and Carlson 1965).

Little inflammation is associated with lymphocystis, but infected cells are often clustered in a collagenous hyalin matrix (Howse and Christmas 1970). Electron micrographs show that inclusion bodies are dense areas of hexagon-shaped virus particles that are transported to different sites by the bloodstream. In most instances, lymphocystis lesions will regress after a period, cells will lyse or slough, and infected areas heal and leave little evidence of infection.

Significance

Lymphocystis virus–infected fish generally do not die as a result of infection but do have reduced economic value. Entire populations of juveniles or adult fish may be rejected by those engaged in sport and commercial fishing if heavily infected with lesions even though lesions are transient.

PILCHARD HERPESVIRUS DISEASE

In 1995, massive mortality occurred in marine Australian pilchard (a sardinelike fish) in Australia and New Zealand and was attributed to a herpesvirus (Whittington et al. 1997). This was thought to have been the most extensive mortality recorded in any fish species in terms of numbers and geographical range affected and serves as an example of how a highly virulent pathogen can affect a dense (schooling) population of aquatic animals.

Geographical Range and Species Susceptibility

The pilchard herpesvirus disease (PHD) was confined to the western, southern, and eastern coasts of Australia and northeastern coast of New Zealand. To date, the virus has only been reported in pilchards.

Clinical Signs and Findings

Pilchards infected with PHD were primarily adults (larger than 11 cm) of both sexes (Hyatt et al. 1997). It was observed that "affected fish left the school, ceased swimming, turned onto their sides, resumed swimming sluggishly, and then died in extremis within a few minutes" (Whittington et al. 1997). Once startled, a flashing behavior began within 15 seconds and death occurred in 3 to 5 minutes. Some fish died with mouth open and flared opercula. Tissue around the eyes of fresh dead and moribund fish was congested and gills were dark red to dark brown, compared to a bright red gill in healthy fish. Internally, the spleen varied from pale brown to dark red or black.

Diagnosis

Pilchard herpesvirus disease is diagnosed by clinical signs and confirmed by histopathology and detection of virus in gill tissues by electron microscopy (Whittington et al. 1997). The virus could not be isolated in several permanent fish cell lines (CHSE-214 [chinook salmon embryo], RTG-2, BF-2, fathead minnow [FHM], or epithelioma papillosum of carp [EPC]) inoculated with filtrates from homogenized gill and internal organ tissue (Hyatt et al. 1997). Bacteria, parasites, and environmental disturbances were ruled out as possible causes of mortality.

Virus characteristics

By electron microscopy, a replicating herpeslike virus was seen in gill tissue cells from all affected fish but from no unaffected fish. The icosahedral virus was 92 nm in diameter with many virions containing an envelope acquired as it passed through the plasma membrane. Viral replication occurred in the nucleus and virus was released into the cytoplasm as the nuclear membrane degenerated. Virus release from cells was associated with lysis of gill epithelial cells.

Epizootiology

Other than observations noted by Whittington et al. (1997), little is known of the epizootiology of PHD. The disease that occurred in Australian and New Zealand waters appeared to be the same, with identical etiologies. Infection progressed from mild to subacute to acute as it moved through the pilchard schools. The epizootic began in March (1995) on the southern coast of Australia, spread to the east and west coast, and then to New Zealand in June, and continued in both localities until September. In Australian waters, the disease spread at about 30 km/d and generally followed the pilchard migration pattern, which was contrary to prevailing currents. Mortality occurred along 5000 km of Australian and 500 km of New Zealand coast, which roughly constitutes the geographical range of the Australian pilchard. Loss estimates were about 10% in some pilchard schools.

Pilchards comprise a large part of the gannet's diet, and because these sea birds were observed eating affected fish, it was speculated that they may have contributed to spread of disease (Whittington et al. 1997).

Pathological Manifestations

Significant lesions were noted only in the gills. Acute to subacute inflammation of these lesions was followed by epithelial hypertrophy and hyperplasia. Lesions were initially focal but became generalized over a 4-day period. The only consistent internal organ virus-related histopathology was engorgement of the spleen with erythrocytes.

Significance

A small pilchard fishery exists in western Australia, New South Wales, and New Zealand, and the economic impact of PHD on this enterprise is yet undetermined (Whittington et al. 1997). A significant loss of pilchards could have an ecological impact because of their importance in the diet of many marine predators.

VIRAL ERYTHROCYTIC NECROSIS

Viral erythrocytic necrosis (VEN) manifests itself as an infection of the erythrocytes and affects a wide variety of marine and anadromous fish species. Viral erythrocytic necrosis was originally known as piscine erythrocytic necrosis (PEN), until Walker and Sherburne (1977) suggested that the nature of the disease would be better reflected if emphasis were placed on blood instead of fish; thus, the name "viral erythrocytic necrosis." Laird and Bullock (1969) proposed a viral etiology for VEN, and Walker (1971) later provided electron microscopic evidence to support that etiology. The etiological agent of VEN is erythrocytic necrosis virus (ENV). Wolf (1988) cites literature from the early 1900s that described the disease, but at that time, its cause was attributed to an intracellular parasite.

Geographical Range and Species Susceptibility

Viral erythrocytic necrosis was first reported in Atlantic cod, Atlantic sea snail, and shorthorn sculpin in coastal waters of the North Atlantic of Canada and the United States (Laird and Bullock 1969). Later, VEN was reported in the United Kingdom (Smail and Egglestone 1980), Portugal (Eiras 1984), Mediterranean Sea (Pinto et al. 1989), and the Pacific region of the United States (Rohovec and Amandi 1981). Essentially, VEN is confined to marine environments, but has been reported in eels in Taiwan (Chen et al. 1985) and freshwater stocks of anadromous salmon (Rohovec and Amandi 1981). Walker and Sherburne (1977) listed 12 species of fish from the Atlantic coast of North America that are affected by VEN, 5 of questionable susceptibility, and 25 that appear to be refractive. Affected species are primarily alewife, rainbow smelt, sea raven, and European seabass (Pinto et al. 1989). Lamas et al. (1995) reported VEN in cultured turbot in the Mediterranean Sea along the Spanish coast. Other economically important susceptible fish species are chum, coho, chinook, and pink salmon, steelhead trout, Atlantic herring, and striped mullet (Sherburne 1977; MacMillan et al. 1980; Rohovec and Amandi 1981; Eiras 1984).

Clinical Signs and Findings

The principal clinical sign of VEN is severe anemia characterized by pale gills, watery colorless blood, and discolored livers (Meyers et al. 1986). Stricken fish will be listless or whirl. Hematocrits will be as low as 2% to 10% (Evelyn and Traxler 1978). Pacific herring infected with ENV did show hemorrhages in the epithelium (Figure 13.3). Walker and Sherburne (1977) reported that captured infected cod were somewhat emaciated and darker than noninfected fish.

Diagnosis

Viral erythrocytic necrosis can only be confirmed by electron microscopy, as virus has not been prop-

FIGURE 13.3. Pacific herring with viral erythrocytic necrosis virus. Note the hemorrhages in the epithelium (arrows). *(Photograph courtesy of J. R. Winton.)*

agated in established cell lines (Evelyn and Traxler 1978); however, Reno and Nicholson (1980) demonstrated that ENV replicated in vitro in cod erythrocytes at 4°C. Viral erythrocytic necrosis infections can be detected by staining blood smears and observing the erythrocytic cytoplasmic inclusion bodies. Nuclei of infected cells may be round, bilobed, or U-shaped (Figure 13.4) (Walker and Sherburne 1977). Infected erythrocytes contained small to large and round to lobate eosinophilic (Giemsa stain) inclusion bodies in the cytoplasm (Figure 13.4). Depending on infection stage and severity, 9% to 100% of erythrocytes will contain these inclusions. Pinto et al. (1989) found erythrocytic inclusions in 2.7% to 25% of VEN-infected European seabass. The inclusion body, which is also Feulgen positive, fluoresces bright green with acridine orange and most likely contains the replicating pool of viral nucleic acid.

Virus Characteristics

Electron microscopy of ENV-infected erythrocytes will reveal hexagonal virions in the cytoplasm, which confirms virus infection (Figure 13.4). Although serological methods for ENV identification are not available, Pinto et al. (1989) did purify viral material from infected erythrocytes of seabass and developed an immunofluorescent reagent that allowed viral differentiation from similar viral fish infections.

The ENV that has been described to date is an icosahedral virion with a DNA genome that is classified as a member of the family Iridoviridae. The size of ENV virions varies with afflicted fish species. Virus described in erythrocytes of affected Atlantic herring and salmonids was 140 to 190 nm (Evelyn and Traxler 1978), but a larger, 300- to 350-nm, virus was reported in infected Atlantic cod (Walker and Sherburne 1977). Without benefit of serological procedures and in vitro culture techniques, it is impossible to determine whether all of these viruses are the same; however, with the exception of particle size, they are very similar.

Epizootiology

An ENV infection may manifest itself in different ways depending on fish species, degree of infection, and clinical signs (Table 13.1). Mean hematocrits decline to less than 10% from a normal of near 40% in chum salmon (MacMillan et al. 1980) and even lower in other species. It is proposed that newly infected circulating erythrocytes are destroyed in the hematopoietic centers of the spleen and anterior kidney, thus facilitating virus release. The host compensates for this anemia by erythroblastosis, but additional infection of erythroblasts with virus may prolong infection.

Smail and Egglestone (1980) recorded a 32% incidence of VEN in young-of-the-year Atlantic cod compared with an almost zero incidence in 3 to 4 year olds, although the older fish may possibly have recovered from an earlier infection. Similarly,

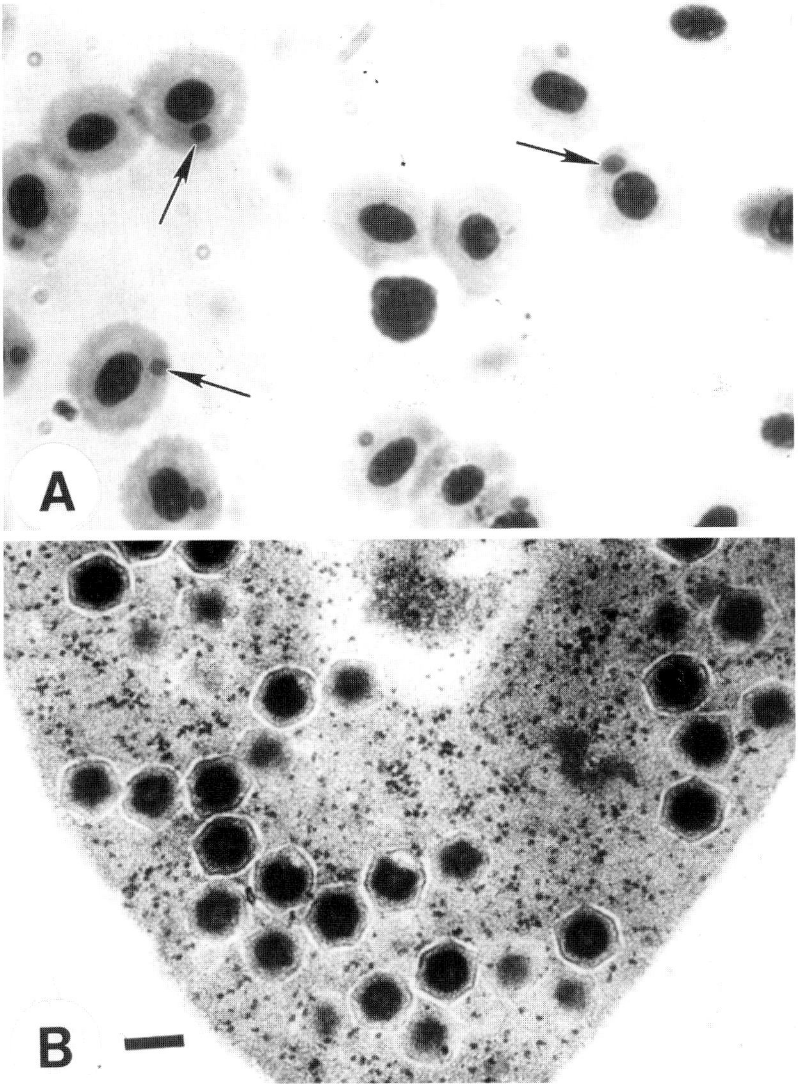

Figure 13.4. Viral erythrocytic necrosis. (A) *Erythrocytes with singular cytoplasmic inclusion bodies* (arrows). (B) *Electron micrograph of the hexagonal viruses in Atlantic cod erythrocytes (*bar = *200 nm). (Photographs courtesy of B. Nicholson.)*

Table 13.1 Summary of pathology of viral erythrocytic necrosis infection in some fish

Fish host	Percent erythrocytes affected	Pathology
Atlantic cod	0.01–99, mature	Skin darkening, no mortality
Common blenny	3–60, mature	None
Alewife	0.17, mature	None
Atlantic herring	90, mature and immature	Hyperemic liver, hemolysed blood, two types of inclusions
Pacific herring	Mature and immature	Erythroblastosis
Chum salmon	80, mature and immature	0.3% mortality, prone to bacterial kidney disease, vibriosis and anemia

Sources: Laird and Bullock (1969); Evelyn and Traxler (1978); Reno and Nicholson (1980); Pinto et al. (1989).

MacMillan and Mulcahy (1979) detected 43% incidence of virus in young Pacific herring, compared with a 4% incidence in 4– to 8-year-old fish. Meyers et al. (1986) found a VEN prevalence of 56% to 100% in 1+ year and 17% to 80% in older Pacific herring in Alaska. Captive European seabass had a 20% to 100% VEN infection, compared with only 6% in wild members of the same species (Pinto et al. 1989). These collective data indicate that younger and confined fish are more susceptible to VEN.

The infection route of ENV is most likely from fish to fish through the water. Experimentally ENV can be transmitted by injection with filtrates derived from infected fish tissue. Eaton (1990) transmitted ENV from Pacific herring to pink, chum, and sockeye salmon, with chum salmon being the most susceptible. These studies show that ENV from one fish species will infect another species. Bloodsucking parasites such as the salmon louse (*Lepeophtheirus* sp.) and gill maggot (*Salmincola* sp.) may also serve as vectors (Smail 1982).

Evelyn and Traxler (1978) found that chum and pink salmon experimentally infected with cell-free filtrates became clinically infected in 12 days (chum) and 3 weeks (pink). Infection rate in chum salmon increased from 50% at 12 days to 100% at 48 days postinoculation. These researchers also reported that at 7 months postinoculation, four infected fish were ENV negative, which indicates that recovery from VEN is possible. Rohovec and Amandi (1981) reported an ENV infection in 20-g coho salmon in fresh water. The fish were experiencing mortalities, but no clinical signs other than hematocrits of less than 10% were noted.

Reno et al. (1985) reported that both Atlantic cod and Atlantic herring exhibited lower hematocrits and erythrocyte counts, but a reduced hemoglobin concentration was reported only in cod. There was no effect on plasma electrolyte or protein concentration in either species. Artificially ENV-infected chum salmon developed lowered hematocrits and hemoglobin concentrations (Haney 1992). The ENV-infected salmon also had higher white blood cell counts and less fragile red blood cells than the controls. Infection progressed through 4 weeks before fish began to recover, making physiological and hematological changes caused by disease transient. However, infected fish recovered more slowly from stress and had greater osmoregulatory difficulties than did control fish. Meyers et al. (1986) also found evidence that ENV-infected Pacific herring suffered osmoregulatory stress that could precipitate death, particularly in young fish.

Survival of ENV-infected fish varies. MacMillan et al. (1980) and Rohovec and Amandi (1981) assayed 35 stocks of sea run salmonids and found 25% to be positive for erythrocytic inclusions. Percentage of ENV-positive fish in the assayed populations ranged from 1.0% to 15.1%. When evaluating the effect VEN had on released salmon, it was found that mortality directly due to the viral disease was limited; however, environmental stressors caused higher mortality in infected fish than in uninfected fish. There is some evidence that VEN is not primarily responsible for deaths of infected fish but renders them more susceptible to environmental insults and secondary bacterial and/or fungal infections.

Pathological Manifestations

Appy et al. (1976) described pathology that takes place in the erythrocytes of cod infected with VEN. The nucleus becomes more rounded and densely stained with the presence of magenta cytoplasmic inclusions that measure 1 to 4 μm in diameter (Figure 13.4). In some cells, inclusions appeared to possess a small central vacuole that increased in size with subsequent degeneration. Infected cells appeared more rounded and irregular than uninfected cells. Some cells fractured and resulted in formation of fragments containing nuclear remains of the inclusion. Cytoplasmic inclusions of infected cells are composed of a pool of icosahedral viral particles (Figure 13.4) with incomplete viruses at the edge of the pool.

Significance

Prevalence of VEN in some fish populations can be high, but its actual importance to these fish stocks is unknown. It does pose a threat to maricultured populations. Although limited mortality in cultured fish does occur, its direct impact on wild populations is unclear. The virus does alter blood parameters and may make infected fish more susceptible to opportunistic bacterial diseases (such as *Vibrio anguillarum*) as well as more sensitive to dissolved oxygen depressions and handling (Reno et al. 1985; Haney et al. 1992).

VIRAL NERVOUS NECROSIS

Viral nervous necrosis is caused by a nodavirus that has been reported from a variety of marine fish species in several geographical locations (Nakai et al. 1995). Of these viruses, striped jack nervous necrosis virus (SJNNV) of hatchery-reared striped jack has received the greatest attention from the research community (Mori et al. 1992).

Species Susceptibility and Geographical Range

Striped jack nervous necrosis virus is currently considered to be a disease of Japan; however, it has also been diagnosed in Greece, Martinique, and France (Breuil et al. 1991; LeBreton et al. 1997). The principal susceptible species is the striped jack during its larval stage, but Le Breton et al. (1997) reported VNN disease in production-size European seabass. Arimoto et al. (1993) showed that larval yellowtail and goldstriped amberjack were refractive to SJNNV, and Muroga (1995) listed 14 other fish species, including barramundi, groupers, other jacks, parrotfish, flounders, and puffers that are affected by VNN.

Clinical Signs and Findings

When larval striped jack are injected with virus, clinical signs indicate that the central nervous system is affected (Arimoto et al. 1994). Infected fish become inappetent, emaciated, and larger larvae develop enlarged swim bladders and vertebral deformities. Larger VNN-infected European seabass exhibited reduced feeding, lethargy, and abnormal swimming (Le Breton et al. 1997). Fish rotated on the long axis, swimming up and down, and moribund fish came to the surface with a curved body and belly up. Fish were darker, with an opaque cornea or bilateral exophthalmia, and became hyperactive when disturbed.

Diagnosis

Striped jack nervous necrosis virus does not replicate in any established fish cell line. Methods of diagnosing the virus in infected larvae include ELISA (Arimoto et al. 1992, Nishizawa et al. 1995), PCR (Nishizawa et al. 1994), and FAT using both monoclonal and rabbit polyclonal antibody (Nguyen et al. 1997). Nishizawa et al. (1995) described an in vitro neutralization test in which larval striped jack were exposed to purified SJNNV virus treated with monoclonal antibodies and then monitored by ELISA using rabbit anti-SJNNV serum.

Virus Characteristics

The causative agent of SJNNV is a nodavirus based on its morphology, structural proteins, and nucleic acids (Mori et al. 1992). The virus is spherical, unenveloped, about 25 to 34 nm in diameter, and contains two single strands of RNA that replicate in cell cytoplasm.

Epizootiology

Striped jack nervous necrosis virus most severely affects larval striped jack when they are 2 to 20 days old (Arimoto et al. 1994). Arimoto et al. (1993) also reported disease occurrence in 3.5- to 4.4-mm–long striped jack larvae that remained susceptible at 78 mm. Striped jack nervous necrosis virus creates a viremia that affects nearly all organs and tissues and is most likely transmitted horizontally.

Using PCR, Mushiake et al. (1994) demonstrated that infected organs provide the opportunity for vertical transmission when they detected a SJNNV-positive gene in spawning striped jack. They found that by examining fish for virus just before spawning, SJNNV-positive fish could be identified and eliminated from spawning activities, thus preventing vertical transmission. Nishizawa et al. (1996) showed, however, that this procedure is not 100% effective in detecting fish with very low levels of SJNNV. Nevertheless, PCR screening of spawner adults and use of only SJNNV-negative fish for spawning has resulted in reduced numbers of virus infected larvae.

With the use of FAT and PCR methods, Nguyen et al. (1997) detected virus in the reproductive fluid of 13-year-old adult (spawners) striped jack but found no virus in fluids of 4-year-old fish that had not spawned. Virus was also detected in the gonad, intestine, stomach, kidney, and liver of a 13-year-old fish by FAT and rabbit anti-SJNNV serum. These data suggest that virus originates in the organs of striped jack spawners and is shed from intestines and gonads, resulting in contaminated eggs.

Mortality of SJNNV-infected striped jack is often near 100% in larvae less than 10 days old, but older fish are less severely affected (Arimoto et al. 1994). Viral nervous necrosis may occur in striped jack at water temperatures from 20 to 26°C but progresses most rapidly at 24°C. Larger European seabass were affected by VNN, but Breuil et al. (1991) noted that mortalities ceased when fish reached 5 g in weight. Le Breton et al. (1997), however, reported VNN in European seabass that were 10 to 45 g and 350 to 580 g, and both groups suffered mortalities of 11% to 60%. These mortalities occurred in July and September when water temperatures were 24.5 to 26.5 °C.

Pathological Manifestations

Naturally (acute and subacute stages of natural infections) and experimentally SJNNV-infected striped jack were studied histopathologically for progression of VNN (Nguyen et al. 1996). General features of pathogenesis in these three groups were similar. Nerve cells in the spinal cord, especially above the swim bladder, were initially affected, and later the brain and retina became necrotic and vacuolated. Deaths occurred 1 to 2 days after lytic degeneration and vacuolation of nervous tissue cells. Results of FAT, histopathology, and electron microscopic observations indicate that SJNNV exhibits a tropism for nerve cells and that initial multiplication occurs in the spinal cord, from which virus spreads to the brain and finally the retina and presumably other tissues and organs. Le Breton et al. (1997) described necrosis of the cornea and frontal part of the head, with blood vessels of the brain showing severe congestion. Large vacuoles were present in the brain and retina.

Significance

In Japan, VNN is most severe in larval striped jack, in which it can be devastating. Its presence in other parts of the world and its potential to affect other fish species make the virus a potentially dangerous pathogen in maricultured facilities.

OTHER VIRUSES OF FISH

During the past 15 years, in addition to the previously discussed iridoviruses, a relatively large number of fish viruses have been isolated and classified as members of the family Iridoviridae (Williams 1996). Heretofore, most viruses in this group had been described in insects, other arthropods, and amphibians, but the number of iridovirus isolations from fish is increasing (Mao et al. 1997). Some iridoviruses from widely different geographical areas and divergent taxonomic fish species are similar (Hedrick et al. 1992). These authors showed that EHNV from redfin perch (Australia), the virus from European catfish (Germany), and black bullhead (France) were similar, in spite of diverse geographical origin. Using molecular characterization and sequence analysis, Mao et al. (1997) determined that nine isolates of iridoviruses from fish, reptiles, and amphibians were more similar to frog virus-3 (genus *Ranavirus*) than to lymphocystis (genus *Lymphocystisvirus*). These authors suggested that the viruses shared features based on geographical origin rather than group of vertebrates from which they were isolated.

The best known iridovirus-induced fish diseases are lymphocystis and VENV. Whittington and Hyatt (1996) listed 10 iridoviruses that cause systemic viremias in fish, and several others have since been described, some of which are briefly discussed here. The pathological effect of fish iridoviruses ranges from insignificant to serious.

Largemouth Bass Iridovirus

No debilitating systemic virus infection had been reported from largemouth bass or any other Centrarchidae, until largemouth bass iridovirus (LMBV) was isolated from adult largemouth bass in Santee Cooper Reservoir, South Carolina (Plumb et al. 1996). However, Hoffman et al. (1969) had reported an incidental reovirus isolation from the epithelium of adult bluegill during epitheliocystis investigations. Also, lymphocystis virus (iridovirus) and the 13p_2 virus (aquareovirus), neither of which are debilitating, are other viruses known to infect centrarchids (Wolf 1988). The LMBV has only been isolated from adult largemouth bass in the South Carolina reservoir (Plumb et al. 1996). In 1997, a virus similar, if not identical, to LMBV was isolated from four additional wild largemouth bass populations in the southeastern United States during routine surveys (unpublished). Bluegills, grass carp, bullfrog tadpoles, and metamorphosing bull frogs were experimentally refractive (unpublished).

Moribund largemouth bass in the reservoir lost equilibrium and floated at the surface, apparently due to enlarged swim bladders, but otherwise appeared normal and had no external lesions. Initial virus isolation came from fish that were frozen in the moribund state and assayed for virus within 48 hours. Gills of these fish remained red but soft, eyes were clear, and body surface and fins retained normal color with no macroscopic lesions. Internally, infected fish looked normal except for the gas gland located on the ventral surface of the sometimes greatly enlarged swim bladder.

Largemouth bass iridovirus disease is diagnosed by isolation of the virus in FHM or BF-2 cells inoculated with filtrates of homogenized internal organs. Initial focal CPE, which may be visible within 24 hours, consists of a few pyknotic cells that form circular, cell-free areas with rounded cells at the margins (Figure 13.5). As CPE progresses, additional cells become pyknotic, rounded, and detached, until eventually the entire cell sheet is affected. Culture media harvested from infected FHM cells contained $10^{8.9}$ $TCID_{50}$/mL.

Icosahedral, unenveloped, cytoplasmic, virus particles in infected FHM cells measured 132 to 145 nm in diameter (Figure 13.5). The enveloped virion, acquired after passing through the plasma membrane, measured about 174 nm in diameter. Infected FHM cell cultures stained with acridine orange developed an apple green cytoplasm indicative of a double-stranded nucleic acid. Based on these data, and that generated by PCR primers for other poikilotherm iridoviruses, this virus is an iridovirus (R. P. Hedrick, University of California, Davis, California, personal communication). There was also a weak but positive fluorescent reaction for tissue culture cells infected with LMBV with indirect FAT against fish iridoviruses. According to Piaskoski (1997), LMBV is more closely related to frog virus-3 than it is to other fish iridoviruses. Recent PCR studies by G. Chinchar (Department of Microbiology, University of Mississippi Medical Center, personal communication) indicate, however, that the DNA sequence from LMBV is identical to an iridovirus isolated from doctorfish.

From July to September of 1995, about 1000 adult largemouth bass, ranging in weight from approximately 2 to 6 kg each, died in Santee-Cooper Reservoir, South Carolina, where surface water temperature approached 30°C during peak mortality. Two large adult largemouth bass (approximately 3 kg each) were necropsied in September and yielded virus, although bacterial cultures and parasite examinations were negative. Two additional adult fish captured in July (1995) and frozen 4 months before being assayed also yielded virus from the liver, spleen, and trunk kidney. Virus assays of smaller moribund largemouth bass (0.5 to 1 kg each) taken from the reservoir in autumn of 1996 yielded virus, but moribund bluegill captured at the same time and from the same location were negative for virus. The isolation of LMBV during a 15-month period suggests that the virus is endemic in the Santee-Cooper Reservoir largemouth bass population (Piaskoski 1997).

Experimental studies show that LMBV can be transmitted to smaller, naive largemouth bass by injection or horizontally by cohabitation. No experimentally infected fish showed signs of disease, however, other than an inflamed lesion that measured about 0.5×1.0 cm at the injection site. Virus was isolated from the swim bladder, liver, spleen, trunk kidney, ovary, blood, normal muscle, and necrotic muscle from the injection site at 8 and 26 days postinjection. Viremia in largemouth bass that were injected with media from virus-infected FHM cells was clearly established, with internal organs and muscle developing virus titers that ranged from $10^{2.3}$ to $10^{7.5}$ $TCID_{50}$/g.

The significance of largemouth bass iridovirus is not clear. Experimental transmission studies indicate that the virus does not appear to be highly pathogenic; therefore, its role in the fish kill in Santee-Cooper Reservoir is inconclusive.

Other Iridoviruses

Iridoviruses have also been isolated from goldfish (Berry et al. 1983), guppy, and doctorfish (Hedrick and McDowell 1995), and from turbot in Denmark (Bloch and Larsen 1993). An iridovirus was isolated from cultured, imported, brown-spotted grouper in Singapore (Chua et al. 1994). This disease, known as sleepy grouper disease, has been responsible for significant economic losses in netcage populations from April to August. Each of these iridovirus diseases have varying significance in terms of their pathological effect on respective hosts.

FISH TUMOR VIRUSES

Fish tumors have received considerable attention recently because of a perceived relationship between the health of fish and the quality of the

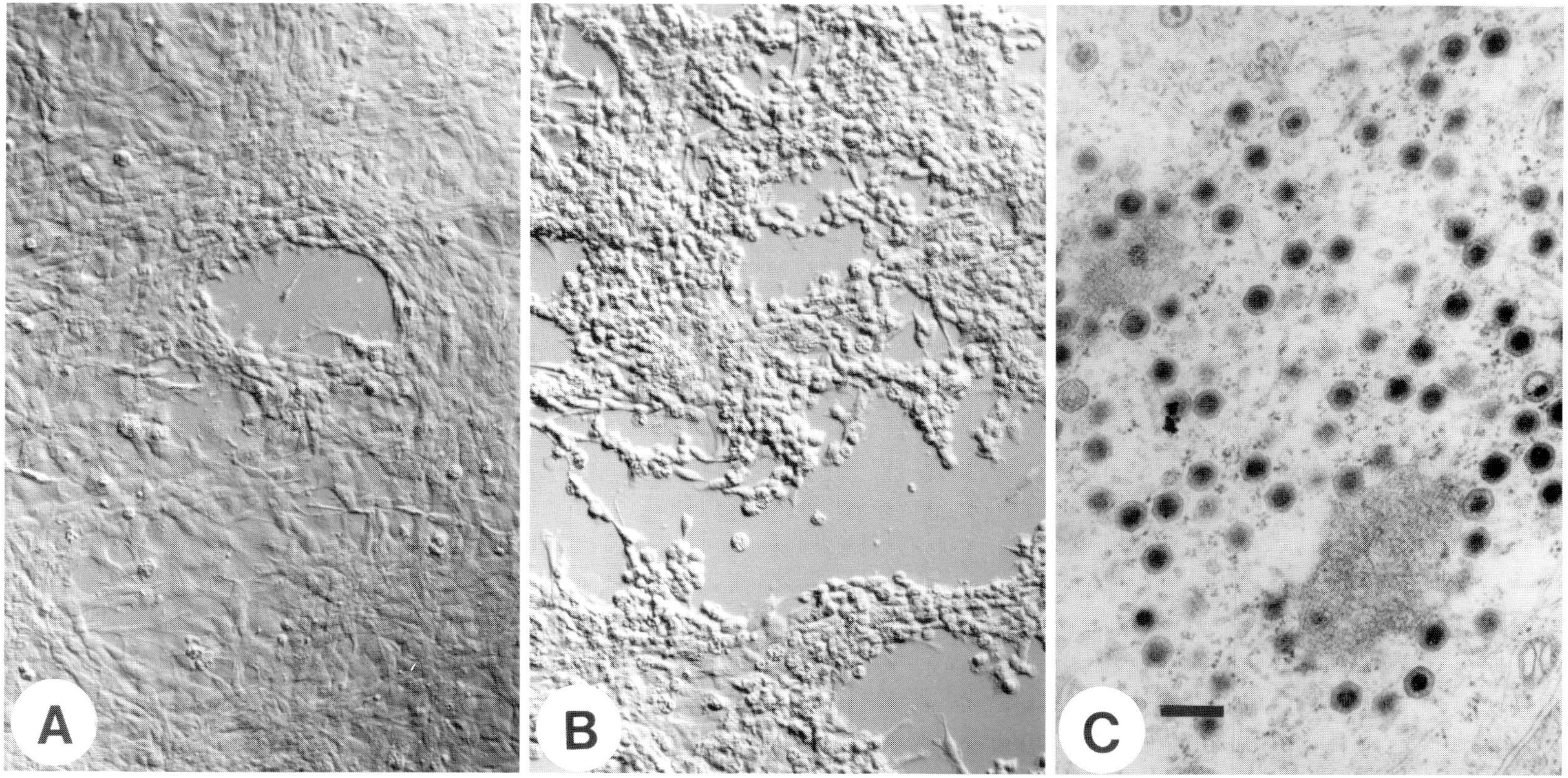

FIGURE 13.5. Largemouth bass iridovirus (LMBV) in fathead minnow (FHM) cells.
(A) *Focal cytopathic effect (CPE) 24 hours postinoculation and incubation at 30°C.*
(B) *Total CPE of FHM cells 36 hours postinoculation and incubation at 30°C.*
(C) *Electron micrograph of LMBV in FHM cytoplasm* (bar = 300 nm).

aquatic environment in which they live. Fish tumors are not a recent occurrence but have been known and recognized for centuries (Anders and Yoshimizu 1994). Although fish skin tumors have a variety of etiologies, many are undetermined. According to Anders and Yoshimizu (1994), there are about 50 types of fish tumors or tumorlike proliferations that range from benign epidermal papillomas to melanomas and carcinomas. About 50% of fish tumors affect the skin, and approximately half of these are in some way virus related. In some fish tumor diseases in which virus is suspected, a viral agent is actually isolated in tissue culture; in others, virus detection is only by electron micrographs of tumors; and still others merely remain suspected of having a viral etiology. Most tumor viruses are known from wild fishes but several play important roles in aquaculture. The following discussion approaches the subject in a general way because several of these diseases have already been discussed in greater detail: fish pox in common carp (Chapter 7), stomatopapilloma of eel (Chapter 8), salmonid herpesvirus type-2 (Chapter 10), and epidermal hyperplasia and discrete epidermal hyperplasia of walleye (Chapter 12).

Viruses Isolated from Tumors

Most isolated viruses known to cause tumors are herpesviruses and have been discussed in previous chapters. Other fish tumors from which viruses have been isolated in cell culture but in which viruses may not actually have caused the tumor are a picornavirus from European smelt (Ahne et al. 1990), rhabdovirus from European eel (Ahne et al. 1987), and birnavirus from North Sea dab (Olesen et al. 1988).

Experimental Transmission But No Virus Isolated

Two herpesviruses and two retroviruses have been involved in experimental tumor transmission from infected fish to noninfected fish via exposure to cell-free filtrates. Epizootic epitheliotropic disease of lake trout caused by a herpesvirus (Chapter 10) and the retrovirus associated with dermal sarcoma of walleye (Chapter 12) have been discussed.

In Japan, several groups of larval Japanese flounder (10 to 30 days old) contracted a tumorous disease that affected the fins and body surface, which were opaque owing to the presence of proliferated epithelial cells (Iida et al. 1989). Mortality reached 80% to 90% in a few weeks at water temperatures of about 20°C. Numerous hexagonal viruslike particles were seen in the nucleus and cytoplasm of the tumor cells when viewed by electron microscopy. The disease was successfully transmitted to naive fish via exposure to 0.45 μm filtrates. Sensitivity of the agent to ether and pH 3, in conjunction with its physical structure, would indicate that the virus was a herpesvirus, but isolation attempts in 33 fish cell lines were unsuccessful.

Other experimentally induced malignant skin tumors of fish associated with viruses are lymphosarcoma of northern pike and muskellunge caused by a retrovirus (Chapter 9) (Sonstegard 1976), squamous cell carcinoma of rudd (Hanjavanit and Mulcahy 1989), and a neurofibromatosis-like disease in damselfish (Schmale and Hensley 1988). Anders and Yoshimizu (1994) stated, "It is notable that successful experimental induction of benign tumors was mostly associated with herpesviruses, whereas, formation of malignant forms seems to be restricted to the involvement of retroviruses of the subfamily Oncovirinae."

Viruses Known by Electron Microscopy

Viruses detected only by electron microscopy of infected tissues have been associated with a large number of fish tumors. According to Anders and Yoshimizu (1994), these include members of the herpesvirus, adenovirus, and retrovirus groups, but some suspected virus particles are questionable. Most fish herpesvirus diseases have been demonstrated by isolation and experimental transmission, but several putative herpesviruses are known only by electron microscopy.

Adenoviruslike particles have been associated with epidermal hyperplasia and papillomas in Baltic cod (Jensen and Bloch 1980) and North Sea dab (Bloch et al. 1986). According to Anders and Yoshimizu (1994), 17 different retroviruslike particles have been found in fish, 10 of which are associated with skin tumors. Some of these have been isolated in cell culture (Frerichs et al. 1991), whereas others have not. The best known of this latter group is walleye dermal sarcoma (Martineau et al. 1991) (Chapter 12). In addition, reverse transcriptase activity in tissue extracts indicated the presence of retrovirus in skin tumors in northern pike (Papas et al. 1976) and stone roller (Sonstegard 1977).

Factors that Affect Virus-Induced Tumors in Fish

A number of factors affect development of fish tumors (Anders and Yoshimizu 1994). The season of the year will affect prevalence of tumors, with highest incidence occurring during spawning season, smoltification, gonadal development, or metamorphosis. During these periods, usually in spring or early summer, endocrine systems and often immune systems are in states of flux, and most likely, this instability plays a role in activation of latent viruses to produce tumors. Interaction between skin, virus, and biological and environmental factors play a major role in epidermal tumor development. This is mediated by immune suppression and increased endocrinological activity that are perceived to be influenced by environmental quality. Environmental parameters that can depress defense mechanisms are fluctuations in water temperature and salinity, high population density, condition factors, high concentrations of certain pollutants, and carcinogens. With regard to fish tumor viruses, these factors could favor virus stimulation and proliferation of epidermal and fibroblastic cells to form tumors (Anders and Yoshimizu 1994). These authors also noted that epidermal fish tumors are induced by herpesviruses, adenoviruses, papovaviruses, and retroviruses, all of which are also responsible for skin tumors in humans.

MANAGEMENT OF OTHER FISH VIRUSES

Control and management of lymphocystis are difficult. When encountered in aquaculture, avoidance of known carrier populations is recommended. When ornamental fish are involved, removal of infected fish before cells can rupture is advised to prevent release of additional virus. Also, replacement of 50% of the water in aquaria every 2 to 3 days will dilute virus and possibly reduce infection incidence. Proper sanitation and disinfection of equipment and holding containers are essential to prevent transmission of the virus to noninfected fish.

In the case of VEN, because fry and young-of-the-year fish seem to be more susceptible, they should be reared in water that is free of infected fish. Adverse environmental conditions, such as low dissolved oxygen and other stressors, should be avoided, especially when disease has occurred.

A management scheme to deal with EHNV by avoidance and possible disease-free inspections was proposed by Whittington and Hyatt (1996). They indicated that EHNV cannot be eradicated from farmed rainbow trout or in water supplies from reservoirs because of carrier wild redfin perch. In an effort to restrict EHNV to those areas of Australia where virus is endemic, stringent guidelines for production farms have been established. Infected farms may sell only killed fish or fillets, and export of live fish from these zones is prohibited. Government-owned farms practice specific pathogen free (SPF) certification for EHNV before distribution of fish or eggs. Additionally, in the event that new infection sites are identified that have a fish-free closed water system, eradication of infected fish is recommended. Equipment and culture units should be completely disinfected with 200 mg of sodium hypochlorite per liter, dried, and organic matter removed (Langdon 1989). Facilities should be repopulated only with stocks certified to be EHNV free.

Control of viral nervous necrosis may vary with geographical region, but absence of stress on susceptible fish is uniformly required (Nakai et al. 1995). The PCR method for detecting SJNNV-positive spawner striped jack combined with hatchery hygiene and husbandry procedures recommended by Nishizawa et al. (1994) can be helpful in preventing and controlling the virus in a particular species. A similar approach to that used for striped jack may be applied to European seabass operations in the Mediterranean region, but Le Breton et al. (1997) suggests that this procedure may not always be effective because of widespread distribution of the pathogen in the region. They also suggested that development of a vaccine may be required to combat the disease.

There are no particular management practices applied to fish tumor viruses other than avoidance of known infected populations. Also, when fish with tumors are found in culture units, they should be removed immediately.

REFERENCES

Ahne, W., I. Schwanz-Pfitzner, and I. Thomsen. 1987. Serological identification of 9 viral isolates from European eels (*Anguilla anguilla*) with stomatopapilloma by means of neutralization tests. *Journal of Applied Ichthyology* 3:30–32.

Ahne, W., K. Anders, M. Halder, and M. Yosimizu. 1990. Isolation of picornavirus-like particles from the European smelt (*Osmerus eperlanus*). *Journal of Fish Diseases* 13:167–168.

Ahne, W., M. Bremont, R. P. Hedrick, A. D. Hyatt, and R. J. Whittington. Iridoviruses associated with epizootic haematopoietic necrosis (EHN) in aquaculture. *World Journal of Microbiology and Biotechnology* 13:367–373.

Alpers, C. E., B. B. McCain, M. S. Myers, and S. R. Wellings. 1977. Lymphocystis disease in yellowfin sole (*Limanda aspera*) in the Bering Sea. *Journal of the Fisheries Research Board of Canada* 34:611–616.

Amin, O. M. 1979. Lymphocystis disease in Wisconsin fishes. *Journal of Fish Diseases* 2:207–217.

Anders, K. 1989. Lymphocystis disease of fishes. In: *Viruses of Lower Vertebrates,* edited by W. Ahne and E. Kurstak, 141–160. Berlin: Springer-Verlag.

Anders, K., and M. Yoshimizu. 1994. Role of viruses in the induction of skin tumors and tumor-like proliferations of fish. *Diseases of Aquatic Organisms* 19:215–232.

Appy, R. G., M. D. B. Burt, and T. J. Morris. 1976. Viral nature of piscine erythrocytic necrosis (PEN) in the blood of Atlantic cod (*Gadus morhu*). *Journal of the Fisheries Research Board of Canada* 33:1380–1385.

Arimoto, M., K. Mushiate, Y. Mizuta, T. Nakai, K. Muroga, and I. Furusawa. 1992. Detection of striped jack nervous necrosis virus (SJNNV) by enzyme-linked immunosorbent assay (ELISA). *Fish Pathology* 27:191–195.

Arimoto, M., K. Mori, T. Nakai, K. Muroga, and I. Furusawa. 1993. Pathogenicity of the causative agent of viral nervous necrosis in striped jack, *Pseudocaranx dentex* (Block & Schnieder). *Journal of Fish Diseases* 16:461–469.

Arimoto, M., K. Maruyama, and I. Furusawa. 1994. Epizootiology of viral nervous necrosis (VNN) in striped jack. *Fish Pathology* 29:19–24.

Berry, E. S., T. B. Shea, and J. Gabliks. 1983. Two iridovirus isolates from *Carassius auratus* (L). *Journal of Fish Diseases* 6:501–510.

Bloch, B., and J. L. Larsen. 1993. An iridovirus-like agent associated with systemic infection in cultured turbot *Scophthalmus maximus* fry in Denmark. *Diseases of Aquatic Organisms* 15:235–240.

Bloch, B., S. Mellergaard, and E. Nielsen. 1986. Adenovirus-like particles associated with epithelial hyperplasias in dab, *Limanda limanda* (L.). *Journal of Fish Diseases* 9:281–285.

Bowden, R. A., D. J. Oestmann, D. H. Lewis, and M. S. Frey. 1995. Lymphocystis in red drum. *Journal of Aquatic Animal Health* 7:231–235.

Breuil, G., J. F. Bonami, and Y. Pichot. 1991. Viral infection (picorna-like virus) associated with mass mortalities in hatchery-reared sea-bass (*Dicentrarchus labrax*) larvae and juveniles. *Aquaculture* 97:109–116.

Bunkley-Williams, L. J. M. Grizzle, and E. H. Williams, Jr. 1996. First report of lymphocystis in the family Gerreidae: Caitipa mojarras *Diapterus rhombeus* from La Parguera, Puerto Rico. *Journal of Aquatic Animal Health* 8:176–179.

Chen, S. N., G. H. Kou, R. P. Hedrick, and J. L. Fryer. 1985. The occurrence of viral infections of fish in Taiwan. In: *Fish and Shellfish Pathology,* edited by A. E. Ellis, 313–319, Orlando, FL: Academic Press.

Chua, F. H. C., M. L. Ng, K. L. Ng, J. J. Loo, and J. Y. Wee. 1994. Investigation of outbreaks of a novel disease, "sleepy grouper disease," affecting the brown-spotted grouper, *Epinephelus louvina* Forskal. *Journal of Fish Diseases* 17:417–427.

Clifford, T. J., and R. L. Applegate. 1970. Lymphocystis disease in tagged and untagged walleyes in a South Dakota lake. *The Progressive Fish Culturist* 32:177.

Colorni, A., and A. Diamant. 1995. Splenic and cardiac lymphocystis in the red drum, *Sciaenops ocellatus* (L). *Journal of Fish Diseases* 18:467–471.

Cook, D. W. 1972. Experimental infection studies with lymphocystis virus from Atlantic croaker. In: *Proceedings of the Third Annual Workshop, World Mariculture Society,* edited by J. W. Avault, E. Boudreaux, and E. Jaspers, 329–335. St. Petersburg, FL: January 26–28.

Dukes, T. W., and A. R. Lawler. 1975. The ocular lesions of naturally occurring lymphocystis in fish. *Canadian Journal of Comparative Medicine* 39:406–410.

Dunbar, C. E., and K. Wolf. 1966. The cytological course of experimental lymphocystis in bluegill. *Journal of Infectious Disease* 116:466d–472.

Durham, P. J. K., and C. D. Anderson. 1978. Lymphocystis disease in imported tropical fish. *New Zealand Vetmedicine Journal* 29:88–91.

Eaton, W. D. 1990. Artificial transmission of erythrocytic necrosis virus (ENV) from Pacific herring in Alaska chum, sockeye, and pink salmon. *Journal of Applied Ichthyology* 6:136–141.

Eiras, J. C. 1984. Virus infection of marine fish: prevalence of viral erythrocytic necrosis (VEN) in *Mugil cephalus* L., *Blennius pholis* L. and *Platichthys flesus* L. in coastal waters of Portugal. *Bulletin of the European Association of Fish Pathologist* 4:52–56.

Evelyn T. P. T., and G. S. Traxler. 1978. Viral erythrocytic necrosis: natural occurrence in Pacific salmon and experimental transmission. *Journal of the Fisheries Research Board of Canada* 35:903–907.

Faisal, M. 1989. Lymphocystis in the Mediterranean golden grouper *Epinephelus alexandrinus* Valenciennes 1828 (Pisces, Serranidae). *Bulletin of the European Association of Fish Pathologist* 9:17.

Frerichs, G. N., D. Morgan, D. Hart, C. Skerrow, R. J. Roberts, and D. E. Onions. 1991. Spontaneously productive C-type retrovirus infection of fish cell lines. *Journal of General Virology* 72:2537–2539.

Gould, A. R., A. D. Hyatt, S. H. Hengstberger, R. J. Whittington, and B. E. H. Coupar. 1995. A polymerase chain reaction (PCR) to detect epizootic haematopoietic necrosis virus and Bohle iridovirus. *Diseases of Aquatic Organisms* 22:211–215.

Haney, D. C., D. A. Hursh, M. C. Mix, and J. R. Winton. 1992. Physiological and hematological changes in chum salmon artificially infected with erythrocytic necrosis virus. *Journal of Aquatic Animal Health* 4:48–57.

Hanjavanit, C., and M. F. Mulcahy. 1989. Squamous cell carcinoma in rudd (Pisces, Cyprinicae). In: *Abstracts of the IVth International Conference of the EAFP*, edited by J. L. Barja and A. E. Toranzo, 89. Santiago de Compostela, Spain: European Association of Fish Pathologists.

Hedrick, R. P., and T. S. McDowell. 1995. Properties of iridoviruses from ornamental fish. *Veterinary Research* 26:423–427.

Hedrick, R. P., T. S. McDowell, W. Ahne, C. Torchy, and P. de Kinkelin. 1992. Properties of three iridovirus-like agents associated with systemic infections of fish. *Diseases of Aquatic Organisms* 13:203–209.

Hengstberger, S. G., A. D. Hyatt, R. Speare, and B. E. H. Coupar. 1993. Comparison of epizootic haematopoietic necrosis and Bohle iridovirus, recently isolated Australian iridoviruses. *Diseases of Aquatic Organisms* 15:93–107.

Hoffman, G. L., C. E. Dunbar, K. Wolf, and L. O Zwillenberg. 1969. Epitheliocystis, a new infectious disease of the bluegill (*Lepomis macrochirus*). *Antonie Van Leeuwenhoek Journal of Microbiology and Serology* 28:146–158.

Howse, H. D., and J. Y. Christmas. 1970. Lymphocystis tumors: histochemical identification of hyaline substances. *Transactions of the American Microscopy Society* 89:276–282.

Hyatt, A. D., and seven others. 1997. Epizootic mortality in the pilchard *Sardinops sagax neopilchard* in Australia and New Zealand in 1995. II. Identification of a herpesvirus within the gill epithelium. *Diseases of Aquatic Organisms* 28:17–29.

Iida, Y., K. Masumura, T. Nakai, M. Sorimachi, H. Matsuda. 1989. A viral disease in larvae and juveniles of the Japanese flounder, *Paralichthys olevaceus*. *Journal of Aquatic Animal Health* 1:7–12.

Jensen, N. J., and B. Bloch. 1980. Adenovirus-like particles associated with epidermal hyperplasia in cod (*Gadus morhua*). *Nordic Veterinary Medicine* 32:173–175.

Laird, M., and W. L. Bullock. 1969. Marine fish hematozoa from New Brunswick and New England. *Journal of the Fisheries Research Board of Canada* 26:1075–1102.

Lamas, J., M. Noya, A. Figueras, and A. E. Toranzo. 1995. Pathology associated with viral erythrocytic infection in turbot, *Scophthalmus maximus* (L). *Journal of Fish Diseases* 18:425–433.

Langdon, J. S. 1989. Experimental transmission and pathogenicity of epizootic haematopoietic necrosis virus (EHNV) in redfin perch, *Perca fluviatilis* L., and 11 other teleosts. *Journal of Fish Diseases* 12:295–310

Langdon, J. S., and J. D. Humphrey. 1987. Epizootic hematopoietic necrosis, a new viral disease of redfin perch, *Perca fluviatilis L.*, in Australia. *Journal of Fish Diseases* 10:289–297.

Langdon, J. S., J. D. Humphrey, L. M. Williams, A. D. Hyatt, and H. A. Westbury. 1986. First virus isolation from Australian fish: an iridovirus-like pathogen from redfin perch, *Perca fluviatilis* L. *Journal of Fish Diseases* 9:263–268.

Langdon, J. S., J. D. Humphrey, and L. M. Williams. 1988. Outbreaks of and EHNV-like iridovirus in cultured rainbow trout, *Salmo gairdneri* Richardson, in Australia. *Journal of Fish Diseases* 11:93–96.

Lawler, A. R., H. D. Howse, and D. W. Cook D. 1974. Silver perch, *Bairdiella chrysura*: new host for lymphocystis. *Copeia* 1974:266–269.

Le Breton, A., L. Girsez, J. Sweetman, and F. Olevier. 1997. Viral nervous necrosis (VNN) associated with mass mortalities in cage-reared sea bass, *Dicentrarchus labrax* (L.). *Journal of Fish Diseases* 20:145–151.

Lorenzen, K., S. A. Clers, and K. Anders. 1991. Population dynamics of lymphocystis disease in estuarine flounder, *Platichthys flesus* (L). *Journal of Fish Biology* 39:577–587.

MacMillan, J. R., and D. Mulcahy. 1979. Artificial transmission to and susceptibility of Puget Sound fish to viral erythrocytic necrosis (VEN). *Journal of the Fisheries Research Board of Canada* 36:1097–1101.

MacMillan, J. R., D. Mulcahy, and M. Landolt. 1980. Viral erythrocytic necrosis: some physiological consequences of infection in chum salmon (*Oncorhynchus keta*). *Canadian Journal of Fisheries and Aquatic Sciences* 37:799–804.

Mao, J., R. P. Hedrick, and V. G. Chinchar. 1997. Molecular characterization, sequence analysis, and taxonomic position of newly isolated fish iridoviruses. *Virology* 229:212–220.

Martineau, D., P. R. Bowser, R. R. Renshaw, and J. W. Casey. 1991. Molecular characterization of a unique retrovirus associated with a fish tumor. *Journal of Virology* 66:596–599.

McCosker, J. E. 1969. A behavioral correlate for passage of lymphocystis disease in three blennioid fish. *Copeia* 1969:636–637.

Meyers, T. R., A. K. Hauck, W. D. Blandenbeckler, and T. Minicucci. 1986. First report of viral erythrocytic necrosis in Alaska, USA, associated with epizootic mortality in Pacific herring, *Clupea harengus pallasi* (Valenciennes). *Journal of Fish Diseases* 9:479–491.

Mori, K., T. Nakai, K. Muroga, M. Arimoto, K. Mushiake, and I. Furusawa. 1992. Properties of a virus belonging to Nodaviridae found in larval striped jack (*Pseudocaranx dentes*) with nervous necrosis. *Virology* 187:368–371.

Muroga, K. 1995. Viral and bacterial diseases in larval and juvenile marine fish and shellfish: a review. *Fish Pathology* 30:71–85.

Mushiake, K., T. Nishizawa, T. Nakai, I. Furusawa, and K. Muroga. 1994. Control of VNN in striped jack: selection of spawners based on the detection of SJNNV gene by polymerase chain reaction (PCR). *Fish Pathology* 29:177–183.

Nakai, T., K. Mori, T. Nishizawa, and K. Muroga. 1995. Viral nervous necrosis of larval and juvenile marine fish. In: *Proceedings of the International Symposium on Biotechnology Applications in Aquaculture.* Asian Fisheries Society Special Publication 10:147–152.

Nigrelli, R. F. 1950. Lymphocystis disease and ergasilid parasites in fishes. *Journal of Parasitology* 36:36. Abstract.

Nigrelli, R. F., and G. D. Ruggieri. 1965. Studies on virus diseases of fishes. Spontaneous and experimentally induced cellular hypertrophy (lymphocystis disease) in fishes of the New York Aquarium with a report of new cases and an annotated bibliography (1874–1965). *Zoologica* 50:83–96.

Nishizawa, T., M. Kise, T. Nakai, and K. Muroga. 1995. Neutralizing monoclonal antibodies to striped jack nervous necrosis virus (SJNNV). *Fish Pathology* 30:111–114.

Nishizawa, T., K. Muroga, and M. Arimoto. 1996. Failure of the polymerase chain reaction (PCR) method to detect striped jack nervous necrosis virus (SJNNV) in striped jack (*Pseudocaranx dentex*) selected as spawners. *Journal of Aquatic Animal Health* 8:332–334.

Nishizawa, T., K. Mori, T. Nakai, I. Furusawa, and K. Muroga. 1994. Polymerase chain reaction (PCR) amplification of RNA of striped jack nervous necrosis virus (SJNNV). *Diseases of Aquatic Organisms* 18:103–107.

Nguyen, H. D., T. Nakai, and K. Muroga. 1996. Progression of striped jack nervous necrosis virus (SJNNV) infection in naturally and experimentally injected striped jack (*Pseudocaranx dentex*) larvae. *Diseases of Aquatic Organisms* 24:99–105.

Nguyen, H. D., K. Mushiake, T. Nakai, and K. Muroga. 1997. Tissue distribution of striped jack nervous necrosis virus (SJNNV) in adult striped jack. *Diseases of Aquatic Organisms* 28:87–91.

Olesen, N. J., P. E. V. Jørgensen, B. Bloch, and S. Mellergaard. 1988. Isolation of an IPN-like virus belonging to the serogroup II of the aquatic birnaviruses from dab, *Limanda limanda* L. *Journal of Fish Diseases* 11:449–451.

Overstreet, R. M. 1988. Aquatic pollution, Southeastern U. S. coasts: histopathologic indicators. *Aquatic Toxicology* 11:213–239.

Overstreet, R. M., and H. D. Howse. 1977. Some parasites and diseases of estuarine fishes in polluted habitats of Mississippi. *Annals of the New York Academy of Science* 298:427–462.

Papas. T. S., J. E. Dahlberg, and R. A. Sonstegard. 1976. Type C virus in lymphosarcoma in northern pike (*Esox lucius*). *Nature* 261:506–508.

Peters, N., and W. Schmidt. 1995. Formation and disintegration of virions in lymphocystis cells of plaice *Pleuronectes platessa. Diseases of Aquatic Organisms* 21:109–113.

Petty, L. L., and J. J. Magnuson. 1974. Lymphocystis in age 0 bluegills (*Lepomis macrochirus*) relative to heated effluent in Lake Monona, Wisconsin. *Journal of the Fisheries Research Board of Canada* 31:189–193.

Piaskoski, T. O. 1997. Characterization of the largemouth bass virus in cell culture and protein profile comparison to frog virus-3 and lymphocystis disease virus. Auburn, AL: Auburn University. MS thesis.

Pinto, R. M., P. Alvarez-Pellitero, A. Bosch, and J. Jafre. 1989. Occurrence of viral erythrocytic infection in the Mediterranean sea bass, *Dicentrarchus labrax* (L.). *Journal of Fish Diseases* 12:185–191.

Plumb, J. A., J. M. Grizzle, H. E. Young, A. D. Noyes, and S. Lamprecht. 1996. An iridovirus isolated from wild largemouth bass. *Journal of Aquatic Animal Health* 8:265–270.

Reddacliff, L. A., and R. J. Whittington. 1996. Pathology of epizootic haematopoietic necrosis virus (EHNV) infection in rainbow trout (*Oncorhynchus mykiss* Walbaum) and redfin perch (*Perca fluviatilis* L). *Journal of Comparative Pathology* 115:103–115.

Reiersen, L. O., and K. Fugelli. 1984. Annual variation in lymphocystis infection frequency in flounder, *Platichthys flesus* (L.). *Journal of Fish Biology* 24:187–191.

Reno, P. W., and B. L. Nicholson. 1980. Viral erythrocytic necrosis (VEN in Atlantic cod (*Gadus morhua*): in vitro studies. *Canadian Journal of Fisheries and Aquatic Sciences* 37:2276–2281.

Reno, P. W., D. V. Serreze, S. K. Hellyer, and B. L. Nicholson 1985. Hematological and physiological effects of viral erythrocytic necrosis (VEN) in Atlantic cod and herring. *Fish Pathology* 20:353–360.

Roberts, B. J. 1975. Experimental pathogenesis of lymphocystis in the plaice (*Pleuronectus platessa*). In:

Wildlife Diseases, edited by L. A. Page, 431–441. New York: Plenum Press.

Robin, J., and L. Bertholimue. 1981. Purification of lymphocystis disease virus (LDV) grown in tissue culture, evidences for the presence of two types of viral particles. *Reviews in Canadian Biology* 40:323–329.

Rohovec, J. S., and A. Amandi. 1981. Incidence of viral erythrocytic necrosis among hatchery reared salmonids of Oregon. *Fish Pathology* 15:135–141.

Ryder, R. A. 1961. Lymphocystis as a mortality factor in a walleye population. *The Progressive Fish-Culturist* 23:183–186.

Schmale, M. C., and G. T. Hensley. 1988. Transmissibility of a neurofibromatosis-like disease in bicolor damselfish. *Cancer Research* 48:3828–3833.

Sherburne, S. W. 1977. Occurrence of piscine erythrocytic necrosis (PEN) in the blood of the anadromous alewife, *Alosa pseudoharengus* from Maine coastal streams. *Journal of the Fisheries Research Board of Canada* 34:281–286.

Smail, D. A. 1982. Viral erythrocytic necrosis in fish: a review. *Proceedings of the Royal Society of Edinburgh (B)* 81:169–175.

Smail, D. A., and S. I. Egglestone. 1980. Virus infections of marine fish erythrocytes: prevalence of piscin erythrocytic necrosis in cod *Gadus morhua* L. and blenny *Blennius pholis* L. in coastal and off shore waters of the United Kingdom. *Journal of Fish Diseases* 3:41–46.

Sonstegard, R. A. 1976. Studies of the etiology and epizootiology of lymphosarcoma in *Esox lucius* L. and *Esox masquinongy*. *Progress in Experimental Tumor Research* 20:141–155.

Sonstegard, R. A. 1977. Environmental carcinogenesis studies in fishes of the Great Lakes of North America. *Annals of the New York Academy of Science* 298:261–269.

Steiner, K. A., R. J. Whittington, R. K. Petersen, C. Hornitzky, and H. Garnett. 1991. Purification of epizootic haematopoietic necrosis virus and its detection using ELISA. *Journal of Virological Methods* 33:199–209.

Walker, R. 1971. PEN, a viral lesion of erythrocytes. *American Zoologist* 11:707. Abstract.

Walker, R., and S. W. Sherburne. 1977. Piscine erythrocytic necrosis virus in Atlantic cod, *Gadus morhua,* and other fish: ultrastructure and distribution. *Journal of the Fisheries Research Board of Canada* 34:1188–1195.

Weissenberg, R. 1965. Fifty years of research on the lymphocystis virus disease of fishes (1914–1964). *Annals of the New York Academy of Science* 126:362–374.

Whittington, R. J., and A. D. Hyatt. 1997. Diagnosis and prevention of epizootic haematopoietic necrosis virus (EHNV) infection. In: *New Approaches to Viral Diseases of Aquatic Animals.* National Research Institute of Aquaculture, 80–93. Nahsei, Mie, Japan.

Whittington, R. J., and A. D. Hyatt. 1996. Contingency planning for control of epizootic haematopoietic necrosis disease. *Singapore Veterinary Journal.* 20:79–87.

Whittington, R. J., and G. L. Reddacliff. 1995. Influence of environmental temperature on experimental infection of redfin perch (*Perca fluviatilis*) and rainbow trout (*Oncorhynchus mykiss*) with epizootic haematopoietic necrosis virus, an Australian iridovirus. *Australian Veterinary Journal* 72:421–424.

Whittington, R. J., A. Philbey, G. L. Reddacliff, and A. R. MacGown. 1994. Epidemiology of epizootic haematopietic necrosis virus (EHNV) infection in farmed rainbow trout, *Oncorhynchus mykiss* (Walbaum): findings based on virus isolation, antigen capture ELISA and serology. *Journal of Fish Diseases* 17:205–218.

Whittington, R. J., C. Kearns, A. D. Hyatt, S. Hengstberger, and T. Rutzou. 1996. Spread of epizootic haematopoietic necrosis virus (EHNV) in redfin perch (*Perca fluviatilis*) in southern Australia. *Australian Veterinary Journal* 73(3):112–114.

Whittington, R. J., J. B. Jones, P. M. Hine, and A. D. Hyatt. 1997. Epizootic mortality in the pilchard *Sardinops sagax neopilchardus* in Australia and New Zealand in 1995. I. Pathology and epizootiology. *Diseases of Aquatic Organisms* 28:1–16.

Williams, T. 1996. The iridoviruses. *Advances in Virus Research* 46:345–412.

Williams, E. H., Jr., J. M. Grizzle, and L. Bunkley-Williams. 1996. Lymphocystis in Indian glassfish *Chanda ranga* imported from Thailand to Puerto Rico. *Journal of Aquatic Animal Health* 8:173–175.

Witt, A., Jr. 1957. Seasonal variation in the incidence of lymphocystis in the white crappie from the Niangua arm of the Lake of the Ozarks, Missouri. *Transactions of the American Fisheries Society* 85:271–279.

Wolf, K. 1962. Experimental propagation of lymphocystis disease of fishes. *Virology* 18:249–256.

Wolf, K. 1988. *Fish Viruses and Fish Virus Diseases.* Ithaca, NY: Cornell University Press.

Wolf, K., and C. P. Carlson. 1965. Multiplication of lymphocystis virus in the bluegill (*Lepomis macrochirus*). *Annals of the New York Academy of Science* 126:414–419.

Wolf, K., M. Gravell, and R. G. Malsberger. 1966. Lymphocystis virus: isolation and propagation in centrarchid fish cell lines. *Science* 151:1004–1005.

Yamamoto, T., R. K. Kelly, and O. Nielsen. 1985. Morphological differentiation of virus associated skin tumors of walleye (*Stizostedion vitreum vitreum*). *Fish Pathology* 20:361–372.

Yoshimizu, M., T. Nomura, T. Awakura, Y. Ezura, and T. Kimura. 1989. Prevalence of pathogenic fish viruses in anadromous masu salmon (*Oncorhynchus masou*) in the northern part of Japan, 1976–1987. *Physiology Ecology of Japan,* 1(Special Volume):559–576.

III
Bacterial Diseases

An aquatic environment contains numerous species of bacteria, many of which are essential to the balance of nature and are of no direct consequence in fish disease. Some 60 to 70 bacterial species, however, are capable of causing disease in aquatic animals, several of which are also infectious to humans.

The aquatic environment, especially aquacultural and eutrophic waters, provides a natural habitat for growth and proliferation of bacteria because of availability of nutrient-rich organic materials that enhance bacterial growth. Bacterial flora of water is influenced by nutrient availability, pH, temperature, and other factors that affect their growth pattern, virulence, and pathogenicity. To grow, some bacteria need organic matter to provide nutrients, some can survive as free living organisms or serve as fish pathogens (facultative), and others are fastidious and survive indefinitely only within a host (obligate pathogens). Also, the level of salinity in water or culture media may affect growth and survival of some bacteria. Generally, an optimum pH 6 to 8 is desirable for growth of most bacteria, and many die if pH is above 11 or below 5. A temperature of 20 to 42°C is necessary for optimum growth of most bacteria, but some will grow at above 50°C (thermophiles), and others can grow at 0°C (psychrophiles). Mesophil bacteria that are most prolific at temperatures of 18 to 45°C include many bacteria that are pathogenic to fish.

Three basic bacterial cell morphologies are spherical (coccus), rod (bacillus), or spiral shaped (spirillum). Bacteria have a particular staining characteristic (Gram's) referred to as Gram positive (blue) or Gram negative (red or pink). Gram-positive organisms may also be acid fast, which relates to presence or absence of mycotic acid in the cell wall. Most bacteria responsible for fish disease are Gram-negative rods, but some pathogens that are Gram-positive rods or cocci and a few that are acid-fast rods also cause disease in aquatic animals.

In terms of pathological capability, there are basically two disease-producing types of bacteria infectious to fish: (1) *Obligate pathogens* and (2) *nonobligate* or *facultative pathogens*. Although there are few true obligate pathogenic bacteria that cause fish diseases, *Renibacterium salmoninarum,* the etiological agent of bacterial kidney disease, and *Mycobacterium* spp. are examples that are rarely found outside of a host. However, facultative bacteria can survive indefinitely in water, and when environmental conditions are conducive, infectious fish diseases may result. *Aeromonas hydrophila,* a primary bacterial species involved in the motile *Aeromonas* septicemia complex, is an example of a facultative bacteria.

Fish bacterial infections can occur as a bacteremia, which implies the presence of bacterial organisms in the bloodstream without clinical infection, or as a septicemia, which indicates that bacteria and toxins are actually present in the circulatory system and usually precipitate disease and clinical signs. Inflammation, hemorrhage, and necrosis are clinical signs associated with septicemia. Pathogenic bacteria can cause disease-producing exotoxins, which is generally a characteristic, but not exclusively, of Gram-positive organisms. Gram-negative bacteria can produce either exotoxins or endotoxins, which consist of proteolytic enzymes that kill host cells and cause necrosis or can make blood vessels more porous and cause hemorrhage.

Bacterial diseases are categorized according to fish species they infect, staining reaction of the cell, or by bacterial family, genus, and species (taxonomic grouping). In this book, bacterial diseases are organized according to family of cultured fishes that they most severely affect; therefore, it should not be concluded that the organism or disease discussed is limited to a particular fish group.

14 Catfish Bacterial Diseases

Channel catfish are the most extensively cultured fish in the United States. Three major bacterial infections that affect these fish in the culture environment are columnaris (*Flavobacterium columnare*), enteric septicemia of catfish (*Edwardsiella ictaluri*), and motile *Aeromonas* septicemia (*A. hydrophila* and related motile aeromonads). Other occasional diseases of catfish are edwardsiellosis (*Edwardsiella tarda*), pseudomonas septicemia (*Pseudomonas fluorescens*), and other incidental infections.

COLUMNARIS

Columnaris, often referred to as cotton wool or mouth fungus, is an acute to chronic infectious skin disease, especially in channel catfish. The disease was first described by Davis (1922), but it was not until 22 years later that the organism was isolated and characterized (Ordal and Rucker 1944). The causative agent of columnaris has gone through several reclassifications and name changes since its original designation as *Bacterium columnaris*. It has been named *Chondrococcus columnaris* (Ordal and Rucker 1944), *Cytophaga columnaris* (Garnjobst 1945), *Flexibacter columnaris,* and again *Cytophaga columnaris.* Bernardet and Grimont (1989) presented DNA relatedness and phenotypic characterization to justify retaining the name *Flexibacter columnaris.* More recently, Bernardet et al. (1996) redescribed this group of bacteria and renamed the organism that causes columnaris in freshwater fish *Flavobacterium columnare,* which is the term used throughout this text. The genus *Flavobacterium,* yellow pigmented bacteria, contains three additional fish pathogens that affect salmonids (Chapter 17): *Flavobacterium psychrophila* [*Flexibacter* (*cytophaga*) *psychrophilum*] (Borg 1960), etiological agent of coldwater disease, and *Flavobacterium branchiophilum* and *F. aquitile,* etiological agents of bacterial gill disease.

Geographical Range and Species Susceptibility

Columnaris disease exists worldwide in fresh water habitats, with channel catfish and other ictalurids being most severely affected (Meyer 1970). No wild or cultured freshwater fish, including ornamental fish in aquaria, are totally resistant to columnaris. Cultured eels (fresh and brackish water) are highly susceptible to *F. columnare* (Wakabayashi et al. 1970) as are salmonids, particularly hatchery reared trout and migrating adult salmon (Becker and Fujihara 1978), cultured centrarchids, common carp in Europe (Bernardet 1989), golden shiners, fathead minnows, and goldfish in the United States, and tilapia wherever cultured. The marine counterpart to fresh water columnaris disease, *Flexibacter maritimus,* occurs only in saltwater fishes (Chapter 20).

Clinical Signs and Findings

Clinical signs of columnaris are easily recognized and differ little between species; however, lesion location will vary from outbreak to outbreak. Disease severity, type and location of lesions, and pathogen virulence may correspond to the particular strain of *F. columnare* responsible for infection.

Clinical signs of columnaris are nearly pathognomonic for the disease, but diagnosis can be complicated by simultaneous viral, bacterial, and/or

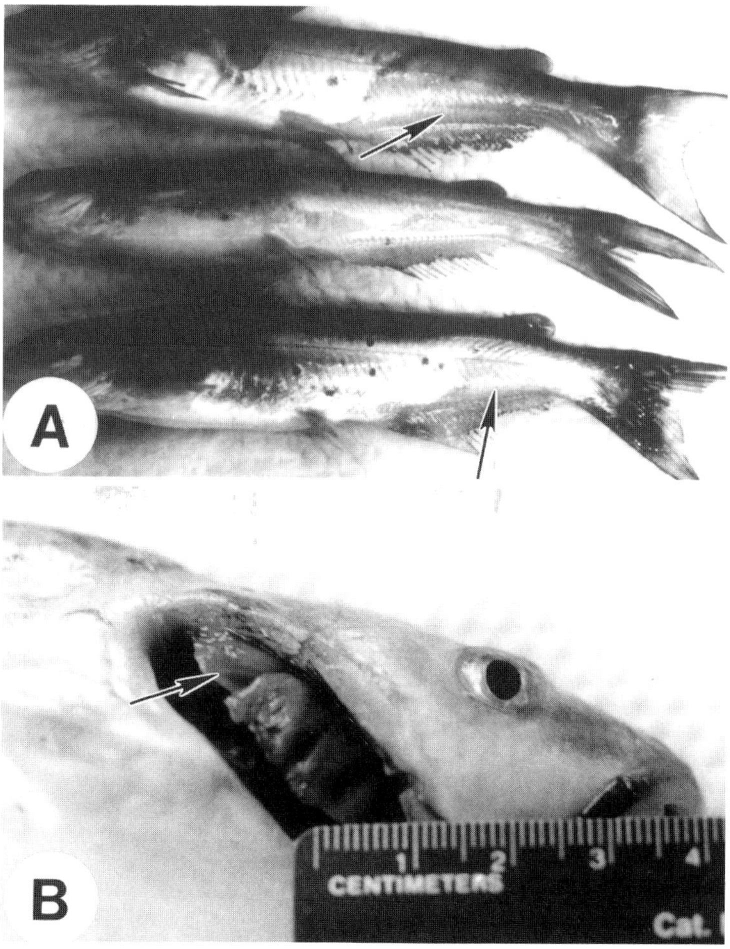

FIGURE 14.1. Columnaris lesion. (A) *Columnaris lesion on caudal peduncle* (arrows) *and frayed caudal fin.* (B) *Lesion on gill* (arrow) *of channel catfish.*

parasitic infections. Columnaris generally begins as an external infection on fins, body surface, or gills. The fins become frayed (necrotic) with grayish to white margins, and initial skin lesions appear as discrete bluish gray areas that evolve into depigmented necrotic lesions, causing fish to lose their metallic sheen (Figure 14.1). Skin lesions have yellowish (mucoid material) or pale margins accompanied by mild inflammation. These same lesion types occur on eel, trout, cyprinids, centrarchids, tilapia, and other fish groups infected with columnaris. Gill lesions appear as white to brown necrotic areas. The color depends on the presence of debris and/or secondary fungus in the lesion (Figure 14.1). Lesions can develop exclusively on the gills, which usually results in subacute disease and mortality.

In some instances columnaris becomes systemic with little or no pathological change occurring in the visceral organs. Whether bacteria isolated from internal organs of systemically infected fish was taxonomically *F. columnare* is not clear, but bacteria can be isolated from kidneys of more than 50% of necropsied catfish infected with epidermal columnaris (Hawke and Thune 1992).

Diagnosis

Columnaris is normally diagnosed by recognition of typical lesions on the body, fins, and gills of diseased fish and presence of long, slender rods in wet mounts made from suspect lesions (Figure 14.2). These nonflagellated bacteria display gliding motility and form "hay stacks" or columns in wet mounts, which is basically confirmatory of the disease.

A moist medium with a low nutrient level and low agar content such as cytophaga (Ordal's)

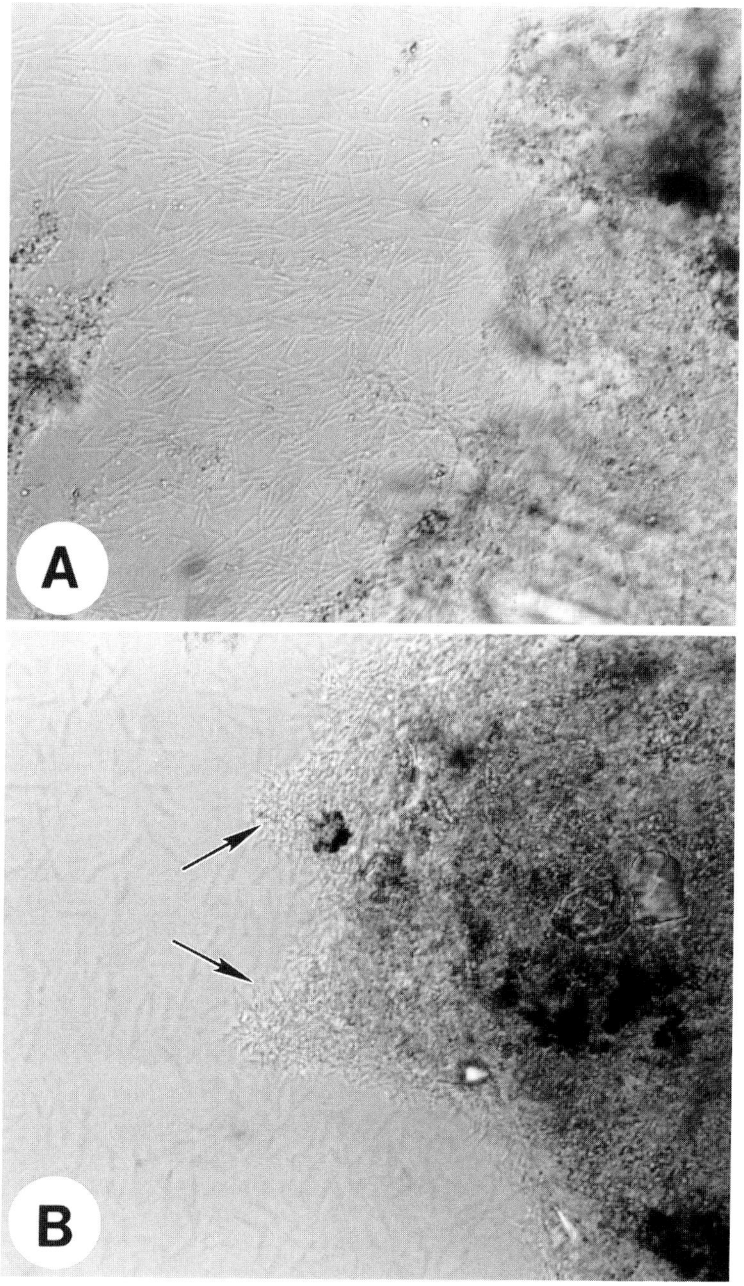

FIGURE 14.2. Wet mounts of Flavobacterium columnare *from channel catfish showing* (A) *long slender, flexing rods and* (B) *the hay-stacking* (arrows) *typical of virulent* F. columnare.

(Anacker and Ordal 1959) and Hsu-Shotts (Shotts 1991) is required for isolation of *F. columnare.* Ordal's and Hsu-Shotts media can also be prepared as broth by eliminating the agar. In broth culture, *F. columnare* forms a distinct yellow, mucoid pellicle at the miniscus. Hawke and Thune (1992) found that a modification of media described by Fijan (1969) worked best for isolation of *F. columnare,* especially in the presence of other bacteria. Song et al. (1988a), however, found that media of Shieh (1980) were superior to any other. The organism grows poorly or not at all on conventional media.

Bacterial Characteristics

When a columnaris isolate is incubated at 25 to 30°C for 48 hours, growth of most isolates will appear as spreading, rhizoid, discrete colonies with yellow centers that adhere tightly to the media. In comparing 27 isolates of *F. columnare* from channel catfish in the Southeastern United States, Davidson (1996) found colonies with subtly different morphologies, which were also described by Shamsudin and Plumb (1996). When grown on Hsu-Shotts media, these isolates produced three colony morphologies: (1) bright yellow, dry, umbonate and spreading with irregular edges (most were of this type); (2) bright yellow, moist, and spreading with uneven edges; and (3) pale yellow, dry, and flat, with uneven edges and more spreading than was noted than in other colony types. Eleven of 27 isolates also grew weakly in the presence of 1% NaCl. An additional 11 isolates studied by Shamsudin and Plumb (1996) from blue catfish, channel catfish, largemouth bass, and fathead minnows had no morphological or physiological differences relative to host. All isolates were confirmed as *F. columnare* based on colonial and cellular morphology and biochemical and physiological characterization, most of which were uniform and similar to the ATCC no. 49512 isolate. Isolation of *F. columnare* may be enhanced by addition of polymyxin B (10 IU/mL) and neomycin (5 µg/mL) to the media because they inhibit growth of noncolumnaris organisms (Fijan 1969). Columnaris is then confirmed by slide agglutination using specific *F. columnare* antisera.

According to Bernardet et al. (1996) the newly emended genus *Flavobacterium* is a Gram-negative rod that measures 2- to 10-µm long and about 0.5 µm in diameter, motile by gliding, produces yellow colonies on agar, is a chemoorganotroph and a facultative anaerobe, and decomposes several polysaccharides but not cellulose. These organisms are widely distributed in soil and fresh water habitats. The G + C contents of *Flavobacterium* DNAs range from 32 to 37 mol%. Major distinguishing factors between *F. columnare, F. branchiophilum, F. psychrophila,* and *F. maritimus* are acid production from glucose, H_2S production, catalase, optimum growth temperature, salinity tolerance, and the presence of chondroitinase (Table 14.1). Griffin (1992) devised a simplified method of identifying *F. columnare* using five characteristics that distinguish it from other yellow pigment producing aquatic bacteria: (1) ability to grow in the presence of neomycin sulfate and polymyxin B; (2) typical thin, rhizoid, yellowish colonies; (3) ability to degrade gelatin; (4) bind congo red; and (5) production of chondroitin lyase.

Using DNA homology, Pyle and Shotts (1981) suggested that strains of columnaris from salmonids and those from warmwater fish were indeed different and that three distinct groups existed within the cold water isolates. Song et al. (1988b) found three distinct groups among 26 *F. columnare* isolates from Canada, Chile, Japan, Korea, Republic of China, and the United States based on DNA homology, although there was diversity in colony morphology and some biochemical characteristics not necessarily related to DNA homology. Twenty isolates in one homologous group included representatives from each country.

Analysis by sodium dodecyl sulfate–polyacrylamide gel electrophoresis (SDS–PAGE) of outer membrane profiles (OMP) of 27 channel catfish *F. columnare* isolates indicated four distinct groups based on OMP molecular mass that ranged between 40 and 60 kD (Davidson 1996). If all isolates examined in these studies were in fact *F. columnare,* different strains of the pathogen apparently exist within the species.

Epizootiology

Columnaris disease may occur as a primary infection without significant predisposing stress to the host, but more commonly, it develops as a secondary infection due to environmental stress or trauma. In either case, the disease can become an acute infection with rapidly developing mortality.

Columnaris may occur as a combination of external and systemic infection of the body, fins, or gills. Hawke and Thune (1992) found that in 53 cases involving *F. columnare* in channel catfish, 11% were solely external, 17% were solely internal, and 72% were a combination of the two.

Columnaris often appears in association with one or more other pathogens and is secondary to a more primary but less lethal organism (external protozoan parasites) or bacteria. Of the 53 *F. columnare* infections studied by Hawke and Thune (1992), 46 involved more than one bacterial infection and most commonly included *E. ictaluri* and *Aeromonas* spp. Marks et al. (1980) were unable to induce a *F. columnare* infection experimentally unless a *Corynebacterium* sp. was also present. Chowdhury and Wakabayashi (1989) reported that *F. columnare* invaded fish when several other

TABLE 14.1. BIOPHYSICAL AND BIOCHEMICAL CHARACTERISTICS OF *FLAVOBACTERIUM COLUMNARE* AND *FLEXIBACTER MARITIMUS*

Characteristic	*Flavobacterium columnare*[a]	*Flexibacter maritimus*[a]
Cell morphology	Long, Gram-negative rods	Long, Gram-negative rods
Colony morphology	Flat, rhizoid, adheres to agar	Flat, irregular
Cell size (μm)	0.3–0.5 × 3–10	0.5 × 2–30
Yellow pigmented colony	+	+
Motility	Gliding	Gliding
Flexirubin pigment	+	–
Binds Congo red	+	+
Resistant to neomycin sulfate, polymyxin B	+	+
Chondroitin lyase	+	–
o-nitrophenyl-ß-D-galactopymanoride	–	–
Growth on peptone	+	+
Glucose source of carbon	–	–
Acid from carbohydrates	–	+
Degredation of		
Gelatin	+	+
Casein	–	+
Starch	–	–
Tyrosine	–	+
Urease	?	+
H_2S	+	–
Nitrate reduced	–	+
Catalase	+	+
Cytochrome oxidase	+	+
Optimum growth at (°C)	25–30	30
Growth tolerance (°C)	10–37	15–34
Growth in 0% SW-HS[b]	+	+
33% SW-HS	–	+
66% SW-HS	–	+
100% SW-HS	–	+
G + C content (mol%)	32–37	33–42
Habitat	Freshwater (saprophytic)	Marine (saprophytic)

Sources: Wakabayashi et al. (1986); Chen et al. (1995); Bernardet et al. (1996).
[a]+ = positive reaction or characteristic; – = negative
[b]SW-HS is Hsu-Shotts made with % of sea water.

bacteria were present but not when *A. hydrophila* or *P. fluorescens* were present. They also showed that *F. columnare* did not survive well in vitro when density of *A. hydrophila* was approximately 100 times higher than that of *F. columnare*. In view of these conflicting reports, the role of *F. columnare* in primary or secondary infections and its relationship with other pathogens are unclear.

In cultured channel catfish populations where no other species are present and the water supply comes from a well, catfish would be considered the pathogen source. Fish living in stream or reservoir water can also serve as a source for the bacteria. It has been proposed by Bullock et al. (1986) that course fish (suckers, carp, and so forth) are reservoirs of *F. columnare*.

Transmission of columnaris is generally from fish to fish via water, but numerous factors can affect its transmission and contraction. The most common precursors to development of columnaris as a secondary infection in channel catfish include handling, seining, transportation, temperature shock, water quality (low dissolved or supersaturation of gasses—oxygen, nitrogen, and so forth), and presence of other infectious diseases (Hanson and Grizzle 1985). Channel catfish are susceptible to columnaris at temperatures from 15 to 30°C, and young fish are more severely affected than adult fish. Centrarchids are especially susceptible when held in abnormally cool water during summer.

F. columnare infections can be chronic and cause lingering, gradually accelerating mortality in channel catfish, but more often, *F. columnare* appears suddenly and accelerates to subacute mortality in a matter of days. Ninety percent mortality in tank-held fingerling channel catfish is not uncommon during optimum disease conditions. Mortality in pond populations is usually lower but may reach 50% to 60%. Several researchers have reported greater than 90% mortality in salmonid and eel populations during columnaris epizootics (Hussain and Summerfelt 1991). Cases have been reported in which columnaris-infected juvenile channel catfish suffered about 50% mortality in a 24-hour period.

Juvenile rainbow trout and other salmonids are more susceptible than are older fish, particularly when held in water at temperatures approaching 20°C. Salmonid fingerlings can suffer up to 100% mortality if injured and held at 18 to 20°C; however, very low mortality occurs in uninjured fish held under the same conditions. Adult migrating salmon also become more susceptible to columnaris the farther they move upstream toward spawning grounds. *F. columnare* infections in trout and salmon are enhanced by low levels of dissolved oxygen and elevated ammonia, which is probably also true in other cultured fish species (Chia-Reiy et al. 1982).

Hussain and Summerfelt (1991) experimentally induced columnaris infections in a group of walleye. Up to 70% of mechanically injured fish contracted the disease, but no infection occurred in noninjured fish. Bait minnows (fathead minnows and golden shiners) held in bait shop tanks are highly susceptible to columnaris.

Although columnaris occurs in every month of the year, it tends to be seasonal, especially in temperate climates where it has two peak periods (Figure 14.3). Infections increase in late March through April; low incidence occurs during summer but increases again in autumn. This pattern probably depends on optimum water temperatures for the pathogen, the presence of greater numbers of susceptible size and age fish in the spring, and movement and transport of fingerlings in the fall. Bowser (1973) found that black bullheads in Clear Lake, Iowa, had widespread columnaris infections in May and June with no reported incidences after July. *F. columnare* infections in Taiwan were reported by Kuo et al. (1981) to be highest in tilapia and eel populations during September and October, lower in March through June, and very low in

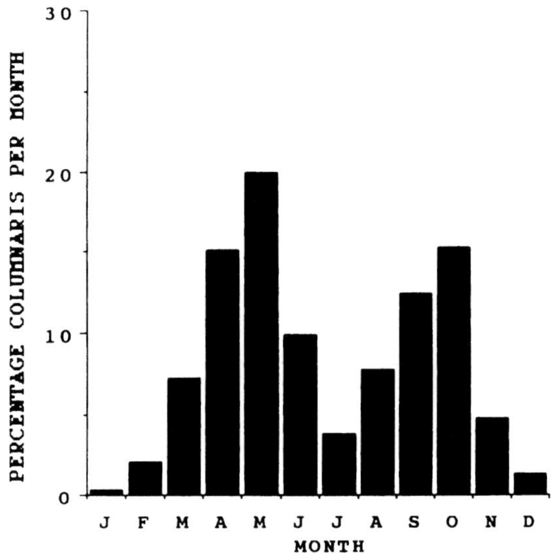

FIGURE 14.3. Monthly distribution of Flavobacterium columnare *infections (all fish species) in Alabama from 1991 through 1996* (N = 1274). *Diagnostic data compiled from diagnostic records at the Southeastern Cooperative Fish Disease Laboratory, Auburn University, Alabama, and the Alabama Fish Farming Center, Greensboro, Alabama.*

January through February and June through August.

The ability of *F. columnare* to survive in water has been studied by several researchers. Fijan (1968) showed that pathogen survival was reduced at pH 7 or less, in waters with hardness less than 50 mg/L $CaCO_3$, and/or in waters containing low organic matter. Chowdhury and Wakabayashi (1988) found that *F. columnare* survival decreased little during 7 days in chemically defined water containing 0.03% NaCl, 0.01% KCl, 0.002% $CaCl_2{\cdot}H_2O$ and 0.004% $MgCl_2{\cdot}6H_2O$, but survival was reduced in water with higher concentrations of these ions. When sterile mud was seeded with *F. columnare,* 62% of the cells survived after 77 hours at 10°C, but only 35% survived at 20°C (Becker and Fujihara 1978). These data imply that under normal conditions, the organism does not survive well in the environment for extended periods without a fish host; however, this possibility has not been totally eliminated.

Pathological Manifestations

Initial infections of *F. columnare* usually result from mechanical or physiological injury or environmental stress but can develop independent of stressors. It has been proposed that lesion severity depends on strain virulence and that necrotic lesions likely result from the organism's proteolytic enzyme activity. Gill lesions start at the margins of filaments, and necrosis progresses toward the gill arch (see Figure 14.1). Necrotic gills become congested, and epithelium separates from the lamellae.

When skin lesions occur, the dermis and underlying musculature becomes necrotic (Bootsma and Clerx 1976). Capillaries become congested and are destroyed, causing hemorrhaging to appear at margins of the ulceration. Once integument is compromised by the bacterium, a systemic infection may follow. Generally, very little pathology is associated with systemic columnaris infections of channel catfish; however, Hawke and Thune (1992) did note swelling of the trunk kidney in some case studies.

Significance

From 1987 to 1989, columnaris was the most frequently reported infectious disease in the catfish industry accounting for 58% of all bacterial cases (Thune 1991), and it still continues to be a major disease factor. The disease's broad geographical range and extensive species susceptibility add to its significance, which has often been overlooked by fish pathologists because of its role as a secondary pathogen. In reality, each year, columnaris is probably responsible for killing as many cultured fish, irrespective of species, as any other bacterial organism.

ENTERIC SEPTICEMIA OF CATFISH

The genus *Edwardsiella* includes two bacterial species that cause major disease problems in fish: *E. tarda* (Ewing et al. 1965) and *E. ictaluri* (Hawke et al. 1981). *E. tarda* is the causative agent of edwardsiella septicemia and is discussed more extensively in the eel disease section (Chapter 16).

E. ictaluri causes enteric septicemia of catfish (ESC), also known as "hole-in-the-head" disease (Hawke 1979; Hawke et al. 1981). This disease and columnaris are the two most important infectious diseases of cultured catfish. Being the most frequently reported infectious fish disease in the southeastern United States from 1986 through 1996, ESC has caused losses in the millions of dollars and continues to be a major problem.

Geographical Range and Species Susceptibility

E. ictaluri has been confirmed in the United States, Thailand (Kasornchandra et al. 1987), and Australia (Humphrey et al. 1986). In the United States, it has been isolated from channel catfish, primarily from Florida to Texas and north to Missouri and Kentucky. The bacterium has been reported among cultured channel catfish in other states including Arizona, California, Idaho, Indiana, Kansas, New Mexico, and Virginia. As channel catfish propagation expands, the dissemination of *E. ictaluri* into new geographical areas will likely occur. There have already been unconfirmed reports of its presence in countries where channel catfish have been introduced.

E. ictaluri has a more narrow host range than most other warmwater fish disease-producing bacteria. Cultured channel catfish are much more susceptible than other ictalurids; however, white catfish, blue catfish, and brown bullhead have on occasion been naturally infected with *E. ictaluri.* Although there have been naturally occurring *E. ictaluri* infections reported in blue catfish, Wolters et al. (1996) showed that under experimental conditions, juvenile blue catfish had the highest rate of survival (90%), channel catfish the lowest (62%), and channel X blue hybrids had intermediate survival (74%) following pathogen exposure. Naturally occurring infections have been reported in walking catfish in Thailand (Kasornchandra et al. 1987) and in two aquarium species, danio (sind) (Waltman et al. 1985) and green knife fish in the United States (Kent and Lyons 1982). Experimental infections were established in chinook salmon and rainbow trout, but because of temperature constraints, the pathogen is unlikely to produce serious problems in salmonid species (Baxa et al. 1990). European catfish (sheatfish) are only slightly susceptible (Plumb and Hilge 1987), whereas several commonly cultured warmwater species were shown to be refractive (Plumb and Sanchez 1983).

Clinical Signs and Findings

Enteric septicemia is a chronic to subacute disease with almost pathognomonic clinical signs in channel catfish (Hawke 1979). Diseased fish hang listlessly at the surface with a "head-up-tail-down" posture, sometimes spinning in circles, followed by morbidity and death. Affected fish have pale gills, exophthalmia, and occasionally enlarged abdomens. Small depigmented lesions of 1 to 3 mm in diameter appear on the flanks and backs of infected fish. Lesions then progress into similar-size inflamed cutaneous ulcers (Figure 14.4). In chronically ill fish, an open lesion may develop along the central skull line between the eyes at the insertion of the two frontal bones, thus the name "hole-in-the-head" disease. Petechia or inflammation in the skin under the jaw and on the operculum and belly may be so extensive that skin will become bright red (paint brush hemorrhage). Hemorrhage may also occur at the base of fins. Internally, the body cavity may contain a cloudy or bloody fluid and on rare occasions a clear yellow fluid; the kidney and spleen are hypertrophied and the spleen is dark red; inflammation occurs in adipose tissue, peritoneum, and intestine; and the liver is either pale or mottled with congestion (Figure 14.4).

Diagnosis

Generally, *E. ictaluri* is isolated from visceral organs or brain of clinically infected fish on brain heart infusion (BHI) agar or trypticase soy agar (TSA), where punctate colonies will form in 36 to 48 hours when incubated at 28 to 30°C. When using BHI agar, careful observation of primary culture plates is essential because of the possibility that other more rapidly growing bacteria, such as *Aeromonas* spp., will over grow the *E. ictaluri*. Shotts and Waltman (1990) developed an *E. ictaluri* isolation media (EIM) that is selective for the bacteria and can distinguish it from other fish bacterial pathogens. On EIM, *E. ictaluri* produces clear greenish colonies, and at the same time, the medium inhibits growth of Gram-positive bacteria. Other Gram-negative fish pathogens will grow on the media but can be separated by colony morphology and color (Figure 14.5). *E. tarda* colonies will appear small with black centers, *A. hydrophila* colonies are brownish and larger than *E. ictaluri*, and *Pseudomonas fluorescens* colonies will appear blackish and punctate. A complete defined medium consisting of 46 individual components was developed for *E. ictaluri* by Collins and Thune (1996). By reducing the essential components from 46 to 8, they also described a "minimum essential medium" on which *E. ictaluri* grew.

Serological methods developed by Rogers (1981) used fluorescent antibody technique (FAT) and enzyme linked immunosorbent assay (ELISA) rabbit anti–*E. ictaluri* sera to identify *E. ictaluri* either in vitro or in vivo. In these tests, no cross-reactivity was detected with *E. tarda, Salmonella* sp., or *A. hydrophila*. A lack of serological cross-reactivity of *E. ictaluri* with *E. tarda* or *A. hydrophila* was also demonstrated by Klesius et al. (1991) using ELISA and monoclonal antibody. The specificity of *E. ictaluri* antibody was further shown by Chen and Light (1994), who found no correlation between agglutinating *E. ictaluri* antibody and *A. hydrophila* antibody. Ainsworth et al. (1986) used monoclonal antibodies in an indirect FAT for diagnosing *E. ictaluri*. When compared with bacterial culture of brain, liver, spleen, and anterior and posterior kidney, 71% of the fish were positive by culture and 62% were positive by FAT using the monoclonal antibody; but by combining the two methods, 90.3% of the fish were *E. ictaluri* positive. Other studies support the time-saving advantage of using immunoassay to obtain results in 2 hours versus 48 hours required for culture; however, pathogen isolation is still the most reliable method of demonstrating its presence in a fish population.

Baxa-Antonio et al. (1992) suggested that detection of humoral antibody in channel catfish could be used to identify populations that have been exposed to *E. ictaluri*. Klesius (1993) further refined the ELISA system by developing Falcon Assay Screen Test (FAST)–ELISA to detect antibodies against *E. ictaluri* "exoantigen" in 30 minutes. Monoclonal antibody in the FAST–ELISA could be valuable in detecting covert carrier fish or predicting overt infection. Tyler and Klesius (1994) showed that recognition of *E. ictaluri* exoantigen by ELISA is highly specific. However, an absence of antibody may not always indicate whether a fish has been exposed to the pathogen, as all individuals do not uniformly produce antibody after bacterial exposure (Vinitnantharat and Plumb 1993).

Earlix et al. (1996) digested homogenized kidney tissue from infected channel catfish in Triton X-100 and filtered it with a 0.45-µm nitrocellulose membrane to trap the bacteria. After culture of the membrane for 24 hours at 30°C on agar, an ELISA application with *E. ictaluri*–specific monoclonal antibody detected an 80% *E. ictaluri* prevalence in asymptomatic channel catfish. This compared with a 24% prevalence by conventional bacteriological

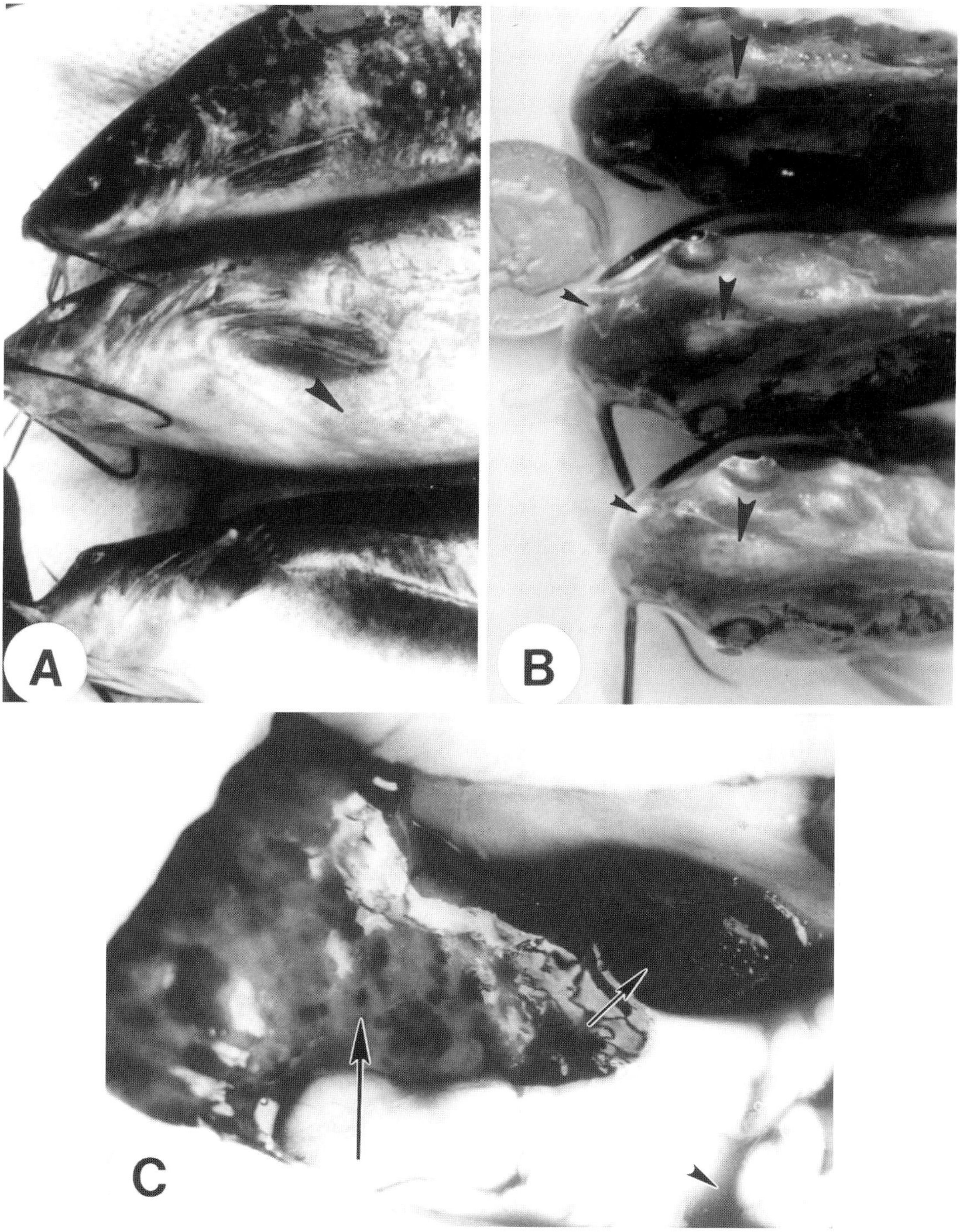

FIGURE 14.4. *Channel catfish infected with* Edwardsiella ictaluri. (A) *Lower fish is not infected, and the upper fish shows the early stages of disease development with depigmented areas* (arrowhead). *The middle fish has petechial hemorrhages* (arrowheads) *in the skin.* (B) *Channel catfish exhibiting open lesions in the cranial region* (large arrowheads), *and inflamed nares* (small arrows) *typical of chronic infection, and exophthalmia.* (C) *Viscera of* E. ictaluri*–infected channel catfish with mottled liver* (large arrow) *and enlarged spleen* (small arrow) *and bloody ascites in body cavity* (arrowhead).

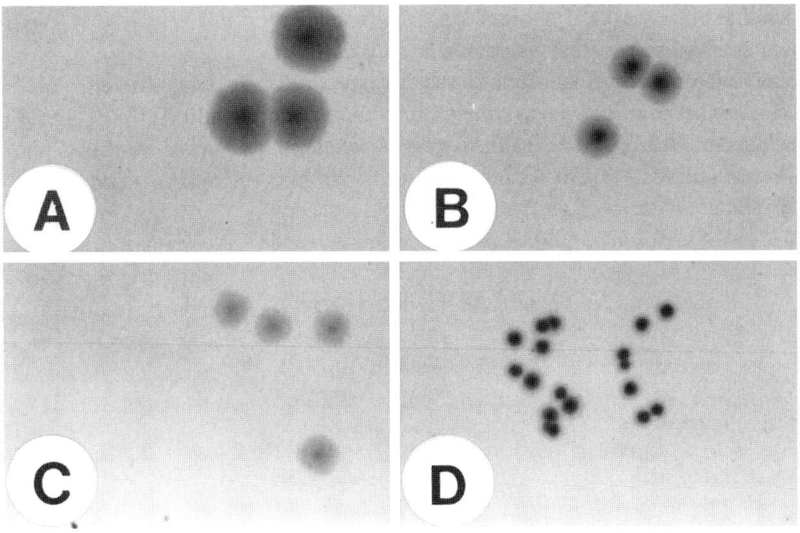

FIGURE 14.5. Colonies of four bacterial pathogens isolated from channel catfish growing on Edwardsiella *isolation media (EIM). All cultures are 48 hours old, incubated at 30°C.* (A) Aeromonas hydrophila *colonies are large and brownish.* (B) Edwardsiella tarda *colonies are medium size, greenish, and have black centers.* (C) Edwardsiella ictaluri *colonies are medium size, pale green with darker green centers.* (D) Pseudomonas fluorescens *are small and black with a pale halo. (Photographs courtesy of D. Earlix.)*

isolation. Implementing these techniques could be useful in determining *E. ictaluri* carrier populations.

Bacterial Characteristics

When biochemical characteristics of *E. ictaluri* (Table 14.2) were described, very little diversity among the many isolates was noted (Hawke et al. 1981; Waltman et al. 1986; Plumb and Vinitnantharat 1989). The organism is a short, Gram-negative rod (0.8 × 1 to 3 µm) that tends to be longer in actively growing cultures. It is cytochrome oxidase negative, weakly motile at 25 to 28°C, and lacks motility at 30°C or above. *E. ictaluri* grows poorly or not at all at 37°C. At 20 to 30°C, it ferments and oxidizes glucose while producing gas. It will not tolerate a NaCl level higher than 1.5% in the media. *E. ictaluri* can easily be separated from *E. tarda* by its indole negative reaction and lack of H_2S production on triple sugar iron (TSI) agar.

Evidence indicates that generally there is only one serological strain of *E. ictaluri* and no cross-reaction occurs with other aquatic bacteria. Antibodies to *E. ictaluri* in naturally infected channel catfish were not removed by adsorption with nine other species of bacteria commonly found in fish intestines and fish ponds (Chen and Light 1994). Also channel catfish immunized with these nine bacterial species failed to develop *E. ictaluri* agglutinating antibodies.

Lobb and Rhoades (1987) and Newton et al. (1988) described one to three plasmids found in *E. ictaluri* isolates, and indications are that all *E. ictaluri* isolates, regardless of origin, possess plasmids. These DNA plasmids are specific enough that Speyerer and Boyle (1987) suggested the possibility of using them as probes to detect *E. ictaluri* in infected fish.

Epizootiology

When first described, *E. ictaluri* was thought to be an obligate pathogen that could not survive for an extended period outside the host (Hawke 1979). It has now been determined that it can survive in pond bottom mud for more than 90 days at 25°C and in water up to 15 days at 25°C or less (Plumb and Quinlan 1986). Survival of *E. ictaluri* in the environment may be influenced by microbial competition because survival is shortened in water or mud that contains other microbes (Earlix 1995). Questions concerning how long *E. ictaluri* remains in epizootic survivors have not been answered; however, Mgolomba and Plumb (1992) found significant numbers of bacteria in blood and all organs of fish 81 days after exposure. Klesius (1992) isolated *E. ictaluri* from channel catfish in a population 280 days after initial pathogen detection. Nearly all *E. ictaluri* epizootics have occurred in channel catfish culture environments, but Chen et al. (1994) found serological evidence that catfish living in natural waters had been exposed to the bacterium.

E. ictaluri is readily phagocytized, but indica-

TABLE 14.2. BIOCHEMICAL AND BIOPHYSICAL CHARACTERISTICS OF *EDWARDSIELLA ICTALURI*, *E. TARDA*, AND *E. HOSHINAE*[ab]

Characteristic	*E. ictaluri*	*E. tarda*	*E. hoshinae*
Cell morphology		short, Gram negative single or paired rods	
Motility			
25°C	+	+	+
37°C	–	+	+
Growth at 40°C	–	+	+
NaCl tolerance			
1.5%	+	+	+
4.0%	–	+	+
Cytochrome oxidase	–	–	–
Indole	–	+	±
Methyl red	–	+	+
Citrate (Christensen's)	–	+	+
H_2S on triple sugar iron	–	+	–
Peptone iron agar	–	+	+
Lysine decarboxylase	+	+	+
Ornithine decarboxylase	+	+	+
Malonate utilization	–	–	+
Gas from glucose	+	+	±
Acid production from			
D-mannose, maltose	+	+	+
D-mannitol, sucrose	–	–	+
Trehalose	–	–	±
L-arabinose	–	–	±
Jordan's tartrate	–	±	–
Nitrate reduced to nitrite	+	+	+
Tetrathionate reductase	?	+	+
On E. ictaluri media (EIM)	Green translucent	Black centers	?
Mol% G + C of DNA	56–57	55–58	53

Sources: Ewing et al. (1965); Grimont et al. (1980); Hawke et al. (1981); Waltman et al. (1986).

[a]All strains of *E. ictaluri, E. tarda,* and *E. hoshinae* tested are negative for Voges-Proskauer, Simmons citrate, urea, phenylalanine deaminase, arginine dihydrolase, gelatin hydrolysis, growth on KCN; acid production from glycerol, salacin, adonitol, D-arabitol, celobiose, dulcitol, erythritol, lactose, L-rhamanose and D-xylose; acid from mucate, esculin hydrolysis, acetate utilization, deoxyribonuclease, lipase, ß-galactosidase (ONPG), pectate hydrolysis, pigment production, tyrosine clearing and oxidase test (Kovacs).

[b]+ = positive for 90–100% of strains; – = negative for 90–100% of strains; ± = mixed reaction.

tions are that bacteria are not destroyed in these cells (Figure 14.6) (Miyazaki and Plumb 1985); phagocytes may contribute to the bacteria's longevity in carrier fish. Transmission from adult channel catfish to offspring at spawning also seems possible, but yet unproved.

The primary mode for *E. ictaluri* transmission is through water via carrier channel catfish that sequester the bacteria in their intestines. Klesius (1994) showed that transmission of *E. ictaluri* from fish that died of ESC to nondiseased contact fish occurred by cannibalism or from bacteria shed by dead fish. The practicality of daily removal of dead fish from a culture unit was shown by Earlix (1995). He found that during an ESC epizootic, a significantly higher concentration of *E. ictaluri* was present in pond water in areas containing large numbers of dead fish than in areas with no carcasses. Because dead fish shed large numbers of bacteria, their removal can reduce bacterial concentrations to which noninfected fish are exposed. It also is possible that cormorants and herons can be vectors of *E. ictaluri* (Taylor 1992). It was demonstrated by ELISA that viable *E. ictaluri* can survive in the intestines of these birds, suggesting that fish-eating birds can be a pathogen source.

Although ESC has been diagnosed during every month of the year and in a wide range of water temperatures, it is considered a seasonal disease occurring primarily when temperatures range from 18 to 28°C in late spring to early summer and again in autumn (Figure 14.7). Mortality in experimentally *E. ictaluri*–infected channel catfish fingerlings was highest at 25°C, slightly lower at 23 and 28°C, and

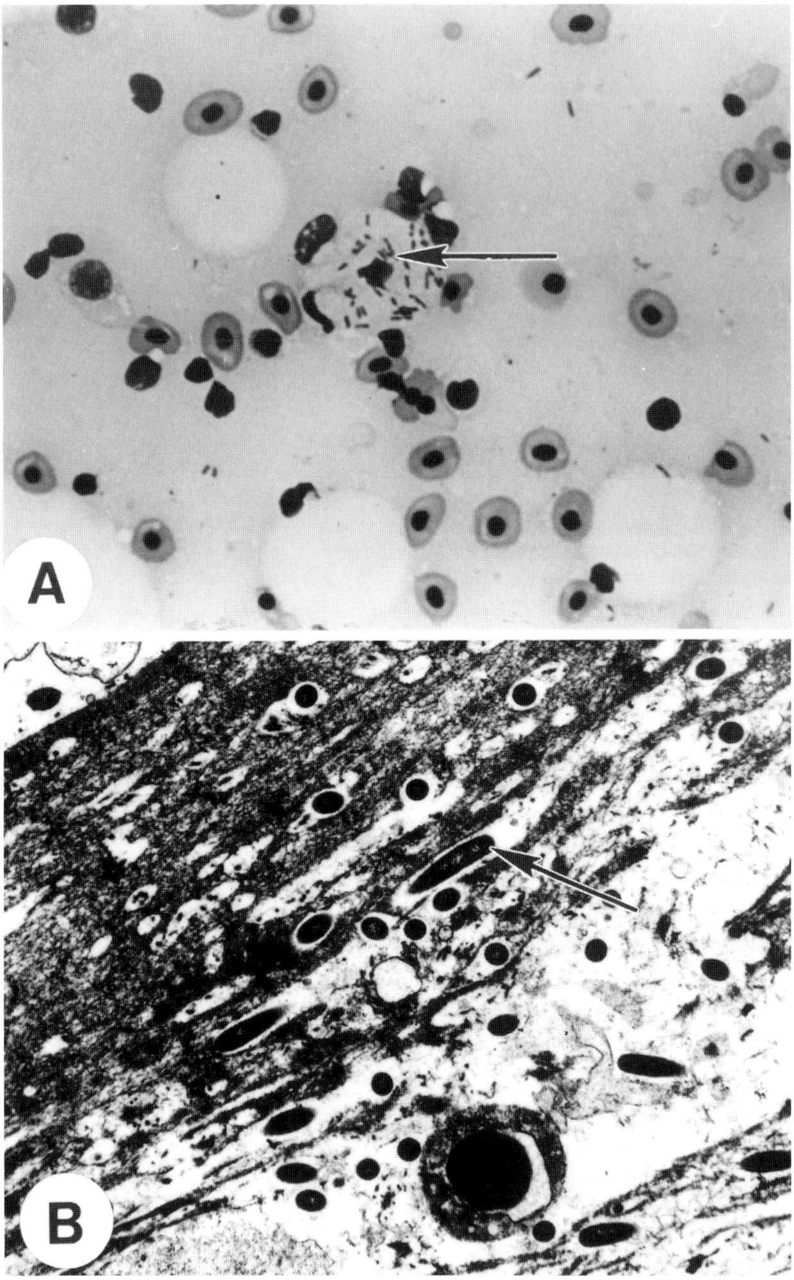

FIGURE 14.6. (A) *Smear from lesion in the skull of channel catfish showing intracellular* Edwardsiella ictaluri (arrow) *in macrophages (H & E stain). (Photograph by T. Miyazaki.)* (B) *Electron micrograph of olfactory organ from channel catfish illustrating* E. ictaluri (arrow) *invading the tissue (×15,000). (Photograph by E. Morrison.)*

no deaths occurred at 17, 21, or 32°C (Francis-Floyd et al. 1987). Baxa-Antonio et al. (1992) found that channel catfish experimentally infected with *E. ictaluri* by immersion experienced 0% mortality at 15°C, 46.6% at 20°C, 97.8% at 25°C, 25% at 30°C and 4% at 35°C. However, an increase of ESC outbreaks during July and August has been noted, which could indicate an expanding temperature tolerance for the pathogen.

Mortality rate in *E. ictaluri*–infected catfish populations varies from less than 10% to more than 50%. The pathogen infects fingerling as well as production-size fish and occurs in ponds, raceways, recirculating systems, and cages. Even though ESC can develop independently of extrinsic influences, adverse environmental circumstances can intensify disease severity. The precise relationship between environmental quality and *E. ictaluri*

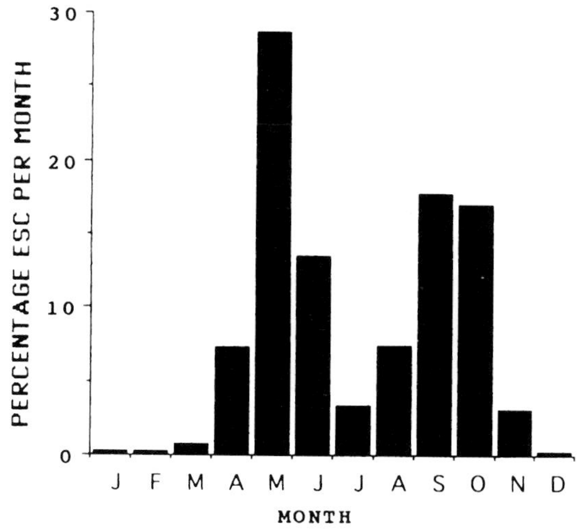

FIGURE 14.7. Seasonal occurrence of Edwardsiella ictaluri *(N = 1252) in Alabama from 1991 through 1996 showing greatest incidence of disease in May–June and September–October when water temperatures are 20 to 28°C. Diagnostic data compiled from diagnostic records at the Southeastern Cooperative Fish Disease Laboratory, Auburn University, Alabama, and the Alabama Fish Farming Center, Greensboro, Alabama.*

outbreaks has not been fully elucidated, but during experimental infections, Wise et al. (1993a) established a correlation between confinement-induced stress and increased susceptibility of channel catfish to the pathogen. They also concluded that stress induced by handling and hauling likely increases mortality due to ESC. The effect of stress on ESC was also shown by Ciembar et al. (1995), who determined that stressed channel catfish suffered higher cumulative infections (47%) 3 weeks after immersion exposure to *E. ictaluri* than did unstressed fish (16%). Injection of the corticosteroid Kenolog increased susceptibility of channel catfish to an initial as well as a second exposure to *E. ictaluri.* It also increased carrier rate of survivors from 1 of 8 in controls to 8 of 8 in Kenolog-injected fish (Antonio and Hedrick 1994). From these studies, there is little doubt that increased stressful conditions play a role in susceptibility and severity of *E. ictaluri.*

Certain dietary nutrients may affect ESC severity. It was shown by Paripatananont and Lovell (1995) that increased dietary zinc (15 to 30 mg Zn/day) enhanced channel catfish resistance to *E. ictaluri,* and organic zinc methionine was of greater value than inorganic zinc sulfate. Diets containing corn materials with fumoninsin (product of the fungus *Fusarium moniliforme*) reduced channel catfish growth and resistance to *E. ictaluri* infections (Lumlertdacha and Lovell 1995). Wise et al. (1993b) noted that addition of 60 mg or more of vitamin E into diets of channel catfish increased agglutinating antibody titers and enhanced ability of macrophages to phagocytize virulent bacteria.

There is no indication that *E. ictaluri* posses a health threat to aquatic animals other than fish. Temperature limitations under which *E. ictaluri* grows most likely preclude the bacterium from being a pathogen for warm-blooded animals.

Pathological Manifestations

The most severely damaged organs in *E. ictaluri*–infected catfish are the trunk kidney and spleen. Both organs develop necrosis, and the liver becomes edematous. Interlamellar gill tissue proliferates, and skin epidermis is destroyed. A mild focal infiltration occurs in the musculature underlying the epidermis, where bacteria apparently colonizes the capillaries causing depigmentation and necrosis (Miyazaki and Plumb 1985; Shotts et al. 1986). Ulcerative head lesions are necrotic and hemorrhaged, whereas systemic infections are associated with necrosis of hepatocytes and pancreatic cells. Intact, apparently dividing, *E. ictaluri* cells are seen within macrophages (see Figure 14.6). This phenomenon was also seen in electron micrographs of macrophages in tissues of the olfactory sac (Morrison and Plumb 1992).

After experimental exposure to 5×10^8 CFU/mL of *E. ictaluri* via immersion, 93% of channel catfish developed acute ESC and 7% developed chronic disease (Newton et al. 1989). Acute disease was characterized by hemorrhage, ulceration, and enteritis, olfactory sacculitis at 2 days postexposure, followed by hepatitis and dermatitis. Chronic ESC, most commonly observed at 3 to 4 weeks postexposure, was characterized by dorsocranial swelling, ulceration, granulomatous inflammation, and meningoencephalitis of the olfactory bulbs, tracts, and lobes of the brain.

Fish can be infected with *E. ictaluri* by injection, waterborne exposure, or ingestion. Waterborne bacteria invades the olfactory organ via the nasal opening, migrates into the olfactory nerve, enters the brain (Shotts et al. 1986) and spreads from the meninges to the skull and skin, which creates a hole-in-the-head condition. Morrison and Plumb (1992) showed that *E. ictaluri* can attach to the olfactory epithelium and migrate into the submucosa (see Figure 14.6). Injury included loss of sensory cilia and microvilli from the olfactory mucosal surface within 1 hour of exposure. Degeneration of olfactory receptors and supporting cells was evident 24 hours postinfection. Electron microscopy confirmed *E. ictaluri* on the mucosal surface and within the epithelium. Host leukocytes migrated through the olfactory epithelium into the interlamellar lumen and phagocytized the bacteria that did not appear to be destroyed. Indications are that the olfactory sac of channel catfish is a primary invasion site for *E. ictaluri* and phagocytes play a role in establishing a septicemia.

When ingested, *E. ictaluri* enters the bloodstream through the intestine and a septicemia results (Newton et al. 1989). Catfish orally exposed to *E. ictaluri* developed enteritis, hepatitis, interstitial nephritis, and myositis within 2 weeks of infection. Baldwin and Newton (1993) showed that at 0.25 hours after intestinal infection, *E. ictaluri* can be found in the kidneys, thus indicating a rapid transmucosal passage. They also suggested that *E. ictaluri* may have invasion and survival potential in the host that is similar to other invasive Enterobacteriaceae.

Little is known about the pathogenic mechanism of *E. ictaluri*. Stanley et al. (1994) demonstrated by electron microscopy a fibril network of connecting cells that were similar to attachment fibrils seen by Morrison and Plumb (1994) when *E. ictaluri* attached to the olfactory mucosa. When comparing virulent to avirulent isolates, Stanley et al. (1994) found that virulent cells had a greater amount of capsular material and surface proteins and demonstrated increased chondroitin degradation.

Significance

Enteric septicemia is the most economically important infectious disease of cultured channel catfish in the United States. *E. ictaluri* infections cost the aquaculture industry millions of dollars annually in killed fish and expenditures for prevention and chemotherapy. The effect of ESC on reduced growth and higher feed conversion ratios is only speculative. Because channel catfish depend on olfactory functions while feeding, any injury to these vital tissues most likely affects this process and would help explain why affected fish do not feed once disease has reached a certain level of severity.

Relatively few cases of ESC were reported immediately after its discovery, but in the early 1980s, the number of *E. ictaluri* isolates began to climb at an alarming rate. In the southeastern United States, 47 cases of ESC were reported in 1981, 1042 in 1985, and 1605 in 1988. Of all reported fish disease outbreaks in the southeast, ESC accounted for about 28% in 1985 and 30.4% in 1988.

MOTILE *AEROMONAS* SEPTICEMIA

Motile *Aeromonas* septicemia (MAS) is associated with infections caused by motile members of the genus *Aeromonas*. Synonyms for this syndrome are hemorrhagic septicemia, infectious dropsy, infectious abdominal dropsy, red pest, red disease, red sore, rubella, and others. The disease, currently known as MAS, became part of modern fish health in the 1930s when in Europe it was named infectious dropsy of carp. In North America, the syndrome was known as hemorrhagic septicemia until the mid-1970s when its name was changed to "motile *Aeromonas* septicemia." At this time, it became apparent that more than one motile member of the *Aeromonas* genus could cause the same disease syndrome. *A. hydrophila* (*punctata, liquefaciens*), *Aeromonas sobria*, and *Aeromonas caviae* are the principal motile species that affect fish (Austin and Austin 1987). Carnahan et al. (1991) listed seven motile species of the genus but found only the three previously mentioned to be regularly associated with diseased fish.

Geographical Range and Species Susceptibility

Motile *Aeromonas* septicemia is a ubiquitous disease that affects fishes found in warm, cool, and cold fresh water around the world and occasionally, those in brackish water especially in Europe, North and South America, and Asia.

Motile *Aeromonas* septicemia is generally, but not exclusively, associated with warmwater fish. Channel catfish, other ictalurids, silurids, clariads, carp and other cyprinids, eels, centrarchids, and true basses (striped bass) are susceptible. Trout and salmon are also susceptible but are usually affected only when water temperatures reach their upper tolerance (stressful) limits (Nieto et al. 1985). As far as is known, no fish species is totally immune or resistant to organisms that cause MAS. Motile aeromonads infect other aquatic animals as well, causing "red-leg" disease in frogs and fatal disease in reptiles (Shotts et al. 1972). They can also produce fatal septicemia and localized infections in humans (Goncalves et al. 1992).

Clinical Signs and Findings

Clinical signs of MAS are varied and diverse. Generally, clinical signs are categorized as either behavioral or external, but internal lesions do occur. Motile *Aeromonas* septicemia–infected fish lose their appetite, become lethargic, and swim lazily at the surface. When diseased fish first appear at the surface, they usually dive when disturbed but eventually lose their equilibrium, return to the surface, and move into shallow water.

External signs of MAS are varied, with none being specific, but slight differences in disease manifestation can be noted between scaled and scaleless fish. In catfish and other scaleless fishes, fins are frayed, hemorrhaged or hyperemic, and congested. Epidermal lesions begin as irregularly shaped depigmented areas that eventually develop into necrotic skin that sloughs, leaving open, ulcerated lesions with exposed muscle (Figure 14.8). Lesion margins are whitish or hemorrhaged. External lesions can occur at any location: caudal peduncle, dorsally, ventrally, laterally, or on top of the head. On scaled fish, lesions that begin as hemorrhages at the base of scales have a red (inflamed) central area surrounded by whitish (necrotic) tissue from which scales have been lost (Figure 14.8). In scaleless and scaled fish, the fungus *Saprolegnia* spp. will often attack necrotic tissue, giving the lesion a fuzzy, brownish appearance. Motile *Aeromonas* septicemia–infected fish may have exophthalmia with hemorrhages or opaqueness of the eye, enlarged abdomens with ascites, pale gills indicative of anemia, edematous musculature, and scales that show lepidorthosis before being lost. Internally, organs are friable, have a generalized hyperemia, swollen kidney and spleen, and the liver is often mottled with hemorrhage interspersed with light areas. The body cavity may contain a clear fluid (ascites), but more often the fluid is bloody and cloudy. The intestine is flaccid, hyperemic, contains yellowish mucus, and is void of food.

Diagnosis

Diagnosis of MAS cannot be based solely on clinical signs because other bacterial organisms (such as *Pseudomonas* sp.) and protozoan parasites (such as *Epistylis* sp. and *Ichthyobodo* sp.) often produce similar clinical signs and identical external lesions. A definitive MAS diagnosis can only be made by a complete necropsy of diseased fish in conjunction with isolation and identification of the causative organism. Primary isolation of motile aeromonads (*A. hydrophila, A. sobria,* or *A. caviae*) can be made on either BHI or TSA. The cultures should be incubated at 25 to 30°C for 24 to 48 hours, at which time entire, slightly convex colonies, which are mucoid and white to yellowish, will be obvious. *A. hydrophila* can be isolated and identified on Rimler-Shotts (RS) selective media, where an orange–yellow colony will form when incubated at 35°C (Shotts and Rimler 1973).

Bacterial Characteristics

Once the bacterium is isolated, it must be determined whether it is a Gram-negative, short, motile rod that is cytochrome oxidase positive and ferments glucose, all of which constitute a presumptive identity for motile aeromonads. These bacteria measure 0.8 to 0.9 × 1.5 μm, are polar flagellated, produce no soluble pigments, and are resistant to vibriostat (0/129) (2, 4-diamino-6, 7-disopropyl pteridine phosphate) and Novobiocin (Table 14.3).

The genus *Aeromonas* is in the family Vibrionaceae; however, Colwell et al. (1986) proposed that the family Aeromonodaceae, which would include *Aeromonas,* be recognized. *Aeromonas hydrophila, A. sobria,* and *A. caviae* can be distinguished by biochemical tests that include gas from glucose, esculin hydrolysis, and acid from arabinose (Table 14.3). Serological identification of motile aeromonads is not a common identification tool because of antigenic diversity and genetic complexity (Janda 1991). This is particularly true of the ubiquitous *A. hydrophila,* which has as many as 12 O-antigens and 9 H-antigens (Sakazake 1987). Historically, monovalent antisera to a particular *A. hydrophila* strain agglutinates only a small percentage of heterologous *A. hydrophila.* A virulent strain of *A. hydrophila* from epizootic ulcerative syndrome (EUS) in Asia was used to produce a monoclonal antibody (designated F26P5C8) that identified viru-

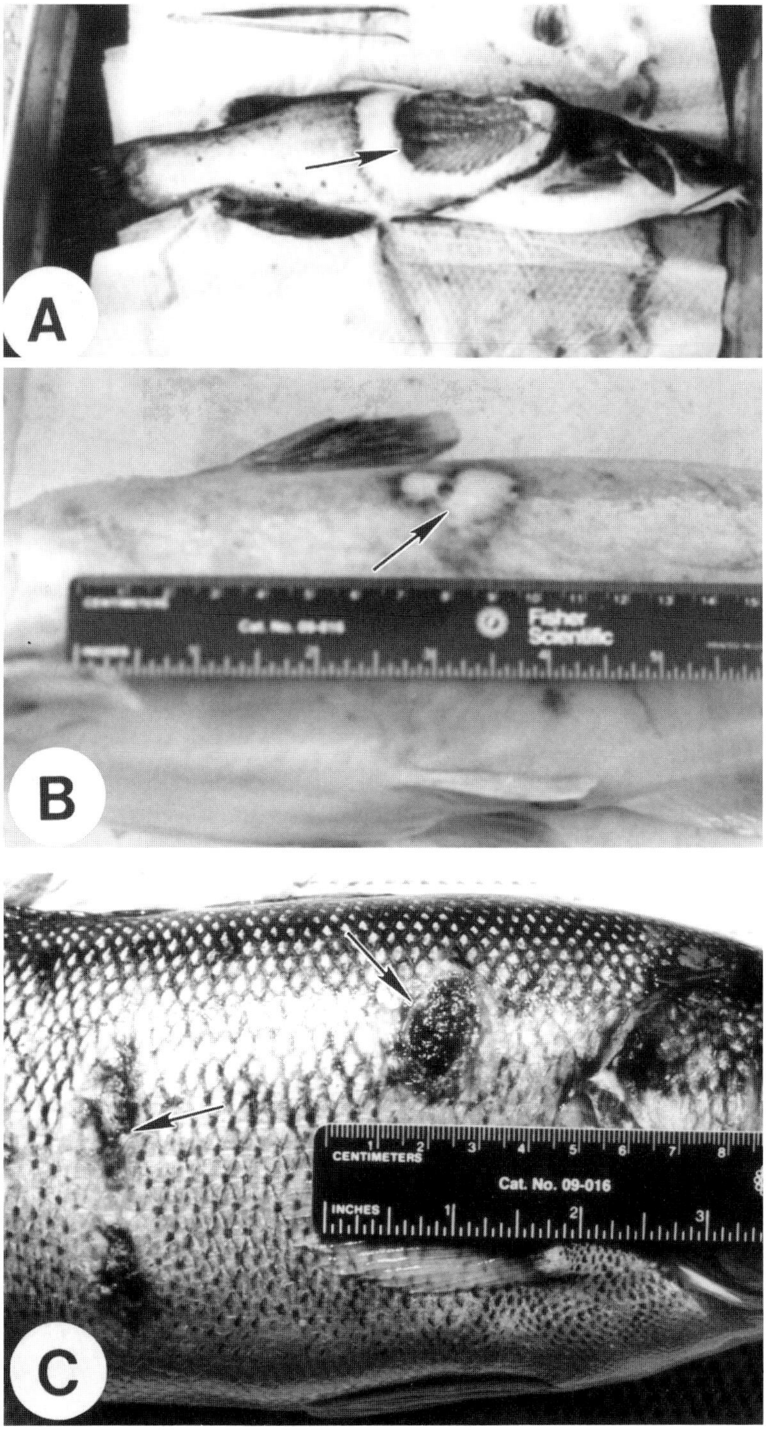

Figure 14.8. Motile Aeromonas *septicemia (*Aeromonas hydrophila*) infected fish.* (A) *Channel catfish with necrotic lesions* (arrows) *with exposed musculature.* (B) *Early stage of* A. hydrophila*–induced lesion with hemorrhage on margin of necrotic tissue* (arrow). (C) *Largemouth bass with ulcerative skin lesions* (arrows).

lent strains of this bacterium from other epizootics as serotype I (Cartwright et al. 1994). This monoclonal antibody also identified a large number of *A. hydrophila* isolates from Australia and Japan, but their virulence was unknown.

Epizootiology

Motile *Aeromonas* septicemia has been one of the most frequently diagnosed bacterial fish diseases since 1972 and was the most severe disease prob-

TABLE 14.3. BIOCHEMICAL AND BIOPHYSICAL CHARACTEISTICS OF *AEROMONAS HYDROPHILA, A. CAVIAE,* AND *A. SOBRIA*

Character/substrate	*A. hydrophila*	*A. caviae*	*A. sobria*
Morphology	Short, Gram negative, single or paired rods		
Cytochrome oxidase	+[b]	+	+
Catalase	+	+	+
Growth on nutrient agar (37°C)	+	+	+
Ornithine decarboxylase	–	–	–
Indole production	+	+	+
Fermentation			
Glucose, sucrose, mannitol	+	+	+
Dulcitol, rhamnose, xylose	–	–	–
Raffinose, inositol, adonitol	–	–	–
NO_3 reduction to NO_2	+	+	+
Growth on peptone without NaCl	+	+	+
0/129 resistant	+	+	+
Hydrolysis of starch, gelatin, ONPG, RNA and DNA hydrolysis	+	+	+
Gas from glucose[b]	+	–	+
Salacin	+	+	–
B-D xylosidase	+	+	–
Arbutine	+	+	–(14%+)
G-glucosidase	+	+	–(29%+)
Acid from arabinose[b]	+	+	–
Voges-Proskauer	+	–	+
Elastase	+	–	–
Esculin	+	+	–
Growth on KCN	+	+	–
L-histidine utilization	+	+	–
L-arginine utilization	+	+	–
H_2S from cystine	+	–	+

Sources: Shotts and Rimler (1973); Popoff (1984).
[a]+ = positive reaction and – negative reaction for substrate.
[b]Characteristics most often used to separate the three species if they are isolated from fish.

lem encountered by catfish farmers until the mid-1980s (A. J. Mitchell, Fish Farming Experiment Station, Stuttgart, Arkansas, personal communication). The disease accounted for as many as 60% of total bacterial cases reported in some years during that period.

Motile *Aeromonas* septicemia is generally a seasonal disease, but it has been diagnosed every month of the year across the southern United States where it peaks in the spring (Figure 14.9). As one moves northward, infections tend to occur in late spring or early summer. This disease pattern can be attributed to lower resistance following winter and spring environmental stress periods, hormonal imbalance during spawning, presence of large numbers of susceptible young fish, and the conduciveness of water temperatures in spring to bacterial infections. As summer progresses, fish tend to develop an immunity and/or natural resistance to *Aeromonas* spp. organisms, and MAS outbreaks decrease, only to rise again in fall when water temperatures are more favorable and juvenile fish are being handled and transported.

Fish mortality associated with motile *Aeromonas* infection is usually chronic with obscure peaks of deaths per day; cumulative mortality can be significant. High mortality may result when a highly virulent bacterial strain is involved, but normally total losses will be less than 50%. Infections of *A. hydrophila* can occur in very young fish to adults. Epizootics of young fish may be subacute, whereas die-off in older fish is more chronic. Kage et al. (1992) found that 90% to 100% of 7-day-old Japanese catfish died as a result of an *A. hydrophila* infection.

Infections may be external, internal (systemic), or more commonly, both. In external infections, skin lesions from which bacteria can be isolated are the only obvious clinical sign of disease, and bacteria cannot be isolated from internal organs. It can

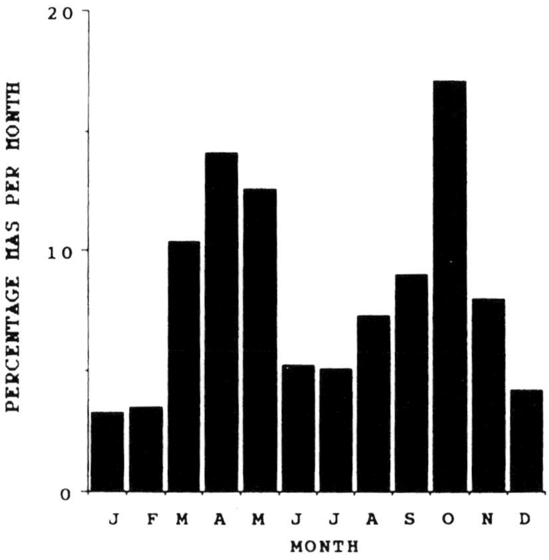

FIGURE 14.9. Monthly distribution of motile Aeromonas *septicemia infections in fish in Alabama from 1991 through 1996 (N = 919). Diagnostic data compiled from diagnostic records at the Southeastern Cooperative Fish Disease Laboratory, Auburn University, Alabama, and the Alabama Fish Farming Center, Greensboro, Alabama.*

be argued that this phase of the syndrome should not be considered motile *Aeromonas* septicemia because there is no septicemia. Systemic infections are characterized by a septicemia, and the causative organism can easily be isolated from any internal organ as well as skin lesions.

Infections of *A. hydrophila* in fish are often associated with some type of predisposing stress such as temperature shock, low oxygen (see Figure 1.5), high ammonia and other adverse water quality problems, trauma from improper handling or hauling, and presence of other disease organisms (Plumb et al. 1976; Grizzle and Kiryu 1992). Peters et al. (1988) found that even social stress of juvenile rainbow trout enhanced ventilation, elevated glucose and leukocyte volume, and increased susceptibility to *A. hydrophila*. Motile aeromonad infections in most animals, including humans, usually occur as a secondary infection when a debilitating condition already exists; however, it can be a primary infection, and if *A. hydrophila* becomes established in fish, it is often the cause of morbidity and death.

A. hydrophila infections are often associated with eutrophic lakes and ponds that receive large amounts of organic enrichment (Shotts et al. 1972), similar to conditions found in catfish culture ponds that receive large daily rations of feed. The relationship between water quality–induced stress and an *A. hydrophila* infection in a eutrophic channel catfish culture pond was shown by Plumb et al. (1976) and later duplicated in the laboratory (Walters and Plumb 1980). An abrupt decrease in the pond's dissolved oxygen concentration resulted in a drop in pH and increased concentrations of ammonia and carbon dioxide, followed in 6 days by an *A. hydrophila* infection (see Figure 1.3).

Susceptibility of some fish species to *A. hydrophila* is also linked to water temperature. Groberg et al. (1978) reported that when exposed to *A. hydrophila,* coho and chinook salmon and steelhead trout suffered 64% to 100% mortality at 18°C, compared with 0% mortality at 9.4°C. Rainbow trout demonstrated an increased susceptibility to *A. hydrophila* when water temperature was raised from 5.5 to 8°C to more than 11°C (Nieto et al. 1985). Algae in unclean holding tanks can also contribute to higher *A. hydrophila* concentrations and incidence of infection (Levanon et al. 1986).

Although rare, *A. hydrophila* has the ability to cause high mortality among cultured fish without presence of severe external (stressful) influences. This inconsistency may result from the presence of *A. hydrophila* strains that possess specific virulent or pathogenic characteristics. There is evidence that the motile *Aeromonas* complex involves secondary and opportunistic pathogens, but *A. hydrophila*'s ability to cause disease and death of fish should not be overlooked because occasionally highly virulent strains emerge. Regardless of whether the organism serves as a primary or secondary invader of stressed fish, it is often the final insult that leads to death.

In Asia, *A. hydrophila* has been closely associated with EUS, a disease that has plagued both wild and cultured fish populations from Indonesia to India since the early 1980s (Boonyaratpalin 1989). The definitive etiology of EUS is uncertain because a virus (snakehead rhabdovirus), fungus (*Aphanomyces invaderis*), and *A. hydrophila* have all been associated with the lesions. In Southeast Asia, *Aphanomyces* was found in most fish lesions associated with EUS (Roberts et al. 1993; Willoughby et al. 1995). The massive tissue damage incurred was blamed on the fungus, but researchers were skeptical as to whether this organism was the primary pathogen. Although clinical signs of EUS and MAS are almost identical, investigators have suggested that *A. hydrophila* is secondary to the primary cause but may have been what actually killed the fish.

A. hydrophila, which adapts readily to its environment, is found in most natural fresh water ponds, streams, reservoirs, and bottom muds where it exists as a facultative organism and utilizes any available organic material as a nutrient source (Hazen et al. 1978). Often, different bodies of water, or watersheds, harbor a unique strain of *A. hydrophila* because of its serological diversity and the widespread facultative nature of the bacterium. If fish that are resistant to a particular strain of the bacterium are moved to a different body of water, the possibility exists they will be exposed to a different *A. hydrophila* strain for which they have no immunity, and they may become infected as a result of handling and transport.

As noted earlier, *A. hydrophila* primarily affects fish, but other aquatic animals and humans can also be infected. In humans, these infections can become quite severe, causing enteritis, meningitis, and localized infections on extremities (Ketover et al. 1973). Puncture wounds inflicted by catfish spines may result in serious *A. hydrophila* infections (Hargraves and Lucey 1990; Murphey et al. 1992). Several human deaths due to *A. hydrophila* have been reported, but often the patient had been debilitated by another ailment. Goncalves et al. (1992) attributed a case of pneumonia in a healthy male to an *A. hydrophila* septicemia when he became ill 3 days after swimming in the ocean, and the author suggested that ocean water was the source. That *A. hydrophila* is a fresh water organism casts some doubt as to the source. King et al. (1992) reported an *Aeromonas* spp. isolation rate in humans of 10.6 cases per one million population in California. Although 2% of the patients died, all had serious underlying medical conditions in addition to the *Aeromonas* infection. They concluded that *Aeromonas* spp. infections in humans is not an important public health problem and is largely unpreventable; however, Aeromonas infections in humans is now a reportable disease in California. Because of an increased incidence of *A. hydrophila* infections in humans, handling infected fish with caution is advisable.

Pathogenic capabilities of *A. hydrophila, A. sobria,* and *A. caviae* are still unclear. Gray et al. (1990) found that in 61 isolates of motile *Aeromonas* spp. taken from pig and cow feces and other environmental sources (none were from fish), 96.4% of *A. hydrophila,* 36.4% of *A. sobria,* and only 13.6% of *A. caviae* were cytotoxic. It was concluded that *A. hydrophila* was highly pathogenic, *A. sobria* was only moderately pathogenic, and *A. caviae* was not pathogenic. This may not hold true for fish isolates because of those taken from MAS-diseased channel catfish in Mississippi, *A. sobria* was more frequently found than *A. hydrophila* (M. Johnson, Mississippi State University, Stoneville, Mississippi, personal communication). In view of this, it is possible that *A. sobria* may be the primary cause of MAS infections in cultured channel catfish but often misidentified as *A. hydrophila.*

Pathological Manifestations

Pathology of MAS is not distinctly different from most other septicemic bacterial fish infections (Miyazaki and Jo 1985; Ventura and Grizzle 1988). Lesions are typical of those caused by bacterially produced protease and hemolysins. Epidermal infections are characterized by necrotic lesions that have spongy centers and hemorrhagic margins. The epidermis adjacent to lesions is edematous and the dermis becomes hemorrhagic with macrophage and lymphocyte accumulation accompanied by severe inflammation. Scale pockets become edematous, causing lepidorthosis. In systemic infections, most internal organs are edematous with diffused necrosis of the liver, kidney and spleen. Inflammation is not apparent in diseased internal organs, but hemorrhage and/or erythema does occur.

Virulence of *A. hydrophila* may be influenced by specific biochemical characteristics or by production of extracellular products that independently produce pathological effects when injected into fish (Santos et al. 1987). It was reported by Wakabayashi et al. (1980) that virulence of *A. hydrophila* is related to proteolytic casein and elastin hydrolysis. By correlating extracellular enzymatic activity of 127 strains of *A. hydrophila,* they showed that elastase-positive strains produced lesions and mortality when injected into channel catfish. It was suggested that other extracellular substances such as hemolysins and proteases may also be involved with *A. hydrophila*'s pathogenic mechanism (Chabot and Thune 1991). Additional factors that may influence pathogenesis of the organism are its ability to adhere to tissue surfaces; resistance to phagocytosis; production of surface proteins, siderophores, and lipopolysaccharide (LPS); presence of pili, S-layer, or outer membrane proteins; and the bacteriostatic activity of host serum. Also, water isolates of the bacterium may be avirulent or less pathogenic to fish than are isolates from diseased fish (de Figueiredo and Plumb 1977). The significance of these potentially pathogenic factors is yet unclear; in the final analysis, the only accurate measure of the organism's virulence and pathogenicity is whether or not it kills fish.

Significance

There is disagreement among fish pathologists as to the significance of MAS. The frequency of its appearance cannot be disputed, but its importance to many fish pathologists is reduced because it is usually a secondary disease. When MAS, especially *A. hydrophila* and *A. sobria,* is present in an aquaculture environment where a high potential for stress exists, the disease cannot be ignored because in most instances, it is often what kills the fish. Also, the fact that the motile *Aeromonas* group occurs worldwide and affects such a large variety of fish species make the disease syndrome an important factor in both cultured and wild fish populations.

OTHER BACTERIAL DISEASES OF CATFISH

Several other bacterial organisms occasionally cause infectious diseases in channel catfish. From a clinical standpoint, *Pseudomonas fluorescens* causes a disease that clinically is nearly identical to MAS and at one time was considered as one cause of "hemorrhagic septicemia." This organism occurs as a facultative bacterium in many water sources and occasionally causes infections in channel catfish during and after environmental stress. Clinical signs, morbidity, and mortality patterns are similar to those of MAS and, when present, will usually cause a septicemia. *P. fluorescens* is isolated from skin, muscle lesions, and internal organs on high nutrient agars (BHI or TSA). The organism is a Gram-negative, slightly motile bacillus that is cytochrome oxidase positive. Colonies of *P. fluorescens* on EIM are small and black (see Figure 14.5).It is distinguished from motile aeromonads and other fish pathogenic bacteria because it is oxidative only in glucose motility deep; it does not produce gas, and it produces a diffusible fluorescent pigment on agar medias.

E. tarda produces a disease in channel catfish known as "emphysematous putrefactive disease of catfish" (EPDC) and is discussed in more detail in Chapter 16 (Meyer and Bullock 1973). This organism is only seen in about 1% to 2% of catfish disease cases, but when present, it can be a major problem. Generally, prevalence of EPDC in an infected population is less than 5%, but when fish are crowded in holding tanks, the incidence can increase to 50% or higher. An *E. tarda* infection in channel catfish is characterized by lethargic swimming and the presence of small, 3- to 5-mm, cutaneous lesions located dorsolaterally on the body (Meyer and Bullock 1973). Within the flank muscles or caudal peduncle, these small lesions progress into larger abscesses and develop obvious convex, swollen areas. The skin loses its pigmentation, and incised lesions emit a foul-smelling gas. As infection progresses, posterior body mobility is lost. Internally, there is a general hyperemia characteristic of a septicemia and the body cavity has a putrid odor. The kidney, in particular, is enlarged, and the liver is mottled or abscessed.

E. tarda infections are usually prevalent in channel catfish approximately 0.4 kg or larger; however, subadult channel catfish are not resistant. In channel catfish experimentally infected with *E. tarda,* pathogenesis and histopathology developed rapidly after skin injury and waterborne exposure, but the process was relatively short lived (Darwish 1997). The organism can be isolated from most internal organs in 2 to 3 days postexposure and disappears in about 8 days. Histopathology consists of necrosis and granulomatous inflammation of the muscle, liver, kidney, and spleen and peaks at 6 days postinfection, with tissue healing being nearly complete at 8 days.

Bacillus mycoides was reported to be a potential pathogen of channel catfish by Goodwin et al. (1994). This organism was isolated from diseased channel catfish during an epizootic in May when water temperatures were 25 to 27°C. No assessment of morbidity or mortality was given. Diseased fish had pale areas and/or open ulcers on their backs and focal necrosis of epaxial muscles. A Gram-positive bacillus was the only pathogen isolated from the fish. The isolation was facilitated by use of tissue explants rather than an inoculating loop. *B. mycoides* produces a distinct whorllike growth on Müller-Hinton agar with no distinct colonies. The organism appeared as long chains of bacteria in diseased tissue and on culture plates. A similar pathological condition was produced by intramuscular injection of the organism into naive channel catfish. Intraperitoneal injection or skin abrasion failed to facilitate infection. Other Gram-positive bacteria that have occasionally been found in diseased channel catfish are *Carnobacterium* spp., *Staphylococcus* sp., and *Streptococcus* sp., but these are considered insignificant in terms of impact on the fish species.

MANAGEMENT OF CATFISH BACTERIAL DISEASES

Controlling bacterial diseases of channel catfish is a combination of wise management, judicious use of available chemotherapeutics, and vaccination when feasible.

Management

Management of any bacterial infection of channel catfish, whether columnaris, enteric septicemia, or motile *Aeromonas* septicemia, entails proper handling, maintaining a quality environment, and reducing environmental stressors. Ideally, fish should never be handled when in a weakened condition or when environmentally stressed. Water temperature control appears to be an important environmental management tool for channel catfish, particularly in tanks, raceways, and aquaria. When running water is used, flows must be sufficient to flush away metabolic waste while optimum oxygen concentrations and other water quality parameters are maintained.

Channel catfish reared in ponds that use artesian well water that contains up to 4000 mg of NaCl per liter seldom contract ESC (G. Whitis, Alabama Fish Farming Center, Greensboro, Alabama, personal communication). Plumb and Shoemaker (1995) exposed a population of juvenile channel catfish, with a 10% carrier rate of *E. ictaluri* at 15°C, to water with concentrations of NaCl of near 0 to 3000 mg/L at 25°C. Fish held in 0 and 100 mg/L NaCl suffered 100% and 96% mortality, respectively. Fish held in concentrations of 1000, 2000, and 3000 mg/L NaCl had 33%, 43%, and 17% mortality. On the basis of these experimental results and personal observations, some channel catfish farmers have added low grade salt to ponds as a preventive measure against ESC. Although field trial results have not been as dramatic as experimental results, the effect of increasing salt concentrations to more than 1000 mg/L has proved beneficial for ESC control and may also be effective in preventing columnaris infections. Disadvantages of adding large amounts of salt to ponds are cost, its corrosive effect on equipment, dilution of salt if ponds are flushed, and possible environmental repercussions.

There is a tendency for the aquaculturist to "do something" when ESC strikes; however, changing the feeding regimen from every day to once every few days, or complete cessation of feeding during epizootics, may be as effective in limiting mortality as applying antibiotics. Cessation of feeding or feeding medicated feed with Romet-30 every third day resulted in higher survival of *E. ictaluri*–infected channel catfish than in fish that received daily feeding with a normal ration (Wise and Johnson 1998). This procedure is practiced by many catfish farmers and, generally, with satisfying results. As previously noted, when evaluating ESC susceptibility of production-size channel catfish, Kim and Lovell (1996) demonstrated that fish held over winter without receiving feed were more resistant to *E. ictaluri* in spring than fish that had received normal or partial feeding during winter months.

Culturing blue catfish or channel X blue hybrids rather than channel catfish in farms where *E. ictaluri* is endemic is a feasible management approach because *E. ictaluri* is generally less pathogenic to blue catfish and channel X blue hybrids than to pure channel catfish strains (Wolters and Johnson 1994; Wolters et al. 1996); however, all are equally susceptible to columnaris and MAS.

Chemotherapy

Bacterial infections in channel catfish vary in response to chemotherapy and drug options are limited because only Terramycin (oxytetracycline) and Romet-30 (sulfadimethoxine–ormetoprim) are Food and Drug Administration (FDA) approved as treatments. Therefore, this discussion is confined to application of FDA-approved drugs, drugs of low regulatory priority, and those chemicals that have been approved by the U.S. Environmental

Protection Agency (EPA) for use in waters containing food fish.

Oxytetracycline is incorporated at 2.2 g of active drug per kilogram of feed (approximately 1 g/lb), fed at 2% of body weight, which gives an application rate of 50 to 75 mg/kg of body weight per day, and is fed for 10 days. A 21-day withdrawal period is required for Terramycin. Romet-30 is also fed at 50 mg/kg of fish per day (Plumb et al. 1987), but for only 5 days, followed by a 3-day withdrawal period. These treatments are effective against ESC and MAS infections, providing treatment is initiated early.

Johnson and Smith (1994) found that pellet size and formulation of Romet-30 medicated feed affected efficacy of feeding the drug. They determined that small pellets formulated at 5 kg of drug per ton of feed resulted in a significantly ($P < .05$) higher survival rate than was noted in fish fed large pellets with the same medicated formulation. The smaller pellets allowed feeding at 3% of body weight and resulted in fish consuming the medicated pellets. These fish gained more weight than did infected nonmedicated controls or fish fed Romet-medicated feed formulated to be fed at 1% of body weight with 15 kg Romet per ton of feed.

Indiscriminate use of any antibiotics should be avoided to reduce potential for antibiotic resistance. Recently, fish disease diagnostic laboratories have reported that more than 45% of *A. hydrophila* isolates are resistant to Terramycin. This acquired resistance is due, in part, to many years of exposure to improper drug application. Also, plasmid-mediated R-factors are involved in antibiotic resistance of *A. hydrophila* (Shotts et al. 1976). It has been noted that *E. ictaluri* has developed a resistance to Terramycin and/or Romet-30 with as many as 10% to 15% of the isolates studied being resistant to one or both drugs. Waltman et al. (1989) found that *E. ictaluri* resistance to Romet-30 was mediated by an R plasmid, and Starliper et al. (1993) showed that the plasmid could be transferred to nonresistant isolates. Cooper et al. (1993) suggested that bacteria present in agricultural runoff are potential sources for the plasmid responsible for *E. ictaluri* being resistant to Romet-30. Feed containing Romet-30 enhances selection for *E. ictaluri* that have received the R plasmid. Although this resistance can be plasmid or genetically induced, the problem is likely exacerbated by improper use of antibiotics. Use of medicated feed when it is not necessary, improper application rates, and/or continuing treatment longer than recommended will promote pathogen conversion from sensitive to resistance.

Prolonged treatments with potassium permanganate and copper sulfate, neither of which are FDA approved for fish disease treatment, are most commonly used for columnaris. Potassium permanganate is used at 2 to 4 mg/L in ponds indefinitely and up to 10 mg/L in tanks for up to 1 hour depending on organic water load (Phelps et al. 1977). Tucker (1984) showed how potassium permanganate demand can be measured colorimetrically, and Lau and Plumb (1981) showed that 2 mg/L above the potassium permanganate demand was necessary to control columnaris. The most effective way to treat a pond of channel catfish infected with columnaris is to apply potassium permanganate in combination with feeding Terramycin-medicated feed. Prophylactic bath treatments of 1% to 3% salt (NaCl) or 4 to 10 mg of potassium permanganate per liter for 1 hour will reduce incidence of posthandling infections.

Vaccination

Vaccination of channel catfish against bacterial diseases has not developed to the point that they are routinely used in channel catfish culture. A growing pool of data indicates that routine vaccination against some bacterial diseases, namely ESC, may be a distinct possibility.

When injected, *E. ictaluri* is a strong immunogen, making it an excellent candidate for vaccine development against ESC. Vaccination studies of *E. ictaluri* have included use of whole cell bacterins and cell extracts such as LPS and purified "immunodominant" antigens as well as live attenuated preparations. Several studies have shown that fractionating *E. ictaluri* and using a purified immunodominant antigen can provide protection against ESC (Plumb and Klesius 1988; Vinitnantharat et al. 1993). Protein with a molecular mass of 36 kD is a primary immunodominant antigen in the cell wall of *E. ictaluri* and is retained by cells when subcultured up to 30 passages (Vinitnantharat et al. 1993). The organism is also antigenic when fish are immersed in a solution containing *E. ictaluri* at a concentration of 1:10 of the vaccine in water with exposure for 1 to 2 minutes (Vinitnantharat and Plumb 1992). For oral application, an optimum concentration of coated *E. ictaluri* vaccine in feed is 1%; 0.5% is effective, and 10% is not immunosuppressive (Plumb et al. 1994). Orally vaccinated fish with 1% vaccine in the feed had 77% survival compared with 23% for unvaccinated fish when experimentally challenged with *E. ictaluri*. It was shown by Ainsworth et al. (1995) that oral application of particulate *E. ictaluri* antigen elicited a humoral response and resulted in antibodies in the mucus, gut washings, and gut content, but titers were not as high as in the serum.

Antibody titers against *E. ictaluri* are easily quantified, but antibody is yet to be proved as a measure of protection. However, channel catfish that survived a natural epizootic of *E. ictaluri,* and had titers greater than 1:512, were protected against subsequent experimental exposure to the pathogen (Vinitnantharat and Plumb 1993). In contrast, Klesius and Sealey (1995) showed that *E. ictaluri*–specific antibodies were not protective in fish, but titers were less than the 1:512 shown to be protective in the previous study.

The effects of immunomodulators on immunity of channel catfish to *E. ictaluri* have been mixed. An extraction from a marine tunicate (*Ecteinascidia turbinata*) actually decreased survival of channel catfish exposed to *E. ictaluri* and did not affect specific antibody response (Stanley et al. 1995). A similar increase in susceptibility of channel catfish to *E. ictaluri* was shown by Tyler and Klesius (1994) when squaline, an oil-based adjuvant, was administered intraperitoneally.

Commercial vaccines for *E. ictaluri* have undergone successful laboratory and field testing by demonstrating an immunity against mass mortalities due to the pathogen in most, but not all, instances. The vaccines are generally applied by an initial immersion of fry (10 to 14 day old) or fingerlings, followed in 1 to 2 months by an oral booster (Plumb and Vinitnanatharat 1993). Fish may also be immunized by using only immersion vaccination or oral application. The efficacy of vaccinating 2- to 3-week-old channel catfish fry by immersion remains unresolved. Nusbaum and Morrison (1996) demonstrated that live radiolabeled *E. ictaluri* passed through gills of channel catfish and soon appeared in internal organs, but dead radiolabeled bacteria did not enter the bodies of fish being treated. Vaccination against ESC with killed vaccines were not widely embraced by the catfish industry because of expense, inconsistent results, and lack of a convenient application method.

Modified live (attenuated) *E. ictaluri* vaccines may be more effective in large-scale vaccination of channel catfish against ESC. Laurence et al. (1997) developed a recombinant *E. ictaluri purA* mutant vaccine designated LSU-E2. Upon waterborne exposure to the LSU-E2 vaccine, or orally fed, 5-g channel catfish produced 100% protection to subsequent challenge with wild-type *E. ictaluri*. The modified vaccine organism was cleared by the immunized fish in 48 hours. A modified live *E. ictaluri* vaccine was also developed and tested on channel catfish by Klesius and Shoemaker (1998). The rifampicin-resistant *E. ictaluri* mutant (RE-33) stimulated immunity against EC in 10-day-old channel catfish fry. An immersion for 2 minutes in water containing RE-33 stimulated strong acquired cellular immunity but low humoral antibody response. Immunized fish cleared the transient infection in 14 days, which was long enough to stimulate protective immunity for at least 6 months. After challenge of vaccinated fish with 20 different *E. ictaluri* isolates, the mean RPS was 65.9 with 75% of the individual RPS values being above 50.

Vaccination of channel catfish against *F. columnare* showed some promise as early as 1972 (Schachte and Mora 1973) when high antibody titers were achieved in channel catfish injected with vaccine. Moore et al. (1990) vaccinated fingerling channel catfish by immersion in a formalin-killed bacterin of *F. columnare* that resulted in better survival and less need for chemotherapy treatment in vaccinated fish than in controls. In another vaccination study, Maas and Bootsma (1982) found that common carp will absorb *F. columnare* by immersion and indicated that this could be a means of vaccinating against columnaris.

Research to develop a vaccine for *A. hydrophila* dates back to the mid-1950s. Since then, very little effort has gone into the development of motile *Aeromonas* vaccines for channel catfish, but studies have explored its application for other fish species. Researchers have used heat- or formalin-killed bacterins and cell extracts such as LPS with some success. Carp immunized by immersion in LPS from *A. hydrophila* were protected by the cell-mediated system (Baba et al. 1988). Rainbow trout fry were protected against challenge after injection of heat-killed *A. hydrophila* (Khalifa and Post 1976), and Japanese eel were protected from the bacterium by injection with attenuated, heat- or formalin-killed bacteria (Song et al. 1976). Neither of the latter studies, however, used heterologous antigen for challenge. Tiecco et al. (1988) vaccinated eel with formalin and Ampicillin inactivated *A. hydrophila* by injection and immersion and both methods provided some protection, but the formalin inactivated bacterin was not as effective as was the Ampicillin preparation.

Although vaccination of channel catfish against bacterial infections may be a future management tool, this method of prevention should not be perceived as absolute. If a disease for which fish have been vaccinated does develop, mortalities will be lower in vaccinated populations; therefore, a manager will need to analyze the cost-to-benefit ratio to determine whether vaccination is feasible. Generally, vaccinated channel catfish grow faster, show a lower feed conversion rate, and have 10% to 30% better survival than do unvaccinated fish.

REFERENCES

Ainsworth, A. J., C. D. Rice, and L. Xue. 1995. Immune responses of channel catfish, *Ictalurus punctatus* (Rafinesque). After oral or intraperitoneal vaccination with particulate or soluble *Edwardsiella ictaluri* antigen. *Journal of Fish Diseases* 18:397–409.

Ainsworth, A. J., G. Capley, P. Waterstrat, and D. Munson. 1986. Use of monoclonal antibodies in the indirect fluorescent antibody technique (IFA) for the diagnosis of *Edwardsiella ictaluri. Journal of Fish Diseases* 9:439–444.

Anacker, R. L., and E. J. Ordal. 1959. Studies on the myxobacterium *Chondrococcus columnaris,* I. Serological typing. *Journal of Bacteriology* 78:25–32.

Antonio, D. B., and R. P. Hedrick. 1994. Effects of the corticosteroid Kenalog on the carrier state of juvenile channel catfish exposed to *Edwardsiella ictaluri. Journal of Aquatic Animal Health* 6:44–52.

Austin, B., and D. A. Austin. 1987. *Bacterial Fish Pathogens: Diseases in Farmed and Wild Fish.* Chichester, UK: Ellis Horwood Ltd.

Baba, T., J. Immura, R. Izawa, and G. Ikeda. 1988. Immune protection in carp, *Cyprinus carpio* L., after immunization with *Aeromonas hydrophila* crude lipopolysaccharide. *Journal of Fish Diseases* 11:237–244.

Baldwin, T. J., and J. C. Newton. 1993. Early events in the pathogenesis of enteric septicemia of channel catfish caused by *Edwardsiella ictaluri*: light and electron microscopic and bacteriologic findings. *Journal of Aquatic Animal Health* 5:189–198.

Baxa, D. V., V. J. Groff, A. Wishkowsky, and R. P. Hedrick. 1990. Susceptibility of nonictalurid fishes to experimental infection with *Edwardsiella ictaluri. Diseases of Aquatic Organisms* 8:113–117.

Baxa-Antonio, D., J. M. Groff, and R. P. Hedrick. 1992. Effect of water temperature on experimental *Edwardsiella ictaluri* infections in immersion exposed channel catfish. *Journal of Aquatic Animal Health* 4:148–151.

Becker, C. D., and M. P. Fujihara. 1978. The bacterial pathogen *Flexibacter columnaris* and its epizootiology among Columbia River fish. *American Fisheries Society Monograph* no. 2.

Bernardet, J. F. 1989. *Flexibacter columnaris*: first description in France and comparison with bacterial strains from other origins. *Diseases of Aquatic Organisms* 6:37–44.

Bernardet, J-F., and P. A. D. Grimont. 1989. Deoxyribonucleic acid relatedness and phenotypic characterization of *Flexibacter columnaris* sp. nov., nom. rev., *Flexibacter psychrophilus* sp. nov., nom. rev., and *Flexibacter maritimus* Wakabayashi, Hikida, and Masumura 1986. *International Journal of Systematic Bacteriology* 39:346–354.

Bernardet, J. F., P. Segers, M. VanCanneyt, F. Berthe, K. Kersters, and P. van Damme. 1996. Cutting the Gordian knot: emended classification and description of the genus *Flavobacterium,* emended description of the family Flavobacteriaceae, and proposal of *Flavobacterium hydatis* nom. nov. (Basonym, *Cytophaga aquatilis* Strohl and Tait 1978). *International Journal of Systematic Bacteriology* 46:128–148.

Boonyaratpalin, S. 1989. Bacterial pathogens involved in the epizootic ulcerative syndrome of fish in Southeast Asia. *Journal of Aquatic Animal Health* 1:272–276.

Bootsma, R., and J. P. M. Clerx. 1976. Columnaris disease of cultured carp *cyprinus carpio* L. characterization of the causative agent. *Aquaculture* 7:371–384.

Borg, A. F. 1960. Studies on myxobacteria associated with diseases in salmonid fishes. *Journal of Wildlife Diseases* 8:1–85.

Bowser, P. R. 1973. Seasonal prevalence of *Chondrococcus columnaris* infection in black bullheads from Clear Lake, Iowa. *Journal of Wildlife Diseases* 9:115–119.

Bullock, G. L., T. C. Hsu, and E. B. Shotts, Jr. 1986. Columnaris disease of fishes. *Fish Disease Leaflet* 72, United States Fish and Wildlife Service, 9 pp.

Carnahan, A. M., S. Behram, and S. W. Joseph. 1991. Aerokey key for identifying clincal *Aeromonas* species. *Journal of Clinical Microbiology* 29:2843–2849.

Cartwright, G. A., D. Chen, P. J. Hanna, N. Gudkous, and K. Tajima. 1994. Immunodiagnosis of virulent strains of *Aeromonas hydrophila* associated with epizootic ulcerative syndrome (EUS) using monoclonal antibody. *Journal of Fish Diseases* 17:123–133.

Chabot, D. J., and R. L. Thune. 1991. Proteases of the *Aeromonas hydrophila* complex: identification, characterization and relation to virulence in channel catfish, *Ictalurus punctatus* (Rafinesque). *Journal of Fish Diseases* 14:171–184.

Chen, M. F., and T. S. Light. 1994. Specificity of the channel catfish antibody to *Edwardsiella ictaluri*. *Journal of Aquatic Animal Health* 6:266–270.

Chen, M. F., and seven others. 1994. Distribution of *Edwardsiella ictaluri* in California. *Journal of Aquatic Animal Health* 6:234–241.

Chen, M. F., D. Henry-Ford, and J. M. Groff. 1995. Isolation and characterization of *Flexibacter maritimus* from marine fishes of California. *Journal of Aquatic Animal Health* 7:318–327.

Chia-Reiy, L., H-Y. Chung, and G-H. Kuo. 1982. Studies on the pathogenicity of *Flexibacter columnaris*–I. Effect of dissolved oxygen and ammonia on the pathogenicity of *Flexibacter columnaris* to eel (*Anguilla japonica*), *CAPD Fish*. Series no. 8:57–61.

Chowdhury, B. R., and H. Wakabayashi. 1988. Effects of sodium, potassium, calcium and magnesium ions on the survival of *Flexibacter columnaris* in water. *Fish Pathology* 23:231–235.

Chowdhury, B. R., and J. Wakabayashi. 1989. Effects of competitive bacteria on the survival and infectivity of *Flexibacter columnaris*. *Fish Pathology* 24:9–15.

Ciembar, P. G., V. S. Blazer, D. Dawe, and E. B. Shotts. 1995. Susceptibility of channel catfish to infection with *Edwardsiella ictaluri*: effect of exposure method. *Journal of Aquatic Animal Health* 7:132–140.

Collins, L. A., and R. L. Thune. 1996. Development of a defined menial medium for the growth of *Edwardsiella ictaluri*. *Applied and Environmental Microbiology*. 62:848–852.

Colwell, R. R., M. T. MacDonnell, and J. DeLey. 1986. Proposal to recognize the family Aeromonadaceae fam. nov. *International Journal of Systematic Microbiology* 36:473–477.

Cooper, R. K., II, C. E. Starliper, E. B. Shotts, Jr., and P. W. Taylor. 1993. Comparison of plasmids isolated from Romet-30-resistant *Edwardsiella ictaluri* and tribrissen-resistant *Escherichia coli*. *Journal of Aquatic Animal Health* 5:9–15.

Darwish, M. 1997. The pathogenesis of experimental *Edwardsiella tarda* infection in channel catfish (*Ictalurus punctatus*). Auburn, AL: Auburn University. Ph.D. dissertation.

Davidson, M. L. 1996. Characterization of the outer membrane proteins of *Flavobacterium columnare* and examination of their antigenicity to channel catfish (*Ictalurus punctatus*). Auburn, AL: Auburn University. Master's thesis.

Davis, H. S. 1922. A new bacterial disease of fresh-water fishes. *U. S. Bureau of Fisheries Bulletin* 38:261–280.

de Figueiredo, J., and J. A. Plumb. 1977. Virulence of different isolates of *Aeromonas hydrophila* in channel catfish. *Aquaculture* 11:349–354.

Earlix, D. J. 1995. Host, pathogen, and environmental interactions of enteric septicemia of catfish. Auburn, AL: Auburn University. Ph.D. dissertation.

Earlix, D., J. A. Plumb, and W. A. Rogers. 1996. Isolation of *Edwardsiella ictaluri* from channel catfish by tissue homogenization, filtration and enzyme linked immunosorbent assay. *Diseases of Aquatic Organisms* 27:19–24.

Ewing, W. H., A. C. McWhorter, M. R. Escobar, and A. H. Lubin. 1965. *Edwardsiella*, a new genus of Enterobacteriaceae based on a new species, *Edwardsiella tarda*. *International Bulletin of Bacterial Nomenclature and Taxonomy* 15:33–38.

Fijan, N. N. 1968. The survival of *Chondrococcus columnaris* in waters of different quality. In: *Symposium II de la Commission de l'Office International Des Epizooties Des Maladies Des Poissons*. Stockholm.

Fijan, N. N. 1969. Antibiotic additives for the isolation of *Chondrococcus columnaris* from fish. *Applied Microbiology* 17:333–334.

Francis-Floyd, R., M. H. Beleau, P. R. Waterstrat, and P. R., Bowser. 1987. Effect of water temperature on the clinical outcome of infection with *Edwardsiella ictaluri* in channel catfish. *Journal of the American Veterinary Medical Association* 191:1413–1416.

Garnjobst, L. 1945. *Cytophaga columnaris* (Davis) in pure culture: a myxobacterium pathogenic to fish. *Journal of Bacteriology* 49:113–128.

Goncalves, J. R., G. Braum, A. Fernandes, I. Biscaia, M. J. S. Correia, and J. Bastardo. 1992. *Aeromonas hydrophila* fulminant pneumonia in a fit young man. *Thorax* 47:482–483.

Goodwin, A. E., J. S. Roy, J. M. Grizzle, and M. T. Goldsby, Jr. 1994. *Bacillus mycoides*: a bacterial pathogen of channel catfish. *Diseases of Aquatic Organisms* 18:173–178.

Gray, S. J., D. J. Stickler, and T. N. Bryant. 1990. The incidence of virulence factors in mesophilic *Aeromonas* species isolated from farm animals and their environment. *Epidemiology and Infections* 105:277–294.

Griffin, B. R. 1992. A simple procedure for identification of *Cytophaga columnaris*. *Journal of Aquatic Animal Health* 4:63–66.

Grimont, P. A. D., F. Grimont, C. Richard, and R. Sakazaki. 1980. *Edwardsiella hoshinae*, a new species of Enterobacteriaceae. *Current Microbiology* 4:347–351.

Grizzle, J. M., and Y. Kiryu. 1993. Histopathology of gill, liver, and pancreas, and serum enzyme levels of channel catfish infected with *Aeromonas hydrophila* complex. *Journal of Aquatic Animal Health* 5:36–50.

Groberg, W. J., Jr., R. H. McCoy, K. S. Pilcher, and J. L. Fryer. 1978. Relation of water temperature to infections of coho salmon (*Oncorhynchus kisutch*), chinook salmon (*O. tshawytscha*), and steelhead trout (*Salmo gairdneri*) with *Aeromonas salmonicida* and *A. hydrophila. Journal of Fisheries Research Board Canada* 35:1–7.

Hanson, L. A., and J. M. Grizzle. 1985. Nitrite-induced predisposition of channel catfish to bacteria diseases. *The Progressive Fish-Culturist* 47:98–101.

Hargraves, J. E., and D. R. Lucey. 1990. *Edwardsiella tarda* soft tissue infection associated with catfish puncture wound. *Journal of Infectious Diseases* 162:1416–1417.

Hawke, J. P. 1979. A bacterium associated with disease of pond cultured channel catfish. *Journal of the Fisheries Research Board of Canada* 36:1508–1512.

Hawke, J. P., and R. L. Thune. 1992. Systemic isolation and antimicrobial susceptibility of *Cytophaga columnaris* from commercially reared channel catfish. *Journal of Aquatic Animal Health* 4:109–113.

Hawke, J. P., A. C. McWhorter, A. C. Steigerwalt, and D. J. Brenner. 1981. *Edwardsiella ictaluri* sp. nov., the causative agent of enteric septicemia of catfish. *International Journal of Systematic Bacteriology* 31:396–400.

Hazen, T. C., C. B. Fliermans, R. P. Hirsch, and G. W. Esch. 1978. Prevalence and distribution of *Aeromonas hydrophila* in the United States. *Applied Environmental Microbiology* 36:731–738.

Humphrey, J. D., C. Lancaster, N. Gudkovs, and W. McDonald. 1986. Exotic bacterial pathogens *Edwardsiella tarda* and *Edwardsiella ictaluri* from imported ornamental fish *Betta splendens* and *Puntius conchonius* respectively; isolation and quarantine significance. *Australian Veterinary Journal* 63:363–369.

Hussain, M., and R. C. Summerfelt. 1991. The role of mechanical injury in an experimental transmission of *Flexibacter columnaris* to fingerling walleye. *Journal of the Iowa Academy of Science* 98:93–98.

Janda, J. M. 1991. Recent advances in the study of the taxonomy, pathogenicity, and infectious syndromes associated with the genus *Aeromonas. Clinical Microbiology Review* 4:397–410.

Johnson, M. R., and K. L. Smith. 1994. Effect of pellet size and drug concentration on the efficacy of Romet-medicated feed for controlling *Edwardsiella ictaluri* infections in channel catfish fingerlings. *Journal of Aquatic Animal Health* 6:53–58.

Kage, T., R. Takahashi, I. Barcus, and F. Hayashi. 1992. *Aeromonas hydrophila,* a causative agent of mass mortality in cultured Japanese catfish larvae (*Silurus asotus*). *Fish Pathology* 27:57–62.

Kasornchandra, J., W. A. Rogers, and J. A. Plumb. 1987. *Edwardsiella ictaluri* from walking catfish, *Clarias batrachus* L., in Thailand. *Journal of Fish Disease* 10:137–138.

Kent, M. L., and J. M. Lyons. 1982. *Edwardsiella ictaluri* in the green knife fish, *Eigemannia virescens. Fish Health News* 2:ii.

Ketover, B. P., L. S. Young, and D. Armstrong. 1973. Septicemia due to *Aeromonas hydrophila;* clinical and immunological aspects. *Journal of Infectious Diseases* 127:284–290.

Khalifa, K. A., and G. Post. 1976. Immune response of advanced rainbow trout fry to *Aeromonas liquefaciens. The Progressive Fish-Culturist* 38:66–68.

Kim, M. K., and R. T. Lovell. 1995. Effect of overwinter feeding regimen on body weight, body composition and resistance to *Edwardsiella ictaluri* in channel catfish, *Ictalurus punctatus. Aquaculture* 134:237–246.

King, G. E., S. B. Werner, and W. Rizer. 1992. Epidemiology of aeromonas infections in California. *Clinical Infectious Diseases* 15:449–452.

Klesius, P. 1992. Carrier state of *Edwardsiella ictaluri* in channel catfish, *Ictalurus punctatus. Journal of Aquatic Animal Health* 4:227–230.

Klesius, P. H. 1993. Rapid enzyme-linked immunosorbent tests for detecting antibodies to *Edwardsiella ictaluri* in channel catfish, *Ictalurus punctatus,* using exoantigen. *Veterinary Immunology and Immunopathology* 36:359–368.

Klesius, P. 1994. Transmission of *Edwardsiella ictaluri* from infected, dead to noninfected channel catfish. *Journal of Aquatic Animal Health* 6:180–182.

Klesius, P., and W. M. Sealey. 1995. Characteristics of serum antibody in enteric septicemia of catfish. *Journal of Aquatic Animal Health* 7:205–210.

Klesius, P. H. and C. A. Shoemaker. 1999. The development and use of modified live *Edwardsiella ictaluri* vaccine against enteric septicemia of catfish. In: *Veterinary Vaccines and Diagnostics,* edited by R. D. Schultz. Vol. 41, Advances in Veterinary Medicine 523–537.

Klesius, P., K. Johnson, R. Durborow, and S. Vinitnantharat. 1991. Development and evaluation of an enzyme-linked immunosorbent assay for catfish serum antibody to *Edwardsiella ictaluri. Journal of Aquatic Animal Health* 3:94–99.

Kuo, S-C., H-Y. Chung, and G-H. Kou. 1981. Studies on artificial infection of the gliding bacteria in cultured fishes. *Fish Pathology* 15:309–314.

Lau, K. J., and J. A. Plumb. 1981. Effects of organic load on potassium permanganate as a treatment for *Flexibacter columnaris. Transactions of the American Fisheries Society* 110:86–89.

Lawrence, M. L., R. K. Cooper, and R. L. Thune. 1997. Attenuation, persistence, and vaccine potential of an *Edwardsiella ictaluri* purA mutant. *Infection and Immunity* 65:4642–4651.

Levanon, N., B. Motro, D. Levanon, and G. Degani. 1986. The dynamics of *Aeromonas hydrophila* in the water of tanks used to nurse elvers of the European eel *Anguilla anguilla. Bamidgeh* 38:55–63.

Lobb, C. J., and M. Rhoades. 1987. Rapid plasmid analysis for identification of *Edwardsiella ictaluri* from infected channel catfish *Ictalurus punctatus. Applied Environmental Microbiology* 53:1267–1272.

Lumlertdacha, S., and R. T. Lovell. 1995. Fumonisin-contaminated dietary corn reduced survival and antibody production by channel catfish challenged with *Edwardsiella ictaluri. Journal of Aquatic Animal Health* 7:1–8.

Maas, M. G., and R. Bootsma. 1982. Uptake of bacterial antigens in the spleen of carp (*Cyprinus carpio* L). *Developments in Comparative Immunology* (Suppl.) 2:47–52.

Marks, J. E., D. H. Lewis, and G. S. Trevino. 1980. Mixed infection in columnaris disease of fish. *Journal of the American Veterinary Medical Association* 177:811–814.

Meyer, F. P. 1970. Seasonal fluctuations in the incidence of diseases on fish farms. In: *A Symposium on Diseases of Fishes and Shellfishes* edited by S. F. Snieszko, 21–29. American Fisheries Society special publication no. 5. Washington, DC: American Fisheries Society.

Meyer, F. P., and G. L. Bullock. 1973. *Edwardsiella tarda,* a new pathogen of channel catfish (*Ictalurus punctatus*). *Applied Microbiology* 25:155–156.

Mgolomba, T. N., and J. A. Plumb. 1992. Longevity of *Edwardsiella ictaluri* in the organs of experimentally infected channel catfish, *Ictalurus punctatus. Aquaculture* 101:1–6.

Miyazaki, T., and Y. Jo. 1985. A histopathological study of motile aeromonad disease in ayu. *Fish Pathology* 20:55–59.

Miyazaki, T., and J. A. Plumb. 1985. Histopathology of *Edwardsiella ictaluri* in channel catfish, *Ictalurus punctatus* (Rafinesque). *Journal of Fish Diseases* 8:389–392.

Moore, A. A., M. E. Eimers, and M. A. Cardella. 1990. Attempts to control *Flexibacter columnaris* epizootics in pond-reared channel catfish by vaccination. *Journal of Aquatic Animal Health* 2:109–111.

Morrison, E., and J. A. Plumb. 1992. The chemosensory system of channel catfish, *Ictalurus punctatus,* following immersion exposure to *Edwardsiella ictaluri* [abstract]. *American Chemosensory Society,* Orlando, Florida, April 8–10.

Morrison, E. E., and J. A. Plumb. 1994. Olfactory organ of channel catfish as a site of experimental *Edwardsiella ictaluri* infection. *Journal of Aquatic Animal Health* 6:101–109.

Murphey, D. K., E. J. Septimus, and D. C. Waagner. 1992. Catfish-related injury and infection: Report of two cases and review of the literature. *Clinical Infectious Diseases* 14:689–693.

Newton, J. C., R. C. Bird, W. T. Blevins, G. R. Wilt, and L. Wolfe. 1988. Isolation, characterization, and molecular cloning of cryptic plasmids isolated from *Edwardsiella ictaluri. American Journal of Veterinary Research* 49:1856–1860.

Newton, J. C., L. G. Wolfe, J. M. Grizzle, and J. A. Plumb. 1989. Pathology of experimental enteric septicemia in channel catfish *Ictalurus punctatus* Rafinesque following immersion exposure to *Edwardsiella ictaluri. Journal of Fish Diseases* 12:335–348.

Nieto, T. P., M. J. R. Corcobado, A. E. Toranzo, and J. L. Barja. 1985. Relation of water temperature to infection of *Salmo gairdneri* with motile *Aeromonas. Fish Pathology* 20:99–105.

Nusbaum, K. E., and E. E. Morrison. 1996. Entry of ^{35}S-labeled *Edwardsiella ictaluri* into channel catfish. *Journal of Aquatic Animal Health* 8:146–149.

Ordal, E. J., and R. R. Rucker. 1944. Pathogenic myxobacteria. *Society of Experimental Biology and Medicine Proceedings* 56:15–18.

Paripatananont, T., and R. T. Lovell. 1995. Responses of channel catfish fed organic and inorganic sources of zinc to *Edwardsiella ictaluri* challenge. *Journal of Aquatic Animal Health* 147–154.

Peters, G., M. Faisal, T. Lang, and I. Ahmed. 1988. Stress caused by social interaction and its effect on susceptibility to *Aeromonas hydrophila* infection in rainbow trout *Salmo gairdneri. Diseases of Aquatic Organisms* 4:83–89.

Phelps, R. P., J. A. Plumb, and H. C. Harris. 1977. Control of external bacterial infections of bluegills with potassium permanganate. *The Progressive Fish-Culturist* 39:142–143.

Plumb, J. A., and V. Hilge. 1987. Susceptibility of European catfish (*Silurus glanis*) to *Edwardsiella ictaluri. Journal of Applied Ichthyology* 3:45–48.

Plumb, J. A., and P. Klesius. 1988. An assessment of the antigenic homogeneity of *Edwardsiella ictaluri* using monoclonal antibody. *Journal of Fish Diseases* 11:499–510.

Plumb, J. A., and E. E. Quinlan. 1986. Survival of *Edwardsiella ictaluri* in pond water and bottom mud. *The Progressive Fish-Culturist* 48:212–214.

Plumb, J. A., and D. J. Sanchez. 1983. Susceptibility of 5 species of fish to *Edwardsiella ictaluri. Journal of Fish Diseases* 6:261–266.

Plumb, J. A., and C. Shoemaker. 1995. Effects of temperature and salt concentration on latent *Edwardsiella ictaluri* infections in channel catfish. *Diseases of Aquatic Organisms* 21:171–175.

Plumb, J. A., and S. Vinitnantharat. 1989. Biochemical biophysical, and serological homogeneity of *Edwardsiella ictaluri. Journal of Aquatic Animal Health* 1:51–56.

Plumb, J. A., and S. Vinitnantharat. 1993. Vaccination of

channel catfish, *Ictalurus punctatus* (Rafinesque), by immersion and oral booster against *Edwardsiella ictaluri. Journal of Fish Diseases* 16:65–71.

Plumb, J. A., J. M. Grizzle, and J. de Figueiredo. 1976. Necrosis and bacterial infection in channel catfish (*Ictalurus punctatus*) following hypoxia. *Journal of Wildlife Diseases* 12:247–253.

Plumb, J. A., G. Maestrone, and E. Quinlan. 1987. Use of a potentiated sulfonamide to control *Edwardsiella ictaluri* infection in channel catfish (*Ictalurus punctatus*). *Aquaculture* 62:187–194.

Plumb, J. A., S. Vinitnantharat, and W. D. Paterson. 1994. Optimum concentration of *Edwardsiella ictaluri* vaccine in feed for oral vaccination of channel catfish. *Journal of Aquatic Animal Health* 6:118–121.

Popoff, M. 1984. Genus III. *Aeromonas* Kluyver and Van Niel. 1936, 398AL. In: *Bergey's Manual of Systematic Bacteriology,* vol. I, edited by N. R. Krieg, 545. Baltimore, MD: Williams and Wilkins.

Pyle, S. W., and E. B. Shotts, Jr. 1981. DNA homology studies of selected flexibacteria associated with fish diseases. *Canadian Journal of Fisheries and Aquatic Sciences* 38:146–151.

Roberts, R. J., L. G. Willoughby, and S. Chinabut. 1993. Mycotic aspects of epizootic ulcerative syndrome (EUS) of Asian fishes. *Journal of Fish Diseases* 16:169–183.

Rogers, W. A. 1981. Serological detection of two species of *Edwardsiella* infecting catfish. *Developments in Biological Standards* 49:169–172.

Santos, Y., A. E. Toranzo, C. P. Dopazo, T. P. Nieto, and J. L. Barja. 1987. Relationships among virulence for fish, enterotoxigenicity, and phenotypic characteristics of motile *Aeromonas. Aquaculture* 67:29–39.

Schachte, J. H., Jr., and E. C. Mora. 1973. Production of agglutinating antibodies in the channel catfish (*Ictalurus punctatus*) against *Chondrococcus columnaris. Journal of the Fisheries Research Board of Canada* 30:116–118.

Shamsudin, M. N., and J. A. Plumb. 1996. Morphological, biochemical, and physiological characterization of *Flexibacter columnaris* isolates from four species of fish. *Journal of Aquatic Animal Health* 8:335–339.

Shieh, H. S. 1980. Studies on the nutrition of a fish pathogen, *Flexibacter columnaris. Microbiological Letters* 13:129–133.

Shotts, E. B. 1991. Selective isolation methods for fish pathogens. *Journal of Applied Bacteriology, Symposium Supplement* 70:75S.

Shotts, E. B., and R. Rimler. 1973. Medium for the isolation of *Aeromonas hydrophila. Applied Microbiology* 26:550–553.

Shotts, E. B., and W. D. Waltman. 1990. An isolation medium for *Edwardsiella ictaluri. Journal of Wildlife Diseases* 26:214–218.

Shotts, E. B., J. L. Gaines, L. Martin, and A. K. Prestwood. 1972. *Aeromonas*-induced deaths among fish and reptiles in an eutrophic inland lake. *Journal American Veterinary Medicine Association* 161:603–607.

Shotts, E. B., V. L. Vanderwork, and L. M. Campbell. 1976. Occurrence of R factors associated with *Aeromonas hydrophila* isolated from aquarium fish and waters. *Journal of Fisheries Research Board of Canada* 33:736–740.

Shotts, E. B., V. S. Blazer, and W. D. Waltman. 1986. Pathogenesis of experimental *Edwardsiella ictaluri* infections in channel catfish (*Ictalurus punctatus*). *Canadian Journal of Fisheries and Aquatic Sciences* 43:36–42.

Song, Y-L., S-N. Chen, and G. Kou. 1976. Agglutinating antibodies production and protection in eel (*Anguilla japonica*) inoculated with *Aeromonas hydrophila* (*A. liquefaciens*) antigens. *Journal of Fisheries Society of Taiwan* 4:25–29.

Song, Y-L., J. L. Fryer, and J. S. Rohovec. 1988a. Comparison of six media for the cultivation of *Flexibacter columnaris. Fish Pathology* 23:91–94.

Song, Y-L., J. L. Fryer, and J. S. Rohovec. 1988b. Comparison of gliding bacteria isolated from fish in North America and other areas of the Pacific rim. *Fish Pathology* 23:197–202.

Speyerer, P. D., and J. A. Boyle. 1987. The plasmid profile of *Edwardsiella ictaluri. Journal of Fish Diseases* 10:461–469.

Stanley, L. A., J. S. Hudson, T. E. Schwedler, and S. S. Hayasaka. 1994. Extracellular products associated with virulent and avirulent strains of *Edwardsiella ictaluri* from channel catfish. *Journal of Aquatic Animal Health* 6:36–43.

Stanley, L. A., S. S. Hayasaka, and T. E. Schwedler. 1995. Effects of the immunomodulator *Ecteinascidia turbinata* extract on *Edwardsiella ictaluri* infection of channel catfish. *Journal of Aquatic Animal Health* 7:141–146.

Starliper, C. E., R. K. Cooper, E. B. Shotts, Jr., and P. W. Taylor. 1993. Plasmid-mediated Romet resistance of *Edwardsiella ictaluri. Journal of Aquatic Animal Health* 5:1–8.

Taylor, P., 1992. Fish-eating birds as potential vectors for *Edwardsiella ictaluri. Journal of Aquatic Animal Health* 4:240–243.

Thune, R. L. 1991. Major infectious and parasitic diseases of channel catfish. *Veterinary and Human Toxicology* 33 (Suppl. 1):14–18.

Tiecco, G., C. Sebastio, E. Francioso, G. Tantilla, and L. Corbari. 1988. Vaccination trials against "red plaque" in eels. *Diseases of Aquatic Organisms* 4:105–107.

Tyler, J. W., and P. H. Klesius. 1994. Decreased resistance

to *Edwardsiella ictaluri* infection in channel catfish intraperitoneally administered an oil adjuvant. *Journal of Aquatic Animal Health* 6:275–278.

Tucker, C. S. 1984. Potassium permanganate demand of pond waters. *The Progressive Fish-Culturist* 46:24–28.

Ventura, M. T., and J. M. Grizzle. 1988. Lesions associated with natural and experimental infections of *Aeromonas hydrophila* in channel catfish, *Ictalurus Punctatus* (Rafinesque). *Journal of Fish Diseases* 11:397–407.

Vinitnantharat, S., and J. A. Plumb. 1992. Kinetics of the immune response of channel catfish to *Edwardsiella ictaluri*. *Journal of Aquatic Animal Health* 4:207–214.

Vinitnantharat, S., and J. A. Plumb. 1993. Protection of channel catfish following exposure to *Edwardsiella ictaluri* and effects of feeding antigen on antibody titer. *Diseases of Aquatic Organisms* 15:31–34.

Vinitnantharat, S., J. A. Plumb, and A. E. Brown. 1993. Isolation and purification of an outer membrane protein of *Edwardsiella ictaluri* and its antigenicity to channel catfish (*Ictalurus punctatus*). *Fish & Shellfish Immunology* 3:401–409.

Wakabayashi, H., K. Kira, and S. Egusa. 1970. Studies on columnaris of eels–I. Characteristics and pathogenicity of *Chondrococcus columnaris* isolated from pond-cultured eels. *Bulletin of the Japanese Society of Scientific Fisheries* 36: 147–155.

Wakabayashi, H., M. Hikida, and K. Masumura. 1986. *Flexibacter maritimus* sp. nov., a pathogen of marine fishes. *International Journal of Systematic Bacteriology* 36:396–398.

Wakabayashi, J., K. Ganai, T. C. Hsu, and S. Egusa. 1980. Pathogenic activities of *Aeromonas* biovar *hydrophila* (Chester) Popoff and Vernon. 1976 to fishes. *Fish Pathology* 15:319–325.

Walters, G. R., and J. A. Plumb. 1980. Environmental stress and bacterial infection in channel catfish, *Ictalurus punctatus* Rafinesque. *Journal Fish Biology* 17:177–185.

Waltman, W. D., E. B. Shotts, and V. S. Blazer. 1985. Recovery of *Edwardsiella ictaluri* from Danio (*Danio devario*). *Aquaculture* 46:63–66.

Waltman, W. D., E. B. Shotts, and T. C. Hsu. 1986. Biochemical characteristics of *Edwardsiella ictaluri*. *Applied and Environmental Microbiology* 51:101–104.

Waltman, W. D., E. B. Shotts, and R. E. Wooley. 1989. Development and transfer of plasmid-mediated antimicrobial resistance in *Edwardsiella ictaluri*. *Canadian Journal of Fisheries and Aquatic Sciences* 46:1114–1117.

Willoughby, L. G., R. J. Roberts, and S. Chinabut. 1995. *Aphanomyces invaderis* sp. Nov., the fungal pathogen of freshwater tropical fish affected by epizootic ulcerative syndrome. *Journal of Fish Diseases* 18:273–275.

Wise, D. J., T. E. Schwedler, and D. L. Otis. 1993a. Effects of stress on susceptibility of naive channel catfish in immersion challenge with *Edwardsiella ictaluri*. *Journal of Aquatic Animal Health* 5:92–97.

Wise, D. J., and M. R. Johnson. 1998. Effect of feeding frequency and Romet-medicated seed on survival, antibody response, and weight gain of fingerling channel catfish (*Ictalurus punctatus*) after natural exposure to *Edwardsiella ictaluri*). *Journal of the World Aquaculture Society* 29:170–176.

Wise, D. J., J. R. Tomasso, T. E. Schwedler, V. S. Blazer, and D. M. Gatlin III. 1993b. Effects of vitamin E on the immune response of channel catfish to *Edwardsiella ictaluri*. *Journal of Aquatic Animal Health* 5:183–188.

Wolters, W. R., and M. R. Johnson. 1994. Enteric septicemia resistance in blue catfish and three channel catfish strains. *Journal of Aquatic Animal Health* 6:329–334.

Wolters, W. R., D. J. Wise, and P. H. Klesius. 1996. Survival and antibody response of channel catfish, blue catfish, and channel catfish female x blue catfish male hybrids after exposure to *Edwardsiella ictaluri*. *Journal of Aquatic Animal Health* 8:249–254.

15 Carp and Minnow Bacterial Diseases

Carp and goldfish culture (family Cyprinidae) is probably the oldest known type of aquaculture. Common carp have been cultured in Europe since the Middle Ages, and the Chinese have grown carp and goldfish in captivity for 2500 years. Currently, golden shiners and fathead minnows are cultured extensively in the United States for bait fish. Two bacterial diseases that are most often associated with cyprinids are "carp erythrodermatitis" (CE) (Fijan 1972) and "ulcer disease of goldfish" (UDG) (Elliott and Shotts 1980). Ironically, both disease syndromes are caused by the same organism, namely, atypical *Aeromonas salmonicida,* also known as *Aeromonas salmonicida* subspecies *achromogens* and atypical nonmotile *Aeromonas.* The following discussion includes CE, UDG, and other bacterial pathogens that cause disease in cyprinids (motile *Aeromonas* septicemia [MAS], *Pseudomonas* spp., and columnaris) (see Chapter 14).

ATYPICAL NONMOTILE *AEROMONAS* INFECTIONS

Atypical *A. salmonicida* was recognized as a fish pathogen by Smith (1963) in Great Britain. Paterson et al. (1980) later showed that the organism was identical to *Haemophilus piscium,* the etiological agent of "ulcer disease" of trout. Since the early 1970s, atypical *A. salmonicida* has been a major disease organism in cultured cyprinids, in which it produces a subacute to chronic infection. It has also been associated with similar diseases in increasing numbers of other fish species. Carp erythrodermatitis was originally considered a stage of infectious dropsy of carp (IDC); however, CE is now a distinct disease (Fijan 1972). The precise etiology of CE was not determined until Bootsma et al. (1977) reported a nonmotile "atypical" *A. salmonicida* (subspecies *achromogens*) from carp at five farms in Yugoslavia and Germany, at which time Koch's postulates were fulfilled. Goldfish ulcer disease followed a similar scenario until Shotts et al. (1980) solved the etiological problem.

Geographical Range and Species Susceptibility

Atypical *A. salmonicida* infections (CE and UDG) have been documented in most European countries, Great Britain, the United States, Canada, Australia, Israel, Japan, and several Asian countries (Fijan 1972; Fijan and Petrinec 1973; McCarthy 1975; Bootsma et al. 1977; Paterson et al. 1980). Cultured carp with clinical signs typical of CE have also been reported in China. Carp erythrodermatitis per se has not been reported in North America because the syndrome is referred to as ulcerative disease of goldfish (Elliott and Shotts 1980; Shotts et al. 1980).

Common carp, including scaled and mirror varieties, and goldfish are the principal species affected by atypical *A. salmonicida,* but similar types of lesions have been described in a variety of other freshwater and marine fish species. Other susceptible freshwater fish are northern pike (Wiklund 1990), eels, roach, silver bream, common bream, perch, and common minnow. Several species of salmonids, including Atlantic, chum, and coho salmon and rainbow, brook, and brown trout (Evelyn 1971; Wichardt 1983) are also susceptible. Marine fish that have been either naturally or experimentally infected with atypical *A. salmonicida* include

European flounder (Wiklund and Bylund 1991), dab, plaice (Wiklund and Dalsgaard 1995), turbot (Pedersen et al. 1994), and Atlantic tomcod (Williams et al., in press). Most of these infections have occurred in either wild, captured, or cultured fishes in the North Sea-Baltic Sea area of Scandinavia or the western north Atlantic Ocean.

Clinical Signs and Findings

Clinical signs of atypical *A. salmonicida* infections are similar in most fish (Fijan 1976; Elliott and Shotts 1980). Moribund CE fish rest near the surface, at the bottom, or close to pond banks or concentrate near fresh water inflow, a behavioral pattern that has not been described for UDG. Carp become darkly pigmented with slightly extended abdomens and slight to extreme exophthalmia.

Deep ulcers in the skin and muscle tissue are the most obvious clinical signs of CE and UDG (Figure 15.1). These lesions begin as small to large hemorrhages in the skin and scale pockets and progress to extensive necrosis in the superficial epithelium. Hemorrhagic inflammation also occurs on the fins. In other fish species, lesions associated with atypical *A. salmonicida* are generally very similar to those described for CE and UDG (McCarthy 1975; Hubert and Williams 1980). Iida et al. (1984) found that the most common sign of disease associated with atypical *A. salmonicida* in eels was a swollen head and presence of ulcerative lesions on the jaws and cheeks. Affected goldfish have pronounced anemia, exhibit body edema along with lepidorthosis, and have proteinaceous material that may collect on lesions, giving the appearance of a fungal infection (Figure 15.1). Internally, a general hemorrhagic appearance, pale liver, and inflamed intestine are noted. Atlantic salmon (30 to 40 g) infected with atypical *A. salmonicida* exhibited clinical signs that closely resembled those of furunculosis of salmonids.

Clinical signs displayed by diseased turbot in salt water tanks in Denmark were lethargy and necrotic lesions that usually started as erosion on the tips of skin nodules (Pedersen et al. 1994). These nodules contained white centers surrounded by a narrow hemorrhagic zone. Some erosions progressed into ulcers of 0.5 to 3 cm in diameter that were distributed over the body. Similar signs were described in atypical *A. salmonicida* infected European flounder, dab, and plaice (Wiklund and Dalsgaard 1995).

Diagnosis

The first step in diagnosing atypical *A. salmonicida* infections is noting the presence of clinical signs. The pathogen can be isolated from necrotic musculature or hemorrhagic skin lesions, seldom from the visceral organs of carp or goldfish. When other fish species are infected, the pathogen is more often isolated from visceral organs. For example, Cornick et al. (1984) isolated atypical *A. salmonicida* from kidneys of 22% of Atlantic cod with skin lesions, and Hubert and Williams (1980), working with roach, isolated the causative organism from 95% of skin ulcers and 67% and 24% of blood and kidney samples, respectively. Wiklund and Dalsgaard (1995) isolated the organism from skin ulcers of only 73% of affected fish and were unable to isolate it from internal organs of European flounder, dab, or plaice. Bootsma et al. (1977) proposed that isolation media be supplemented with tryptone and/or serum, but Paterson et al. (1980) found that blood agar produced satisfactory isolation results. Atypical *A. salmonicida* is fastidious, especially on primary isolation, but growth capability improves with repeated subcultivation. On isolation, slow-growing pinpoint, nonpigmented, cream-colored colonies appear after 24 to 72 hours at 20°C. Muscle lesions are often contaminated with other, less fastidious bacteria (*Aeromonas hydrophila* or *Vibrio anguillarum*) that grow on media more rapidly than does atypical *A. salmonicida*.

Bacterial Characteristics

Atypical *A. salmonicida* is a short Gram-negative, nonmotile bacillus that produces no diffused brown pigment at temperatures below 25°C (Shotts et al. 1980). The pathogen may be either cytochrome oxidase positive, as is the case in isolates from carp and goldfish, or cytochrome oxidase negative, as in isolates from coho salmon (Chapman et al. 1991), European flounder, dab, plaice, and turbot (Pedersen et al. 1994). Detection of the cytochrome oxidase-negative reaction of atypical *A. salmonicida* is relatively new (Chapman et al. 1991). Atypical *A. salmonicida* cells are about 0.5-µm wide and 0.7- to 1.4-µm long. Optimum growth temperature is 27°C, with no growth taking place at 37°C. As suggested by Pedersen et al. (1994) and Wiklund and Dalsgaard (1995), it appears that atypical *A. salmonicida* isolated in the

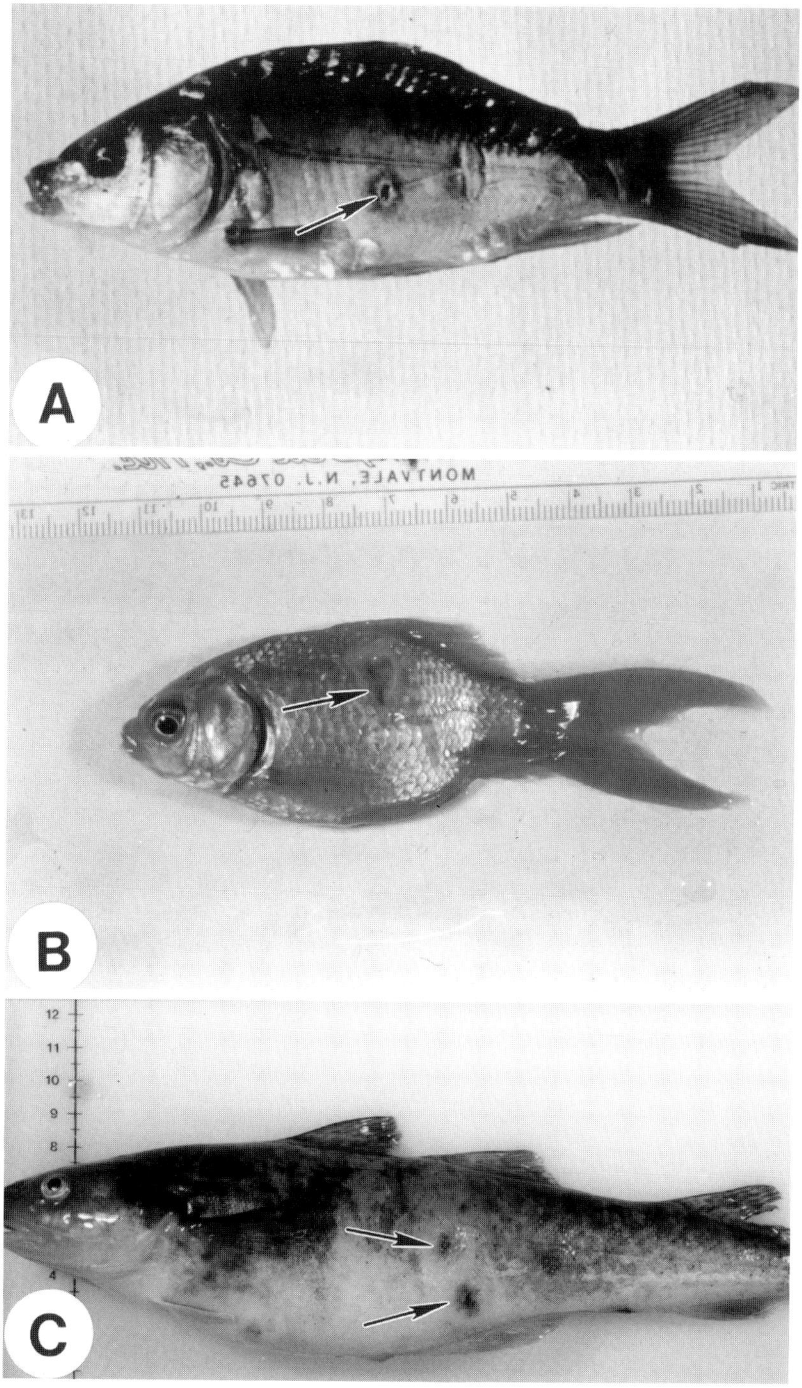

FIGURE 15.1. Diseases caused by atypical Aeromonas salmonicida. (A) *Ulcerative lesion* (arrow) *on the skin of common carp with carp erythrodermatitis. (Photograph courtesy of N. Fijan.)* (B) *Ulcerative lesion* (arrow) *on the skin of goldfish with ulcer disease of goldfish.* (C) *Ulcerative lesion* (arrows) *on lateral surface of Atlantic tomcod. (Photograph courtesy of S. Bastien-Daigle and P. J. Williams and with permission of the American Fisheries Society.)*

North Sea-Baltic Sea region from European flounder, turbot, and so forth, forms a distinct group from isolates of carp, goldfish and other species (Shotts et al. 1980). Major differences in the isolates include fastidiousness, cytochrome oxidase reaction, degradation of casein, esculin hydrolysis, and susceptibility to ampicillin and cephalothin.

All atypical *A. salmonicida* isolates are less biochemically reactive than other bacterial isolates from fish (Table 15.1). Restriction endonuclease analysis

TABLE 15.1. BIOCHEMICAL AND BIOPHYSICAL CHARACTERISTICS OF NONMOTILE AEROMONADS (TYPICAL *AEROMONAS SALMONICIDA SALMONICIDA*, ATYPICAL *A. SALMONICIDA* (SUBSP. *ACHROMOGENS*), AND ATYPICAL *A. SALMONICIDA* FROM NORTH SEA AND BALTIC SEA)

Characteristic/reaction	Typical *A. salmonicida*[a]	Atypical *A. salmonicida*[a]	North, Baltic Sea *A. salmonicida*[a]
Cell morphology		short, Gram-negative rods	
Agglutinates antitypical *A. salmonicida* serum	+	+	+
Motility	–	–	–
Cytochrome oxidase	+	+	±
Oxidation/fermentation glucose	+/+	+/+	+/+
Gas from glucose	+	–	–
Growth in or at			
0% NaCl	+	+	+
3% NaCl	+	+	+
4% NaCl	+	+	+(V)[b]
5°C	+	+	+
30°C	+	+	+
37°C	–	–	–
Degradation of			
Casein	+	+	+
Esculin	+	–	+
Gelatin	+	–	+
Starch	+	+	+
Voges-Proskauer	–	+	+
ß-galactosidase	–	–	+(V)
Indole	–	+	–
Susceptibility to			
Ampicillin (33 µg)	S	R	S
Cephalothin (66 µg)	S	R	S
Acid from			
Glycerol	+	+	–
L-arabinose	+	–	–
Ribose	+	+	–(V)
Galactose	+	+	–(V)
Fructose	+	+	+
D-mannose	–	+	V
Mannitol	+	+	–
N-acetylglucosamine	+	–	–(V)
Arbutine	+	–	–
Salicine	+	–	V
Cellobiose	–	–	–(V)
Maltose	+	–	–(V)
Saccharose	–	+	V
Trehalose	–	+	+(V)
Glycogen	+	+	–(V)

Sources: Kimura (1969); McCarthy (1975); Munroe and Håstings (1993); Wicklund and Dalsgaard (1995).
[a]+ = positive; – = negative; V = most strains variable; R = resistant; S = susceptible.
[b](V) = variable

of atypical *A. salmonicida* isolates in Australia indicates there are distinctly different strains of the organism (Whittington et al. 1995). They showed that an isolate from silver perch was the same as that from Australian goldfish on the same farm and similar to a goldfish isolate in the United States, but it differed from an isolate from goldfish in Singapore. An isolate from farmed greenback flounder in Australia was a completely different strain and was probably brought to the farm by wild-caught marine fish.

Epizootiology

The epizootiology of atypical *A. salmonicida* is similar in most fish species, especially in terms of seasonal and temperature relationship, age susceptibility, mortality patterns, and location of the causative organism in infected fish. Carp erythrodermatitis generally occurs in spring following stocking of production ponds, with less severe outbreaks being reported in autumn. Negligible losses from CE occur in summer. Atypical *A. salmonicida* infections in spring are associated with optimum water temperatures (15 to 20°C) and with spawning activities that can lead to superficial skin injury. During periods when water temperatures fluctuate between 6 and 20°C, carp losses due to CE can be greater than 50% (Fijan and Petrinec 1973). The disease in other fish species also tends to be more prevalent in spring and fall, with little evidence of its presence in summer (McCarthy 1975; Cornick et al. 1984). Atypical *A. salmonicida* can be experimentally transmitted by cohabitation of healthy and diseased fish or inoculation of infected skin material into dermis or scale pockets of naive fish. The most consistent transmission method is to rub infected material onto superficially scarified skin (Fijan 1976; Bootsma et al. 1977; Wiklund 1995a).

Young carp are highly susceptible to CE, but disease severity changes from an acute infection in young fish to a subacute and/or chronic infection as fish become older. A 15% survival was noted in 3- and 5-month-old carp, but when these fish reached 10 months of age, survival was 60% (Wiegertjes et al. 1993). It was suggested that resistance may be an inherited trait because mortality of offspring from one female was 100% when challenged, whereas a companion population from a second female had 25% to 50% mortality.

The source of atypical *A. salmonicida* is presumably carrier fish, although under certain conditions, the pathogen may survive for a period of weeks in the environment (Wiklund 1995b). In sterilized brackish water, the pathogen will survive for less than 14 days, but if sediment is added to the water, it will survive for up to 63 days. Atypical *A. salmonicida* was recoverable for a shorter time in small vessels containing sterile fresh water, and survival was better at 4 than 15°C. Conclusions drawn from these data are that atypical *A. salmonicida* may survive in bottom sediment of brackish water environments for a long time and that this substrate could serve as a pathogen reservoir. Atlantic tomcod captured in the Atlantic Ocean and held in tanks for experimental purposes invariably developed atypical *A. salmonicida* infections, displaying epithelial and muscular lesions in about 10 days after confinement, suggesting that fish was the source. Mortality occurred at 25 days when held at 11°C, and incubation time was doubled at 6°C (Williams et al. 1997).

Ulcerative disease of goldfish is a chronic condition that primarily affects cultured fish that are 1 year or older, including brood stock. The disease usually occurs in early spring following stocking of spawning ponds with adult fish and when water temperatures range between 18 and 25°C. Losses of larger subadult and adult goldfish during this time may reach 45% to 90%. Epizootic survivors appear to have no long-lasting disease effects with the exception of reduced egg yield. Saleable size goldfish (5 to 10 cm) become more susceptible as fish density, parasite load, and environmental stressors increase. Other potential pathogenic bacteria such as *Flavobacterium columnare, A. hydrophila,* and *V. anguillarum* (in marine fish) are often found in skin lesions in association with atypical *A. salmonicida* in goldfish.

In Denmark, an epizootic of atypical *A. salmonicida* in turbot revealed some interesting data (Pedersen et al. 1994). Onset and progression of disease paralleled an increase in water temperature from about 5 to 15°C, during which time mortality reached 15% to 20%. After initial outbreak and drug treatment, a secondary *V. anguillarum* infection occurred.

Atypical *A. salmonicida* isolated from European flounder showed low virulence in rainbow trout, whereas infectious experiments in roach and bleak were inconclusive (Wiklund 1995a). When, however, European flounder were exposed to the bacteria by ingestion or water following a skin abrasion, skin lesions resulted. Although it was concluded that the atypical *A. salmonicida* from European flounder was not a threat to rainbow trout, it was noted by Morrison et al. (1984) that Atlantic cod could become carriers of the bacterium and serve as reservoirs. Also, in experimental studies by Carson and Handlinger (1988), atypical *A. salmonicida* isolated from goldfish was pathogenic to Atlantic salmon, brown, rainbow and brook trout, whereas isolates from marine fish in Scandinavia were not pathogenic to rainbow trout.

Pathological Manifestations

Infections of atypical *A. salmonicida* are generally localized in the skin, where small or large hemorrhagic, inflamed lesions occur. At the center of an inflamed lesion, necrosis will develop on the outer layer of the skin (Fijan 1976). Even though lesions may disappear, infection can remain and result in hydropsy, ascites, exophthalmia, and anemia. It was noted by Bootsma et al. (1977) that bacteria responsible for CE could only be found in dermal and subdermal lesions, but nevertheless a generalized edema did occur. As previously discussed, atypical *A. salmonicida* infections in other fish species (cod and Atlantic salmon) may be systemic, but still the skin will be most severely affected. It was theorized by Pol et al. (1980) that inflammation, tissue necrosis, and possibly generalized edema present in CE-infected fish could be caused by bacterial exotoxins. Histopathology of atypical *A. salmonicida* in Atlantic cod showed that the host developed a defined reaction including encystment of the bacteria (Morrison et al. 1984). The spleen and kidney showed degenerative changes, leukocyte accumulation, and cyst formation.

Significance

In the last 20 years, atypical *A. salmonicida* infections have assumed a more important role in aquaculture and wild fish populations because of expanded species susceptibility. That it occurs in a variety of wild and cultured fish in fresh and salt waters, particularly tomcod, flounder, and halibut, tends to enhance its importance. Under certain conditions, CE can be a very serious disease in European carp culture ponds. Reduced egg production and loss of yearling and adult fish have resulted from UDG infections, and this problem has been exacerbated by the disease's usually poor response to conventional chemotherapeutics.

OTHER BACTERIAL DISEASES OF CYPRINIDS

Motile aeromonads can cause disease problems in cultured carp (common, grass, silver, and bighead), golden shiners, fathead minnows, and goldfish, especially following handling, stocking, or transport. Generally *A. hydrophila,* the principal bacterial organism in infectious dropsy of carp (IDC) syndrome in Europe, is the etiological agent. The bacteria has also been implicated as the etiological agent in a variety of infections and mortalities found in natural cyprinid populations. Clinical signs of *A. hydrophila* in cyprinids are necrotic (frayed) fins, hemorrhaged scale pockets, scale loss, and development of necrotic skin lesions. Infections become systemic and produce edema, anemia, and hyperemia of internal organs. Carp with IDC appear swollen with fluid in the muscles, scales exhibit lepidorthosis, and bloody fluid is present in the body cavity that produces dropsy. Cyprinid mortalities caused by *A. hydrophila* are usually chronic and seldom subacute. *Pseudomonas* spp. and *Pseudomonas fluorescens* produce clinical signs similar to those observed in MAS in cyprinids.

Cyprinids, especially shiners, fathead minnows, and the Asian carp, are extremely susceptible to infections of columnaris. This organism can be responsible for chronic losses in ponds following stocking and other periods of stress. Shiners and fathead minnows crowded into holding tanks while awaiting sale or shipment are very susceptible to columnaris, and mortalities may approach 100% if untreated. Cyprinids infected with columnaris display frayed fins, open ulcerative skin lesions where scales have been sloughed, and grayish color as a result of injury to the mucous layer and epithelium; hemorrhages appear at the base of fins and around the opercle and mouth. Columnaris lesions, particularly around the mouth, may be yellowish owing to the presence of large numbers of bacteria. Also, lesions may contain a variety of other waterborne bacteria including *A. hydrophila, P. fluorescens,* and protozoan parasites.

MANAGEMENT OF CYPRINID BACTERIAL DISEASES

To avoid atypical *A. salmonicida* infections in carp, goldfish, and other susceptible cyprinid species, handling, high fish densities, and skin injuries should be kept to a minimum during critical periods. Brood goldfish, stocked into ponds in early spring just prior to spawning, often contract atypical *A. salmonicida* infections owing to stress. When brood fish are stocked into spawning ponds in late fall or early winter, spring prespawning stress can be eliminated and UDG generally will not occur; therefore, fall brood fish stocking is common on goldfish farms where atypical *A. salmonicida* is endemic.

Fijan (1976) recommended that general prophylactic measures be taken on farms to reduce occurrence and/or severity of CE and UDG. Antibiotic baths should be applied to carp and goldfish after handling. Medicated feed containing oxytetracycline can be fed at 50 mg/kg of fish daily for 10 days to diseased fish; however, success of this procedure has been limited. A high rate of drug resistance for atypical *A. salmonicida* was reported by Barnes et al. (1994), who found that essentially all isolates from Scotland were resistant to amoxicillin (>500 mg/L minimum inhibitory concentration); this resistance was chromosomal rather than plasmid mediated.

In the Canadian Maritime Providences, an intraperitoneal injection of enrofloxacin was used to prevent atypical *A. salmonicida* infections in captured Atlantic tomcod (Williams et al. 1997). An injection of 5 mg/kg of the drug at time of capture, followed by additional injections at 10 and 45 days postcapture, successfully prevented formation of skin and muscle lesions and mortality due to the pathogen.

Several antigen preparations from atypical *A. salmonicida* were tested for their immunogenic potential to protect carp against CE (Evenberg et al. 1988). An injected, formalin-killed, whole cell bacterin provided consistent but moderate protection. However, a concentrated, deactivated culture supernatant afforded better protection against subsequent lethal challenge with the pathogen. A bath challenge system was used by Daly et al. (1994) to demonstrate the efficacy of vaccination with live *A. salmonicida*. Exposure to high concentrations of bacteria resulted in carp developing infection and typical clinical signs, but carp exposed to a sublethal dose of the bacteria developed skin lesions only, which healed after 1 week, followed by recovery. When challenged by subcutaneous injection, 100% of naive carp died, fish exposed to high bacterial concentrations suffered 40% to 60% mortality, but all fish exposed to the sublethal dose of live bacteria recovered. Protection of these fish lasted for 5 months. These data and those of previous studies indicate that extracellular products of atypical *A. salmonicida* are involved in CE and that vaccination of carp with sublethal concentrations of live bacteria against the disease is feasible. Infections of MAS, *Pseudomonas* sp., and columnaris are treated the same in all fish species; application of medicated feed with Terramycin or a bath treatment with potassium permanganate. Holding tanks of bait fish are often given prolonged treatments of commercially prepared concoctions that contain a variety of chemicals for columnaris. When used according to instructions, these compounds usually have some beneficial effect.

REFERENCES

Barnes, A. C., T. S. Håstings, and S. G. B. Amyes. 1994. Amoxycillin resistance of Scottish isolates of *Aeromonas salmonicida*. *Journal of Fish Diseases* 17:357–363.

Bootsma, R., N. Fijan, and J. Blommaert. 1977. Isolation and preliminary identification of the causative agent of carp erythrodermatitis. *Veterinarski Arhives* 47:291–302.

Carson, J., and J. Handlinger. 1988. Virulence of the aetiological agent of goldfish ulcer disease in Atlantic salmon, *Salmo salar* L. *Journal of Fish Diseases* 11:471–479.

Chapman, P. F., R. C. Cipriano, and J. D. Teska. 1991. Isolation and phenotypic characterization of an oxidase-negative *Aeromonas salmonicida* causing furunculosis in coho salmon (*Oncorhynchus kisutch*). *Journal of Wildlife Diseases* 27:61–67.

Cornick, J. W., C. M. Morrison, B. Zwicker, and G. Shum. 1984. Atypical *Aeromonas salmonicida* infection in Atlantic cod, *Gadus morhua* L. *Journal of Fish Diseases* 7:495–499.

Daly, J. G., G. F. Wiegertjes, and W. B. Van Muiswinkel. 1994. Protection against carp erythrodermatitis following bath or subcutaneous exposure to sublethal numbers of virulent *Aeromonas salmonicida* susp. nova. *Journal of Fish Diseases* 17:67–75.

Elliott, D. G., and E. B. Shotts, Jr. 1980. Aetiology of an ulcerative disease in goldfish *Carassius auratus* (L.): microbiological examination of diseased fish from seven locations. *Journal of Fish Diseases* 3:133–143.

Evelyn, T. P. T. 1971. An aberrant strain of the bacterial fish pathogen *Aeromonas salmonicida* isolated from a marine host, the Sable fish (*Anoplopoma finbria*), and from two species of cultured Pacific salmon. *Journal of the Fisheries Research Board of Canada* 28:1629–1634.

Evenberg, D., P. DeGraaff, B. Lugtenberg, and W. B. Van Muiswinkel. 1988. Vaccine-induced protective immunity against *Aeromonas salmonicida* tested in experimental carp erythrodermatitis. *Journal of Fish Diseases* 11:337–350.

Fijan, N. 1972. Infectious dropsy in carp—a disease complex. *Symposium of the Zoological Society of London* 30:39–51.

Fijan, N. 1976. Diseases of cyprinids in Europe. *Fish Pathology* 10:129–134

Fijan, N., and Z. Petrinec. 1973. Mortality in a pond caused by carp erythrodermatitis. *Rivista Italiana Pis-*

cicoltura E Ittiopathologia 8:45–49.

Hubert, R. M., and W. P. Williams. 1980. Ulcer disease of roach, *Rutilus rutilus* (L). *Bamidgeh* 32:46–52.

Iida, T., K. Nakakoshi, and H. Wakabayashi. 1984. Isolation of atypical *Aeromonas salmonicida* from diseased eel, *Anguilla japonica. Fish Pathology* 19:109–112.

Kimura, T. 1969. A new subspecies of *Aeromonas salmonicida* as an etiological agent of furunculosis on "Sakuramasu" (*Oncorhynchus masou*) and pink salmon (*O. gorbuscha*) rearing for maturity, Part 1. On the morphological and physiological properties. *Fish Pathology* 3:34–44.

McCarthy, D. H. 1975. Fish furunculosis caused by *Aeromonas salmonicida* var. achromoqens. *Journal of Wildlife Diseases* 11:489–493.

Morrison, C. M., J. W. Cornick, G. Schum, and B. Zwicker. 1984. Histopathology of atypical *Aeromonas salmonicida* infection in Atlantic cod, *Gadus morhua* L. *Journal of Fish Diseases* 7:477–494.

Munroe, A. L. S., and T. S. Håstings. 1993. Furunculosis. In: *Bacterial Diseases of Fish,* edited by V. Inglis, R. J. Roberts, and N. R. Bromage, 122–142. Oxford, UK: Blackwell Press.

Paterson, W. D., D. Douey, and D. Desautels. 1980. Isolation and identification of an atypical *Aeromonas salmonicida* strain causing epizootic losses among Atlantic salmon (*Salmo salar*) reared in a Nova Scotia Hatchery. *Canadian Journal of Fisheries and Aquatic Sciences* 37:2236–2241.

Pedersen, K., H. Kofod, I. Dalsgaard, and J. L. Larsen. 1994. Isolation of oxidase-negative *Aeromonas salmonicida* from diseased turbot *Scophthalmus maximus. Diseases of Aquatic Organisms* 18:149–154.

Pol, J. M. A., R. Bootsma, and J. M. Berg-Blommaert. 1980. Pathogenesis of carp erythrodermatitis (CE): role of bacterial endo- and exotoxin. In: *Fish Diseases Third COPRAQ-Session,* edited by W. Ahne, 120–125, Berlin, Springer-Verlag.

Shotts, E. B., Jr., F. D. Talkington, D. G. Elliott, and D. H. McCarthy. 1980. Aetiology of an ulcerative disease in goldfish, (*Carassius auratus* L.): characterization of the causative agent. *Journal of Fish Diseases* 3:181–186.

Smith, I. W. 1963. The classification of "Bacterium salmonicida." *Journal of General Microbiology* 33:263–274.

Whittington, R. J., S. P. Djordjevic, J. Carson, and R. B. Callinan. 1995. Restriction endonuclease analysis from goldfish *Carassius auratus,* silver perch *Bidyanus bidanus,* and greenback flounder *Rhombosolea tapirina* in Australia. *Diseases of Aquatic Organisms* 22:185–191.

Wichardt, U-P. 1983. Atypical *Aeromonas salmonicida*-infection in sea-trout (*Salmo trutta* L.) I. Epizootiological studies, clinical signs and bacteriology. *Lasforskningsinstitutet Meddelande* (Salmon Research Institute Report) 6:1–10.

Wiegertjes, G. F., J. G. Daly, and W. B. Muiswinkel. 1993. Disease resistance of carp, *Cyprinus carpio* L.: identification of individual genetic differences by bath challenge with atypical *Aeromonas salmonicida. Journal of Fish Diseases* 16:569–576.

Wiklund, T. 1990. Atypical *Aeromonas salmonicida* isolated from ulcers of pike, *Esox lucius* L. *Journal of Fish Diseases* 13:541–544.

Wiklund, T. 1995a. Virulence of "atypical" *Aeromonas salmonicida* isolated from ulcerated flounder *Platichthys flesus. Diseases of Aquatic Organisms* 21:145–150.

Wiklund, T. 1995b. Survival of "atypical" *Aeromonas salmonicida* in water and sediment microcosms of different salinities and temperatures. *Diseases of Aquatic Organisms* 21:137–143.

Wiklund, T., and G. Bylund. 1991. A cytochrome oxidase negative bacterium (presumptively and atypical *Aeromonas salmonicida*) isolated from ulcerated flounders (*Platichthys flesus* (L.)) in the northern Baltic Sea. *Bulletin of the European Association of Fish Pathologists* 11:74–76.

Wiklund, T., and I. Dalsgaard. 1995. Atypical *Aeromonas salmonicida* associated with ulcerated flatfish species in the Baltic Sea and the North Sea. *Journal of Aquatic Animal Health* 7:218–224.

Williams, P. J., S. C. Courtenay, and C. Vardy. 1997. Use of enrofloxacin to control atypical *Aeromonas salmonicida* in the Atlantic tomcod (*Microgadus tomcod* Walbaum). *Journal of Aquatic Animal Health* 9:216–222.

16 Eel Bacterial Diseases

Generally, eels contract the same types of bacterial diseases as do other warmwater fishes; however, two infectious diseases, edwardsiellosis (*Edwardsiella tarda*), also known as hepatonephritis, and red spot disease (*Pseudomonas anguilliseptica*), seem to have a greater affinity for cultured eel than for other fish species. Other bacterial fish pathogens; *Flavobacterium columnare,* motile *Aeromonas* septicemia (Chapter 14), atypical *Aeromonas salmonicida* (Chapter 15), and *Vibrio anguillarum* (Chapter 17), also occur in eel.

EDWARDSIELLOSIS

Edwardsiellosis, a subacute to chronic disease, affects a variety of fish species especially cultured eel in Asia (Hoshina 1962) and to a lesser degree channel catfish (Meyer and Bullock 1973) in the United States. The disease is also known as hepatonephritis in eel and in channel catfish emphysematous putrefactive disease or fish gangrene because of the foul odor emitted from gas-filled pockets in necrotic muscle. The etiology of edwardsiellosis is *E. tarda,* which was earlier called *Paracolobacterum anguillimortifera* (Wakabayashi and Egusa 1973).

Geographical Range and Species Susceptibility

E. tarda is found in fresh water culture facilities and to some extent in marine environments. It has been reported from 25 countries in North and Central America, Europe, Asia, Australia, Africa, and the Middle East (Austin and Austin 1987).

Although the most prominent fish species infected by *E. tarda* are eels (American, European, and Japanese) and channel catfish, the organism has been isolated from more than 20 freshwater and marine fish species including carp, largemouth bass, striped bass, red seabream, flounder, tilapia, and yellowtail (Hoshina 1962; Meyer and Bullock 1973; Austin and Austin 1987; Francis-Floyd et al. 1993). *E. tarda* has also been isolated from intestinal content of apparently healthy fish during routine bacteriological surveys (Wyatt et al. 1979; van Damme and Vandepitte 1980). Generally considered to be a warm water inhabitant, *E. tarda* has on at least two occasions been implicated in salmonid infections. It was isolated from migrating adult chinook salmon in the Rogue River in Oregon when water temperatures were above normal (Amandi et al. 1982) and again in Nova Scotia, Canada, where it was isolated from a small number of migrating adult Atlantic salmon (Martin 1983).

E. tarda infections are not limited to fish; the pathogen often exist as part of normal intestinal microflora and may be found in fish-eating birds and mammals, reptiles, amphibians, marine mammals, as well as cattle, and swine (Ewing et al. 1965; White et al. 1969; Coles et al. 1978). Humans have also been known to become infected with *E. tarda* (Heargraves and Lucey 1990).

Clinical Signs and Findings

Clinical signs of *E. tarda* infections differ slightly between locale and fish species. Infected eels exhibit lethargic swimming, tend to float at the surface, fins become congested and hyperemic, petechial hemorrhages develop on the underside, the anal region is swollen and hyperemic, and gas pockets may develop between dermis and muscle

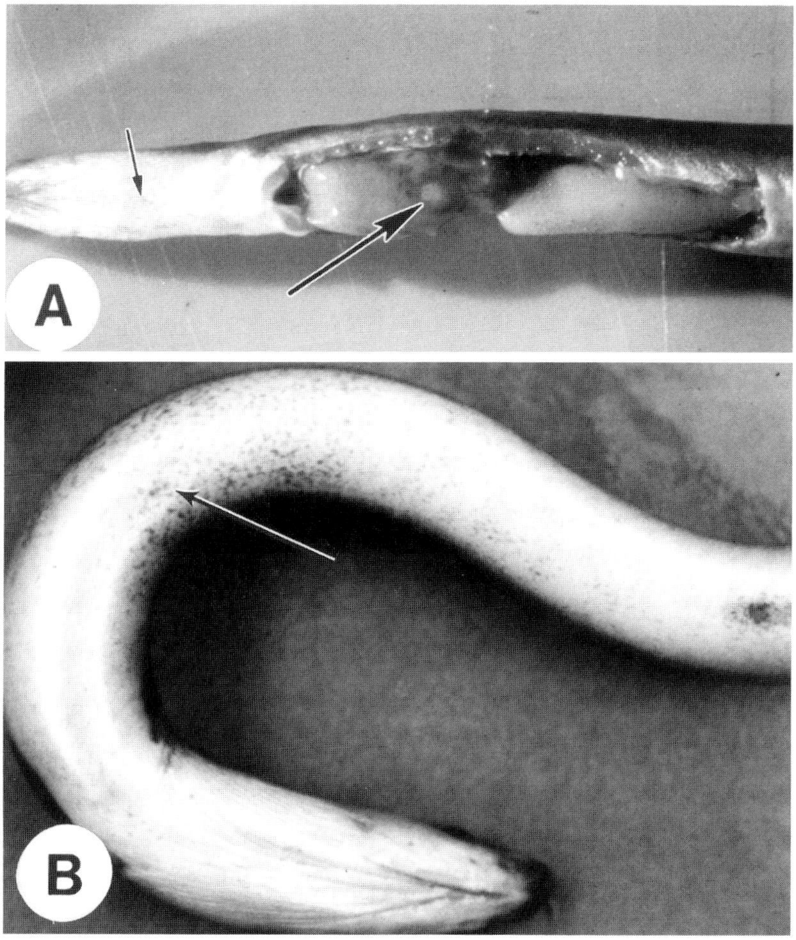

FIGURE 16.1 (A) *Japanese eel infected with* Edwardsiella tarda. *Note the pale, mottled liver with an abscess* (large arrow) *and slight petechiae on the throat* (small arrow). *(Photograph by E. B. Shotts.)* (B) *Japanese eel infected with* Pseudomonas anguilliseptica *showing petechia* (arrow) *(Photograph by T. Miyazaki.)*

(Figure 16.1). Internally, the liver, kidney, and spleen have a whitish appearance and may develop abscesses. Infected channel catfish develop small to large abscesses in the lateral musculature that emit a foul odor when incised. A variety of clinical signs occur in other fish species including exophthalmia, opaqueness of the eye, swollen and necrotic skin and muscle lesions, rectal protrusion, and abscesses in internal organs.

Diagnosis

E. tarda is easily isolated from internal organs and muscle lesions of clinically infected fish and identified by conventional bacteriological or serological methods. When working with eels, it is necessary to isolate bacteria from diseased fish for diagnostic purposes because clinical signs of *E. tarda* and other bacterial septicemias cannot be differentiated (Nichibuchi et al. 1980). *E. tarda* is not fastidious; therefore, isolation is easily made on brain heart infusion (BHI) agar or trypticase soy agar (TSA), or any other general purpose medium (Meyer and Bullock 1973; Amandi et al. 1982). Incubation at 26 to 30°C for 24 to 48 hours yields small, circular, convex, transparent colonies approximately 0.5 mm in diameter. The incidence of *E. tarda* isolation from chinook salmon brain was improved from 2% to 19% by inoculating thioglycolate media and then transferring the inoculum to BHI agar rather than inoculating directly onto BHI agar (Amandi et al. 1982). *E. tarda* form small, green colonies with black centers on *Edwardsiella ictaluri* isolation media (EIM) (see Figure 14.5) (Shotts and Teska 1989).

Bacterial Characteristics

E. tarda, a member of the family Enterobacteriaceae, is a small, straight rod about 1 μm in diameter and 2 to 3 µm in length (Farmer and McWhorter 1984). It is Gram-negative, facultatively anaerobic, with peritrichous flagella. Key presumptive characteristics of *E. tarda* are its motility, indole production in tryptone broth, H_2S production on triple sugar iron (TSI) slants, its tolerance of up to 4% salt, production of gas during glucose fermentation, and growth at 40°C. It is catalase positive, cytochrome oxidase negative, ferments glucose, reduces nitrate to nitrite, and is lactose negative (see Table 14.2). Positive identification of *E. tarda* can be made by specific serum agglutination or fluorescent antibody tests because no serological cross-reactivity occurs between it and *E. ictaluri* or other enteric bacteria. *E. tarda* is phenotypically a homogeneous organism. Waltman et al.(1986), as well as others, found only slight variations in biochemical and biophysical characteristics of 116 *E. tarda* isolates from the United States and Taiwan. Using the Minitek Numerical Identification System, Taylor et al. (1995) correctly identified *E. tarda* 100% of the time compared with 83% positive identification with the API 20E system.

Farmer and McWhorter (1984) identified a wild-type *E. tarda* and a second biogroup 1, both of which are easily distinguished from *E. ictaluri* and *Edwardsiella hoshinae* (a pathogen of reptiles and amphibians) (Grimont et al. 1980). Park et al. (1983) identified four *E. tarda* serotypes (A, B, C, and D) among 445 isolates collected from fish and other environmental sources. Seventy-two percent of fish isolates were serotype A, indicating that it may be the predominant disease-causing type in fish. Also, 28 strains of *E. tarda* isolated from diseased flounder were identical to serotype A (Rashid et al. 1994a).

Epizootiology

Edwardsiellosis of catfish occurs most often in summer when water temperatures are higher than 30°C, which agrees with infections in catfish, but the disease is not universally restricted to warm temperatures. According to Egusa (1976), *E. tarda* infections of Japanese eels were more prevalent in summer when water temperatures were highest. This trend was also reported by Kuo et al. (1977) who found that the optimum water temperature for an edwardsiellosis outbreak in eel populations was 30°C. However, Liu and Tsai (1982) found that eel infections in Taiwan were most common during January to April when temperatures fluctuated. In the United States, Amandi et al. (1982) isolated *E. tarda* from 19% of dead or dying Rogue River chinook salmon in Oregon when water temperatures rose from 17 to 20°C. The various temperatures at which *E. tarda* can cause disease may reflect an opportunistic capability that could be more closely associated with increased species susceptibility during temperature-related stress than to pathogenicity of the organism. However, Suprapto et al. (1995) showed that under similar conditions at 20°C, Japanese flounder and eel reacted very differently to *E. tarda* endotoxins. A heat labile endotoxin from virulent *E. tarda* was 15 times more toxic to Japanese flounder than to Japanese eel. They also showed that the LD50 was $10^{1.1}$ CFU/g of flounder and 10^6 CFU/g of eel.

The main source of *E. tarda* is presumably intestinal content of carrier aquatic animals, primarily catfish and eels; however, amphibians and reptiles must also be considered bacterial sources. *E. tarda* is probably released from an infected animal into the water or is naturally present in the mud or water. It has been reported that cell counts of *E. tarda* in eel pond water are higher when clinical disease is present. Hidaka et al. (1983) showed that *E. tarda* cell counts at 22 to 26°C were four times higher when clinical disease was present, which agrees with observations made in *E. ictaluri*–infected channel catfish ponds (Earlix 1995).

Fish are experimentally infected with *E. tarda* by intraperitoneal (IP) or intramuscular (IM) injection, stomach gavage, or immersion in water containing the pathogen; however, infection by immersion is not guaranteed. Huang and Liu (1986) killed 100% of IP injected eels when *E. tarda* was mixed with *Aeromonas hydrophila* but waterborne exposure to the same mixture failed to induce mortality without sublethal concentrations of nitrogenous compounds. Darwish (1997) had little success in infecting channel catfish by immersion in an *E. tarda* bath unless the skin was first superficially abraded. Mekuchi et al. (1995) successfully infected Japanese flounder with *E. tarda* by IM and IP injection, immersion, and orally. Before introducing *E. tarda* into the lumen via silicon tube, Miyazaki et al. (1992) injured the intestine with hydrogen peroxide. Infected fish developed pathological lesions in the kidney and liver that were nearly identical to those in naturally occurring infections and resulted in death 5 to 23 days postinfection.

E. tarda is a hardy, facultative bacterium that can survive for extended periods in water at temperatures of 15 to 45°C (optimum, 30 to 37°C), pH 4.0 to 10.0 (optimum, pH 7.5 to 8.0), and in salt solutions of 0% to 4% (optimum, 0.5 to 1.0% NaCl) (Chowdhury and Wakabayashi 1990). Isolation of *E. tarda* for longer than 76 days from 20°C pond water indicates that water can serve as a pathogen source for extended periods. The organism also survives for days in net fowling material (Wyatt et al. 1979).

Because *E. tarda* is a common inhabitant of catfish pond waters, its presence poses a constant threat of disease. In *E. tarda*–positive catfish ponds, the organism was isolated from 75% of pond water samples, 64% of mud samples, and 100% of apparently healthy frog, turtle, and crayfish samples (Wyatt et al. 1979). Also, *E. tarda* was isolated from the flesh of as many as 88% of dressed domestic catfish but from only 30% of imported dressed fish. In an ecological study, Rashid et al. (1994b) isolated *E. tarda* in 86% of water samples from one salt water flounder pond and in 22% from a second pond. In the first pond, *E. tarda* was present in 44% of sediment samples and 14% of asymptomatic fish but was found in 0% of sediment and only 2% of flounder from a second pond. Ironically, no clinical disease, morbidity, or mortality occurred in either pond.

An environmental stressor may not be essential for *E. tarda* fish infections, but fluctuating and high water temperatures, high water organic content and generally poor water quality, and crowded conditions can contribute to onset and severity of the disease (Meyer and Bullock 1973; Walters and Plumb 1980). When exposed to environmental stressors, 25% to 50% of juvenile channel catfish experimentally infected with *A. hydrophila* also developed an *E. tarda* infection (Walters and Plumb 1980). Only 4% to 13% of nonstressed *A. hydrophila*–uninfected fish developed *E. tarda* infections. These data indicate that environmental stress, in conjunction with other bacterial infections, can predispose channel catfish to disease from naturally present *E. tarda*. Also, sublethal concentrations of copper (100–250 µg/L) in the water reduced resistance of Japanese eels to *E. tarda* infections (Mushiake et al. 1984). Its transmission in catfish, eels, and most other fish species appears to be enhanced at water temperatures from 20 to 30°C. Japanese flounder were most susceptible at 20 to 25°C by IM injection in which an LD_{50} of 7.1×10^1 CFU was established (Mekuchi et al. 1995). This compared with an LD_{50} of 1.7×10^2 CFU for IP injection and 3.6×10^6 CFU/mL and 1.3×10^6 CFU per fish for immersion and oral exposure, respectively, at the same temperature. On several occasions, the author has observed intensively cultured Nile tilapia that were environmentally stressed or had a moderate to heavy *Trichodina* infection (protozoan ectoparasite) become clinically infected with *E. tarda* and/or *Streptococcus* spp. (unpublished). While stressed, neither bacterial infection responded to chemotherapy, but as soon as the stressor was relieved and parasites eliminated, bacterial infections disappeared without antibiotic medication. In experimental infectivity studies, 100-g eels were immersed in *E. tarda* and 60% mortality occurred (Liu and Tsai 1982). Kodoma et al. (1987) reported that naturally occurring *E. tarda* infections can cause an 80% loss in eels.

E. tarda can be a health threat to humans, in whom it usually manifests itself as gastroenteritis and diarrhea. Extraintestinal infections can, however, produce a typhoidlike illness, meningitis, peritonitis with sepsis, cellulitis, and hepatic abscess (Clarridge et al. 1980). The organism has been associated with wound infections precipitated by fish hooks or catfish spines (Hargreaves and Lucey 1990). It has also been implicated in diarrhea associated with an intestinal infection of *Entamoeba histolytica,* with other tropical diarrheas, and from consumption of contaminated freshwater fish (Gilman et al. 1971). According to Janda and Abbott (1993), in English-language literature concerning *E. tarda,* there have been at least 250 reported cases of *E. tarda* infections in humans. The ability of *E. tarda* to infect warm-blooded animals, and humans in particular, was shown by Janda et al. (1991) when they demonstrated its ability to invade HEp-2 cell monolayers, produce cell-associated hemolysin and siderophores, and express mannose-resistant hemagglutination against guinea pig erythrocytes. Janda and Abbott (1993) also showed that strains of *E. tarda* produced 30% to 40% higher levels of cell associated hemolytic activity (hemolysins) than did strains of *E. ictaluri* that are known only to affect fish. This increased hemolytic activity could contribute to *E. tarda* pathogenicity to humans. Infections in humans are usually successfully treated with antibiotics, but deaths have been reported.

Pathological Manifestations

In eels, *E. tarda* causes either a nephritic or hepatic histopathology (Miyazaki and Kaige 1985). In the more common nephritic form, the kidney is enlarged and contains variously sized abscesses that initially occur in the sinusoids of the hematopoietic tissue. This tissue becomes swollen, and foci of infection develops into abscesses that progress into cavities filled with dark red, odiferous, purulent matter. These large abscesses that contain neutrophils, some of which phagocytize bacteria, are walled off by fibrin. Cell liquefaction follows, and small abscesses can be found in other organs and the lateral musculature that adjoins the kidney.

In the hepatitis form of disease, livers are usually enlarged and contain variously sized abscesses, some of which leak fluid into the body cavity. Enlarged abscesses involve hepatocytes and blood vessels which form puslike emboli and pyemia (Egusa 1976) with extensive bacterial multiplication. At least some *E. tarda* isolates produce toxic extracellular products (ECP). Ullah and Arai (1983) found evidence that *E. tarda* does not form endotoxins as do other Gram-negative bacteria but produces two exotoxins that may contribute to pathogenicity. Factors that regulate pathogenicity of *E. tarda* are unclear. Suprapto et al. (1995) detected a heat-labile ECP in a serogroup A strain of *E. tarda* that was lethal to Japanese eel and Japanese flounder. The optimum incubation temperature for ECP production was 25 to 30°C, which coincides with most reports of optimum temperature for fish susceptibility. The ECP production peaked at 72 to 120 hours after media inoculation. An intracellular component (ICC), which occurred from 24 to 120 hours after media inoculation, was also associated with lysis of *E. tarda* cells. Results of these studies suggest that toxin produced by *E. tarda* plays an important role in its virulence.

Significance

Significance of *E. tarda* depends largely on fish species affected and presence of environmental stressors. In Taiwan and Japan, it is one of the most serious bacterial diseases to affect cultured eels. In the United States, unless channel catfish are crowded, the disease is of minor consequence. A subclinical infection in market-size fish will affect quality at slaughter and can cause contamination problems during dressing (Meyer and Bullock 1973). When this occurs, fish processing lines are shut down, cleaned, and disinfected. In other fish species, including intensively cultured tilapia, *E. tarda* may be responsible for some losses but usually they are chronic rather than acute. Caution must always be taken when handling *E. tarda*–infected fish because of potential infectivity to humans.

RED SPOT DISEASE

Red spot disease is a mild to severe bacterial infection that mainly affects cultured eels, but recently, its species susceptibility has expanded. The disease, known as "Sekiten-byo" in Japan, is caused by *P. anguilliseptica* (Wakabayashi and Egusa 1972).

Geographical Range and Species Susceptibility

Red spot disease primarily affects cultured Japanese and European eels. Although most prevalent in Japan, the disease and its etiological agent have been identified in Taiwan, Malaysia, Scotland, and, more recently, in Finland and France (Nash et al. 1987; Stewart et al. 1983; Wicklund and Bylund 1990). Although Japanese eel appear to be the most susceptible fish species, it has occurred in black seabream, giant seaperch, and estuarine grouper in Malaysia (Nash et al. 1987). The organism has also been isolated from Atlantic salmon, brown and rainbow trout, and whitefish (Wicklund and Bylund 1990). Lönnström et al. (1994) isolated *P. anguilliseptica* from wild Baltic herring along the southwest coast of Finland. The organism was also isolated from maricultured gilthead seabream, seabass, and turbot in France (Berthe et al. 1995) and striped jack in Japan (Kusuda et al. 1995). Disease has been produced by experimental infections in ayu (Nakai et al. 1985a), but an isolate from Baltic herring was avirulent to rainbow trout (Lönnström et al. 1994).

Clinical Signs and Findings

Japanese eels infected with *P. anguilliseptica* develop petechial hemorrhage in the subepidermal layer of the jaws, underside of head, and along the ventral body surface (see Figure 16.1). Internally, petechia can occur on the peritoneum, the liver is swollen, and the spleen and kidney become atrophied (Jo et al. 1975). European eels have similar pathology, but it is less severe than in Japanese eels (Ellis et al. 1983).

Clinical signs in nonanguillid species vary. In Baltic herring, hemorrhages were present in the eyes, fins, and on the heads; the cornea was ulcerated, and blood-stained ascites were seen in several fish (Lönnström et al. 1994). Berthe et al. (1995) described petechial hemorrhages in the skin and livers of farmed gilthead seabream and moribund fish exhibited abdominal distension causing a "belly-up" syndrome. Skin lesions were also observed on these fish but lesions and histopathology were less severe than in eels (Miyazaki and Egusa 1977; Ellis et al. 1983). Kusuda et al. (1995) reported hemorrhages in the mouth, nose, brain, and on the operculum of infected striped jack.

Diagnosis

Red spot disease of eels is diagnosed by isolation of *P. anguilliseptica* from internal organs. The slow-growing organism is cultured on general purpose media (BHI or TSA), where it forms 1-mm diameter colonies in 3 to 4 days at 25°C.

Bacterial Characteristics

P. anguilliseptica measures 0.8 μm in diameter by 5 to 10 μm in length. Growth takes place at 5 to 30°C but not at 37°C. Presumptively, *P. anguilliseptica* is motile (more so at 15 than at 25°C), Gram-negative, cytochrome oxidase positive, and negative for soluble pigments, H_2S production, and indole production (Table 16.1). The organism is not reactive on glucose oxidation-fermentation tests; however, it is positive for catalase, degradation of gelatin, and grows on media with 0 to 4% NaCl. The bacterium has low metabolic reactivity for many sources of carbon (Stewart et al. 1983).

Thirteen isolates of *P. anguilliseptica* from France were phenotypically identical to the type strain NCIMB 1949 (Berthe et al. 1995) and to isolates from Finland (Lönnström et al. 1994). Confirmation of the French isolates was achieved by agglutination with antisera of the type strain. There are two antigenic groups of *P. anguilliseptica* (Nakai et al. 1981): type I antigen is thermolabile (121°C for 30 min), and type II is heat stable. The heat-labile type from Japan possesses an R antigen and is pathogenic only to eels. Based on precipitin and agglutination test, *P. anguilliseptica* isolates from Taiwan are identical to the Japanese R antigen type (Nakai et al. 1985b).

Epizootiology

Epizootiological characteristics of red spot were summarized by Muroga (1978) as follows: (1) prevailed in brackish water ponds, (2) prevailed when water temperature was below 20°C and disappeared when it was above 27°C, and (3) prevailed among Japanese eels. In Japan, red spot disease of eels occurs primarily in April and May, becomes less apparent during summer, and reoccurs to a mild degree in October (Muroga et al. 1973). Epizootics appear to coincide with water temperatures of approximately 20°C, and mortalities decline as temperatures rise above 25°C. Experimental infection of Japanese eels confirmed a relationship between water temperature and *P. anguilliseptica* epizootics (Muroga et al. 1975). Nearly all infected Japanese eels held below 20°C died, and 71% of European eels died at 21°C, although few fish of either species held above 27°C succumbed to the disease. In France, disease in gilthead seabream occurred below 16°C, with the highest mortality occurring in the 9 to 16°C range. Striped jack in Japan were found to be infected with *P. anguilliseptica* at 14 to 16°C (Kusuda et al. 1995).

Because *P. anguilliseptica* tolerates up to 4% NaCl, infections may be as severe in brackish and salt water as in fresh water ponds. In Malaysia, the bacterium occurred in giant seaperch and estuarine grouper in offshore cages; mortality was 20% to 60% during the cooler season of November to December and February to March, which also coincided with poor water quality (Nash et al. 1987). It was speculated that the bacterial infection was secondary to stressful environmental conditions and poor nutrition. Mortality among 300-g striped jack infected with *P. anguilliseptica* was 0.1% to 1% daily with a cumulative mortality of 30% from February through April (Kusuda et al. 1995). Mushiake et al. (1984) noted that concentrations of 25 to 100 μg of Cu/L in water reduced the Japanese eels' resistance to *P. anguilliseptica* by adversely affecting phagocytosis and other cell-mediated defense mechanisms. Stewart et al. (1983) demonstrated the affinity of the organism for young eels when 96% of elvers and only 3.9% of adults died following experimental infection.

TABLE 16.1. BIOPHYSICAL AND BIOCHEMICAL CHARACTERISTICS OF *PSEUDOMONAS FLUORESCENS* AND *P. ANGUILLISEPTICA*

Characteristic	*P. fluorescens*[a]	*P. anguilliseptica*[a]
Cell size (μm)	0.8 × 2.3–2.8	0.4 × 2–5
Gram stain	–	–
Motility	+ (weak)	+ (15°C best)
Fluorescent pigment on agar	+	–
Optimum temperature (°C)	25–30	25
Growth at		
4°C	+	+
30°C	+	+
40°C	–	–
Glucose motility deep	Oxidative	?
Growth in 4% NaCl	–	+
Catalase	+	+
Cytochrome oxidase	+	+
Hydrolysis of		
Arginine	+	–
Gelatin	+	+
Starch	+	–
Indole production	–	–
Nitrate reduction	+	–
ß-galactosidase	–	–
Acid production from		
Arabinose	+	–
Arginine	+	–
D-alanine	+	?
Fructose	+	–
Galactose	+	–
Glucose	+	–
Glycerol	+	–
Inositol	+	–
Lactose	–	–
Maltose	+	–
Mannitol	+	–
Mannose	+	–
Raffinose	?	–
Salicin	–	–
Sucrose	+	–
Trehalose	+	?
Xylose	+	–

Sources: Wakabayashi and Egusa (1972); Palleroni (1984).
[a]+ = positive; – = negative; ? = unknown.

After describing the organism in eye lesions of Baltic herring, Lönnström et al. (1994) suggested that *P. anguilliseptica* had been present in that region for a long time as Bylund (1983) had noted similar lesions and Gram-negative bacteria in smears from ulcerated fish eyes. The primary pathogenicity of *P. anguilliseptica* was questioned by Lönnström et al. (1994) because the organism was isolated from eyes or kidneys of only about 43% of Baltic herring assayed and from no livers or spleens. It was also noted that *P. anguilliseptica* had been isolated during disease outbreaks in rainbow trout farms along the Finnish coast and that Baltic herring may have been the disease reservoir. Even though the bacterium showed low pathogenicity to trout in the laboratory, stress factors present in aquaculture could result in significant disease.

Pathological Manifestations

In histopathological studies of red spot disease in Japanese eels, Miyazaki and Egusa (1977) showed that lesions appeared initially in the dermis, subcutaneous adipose tissue, interstitial musculature tissue, vascular walls, bulbous arteriosus, and heart.

Bacteria multiplied in the vascular walls, producing inflammation with serous exudation and cellular infiltration. Many small hemorrhages occurred in the dermal connective tissue; there was congestive edema of visceral organs and fatty degeneration of liver hepatic cells, serous exudation and cellular proliferation of the spleen, glomerulitis, activation of reticuloendothelial cells lining the sinusoids, and atrophy of hematopoietic tissue of the kidney.

Several days postinjection with *P. anguilliseptica,* eels became moribund, demonstrating clinical signs similar to naturally infected fish, and mortality followed at 6 to 10 days postinfection (Wakabayashi and Egusa 1972). No exotoxin production by *P. anguilliseptica* has been demonstrated, but it is speculated that pathogenicity results from presence of these substances. Nakai et al. (1985b) traced the progression of a Japanese eel red spot infection after fish were injected IM with *P. anguilliseptica* and held at 12, 20, and 28°C. Viable bacterial cells in the blood decreased for 1 to 12 hours postinjection, followed by a stationary period when cells remained constant or increased slightly. Bacterial cell counts then increased to 10^8 to 10^{10} CFU/g of tissue or per milliliter of blood, thus establishing a septicemia that persisted until death. Bacterial growth from injection to maximum concentration took approximately 12 days at 12°C and 5 days at 20°C. When inoculated fish were held at 28°C, bacterial cells disappeared from internal organs within 2 days. Experimental exposure of 40-g seabream by IP and subcutaneous injection and skin swabs after abrasion resulted in 100% mortality, and pathogen isolation was made from all carcasses (Berthe et al. 1995). All control fish survived.

Significance

Red spot is one of the most serious diseases of cultured eels in Japan, and with its appearance in Europe, it has now become a global concern. The ability of *P. anguilliseptica* to infect other fish species and its presence in wild populations that may serve as pathogen reservoirs add to its importance. The potential for economic loss to eel producers caused by *P. anguilliseptica* because of its expanding geographical range and wider species susceptibility was emphasized by Berthe et al. (1995).

OTHER BACTERIAL DISEASES OF EEL

A variety of other bacterial diseases affect eels. Columnaris (Chapter 14) infects most all freshwater fish including cultured eels in freshwater ponds. Fins of infected fish turn white and become necrotic and frayed, and affected body areas lose pigmentation, becoming white when the outer layer of skin is destroyed. Mortalities due to columnaris are usually chronic, but under stressful conditions, can be acute.

Motile *Aeromonas* septicemia (*A. hydrophila*) eel infections are not unusual in fresh water (Chapter 14). Necrotic epidermal lesions develop on the head or opercle or in the lateral musculature in fresh water. This disease, which results in chronic losses, is usually associated with some type of environmental stress, handling, or transport.

Infections of atypical *A. salmonicida* have been infrequently reported in cultured eel (Chapter 15). The disease follows a similar pattern to that which occurs in carp and goldfish; ulcers are generally confined to the skin, and losses are more or less chronic.

Vibriosis has been a recognized disease of eels in brackish waters for many years (Chapter 17). In fact, the specific epithet of *V. anguillarum* was derived from a disease description in eels in the 19th century in Italy. Infection produces shallow ulcerative lesions at any body location while eels are in salt or brackish water.

MANAGEMENT OF EEL BACTERIAL DISEASES

Control of bacterial diseases of cultured eels revolves around good aquaculture management, use of chemotherapy, and vaccination when indicated.

Management

Maintaining a high-quality environment, keeping organic enrichment to a minimum, and preventing shock and extreme temperature fluctuations are positive approaches to management of eel diseases. Highly enriched pond waters appear to enhance occurrence of edwardsiellosis in eels. Because fish, turtles, snakes, and so forth may carry *E. tarda* in their intestines, it is desirable to prevent contact with these animals.

Because red spot disease has a tendency to occur in brackish water ponds at temperatures between 15 and 20°C, Muroga (1978) proposed the following management practices for controlling the disease in cultured Japanese eels: (1) maintain water temperature above 27°C, (2) culture eels in freshwater ponds, and (3) culture less-susceptible European eels in which disease is endemic.

Chemotherapeutics

Application of medicated feed is the most commonly used control for bacterial eel infections. In Taiwan, the drug of choice for treating *E. tarda* eel infections is nalidixic acid because a large percentage of isolates are resistant to most other antibiotics including Terramycin (Liu and Wang 1986). In the United States, Terramycin (at a rate of 50 to 100 mg/kg of fish per day for 10 days) is the drug of choice for treating *E. tarda* but Romet-30 is also widely used and is fed at 50 mg/kg of fish per day for 5 days; however, neither of these drugs are Food and Drug Administration (FDA) approved for treating *E. tarda* infections. For fish not skinned during dressing, a withdrawal period for Terramycin is 21 days, and 42 for Romet-30. Norfloxacin was shown to be highly effective against *E. tarda* in vitro (minimum inhibitory concentration of 0.016–0.031 µg/mL) and when incorporated into feed of naturally *E. tarda* infected Japanese flounder for 3 days at 100 mg/kg of body weight (Heo and Kim 1996).

Antibiotic resistance of *E. tarda* has been studied by several researchers, and most concur that it is a significant problem. Fifty-four strains of *E. tarda* from cultured eels in Taiwan (Chen et al. 1984) were found to have a high rate of susceptibility to gentamicin and sulfa-drugs, and Waltman and Shotts (1986) also found that a high percentage of *E. tarda* isolates from Taiwan were resistant to numerous antimicrobials. Liu and Wang (1986) reported that nearly 92% of *E. tarda* isolates taken from water in eel culture ponds demonstrated some degree of antibiotic resistance. Antibiotic resistance of *E. tarda* in the United States has not yet become a significant problem (Plumb et al. 1995).

Medicated feed is also used to control *P. anguilliseptica* infections in eels and other fish. Wicklund and Bylund (1990) found that oxytetracycline was not effective when used in *P. anguilliseptica*–infected salmon, but Romet-30 was effective. Red spot disease of Japanese eels was controlled by 9 consecutive days of baths in oxolinic acid or nalidixic acid (2 to 10 mg/L) with 0 mortalities, compared with 100% mortality in nontreated controls (Jo 1978). Also, when fish were fed oxolinic acid at 5 to 20 mg/kg of fish for 3 consecutive days, 100% survived.

Vaccination

Development of a vaccine using whole cells, disrupted cells, or cell extracts as immunogens has been pursued in Japan and Taiwan (Song et al. 1982). All of these preparations were immunogenic, especially by injection, but a practical and commercially available vaccine has not yet been developed. Salati (1988) reviewed techniques and procedures for vaccinating eels against *E. tarda* and found that the two basic types of vaccines used were whole cell bacterins and bacterial extracts. Song and Kou (1981) reported that a single immersion of 6-g elvers was effective against *E. tarda*, but two or three vaccine exposures proved more effective in eliciting immunity that lasted for 10 weeks. Following eel immunization with *E. tarda* lipopolysaccharide (LPS), culture filtrates, and formalin-killed cells, Salati et al. (1983) concluded that LPS was the immunogenic component. It was later shown that polysaccharide, without the lipid component, was more antigenic than other preparations (Salati and Kusuda 1985). In a more recent study, eels immunized by injection with formalin-killed cells and LPS from *E. tarda* showed only slight to moderate protection from injection challenge with virulent *E. tarda* 21 days postvaccination (Gutierrez and Miyazaki 1994).

Vaccination studies in Japanese flounder have shown marginal success when extracellular and intracellular products were used in injection, immersion, and orally administered vaccines (Rashid et al. 1994a). Although serum agglutinating antibodies were increased after injection and oral exposure, no clear protection was conveyed. With regard to *E. tarda* vaccine, Salati (1988) concluded that bacterins that are simple and cheap to produce on a large scale do provide protection to anguillettes, and although protection is not complete, it may be feasible for use on farms that have severe *Edwardsiella* problems.

Eel vaccination experiments with *P. anguilliseptica* have shown that formalin-killed bacterins are effective immunogens when applied by injection (Nakai et al. 1982). Injected vaccines stimulated high agglutinating antibodies and were highly protective but probably not practical. Immersion apparently had no immunological effect.

REFERENCES

Amandi, A., S. F. Hiu, J. S. Rohovec, and J. L. Fryer. 1982. Isolation and characterization of *Edwardsiella tarda* from fall chinook salmon (*Oncorhynchus tshawytscha*). *Applied Environmental Microbiology* 43:1380–1384.

Austin, B., and D. A. Austin. 1987. *Bacterial Fish Pathogens: Disease in Farmed and Wild Fish*. Chichester, UK: Ellis Horwood Ltd.

Berthe, F. C. J., C. Miche., and J. F. Bernardet. 1995. Identification of *Pseudomonas anguilliseptica* isolated from several fish species in France. *Diseases of Aquatic Organisms* 21:145–150.

Bylund, G. 1983. The health of Baltic herring and the symptoms of eye disease in fish in Pori coastal waters. *Meri* 12:203–212.

Chen, S. C., M. C. Tung, and S. T. Huang. 1984. Sensitivity in vitro of various chemotherapeutic agents to *Edwardsiella tarda* of pond-cultured eel. *COA Fisheries Series No. 10, Fish Disease Research* 6:135–141.

Chowdhury, M. B. R., and H. Wakabayashi. 1990. Survival of four major bacterial fish pathogens in different types of experimental water. *Bangladesh Journal of Microbiology* 7:47–54.

Clarridge, J. E., D. M. Musher, V. Fainstein, and R. J. Wallace. 1980. Extraintestinal human infection caused by *Edwardsiella tarda*. *Journal of Clinical Microbiology* 11:511–514.

Coles, B. M., R. K. Stroud, and S. Sheggeby. 1978. Isolation of *Edwardsiella* from 3 Oregon sea mammals. *Journal of Wildlife Diseases* 14:339–341.

Darwish, M. 1997. The pathogenesis of experimental *Edwardsiella tarda* infection in channel catfish (*Ictalurus punctatus*). Auburn, AL: Auburn University. Ph.D. dissertation.

Earlix, D. 1995. Host, pathogen, and environmental interactions of enteric septicemia of catfish. Auburn, AL: Auburn University. Doctoral dissertation.

Egusa, S. 1976. Some bacterial diseases of freshwater fishes in Japan. *Fish Pathology* 10:103–114.

Ellis, A. E., G. Dear, and D. J. Steward. 1983. Histopathology of Sekiten-byo caused by *Pseudomonas anguilliseptica* in the European eel, *Anguilla anguilla* L., in Scotland. *Journal of Fish Diseases* 6:77–79.

Ewing, W. H., A. C. McWhorter, M. R. Escobar, and A. H. Lubin. 1965. *Edwardsiella*, a new genus of Enterobacteriaceae based on a new species, *Edwardsiella tarda*. *International Bulletin of Bacteriological Nomenclature and Taxonomy* 15:33–38.

Farmer, J. J., and A. L. McWhorter. 1984. Genus X *Edwardsiella* Ewing and McWhorter 1965, 37AL. In Bergey's Manual of Systematic Bacteriology, vol. I, 486–491. Baltimore, MD: Williams and Wilkins.

Francis-Floyd, R., P. Reed, B. Bolon, J. Estes, and S. McKinney. 1993. An epizootic of *Edwardsiella tarda* in largemouth bass (*Micropterus salmoides*). *Journal of Wildlife Diseases* 29:334–336.

Gilman, R. H., M. Madasamy, E. Gan, M. MaRiappan, E. Davis, and K. A. Kyser. 1971. *Edwardsiella tarda* in jungle diarrhea and a possible association with *Entamoeba histolytica*. *Southeast Asian Journal of Tropical Medicine and Public Health* 2:186–189.

Grimont, P. A. D., F. Grimont, C. Richard, and R. Sakazaki. 1980. *Edwardsiella hoshinae,* a new species of Enterobacteriaceae, *Current Microbiology* 4:347–351.

Gutierrez, M. A., and T. Miyazaki. 1994. Responses of Japanese eels to oral challenge with *Edwardsiella tarda* after vaccination with formalin-killed cells or polysaccharide of the bacterium. *Diseases of Aquatic Organisms* 6:110–117.

Hargreaves, J. E., and D. P. Lucey 1990. *Edwardsiella tarda* soft tissue infection associated with catfish puncture wound. *Journal of Infectious Diseases* 162:1416–1417.

Heo, G. J., and J. H. Kim. 1996. Efficacy of norfloxacin for the control of *Edwardsiella tarda* infection in the flounder *Paralichthys olivaceus*. *Journal of Aquatic Animal Health* 8:255–259.

Hidaka, T., Y. Yanohara, and T. Shibata. 1983. On the causative bacteria of head ulcer disease in cultured eels *Anguilla japonica*. *Memorandum of the Faculty of Fisheries Kagoshima University* 32:147–166.

Hoshina, T. 1962. On a new bacterium, *Paracolobactrum anguillimortiferum* sp. *Bulletin of the Japanese Society of Scientific Fisheries* 28:162–164.

Huang, S. T., and C. I. Liu. 1986. Experimental studies on the pathogenicity of *Edwardsiella tarda* and *Aeromonas hydrophila* in eel, *Anguilla japonica*. *COA Fisheries Series, No. 8, Fish Disease Research* 4:40–55.

Janda, J. M., and S. L. Abbott. 1993. Expression of an iron regulated hemolysin by *Edwardsiella tarda*. *FEMS Microbiology Letters* 111:275–280.

Janda, J., S. L. Abbott, S. Kroske-Bystrom, W. K. Powers, R. P. Koka, and K. Tamura. 1991. Pathogenic properties of *Edwardsiella* species. *Journal of Clinical Microbiology* 29:1997–2001.

Jo, Y. 1978. Therapeutic experiments on red spot disease. *Fish Pathology* 13:41–42.

Jo, Y., R. Muroga, and K. Onishi. 1975. Studies on red

spot disease of pond cultured eels—III. A case of the disease in the European eels (*Anguilla anguilla*) cultured in Tokushima Prefecture. *Fish Pathology* 9:115–118.

Kodoma, H., T. Murai, Y. Nakanishi, F. Yamamoto, T. Mikami, and H. Izawa. 1987. Bacterial infection which produces high mortality in cultured Japanese flounder *Paralichthys olivaceus* in Hokkaido Japan. *Japanese Journal of Veterinary Research* 35:227–234.

Kuo, S-C., H-Y. Chung, and G-H. Kou. 1977. *Edwardsiella anguillimortifera* isolated from edwardsiellosis of cultured eel (*Anguilla japonica*). *JCRR Fisheries Series No. 29, Fish Disease Research* 1:1–6.

Kusuda, R., N. Dohata, Y. Fukuda, and K. Kawai. 1995. *Pseudomonas anguilliseptica* infection of striped jack. *Fish Pathology* 30:121–122.

Liu, C. I., and S. S. Tsai. 1982. Edwardsiellosis in pond-cultured eel in Taiwan. *CAPD Fisheries Series No. 8, Report on Fish Disease Research* 4:92–95.

Liu, C. I., and J. H. Wang. 1986. Drug resistance of fish-pathogenic bacteria-II. Resistance of *Edwardsiella tarda* in aquaculture environment. *COA Fisheries Series No. 8, Fish Disease Research* 8:56–67.

Lönnström, L., T. Wicklund, and G. Bylund. 1994. *Pseudomonas anguilliseptica* isolated from Baltic herring *Clupea harengus membras* with eye lesions. *Diseases of Aquatic Organisms* 18:143–147.

Martin, J. D. 1983. Atlantic salmon and alewife passage through a pool and weir fishway of the Mayaguadaric River, New Brunswick, Canada during 1983 [abstract]. *Canadian Annual Report on Fisheries and Aquatic Science.*

Mekuchi, T., T. Kiyokawa, K. Honda, T. Nakai, and K. Muroga. 1995. Infection experiments with *Edwardsiella tarda* in the Japanese flounder. *Fish Pathology* 30:247–250.

Meyer, F. P., and G. L. Bullock. 1973. *Edwardsiella tarda,* a new pathogen of channel catfish (*Ictalurus punctatus*). *Applied Microbiology* 25:155–156.

Miyazaki, T., and S. Egusa. 1977. Histopathological studies of red spot disease of the Japanese eel (*Anguilla japonica*)—I. Natural infection. *Fish Pathology* 12:39–47.

Miyazaki, T., M. A. Gutierrez, and S. Tanaka. 1992. Experimental infection of *Edwardsiellosis* in the Japanese eel. *Fish Pathology* 27:39–47.

Miyazaki, T., and N. Kaige. 1985. Comparative histopathology of edwardsiellosis in fishes. *Fish Pathology* 20:219–227.

Muroga, R. 1978. Red spot disease of eels. *Fish Pathology* 13:35–39.

Muroga, K., Y. Jo, and M. Yano. 1973. Studies on red spot disease of pond-cultured eels-I. The occurrence of the disease in eel culture ponds in Tokushima Prefecture in 1972. *Fish Pathology* 8:1–9.

Muroga, K., Y. Jo, and T. Sawada. 1975. Studies on red spot disease of pond-cultured eels-II. Pathogenicity of the causative bacterium, *Pseudomonas anguilliseptica. Fish Pathology* 9:107–114.

Mushiake, K., K. Muroga, and T. Nakai. 1984. Increased susceptibility of Japanese eel *Anguilla japonica* to *Edwardsiella tarda* and *Pseudomonas anguilliseptica* following exposure to copper. *Bulletin of the Japanese Society of Scientific Fisheries* 50:1797–1801.

Nakai, T., K. Muroga, and H. Wakabayashi. 1981. Serological properties of *Pseudomonas anguilliseptica* in agglutination. *Bulletin of the Japanese Society of Scientific Fisheries* 47:699–703.

Nakai, T., K. Muroga, R. Ohnishi, Y. Jo, and H. Tanimoto. 1982. Studies on red spot disease of pond-cultured eels—IX. A field vaccination trial. *Aquaculture* 3:131–135.

Nakai, T., H. Hanada, and K. Muroga. 1985a. First records of *Pseudomonas anguilliseptica* infection in cultured ayu *Plecoglossus altivelis. Fish Pathology* 20:481–484.

Nakai, T., K. Muroga, J-Y. Chung, and G-H. Kou. 1985b. A serological study on *Pseudomonas anguilliseptica* isolated from diseased eels in Taiwan. *Fish Pathology* 19:259–261.

Nash, G., I. G. Anderson, M. Schariff, and M. N. Shamsudin. 1987. Bacteriosis associated with epizootic in the giant sea perch, *Lates calcarifer,* and the estuarine grouper, *Epinephelus tauvine,* cage cultured in Malaysia. *Aquaculture* 67:105–111.

Nichibuchi, M., K. Muroga, and Y. Jo. 1980. Pathogenic *Vibrio* isolated from eels, Diagnostic tests for the disease due to present bacterium. *Fish Pathology* 14:125–131.

Palleroni, N. J. 1984. Family I. Pseudomonadaceae Winslow, Broadhurst, Buchanan, Krumwiede, Rogers and Smith 1917, 555. In: *Bergey's Manual of Systematic Bacteriology,* vol. I, edited by N. R. Krieg and J. G. Holt, 141–199. Baltimore, MD: Williams and Wilkins.

Park, S., H. Wakabayashi, and Y. Watanabe. 1983. Serotype and virulence of *Edwardsiella tarda* isolated from eel and their environment. *Fish Pathology* 18:85–89.

Plumb, J. A., C. C. Sheifinger, T. R. Shryock, and T. Goldsby. 1995. Susceptibility of six bacterial pathogens of channel catfish to six antibiotics. *Journal of Aquatic Animal Health* 7:211–217.

Rashid, M. M., T. Mekuchi, T. Nakai, and K. Muroga. 1994a. A serological study on *Edwardsiella tarda* strains isolated from flounder (*Paralichthys olivaceus*). *Fish Pathology* 29:221–227.

Rashid, M. M., K. Honda, T. Nakai, and K. Muroga. 1994b. An ecological study on *Edwardsiella tarda* in flounder farms. *Fish Pathology* 29:221–227.

Salati, F. 1988. Vaccination against *Edwardsiella tarda.* In *Fish Vaccination,* edited by A. E. Ellis, 135–151. London: Academic Press.

Salati, F., and R. Kusuda. 1985. Chemical composition of the lipopolysaccharide from *Edwardsiella tarda. Fish Pathology* 20:187–191.

Salati, F., K. Kawai, and R. Kusuda. 1983. Immunoresponse of eel against *Edwardsiella tarda* antigens. *Fish Pathology* 18:135–141.

Shotts, E. B., and J. D. Teska. 1989. Bacterial pathogens of aquatic vertebrates. In: *Methods for Microbiological Examination of Fish and Shellfish,* edited by B. Austin and D. A. Austin, 164–186. Chichester, UK: Ellis Horwood Ltd.

Song, Y. L., and G. H. Kou. 1981. The immuno-responses of eel (*Anguilla japonica*) against *Edwardsiella anguillimortifera* as studied by the immersion method. *Fish Pathology* 15:249–255.

Song, Y. L., G. H. Kou, and K. Y. Chen. 1982. Vaccination conditions for the eel (*Anguilla japonica*) with *Edwardsiella anguillimortifera* bacterin. *Journal of Fisheries Society of Taiwan* 4:18–25.

Stewart, D. J., R. Woldemariam, G. Dear, and F. M. Mochaba 1983. An outbreak of "Sekiten-byo" among cultured European eels, *Anguilla anguilla* L., in Scotland. *Journal of Fish Diseases* 6:75–76.

Suprapto, H., T. Nakai, and K. Muroga. 1995. Toxicity of extracellular products and intracellular components of *Edwardsiella tarda* in the Japanese eel and flounder. *Journal of Aquatic Animal Health* 7:292–297.

Taylor, P. W., J. E. Crawford, and E. B. Shotts, Jr. 1995. Comparison of two biochemical tests systems with conventional methods for identification of bacteria pathogenic to warmwater fish. *Journal of Aquatic Animal Health* 7:312–317.

Ullah, A., and T. Arai. 1983. Pathological activities of the naturally occurring strains of *Edwardsiella tarda. Fish Pathology* 18:65–70.

van Damme, L. R., and J. Vandepitte. 1980. Frequent isolation of *Edwardsiella tarda* and *Plesiomonas shigelloides* from healthy Zairese freshwater fish: a possible source of sporadic diarrhea in the tropics. *Applied Environmental Microbiology* 39:475–479.

Wakabayashi, H., and S. Egusa. 1972. Characteristics of a *Pseudomonas* sp. from an epizootic of pond-cultured eels (*Anguilla japonica*). *Bulletin of the Japanese Society of Scientific Fisheries* 38:577–587.

Wakabayashi, H., and S. Egusa. 1973. *Edwardsiella tarda* (*Paracolobactrum anguillimortifera*) associated with pond-cultured eel disease. *Bulletin of the Japanese Society of Scientific Fisheries* 39:931–936.

Walters, G., and J. A. Plumb. 1980. Environmental stress and bacterial infection in channel catfish, *Ictalurus punctatus* Rafinesque. *Journal of Fish Biology* 17:177–185.

Waltman, W. D., and E. B. Shotts. 1986. Antimicrobial susceptibility of *Edwardsiella tarda* from the United States and Taiwan. *Veterinary Microbiology* 12:277–282.

Waltman, W. D., E. B. Shotts, and T. C. Hsu. 1986. Biochemical and enzymatic characterization of *Edwardsiella tarda* from the United States and Taiwan. *Fish Pathology* 21:1–8.

White, F. H., F. C. Neal, C. F. Simpson, and A. F. Walsh. 1969. Isolation of *Edwardsiella tarda* from an ostrich and an Australian skink. *Journal of the American Veterinary Medical Association* 155:1057–1058.

Wicklund, T., and G. Bylund. 1990. *Pseudomonas anguilliseptica* as a pathogen of salmonid fish in Finland. *Diseases of Aquatic Organisms* 8:13–19.

Wyatt, L. E., R. Nickelson II, and C. Vanderzant. 1979. *Edwardsiella tarda* in freshwater catfish and their environment. *Applied Environmental Microbiology* 38:710–714.

17 Salmonid Bacterial Diseases

A certain mystique is associated with salmonids that no other group of fishes enjoy, and this in part explains the attention they receive from environmentalists, aquaculturists, fish pathologists, and conservation agencies. Moreover, the high economic value of both wild and cultured salmonids demands that considerable attention be directed toward their health management. Bacterial infections that affect salmonids occur worldwide and can have significant impact on cultured and occasionally wild populations. These diseases include furunculosis (*Aeromonas salmonicida*), vibriosis (*Vibrio anguillarum* and related species), enteric redmouth (ERM; *Yersinia ruckeri*), coldwater vibriosis (CV; *Vibrio salmonicida*), bacterial gill disease (BGD; *Flavobacterium branchiophilum* and *F. aquatile*), bacterial coldwater disease (*Flavobacterium psychrophilum*), bacterial kidney disease (*Renibacterium salmoninarum*), and several other less-specific diseases.

FURUNCULOSIS

Furunculosis is one of the oldest and best known bacterial fish diseases. It is generally considered to be a disease of salmonids but is increasingly associated with other cool and occasionally warmwater fishes in both fresh and salt water (Austin and Austin 1987). Its etiological agent is *A. salmonicida* subsp. *salmonicida,* which is also known as "typical *A. salmonicida*" (McGraw 1952; Schubert 1967; Shotts and Bullock 1975).

Geographical Range and Species Susceptibility

Furunculosis of salmonids is found in most regions of the world where trout occur, including Australia, Europe, Japan, Korea, North America, South Africa, and the United Kingdom (McCarthy 1975; Trust et al. 1980; Wichardt et al. 1989). Brook trout are generally considered to be the most susceptible trout species; however, there are resistant strains within this species. All salmonids, especially brown, rainbow, and lake trout, Atlantic salmon, and other anadromous salmonids are susceptible to the disease. Wichardt et al. (1989) indicated that in Sweden, acute typical *A. salmonicida* infections also occur in Arctic char. Nomura et al. (1992) noted that in Japan, rainbow trout are not severely affected by *A. salmonicida,* but cultured amago and masu salmon are susceptible, and as culture of the latter species increases, *A. salmonicida*'s impact on them is likely to also increase. However, in experimental *A. salmonicida* infections of different pink salmon populations, no difference in susceptibility was evident (Beacham and Evelyn 1992a).

A. salmonicida may occur in species of cyprinids, pike, perch, bullheads, and experimentally in some nonsalmonid marine fish species such as sac-larvae turbot and halibut (Fryer and Rohovec 1984; Ostland et al. 1987; Bergh et al. 1997). When nonsalmonids become infected, they often have been in close proximity to *A. salmonicida*–infected salmonids. Naturally occurring and experimental *A. salmonicida* transmission to gold sing wrasse was reported by Bricknell et al. (1996) but wrasse were not as susceptible to the disease as were Atlantic salmon.

Clinical Signs and Findings

Furunculosis is classified into four categories on the basis of severity: acute, subacute, chronic, or

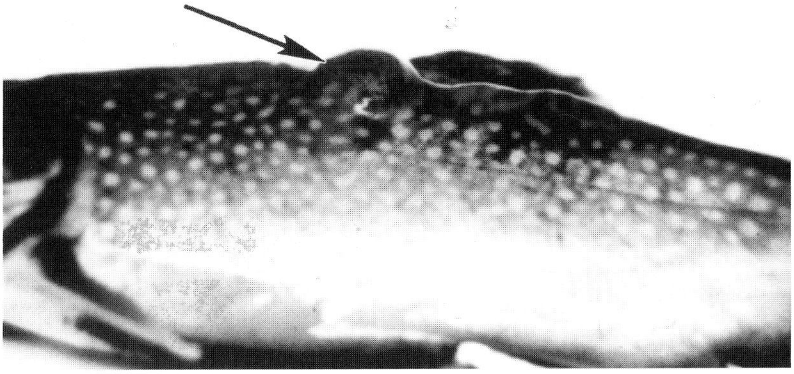

FIGURE 17.1. Brown trout with furuncle (arrow), *caused by* Aeromonas salmonicida. *(Photograph courtesy of the U. S. Fish and Wildlife Service.)*

latent (Herman 1968). Generally, fish infected with *A. salmonicida* become listless and drift downstream or lie on the bottom of ponds, tanks, or net cages, where they respire weakly before death. Few external clinical signs accompany acute furunculosis, but fish will exhibit darkened skin pigmentation, lethargic swimming, and anorexia. Discrete petechia may occur at the base of fins. Internally, the spleen may be enlarged, and septicemia develops, with high mortalities occurring within 2 to 3 days.

Subacute infections develop more gradually but are associated with high morbidity accompanied by a consistent rise in deaths. Various clinical signs are apparent including dark pigmentation, "furuncles" or boillike lesions on the body surface, and hemorrhage at the base of frayed fins (Figure 17.1). In coho salmon, subdermal hemorrhaging that resembles a bruise (hematoma) with a purplish color occurs, but the skin is not ruptured (K. Amos, Washington Department of Fisheries, personal communication). Internally, the peritoneal cavity may contain a bloody, cloudy fluid and the intestine is flaccid, inflamed, and filled with a bloody fluid that exudes from a red, prolapsed anus. The kidney is generally edematous and hemorrhagic, the spleen is usually enlarged and dark red, and the liver is pale. A general hyperemia is evident throughout the viscera.

Chronic furunculosis results in low-grade, but consistent, mortality for an extended period. Clinical findings described for subacute infections are also dramatically apparent in chronically infected moribund fish, but prevalence throughout the population is lower. An inflamed intestine is often the only internal pathological manifestation.

In the latent form of furunculosis, *A. salmonicida* may lie dormant in a fish population, and fish will show no clinical or internal pathological signs of disease; however, they are still considered to be disease carriers. Nomura et al. (1992) indicated that salmon without frank infection had *A. salmonicida* present in kidney and coelomic fluid, but Cipriano et al. (1992) successfully isolated the organism from skin mucus in a higher percentage of fish than that from which it was isolated from internal organs.

Diagnosis

When diagnosing *A. salmonicida,* basic external and internal findings can be helpful, but isolation of the bacterium from diseased fish is essential for positive identification. These isolates are taken from muscle lesions, kidney, spleen, or liver; however, Cipriano et al. (1992) showed that *A. salmonicida* was isolated from skin mucus (a nonlethal sampling procedure) in 56% of infected lake trout, compared with only 6% isolation from kidneys. A similar ratio was seen in mucus versus kidney of Atlantic salmon. *A. salmonicida* will grow on trypticase soy agar (TSA) and brain heart infusion (BHI) agar or on most other general bacteriological medias when incubated at less than 25°C; preferably at 20°C. It grows poorly at 30°C and not at all at 37°C. Upon initial isolation, most colonies of *A. salmonicida* are hard and friable with some being smooth, soft, and glistening. Bacteria in hard colonies (rough) will autoagglutinate in broth and are virulent, whereas bacteria in smooth colonies do not autoagglutinate and are avirulent (Munro and Hastings 1993). Autoagglutination and virulence are associated with the presence of an additional cell surface protein called the "A layer" (Udey and Fryer 1978), which is discussed later.

Bacterial Characteristics

Typical *A. salmonicida,* grown on media with tyrosine, will produce a brown, water-soluble pigment that diffuses throughout the media in 2 to 4 days when incubated below 25°C. Presumptive characteristics of typical *A. salmonicida* are brown, water-soluble pigment and small (0.9 μm × 1.3 μm) Gram-negative, nonmotile rods that ferment and oxidize glucose and are catalase and cytochrome oxidase positive (Shotts and Bullock 1975). Definitive identification is achieved by additional biochemical reactions (see Table 15.1). A comparison of the biochemical reaction between typical *A. salmonicida* (*salmonicida*) and atypical *A. salmonicida* (*achromogens*) using API 50E and conventional tube media was made by Hirvelå-Koski et al. (1994). Distinct differences were noted in several key reactions. Typical *A. salmonicida* (*salmonicida*) is a morphologically and biochemically homogeneous species with several exceptions, the most notable being a variation in pigment production, Voges-Proskauer test results, and its ability to ferment selected sugars (Paterson et al. 1980b; Popoff 1984; Austin and Austin 1987). Some isolates of *A. salmonicida* lose their ability to produce brown pigment after numerous transfers on media. Some motile *A. hydrophila* isolates acquire the ability to produce a similar pigment after numerous subcultures, but color does not develop as rapidly or intensely as in *A. salmonicida.*

Identification of *A. salmonicida* can be made more rapidly with serological rather than cultural procedures by using serum agglutination, fluorescent antibody, or enzyme linked immunosorbent assay (ELISA) on either infected tissue or cultured bacterium (Austin et al. 1986). A specific DNA probe was developed with polymerase chain reaction (PCR) by Mooney et al. (1995), and by this method, they were able to detect *A. salmonicida* DNA in 87% of wild Atlantic salmon in three Irish river systems.

There appears to be only one serological strain of *A. salmonicida salmonicida.* McCarthy and Rawle (1975) showed that a strong serological homogeneity of typical *A. salmonicida* strains exists but that these strains will not cross-react to other aeromonads. Rockey et al. (1988) used monoclonal antibody against *A. salmonicida salmonicida* lipopolysaccharides to demonstrate the homogeneity of this antigenic property. Aoki and Holland (1985) found that at least three major proteins of the outer membranes of different *A. salmonicida* strains were similar.

Epizootiology

Since early descriptions of furunculosis, *A. salmonicida* has been considered to be an obligate pathogen that was passed from fish to fish or generation to generation through carrier fish (Popoff 1984). It was believed to have only a short life span outside its host; however, Michel and Dubois-Darnaudpeys (1980) demonstrated that the organism will survive and retain its pathogenicity for 6 to 9 months in pond bottom mud. Allen-Austin et al. (1984) reported that the bacterium will remain viable in fresh water at undetectable levels for more than 1 month, but Rose et al. (1990) showed that *A. salmonicida* survived for less than 10 days in seawater. Nomura et al. (1992) maintained viable *A. salmonicida* for 60 days in sterilized fresh water but for only 4 days in unsterile water, concluding that the organism could survive in water long enough to infect other fish. *A. salmonicida* can survive for a short time in the marine environment; according to colony forming units on culture media, the bacteria disappeared in about 10 days after seeding into a sterilized environment at 25 ppt salinity and 16 to 22°C (Effendi and Austin 1994). By a variety of direct counts and other detection methods, they noted that the organism did not become less numerous. Also, in unsterile environments and substrates, the organism did not survive much past 10 days, but on the same sterile microcosms, there was very little loss of cell numbers at 20 days. The important question of whether these remaining cells were viable and pathogenic was not answered. In view of the discrepancies in these studies, and in spite of data supporting its extended survival outside of the host, *A. salmonicida* is still considered to be an obligate pathogen.

McCarthy (1980) reviewed the literature on transmission of furunculosis and concluded that infection occurred by fish-to-fish contact either through the skin or by ingestion. Skin lesions caused by injury or parasites will enhance bacterial invasion. Cipriano (1982) demonstrated contact exposure of furunculosis in brook trout. Carrier fish that show no overt signs of disease and from which bacteria cannot be easily isolated can still be implicated in disease transmission. Nonsalmonid *A. salmonicida*–infected fish can serve as pathogen reservoirs. Ostland et al. (1987) demonstrated transmission of *A. salmonicida* from infected common shiners to coho salmon and brook trout.

Contaminated water and/or equipment can also serve as a source of disease transmission. Wichardt et al. (1989) pointed out that in Sweden, it had been documented that fish at several trout and salmon farms became infected after *A. salmonicida*–infected fish were released into upstream water supplies. The role of aerosol transmission of *A. salmonicida* was investigated by Wooster and Bowser (1996), who found that the organism could travel at least 104 cm from its source through the air and then be recovered from water. This airborne potential makes management of *A. salmonicida* in closely associated culture units more critical.

Hjeltnes et al. (1995) experimentally transmitted *A. salmonicida* to cultured cod, halibut, and wrasse from Atlantic salmon; however, this occurred only rarely and did not result in major disease in these species. Experiments by Bricknell et al. (1996) indicated that the pathogen could naturally infect wrasse but was unlikely to cause a primary infection; however, they could serve as pathogen reservoirs. According to Pérez et al. (1996), 30-g turbot in seawater were more susceptible to *A. salmonicida* than were 25-g rainbow trout in fresh water. Intragastric inoculation failed to result in infection in either species.

Bullock and Stuckey (1975) demonstrated that rainbow trout can carry *A. salmonicida* after initial infection for up to 2 years without reexposure. They also noted that when rainbow trout were stressed by water temperature of 18°C and injected with an immunosuppressant (Kenolog 40—triamcinolone acetonide), latently infected fish shed *A. salmonicida*. Following immunosuppression with Kenolog 40, 33% of the trout exhibited overt disease, compared with 4% of fish stressed only by high water temperatures. There was also a 73% mortality in chemically immunosuppressed fish, compared with 33% mortality in those stressed solely by heat. In naturally occurring trout infections, mortality can reach 5% to 6% per week and a total mortality of 85% has been reported in untreated populations.

Nomura et al. (1992) isolated *A. salmonicida* from the kidney and coelomic fluid of chum, pink, and masu salmon that showed no frank signs of disease after entering fresh water. The ability of typical *A. salmonicida* to occur and survive in wild fish populations was shown by Mooney et al. (1995). By means of PCR, they detected an 87% carrier rate in wild Atlantic salmon in Irish rivers. Although pathogen levels were extremely low in each fish, they speculated that the bacterium is widespread among wild salmon populations in Irish rivers.

When Cipriano et al. (1996a) examined an asymptomatic rainbow trout population in a recirculating water system that used a fluidized sand biofilter, they isolated the organism from only 1% of spleens and from no kidneys or livers of these salmonids. However, *A. salmonicida* was isolated from 15% of gill and 19% of mucus samples. The pathogen was not isolated during repeated examinations of the fluidized biofilters and tank water, thus indicating that the pathogen did not become established in the recirculation system.

Risk factors involved in post-smolt Atlantic salmon contracting furunculosis after moving into sea cages were related to location, feeding, rearing, and presence of other disease organisms (Jarp et al. 1995). When a total of 116,480 fish died at 124 sea sites, 54% of the deaths resulted from furunculosis, 10.5% from vibriosis, and 39.5% from infectious pancreatic necrosis virus (IPNV).

According to McCarthy (1980), when an infected trout population overcomes clinical furunculosis, either as a result of natural defenses or chemotherapy, at least some survivors will become carriers. In spite of extensive research in this area, the carrier state of *A. salmonicida* is still controversial. For example, the bacterium was not isolated from any tissue of asymptomatic Atlantic salmon by Hiney et al. (1994). Upon stressing these same fish, *A. salmonicida* was isolated from the mucus, fins, and gills of about 50% of the fish and from about 10% of blood samples. Frequency of stress-induced infections was about 65%. Before fish were stressed, an ELISA was used to detect whether *A. salmonicida* was present. It was found in 50% of intestinal samples but not in any kidney or mucus samples. Application of ELISA on post-stressed fish indicated a positive reaction in about 25% of kidney samples, and when results of kidney, intestinal content, and mucus were combined, 45% of the fish were positive. Hiney et al. (1994)

suggested that intestines may be the primary location of *A. salmonicida* in carrier fish and that stress induces bacterial colonization of mucus on fins and gills. In comparative carrier experiments, Pérez et al. (1996) showed that rainbow trout that survived an *A. salmonicida* infection became pathogen carriers, but infected turbot did not.

In *A. salmonicida* carrier populations, clinical furunculosis can appear without any apparent environmental insult to fish, but usually its recrudescence is stress mediated. If fish are stressed as a result of handling, elevated water temperatures, low oxygen, or other adverse environmental conditions, a latent infection can become chronic, subacute, or acute.

Kingsbury (1961) correlated outbreaks of furunculosis with depressed oxygen levels. When oxygen concentrations dropped to less than 5 mg/L during the night, losses due to *A. salmonicida* increased. Nomura et al. (1992) found that 12.4% of chum salmon held at high density (14.7 fish per square meter) were infected with *A. salmonicida*, whereas lightly stocked fish (4.9 fish per square meter) were disease free. They also found that *A. salmonicida* prevalence was significantly higher in fish held in water with low dissolved oxygen (6 to 7 mg/L) than in fish held in water with 10 mg of oxygen per liter. Survival in high density–low oxygen water was approximately 40% less than in low density–high oxygen conditions. Water temperatures can also play an important role in furunculosis outbreaks. In trout, the optimum water temperature for disease development is 15 to 20°C. When infected fish are held in water at 20°C, *A. salmonicida* in the blood will increase dramatically within 48 hours, but at 10°C, 168 hours are required for bacteria to reach the same concentrations.

Transmission of *A. salmonicida* during spawning has been thought to be important, as the organism can be readily isolated from ovaries, eggs, and ovarian fluid at this time (McCarthy 1980). Bullock and Stucky (1975) were unable to affect spawning related transmission in four attempts and concluded that vertical transmission does not occur. Nomura et al. (1992) showed that *A. salmonicida* on chum salmon eggs was naturally reduced by a log of about 10^6 colony forming units in 48 hours. They also showed that the organism could not be isolated from eggs 5 days after onset of incubation, they therefore also concluded that vertical transmission of *A. salmonicida* was unlikely.

Furunculosis is most prevalent during late spring and summer when subacute epizootics often occur; however, the disease can occur at any time. Piper et al. (1982) noted that twice as many furunculosis cases occurred in July than in any other month.

Some populations of brook and rainbow trout have been selected for their resistance to furunculosis, and the use of these strains have helped reduce the impact of disease in certain areas (Ehlinger 1977). Naturally inherited defense mechanisms of trout account for part of their resistance to furunculosis, but serum bactericidal activity is also important (Hollenbeq et al. 1995).

Pathological Manifestations

Death of trout infected with *A. salmonicida* is attributed to a massive septicemia and toxic extracellular products (ECP) produced by the organism that interfere with the host's blood supply and result in massive tissue necrosis. Skin lesions are characterized by loss of scales, necrosis of epithelium and muscle, capillary dilatation, and peripheral hemorrhage of the lesion (McCarthy and Roberts 1980). Klontz et al. (1966) experimentally infected rainbow trout with *A. salmonicida* and found that the most consistent pathological aberration was an enlarged spleen that appeared 16 hours after injection and persisted throughout the infection. Sinusoids of the spleen became congested and engorged with erythrocytes. Bacteria could be isolated at 8 hours postinjection but not during the ensuing 48 hours, after which time a septicemia developed. Extensive inflammation and muscle necrosis developed at the injection site.

Histopathological changes were similar to those observed in naturally infected trout, especially involving the hematopoietic organs. Hematopoietic tissue of the kidney was most severely affected by an increase in lymphoid hemoblasts, macrophages, neutrophils, and lymphocytes until finally hematopoiesis in the kidney ceased. Renal tissues were largely spared from injury.

Pathogenicity of *A. salmonicida* was initially correlated with colonial morphology on solid bacteriological media. Organisms from "smooth" colonies were considered pathogenic and those from "rough" colonies were not (Anderson 1972); however, this theory has been reversed and pathogenicity clarified. Udey and Fryer (1978) found that *A. salmonicida* virulence was associated with presence of an A layer (additional layer) and these cells actually formed the rough colonies but were absent in bacteria in smooth nonvirulent strains that autoagglutinated. This aspect of *A. salmonicida* has been studied by other investigators who concur that the A layer (A protein monomer) correlates with virulence (Evenberg and Lugtenberg 1982; Trust et al. 1983). The surface protein seems to protect *A. salmonicida* from its host's natural defense mechanisms.

The ECPs (toxins and enzymes) produced by *A. salmonicida* include several proteases, haemolysins, and a leukocidin (Munro et al. 1980; Titball and Munn 1981; Cipriano et al. 1981). *Aeromonas salmonicida* ECPs are capable of producing a syndrome similar to chronic furunculosis, which manifests itself by muscle necrosis and edema at the injection site. It appears that pathological manifestation of *A. salmonicida* can be attributed to a combination of cellular and extracellular components with the precise mechanisms yet to be fully understood.

Significance

Its severity, host susceptibility, and wide geographic range makes furunculosis a significant disease in salmonid aquaculture. In some geographical areas where culture of trout and other salmonids is intensive, furunculosis may be the most serious infectious disease to occur, particularly on salmon farms in the United Kingdom and Scandinavia (Munro and Hastings 1993). *Aeromonas salmonicida*'s significance is also enhanced by its capability of causing infections and some mortality in nonsalmonid fishes, thus producing possible carrier fish. Increased stocking rates in culture units, which leads to a general degradation of environmental conditions, has also contributed to the disease severity.

VIBRIOSIS

Since the mid–18th century it has been speculated that fish, particularly those of marine origin, have been infected by members of the bacterial genus *Vibrio* (Colwell and Grimes 1984). Up to 13 *Vibrio* species have been reported to cause fish infections and include *V. alginolyticus, V. anguillarum, V. carchariae, V. cholerae, V. damsela, V. harvei, V. ordalii, V. parahaemolyticus, V. mimicus, V. vulnificus,* and *V. salmonicida* (Colwell and Grimes 1984; Egidius et al. 1986; Saeed 1995). A new species, *Vibrio ichthyoenteri,* has been proposed by Ishimoru et al. (1996) and *V. splendidus* and *V. pelagius* have been isolated from cod (Santos et al. 1997). *V. anguillarum, V. ordalii,* and *V. salmonicida* are best noted for their pathogenicity to fish. Others have been sporadically associated with fish infections and are not discussed at length here. The following discussion includes diseases caused by *V. anguillarum* and *V. ordalii,* which are collectively referred to as "vibriosis." Both organisms produce hemorrhagic septicemias, but the diseases they cause differ slightly. Coldwater vibriosis (Hitra disease) caused by *V. salmonicida* is discussed separately. *Vibrio* spp. infections have a variety of common names: vibriosis, red pest, saltwater furunculosis, boil disease, and ulcer disease.

Geographical Range and Species Susceptibility

Although vibriosis is considered to be primarily a saltwater fish disease, it does occasionally occur in freshwater fish (Muroga 1975; Tajima et al. 1985). *V. anguillarum* is found worldwide but is particularly significant along the Gulf of Mexico, Pacific and Atlantic coasts of North America, the North Sea, Atlantic and Mediterranean coasts of Europe and North Africa, the Middle East, and throughout Asia. *V. ordalii* occurs along the Northwest Pacific coast of North America and in Japan (Schiewe 1981; Muroga 1992). Other *Vibrio* spp. are not associated with any particular geographical region.

Anderson and Conroy (1970) listed approximately 50 fish species, mostly marine, from which *V. anguillarum* has been isolated, and more recently, numerous species have been added to the list. Colwell and Grimes (1984) listed 12 families and 42 species of marine and freshwater fishes affected by *Vibrio* spp.; *V. anguillarum* was the most frequently isolated, and salmon the fish group most often affected. *V. anguillarum* occurs in wild salmon populations, but most significant infections occur in culture populations, especially in coho, chinook, and sockeye salmon. Essentially, all salmonids can become infected, particularly anadromous species cultured in salt water cages or reared to smolts in fresh water and released into salt water.

Often, the most severely affected nonsalmonid

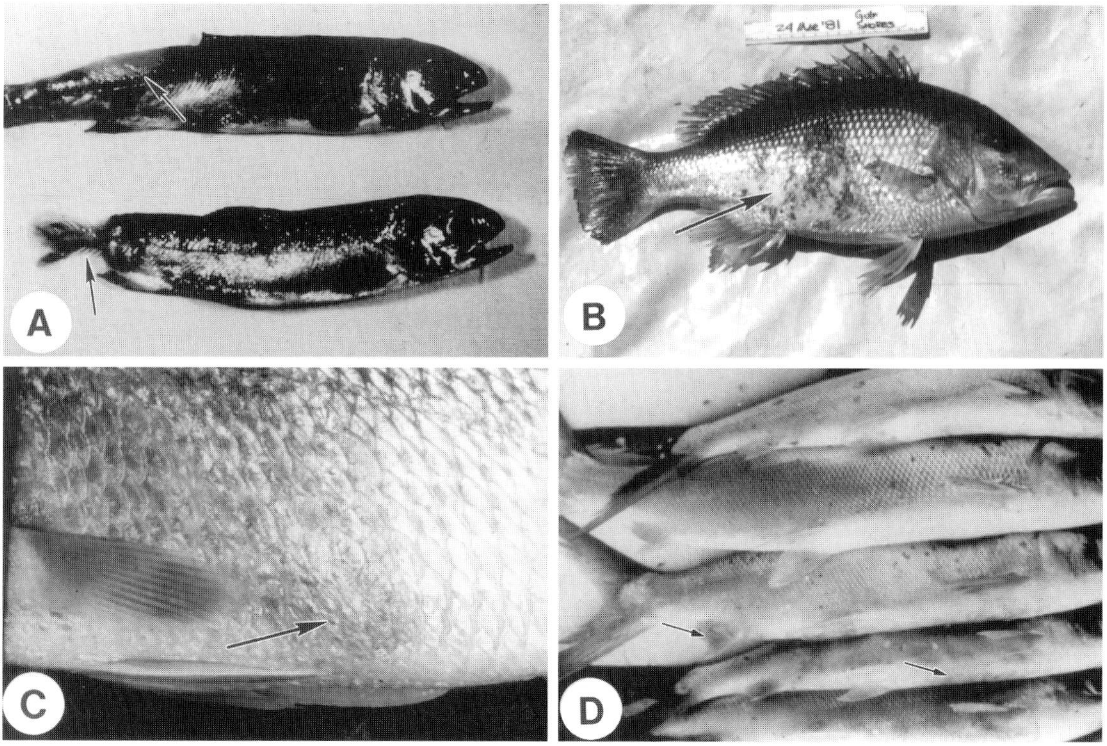

FIGURE 17.2. Clinical vibriosis. (A) *Pacific salmon with epidermal lesion on the back and caudal peduncle* (arrows). *(Photograph courtesy of J. Rohovec).* (B) *Red grouper with hemorrhagic lesion* (arrow) *on the skin.* (C) *Red drum with diffused hemorrhage* (arrow). *(Photograph by J. Hawke.)* (D) *Milkfish with hemorrhage* (arrows) *in the scale pockets and epidermis.*

cultured species are maricultured Atlantic cod, striped bass, European eel, Japanese eel, milkfish grown in brackish water ponds, and ayu grown in freshwater ponds. Other commonly affected families include flounder, sole and turbot, mullets, seabasses, and groupers. As mariculture of these fisheries expands worldwide, outbreaks of *V. anguillarum* will most likely become more significant. It is unlikely that any group of marine or brackish water fish is resistant to *V. anguillarum,* especially in mariculture. Ornamental fish in home and commercial aquaria are also vulnerable to the bacterium. *V. ordalii* is specific for anadromous salmon and freshwater trout (Schiewe 1981).

Clinical Signs and Findings

Clinical signs of vibriosis are similar to those of motile *Aeromonas* septicemia, as both cause the same type of hemorrhagic syndrome (Figure 17.2). Most vibrio-infected salmonids, as well as other infected species, exhibit diminished feeding activity and swim lethargically along the edges of ponds, raceways, or cages, accompanied by intermittent erratic or spinning patterns (Novotny and Harrell 1975). Infected fish also display pale gills (anemic) and skin discoloration and have hemorrhagic, ulcerative necrotic skin lesions that may expose underlying muscle. These lesions can develop on the head, within the mouth, around the vent, and at the base of fins. The vent is often red and swollen and exudes a bloody discharge. Abdominal distension and exophthalmia will occur in many fish. Hemorrhage in the eye is a common clinical sign in coho salmon and sometimes may be the only gross pathological sign in affected fish.

Internal gross pathology might or might not be striking; however, a generalized hyperemia can occur, body cavities often contain bloody ascitic fluid, intestines are inflamed and flaccid, and petechia are present in the liver, kidney, and adipose tissue and on the peritoneum. Evelyn (1971a) re-

TABLE 17.1. BIOCHEMICAL AND BIOPHYSICAL CHARACTERISTICS OF *VIBRIO ANGUILLARUM*, *V. ORDALII*, AND *V. SALMONICIDA*[a]

Test/characteristic	*V. anguillarum*	*V. ordalii*	*V. salmonicida*
Cell morphology	short, sometimes curved, Gram-negative rods		
Growth on			
0% NaCl	±	–	–
3% NaCl	+	+	+
Growth at 37°C	+	–	–
Voges-Proskauer	+	–	–
Arginine	+	–	–
Christiensens citrate	+	–	–
Starch hydrolysis	+	–	–
Gelatin hydrolysis	+	+	–
ONPG	+	–	–
Lipase	+	–	–
Indole from tryptone	+	–	–
Glycogen	+	–	–
Nitrate reduction	+	(+)	–
Acid from			
Celobiose	+	–	(+)
Fructose	+	–	–
Galactose	+	–	–
Glycerol	+	–	(+)
Glucose	+	+	+
Sorbitol	+	–	–
Trehalose	+	–	+
Maltose	+	+	–
Sucrose	+	+	–
Gentiobiose	–	–	–
Gluconate	+	–	+
D-mannose	+	–	+
Sensitive to 0/129	+	+	+

Sources: Pacha and Kien (1969); Colwell and Grimes (1984); Holm et al. (1985); Dalsgaard et al. (1988).
[a]+ = positive; – = negative; (+) = weakly positive.

ported that death in Pacific salmon usually occurred before large necrotic muscle lesions could develop or massive changes could occur within the internal organs. However, Novotny and Harrell (1975) noted that the spleen can sometimes be two to three times larger than normal.

Diagnosis

Vibriosis is diagnosed by isolating the pathogen on general laboratory media (BHI or TSA and so forth). Isolation can be enhanced by addition of 0.5% to 3.5% NaCl to the media (Tajima et al. 1981). Santos et al. (1996) found that of 45 *V. anguillarum* or related isolates from around the world, none would grow in media with 0% salt. Media inoculated with *V. anguillarum* is incubated at 25 to 30°C for 24 to 48 hours at which time round, raised, entire, shiny, cream-colored colonies which measure about 1 to 2 mm in diameter will be obvious (Colwell and Grimes 1984; Austin and Austin 1987). Isolates from fish with systemic infection can be obtained from any internal organ, especially those with copious blood supplies (Tajima et al. 1981; Ransom et al. 1984).

Isolation of *V. ordalii* is greatly enhanced by using seawater agar with up to 3% NaCl. Incubation at 15 to 25°C for 7 days will produce colonies that are not dissimilar from those of *V. anguillarum* (Schiewe et al. 1981). Although *V. ordalii* and *V. anguillarum* share numerous biophysical and biochemical characteristics, as well as some serological cross reactivity (Chart and Trust 1988), there are distinct taxonomical differences between these two pathogens and *V. salmonicida* (Table 17.1).

Bacterial Characteristics

V. anguillarum is a slightly curved, Gram-negative rod that measures about 0.5 × 1.5 μm. The organism is motile, cytochrome oxidase positive, ferments carbohydrates without gas production, and is sensitive to novobiocin and vibriostat 0/129 (2-4-diamino-6, 7-disopropyl pteridine phosphate) (Table 17.1). Biochemically, *V. anguillarum* is a heterogeneous organism with at least five different described biotypes (Schiewe et al. 1977; Austin and Austin 1987).

V. anguillarum is further fractionated by serological diversity. Initially, three salmonid serotypes were identified in the northwestern United States and Europe. Currently, serological groupings number as many as six serotypes; therefore, serology can be used to separate and identify *V. anguillarum.* Using slide agglutination, Santos et al. (1996) serologically characterized 45 *V. anguillarum* and related isolates taken from fish, shellfish, and the environments of Denmark, Spain, Chile, Norway, and the United States. Some phenotypic and serological relatedness was found among the strains, but 22 (49%) were typed among seven 0-antigen serogroups. They also determined that isolates from turbot and cod were serologically different from the others. Knappskog et al. (1993) found, however, that all 14 isolates from cod in Norway were *V. anguillarum* serotype 02. Clearly, *V. anguillarum* is a diverse species, which may help explain its wide geographical distribution and multiple fish species susceptibility. For our purposes, biogroups I and II, as proposed by Håstein and Smith (1977), and biogroup III, are discussed here. Most infections in salmon are caused by members of biogroup I. A serologically and biochemically distinct species, *V. ordalii,* constitutes biogroup III.

Epizootiology

According to West and Lee (1982), *V. anguillarum* is a marine microbe that may be found free living or in fish; the species is quite varied in its habitat requirements. Isolates from fish may possess low to high virulence. It is not known whether different strains are obligate pathogens while others are facultative, but Larsen (1983) showed that environmental strains of *V. anguillarum* were less pathogenic than were reference strains from fish. Vibriosis in salt water is equivalent to motile *Aeromonas* septicemia in fresh water, with the brackish water environment serving as a transitional zone where either may occur. *V. anguillarum* causes infection in wild fish but has its greatest impact on cultured fish, often during periods of environmentally induced stress. The disease is more prevalent during summer when water temperatures are high and dissolved oxygen is at its lowest. High population density or poor hygiene will contribute to epizootics. When fish are exposed to only a low number of highly virulent isolates, disease can occur without presence of exogenous stressors.

V. ordalii is not as widespread as *V. anguillarum,* having been isolated only from fish and not from water or bottom sediments. This organism also preferentially infects the skeletal muscle, heart, gills, and intestinal tract of salmonids and does not produce as extensive a septicemia as does *V. anguillarum.*

V. anguillarum and *V. ordalii* are both capable of infecting and killing large numbers of cultured fishes. Thorburn (1987) noted that mortalities on various Swedish rainbow trout farms ranged from 0% to 17% and were correlated to transport of fish, farm size, and number of years the farm had been in production. Less severe effects occurred at older and larger farms, suggesting that experience in management may have had a positive influence on disease outcome. Up to 100% mortality has been reported in Atlantic salmon (Sawyer et al. 1979) after experimental exposure to *V. anguillarum.* Experimental waterborne challenges in Atlantic cod produced mortality of 4% to 36% using 13 different isolates from cultured cod fry (Knapposkog et al. 1993). Some less-common vibrios may show differential fish species pathogenicity: *V. splendidus* and *V. pelagiius,* which were isolated from cod in Spain and Denmark, respectively, were more pathogenic to rainbow trout than to cod (Santos et al. 1997).

Increased water temperature has a detrimental effect on mortality of *V. anguillarum*–infected salmonids. For example, coho salmon suffered 58% to 60% mortality when held at 18 to 20°C, 40% at 15°C, 28% at 12°C, and only 4% at 6°C (Groberg et al. 1983).

Salmon smolts held in fresh water have little, if any, exposure to *V. anguillarum;* therefore, when they reach salt water they are highly vulnerable to infection. In a vaccination experiment, Harrell (1978) found that more than 90% of unvaccinated 0-age sockeye salmon were killed by *V. anguillarum* during their first 50 days in saltwater cages and that mortalities of 60% to 90% are common among young salmon when stocked into salt water.

In Norway, mortality of *V. anguillarum*–infected Atlantic salmon transferred from fresh water to salt water was influenced by time of transfer; fish moved late in the growing season suffered higher mortality than those that were moved early (Groberg et al. 1983).

In most instances, *V. anguillarum* is transmitted through the water column and by contact, but Grisez et al. (1996) conclusively transmitted the disease via the oral route in turbot. *Artemia* nauplii were incubated in a suspension of *V. anguillarum* and then fed to turbot, which showed clinical signs of disease 24 hours after feeding. All fish that had been fed contaminated nauplii died within 4 days, and the pathogen was detected in all exposed fish by immunohistochemistry. It was concluded that *V. anguillarum* was first transported through the intestinal epithelium by endocytosis, released into the lamina propria, transported by the blood to body organs, and resulted in septicemia and death.

Salmonids have been the recipients of the majority of vibriosis research. Wild and cultured flounder, eels, and milkfish are susceptible to the pathogen and can develop acute or chronic infections. In flounder, acute vibriosis occurs year-round, with onset of clinical signs appearing 12 to 24 hours postinoculation, and death within 2 to 4 days (Watkins et al. 1981). Clinical signs in acute vibriosis are not as evident as in the chronic form. Chronic vibriosis appears 1 to 4 days postexposure, with a duration of 2 to 6 weeks, during which death or recovery of some fish may occur.

In Japan, juvenile ayu are susceptible to *V. anguillarum*. The problem is more severe in sea run fish, in which incidences can be as high as 17%, compared with less than 1% after stocking into fresh water (Muroga et al. 1984). Also, diseased ayu from salt water environments have higher numbers of bacteria in their tissues. A similar pattern is true for cultured eels in Japan, where *V. anguillarum* is a greater problem in brackish water ponds than in fresh water ponds. These observations indicate that *V. anguillarum* is introduced into fresh water via infected fish that are captured in salt water and transferred to fresh water.

Pathological Manifestations

It has been proposed that *V. ordalii* invades the rectum and posterior gastrointestinal tract or possibly the integument (Ransom 1978). Its pathology differs from that of *V. anguillarum* in that *V. ordalii* is not dispersed evenly in various organs and tissues but produces microcolonies in skeletal and heart muscle, gill tissue, and throughout the gastrointestinal tract (Ransom et al. 1984). Also, a lower number of *V. ordalii* cells are present in the blood during an infection than in *V. anguillarum* infections, possibly because the bacteremia develops later in the disease cycle.

Experimental oral *V. anguillarum* infections in turbot showed that 50% had bacterial cells in the spleen and 80% of bacterial cells in the gut remained attached to the intestinal wall following washing (Olsson et al. 1996). These authors proposed that the intestine is the portal of entry for *V. anguillarum* and that bacteria penetrates the intestinal mucus that overlays the epithelial cells and then enters the bloodstream.

Pathogenesis of *V. anguillarum* was reported to be facilitated by an ECP that is heat stable at 100°C for 15 minutes (Umbreit and Tripp 1975). Live and heat-killed bacteria and heat-treated cell-free culture fluid killed 53% to 80% of injected goldfish. Conversely, no evidence was found by Harbell et al. (1979) that endotoxin, culture supernatant, or cell lysate caused pathology or death when injected into coho salmon.

Significance

That vibriosis occurs worldwide and affects a wide variety of fish species makes it an important fish disease, especially in marine and brackish water fishes. Vibriosis is considered to be the most common bacterial disease of marine species in general, and it is probably the most important disease of cultured marine fishes, particularly salmonids in the United States (Novotny and Harrell 1975). Muroga (1975) also stated that vibriosis is the most important infectious disease of cultured ayu in Japan. Although most reports of vibriosis involve cultured fish, wild marine fish and ornamental fish are also susceptible.

COLDWATER VIBRIOSIS

Coldwater vibriosis, also known as Hitra disease and hemorrhagic syndrome, is a disease primarily of the Norwegian Atlantic salmon farming industry (Hjeltnes and Roberts 1993). The disease was

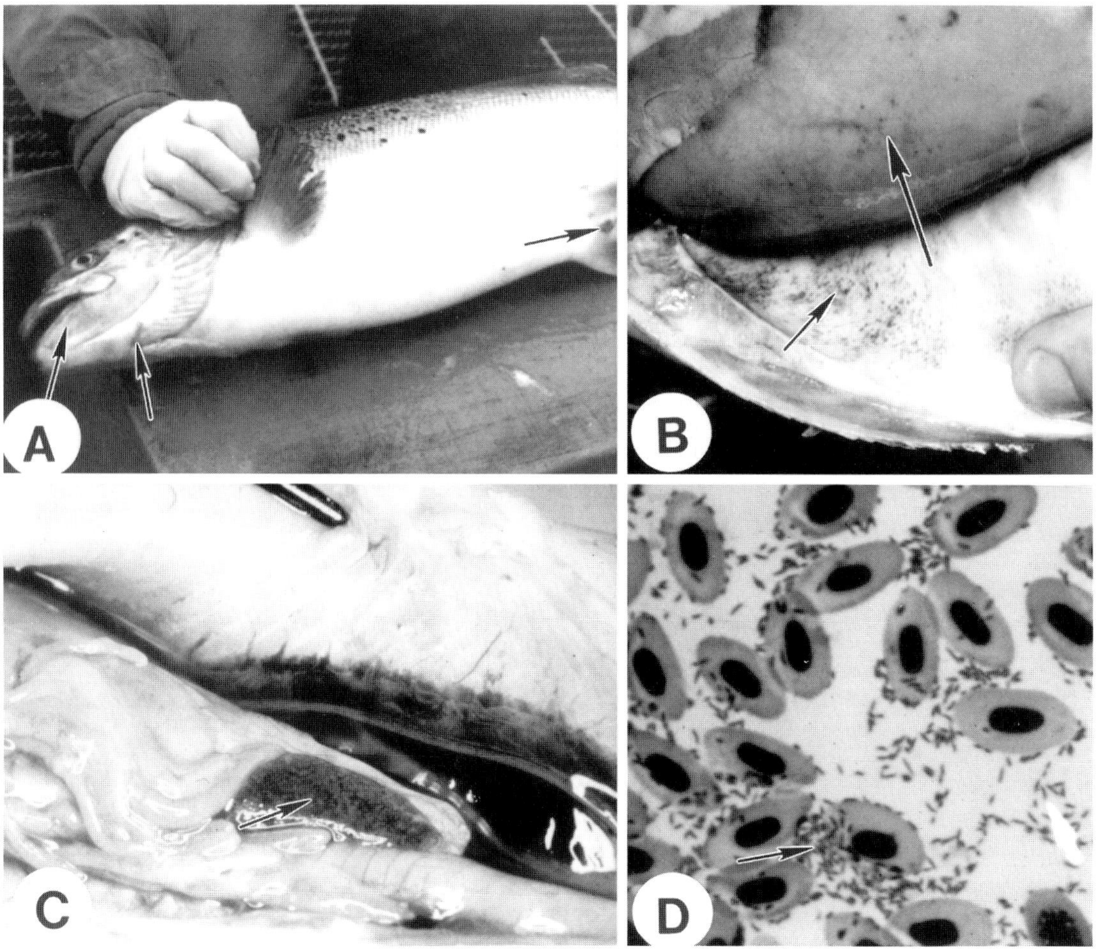

FIGURE 17.3. Atlantic salmon infected with Vibrio salmonicida. (A) *Hemorrhage* (arrows) *on the skin. (Photograph by H. Mitchell.)* (B) *Petechia* (large arrow) *on pale liver and peritoneum* (small arrow). (C) *Granular-appearing spleen.* (D) *Blood smear of Atlantic salmon infected with coldwater vibriosis and a high number of* V. salmonicida *in the blood. (Photographs B, C, and D courtesy of B. Hjeltness.)*

first described by Egidius et al. (1981) on the island of Hitra, and since that time, it has been reported annually throughout Norway and elsewhere. The causative agent of CV was described by Holm et al. (1985) and Egidius et al. (1986), who proposed a new species, *V. salmonicida*. The etiology was disputed by Fjølstad and Heyeroaas (1985) and Poppe et al. (1985), who hypothesized that Hitra disease is a multifactorial disease that could include a nutritional deficiency (vitamin E), as well as environmental stress. Generally, however, *V. salmonicida* is accepted as the etiological agent of Hitra disease (coldwater vibriosis) (Bruno et al. 1986; Hjeltness et al. 1987a).

Geographical Range and Species Susceptibility

Coldwater vibriosis has been reported in cultured Atlantic salmon along the coast of Norway, in the Shetland Islands of northern Scotland, and in the Faroe Islands (Hjeltnes et al. 1987a; Dalsgaard et al. 1988). To date, CV has not been reported from other parts of Europe but was detected by H. Mitchel (Connors Aquaculture, Inc., Eastport, Maine, personal communication) in eastern Canada and the northeastern United States in 1989 and again in 1993. The disease is found primarily in salt water or brackish water net-cages.

Atlantic salmon is the most susceptible fish species to CV; rainbow trout can be infected but are far less susceptible (Egidius et al. 1984). *V. salmonicida* was also isolated from highly stressed wild juvenile cod, but it was thought that this incidence was unusual and cod should not be considered a likely host (Jørgensen et al. 1989).

Clinical Signs and Findings

The initial clinical sign of CV is the tendency for infected fish to swim on their sides near the surface of cages (Holm et al. 1985; Bruno et al. 1986). Infected fish, which generally appear well nourished, might or might not exhibit extensive skin hemorrhage depending on stage of disease. When hemorrhages do appear, they are located at the base of fins and on the abdominal region (Figure 17.3). Gills are generally pale, but the opercular cavity may be hemorrhaged. The anus may be reddish and prolapsed. Internally, a bloody fluid accumulates in the peritoneal cavity, with hemorrhages on tissue surfaces ranging from petechia to ecchymosis (Figure 17.3). The spleen is grayish, the liver is usually yellowish and anemic, and hemorrhages occur on the air bladder, in fatty tissue around the pyloric cecae, throughout the visceral cavity, and on the peritoneal wall. The gut, particularly the posterior region, is hemorrhagic and the lumen contains a watery, bloody fluid. Occasionally, hemorrhages are found in the muscle. Stained blood smears show bacteria in the plasma (Figure 17.3).

Diagnosis

Although *V. salmonicida* is not easily isolated on culture media, it can be recovered from internal organs and blood of infected fish. Initially, *V. salmonicida* was isolated on nutrient agar with 5% human blood and 1.5% to 2% NaCl (Egidius et al. 1984), but either TSA or BHI is a suitable medium for isolation if it contains salt. The bacterium prefers 0.5% to 4% NaCl, grows optimally at 15 to 17°C, and will grow at 1 to 22°C but not at 26°C. At 15°C, slight growth is evident in 24 hours, with more pronounced growth at 72 hours. Colonies are smooth, grayish, opaque, slightly raised with entire margins, and range in size from punctate to 1 or 2 mm in diameter.

Espelid et al. (1988) used a monoclonal antibody against the surface antigen of *V. salmonicida* to identify the organism by ELISA. All isolates possessed a common outer membrane antigen designated as VS-P1, which is a protein–lipopolysaccharide complex. Immunohistochemistry, based on the avidin–biotin complex, was used by Evensen et al. (1991) to identify *V. salmonicida* in formalin fixed and histologically prepared tissues from Atlantic salmon. This method was also used to identify the organism successfully in preserved tissue from the original Hitra disease episode in 1977.

Bacterial Characteristics

V. salmonicida is Gram negative, cytochrome oxidase positive, motile, and sensitive to vibriostat 0/129 (see Table 17.1). The bacterium is a slightly curved rod that measures 0.5 × 2–3 μm and may be pleomorphic on initial isolation (Holm et al. 1985; Egidius et al. 1986). It is not hemolytic to human erythrocytes.

Holm et al. (1985) determined that *V. salmonicida* is a serologically uniform species that is distinct from *V. anguillarum* and *V. ordalii*. Jørgensen et al. (1989) determined by DNA analysis that isolates from Atlantic salmon and Atlantic cod were identical. Wiik et al. (1989) identified four plasmid profiles among 32 isolates, which would indicate possible species diversity, and Schroder et al. (1992) found serological diversity among *V. salmonicida* isolates from Atlantic salmon and Atlantic cod. In spite of possible serological and plasmid differences, all strains of *V. salmonicida* are biochemically similar (see Table 17.1).

Epizootiology

Although there has been some disagreement about the etiology of Hitra disease, it has been shown that it is synonymous with CV. Fjølstad and Heyeroaas (1985) found the pathology of Hitra disease to be similar to that of a vitamin E deficiency in higher vertebrates, and because a pathogenic organism could not consistently be isolated, they proposed a nutritional etiology. Observations by Poppe et al. (1985) cannot be ignored, but most scientists working with *V. salmonicida* concur that CV and Hitra disease are the same. Hjeltnes et al. (1987a) and Hjeltnes and Julshamn (1992) clearly correlated Hitra disease and CV to *V. salmonicida;* however, they recognize that environmental conditions and nutrition may also play an important role in the disease.

Coldwater vibriosis can be transmitted to noninfected Atlantic salmon and to a lesser degree to rainbow trout. Hjeltnes et al. (1987a) fulfilled Koch's postulates for *V. salmonicida* and showed that it can be transmitted through water via infected fish. Superficial injury will enhance infectivity, but primarily, the pathogen enters fish through the gills. Atlantic salmon infected via gills began showing behavioral and clinical signs after 1 week and started to die 2 weeks postexposure.

Mortality in *V. salmonicida*-infected yearlings to market size fish can be high. In naturally occurring infections, acute mortality rates of 5% per day have been noted. Injured fish exposed to waterborne bacteria suffered 80% to near 100% mortality in about 20 days. Fish challenged more naturally by cohabitation suffered 24% and 46% mortality in 24 and 35 days, respectively. Experimental waterborne exposure of the more susceptible Atlantic salmon resulted in 90% mortality for a 45-day period, compared with 40% mortality in similarly infected cod (Schroder et al. 1992). Rainbow trout, infected by injection with an unquantitated number of bacteria showed initial signs of disease 9 days postinfection and suffered 80% to 100% mortality after 14 days when held at 9°C (Egidius et al. 1981). These results may be a little misleading because of the high number of bacteria injected; however, it illustrates the disease's potential for severity at this temperature. Generally, cod are not susceptible to *V. salmonicida* but highly stressed, naturally infected juvenile cod on 15 farms suffered 10% to 90% (50% average) mortality from the pathogen. These fish were stocked into net-cages in close proximity to infected Atlantic salmon (Jørgensen et al. 1989). The phenomenon was considered unusual.

The majority of *V. salmonicida* infections have occurred during autumn and winter when water temperatures ranged between 4 and 9°C (Egidius et al. 1981); however, H. Mitchell (personal communication) noted 0.5% mortality per day in some cages in Maine when water temperatures were 1°C. Between 1977 and 1981, the maximum number of CV cases diagnosed in 1 year at the National Veterinary Institute, Oslo, Norway, constituted only a small percentage of the total case load. In 1982, the number of CV cases increased to greater than 70% of total disease submissions, and during ensuing years, the disease caused increasing losses. In recent years, more cases have been reported during the warmer months (B. Hjeltnes, Institute of Marine Research, Bergen, Norway, personal communication), a pattern that has also been noted in North American epizootics. Disease severity seems to increase during May and June when water temperatures rise to 5 to 6°C.

Pathological Manifestations

Bruno et al. (1986) found the histopathology of CV in experimentally infected Atlantic salmon to be similar to that of naturally infected fish. The most significant pathology occurred in the pyloric cecae and mid- and hind-gut where necrotic tissue sloughed into the lumen. Blood vessels were vasodilated and congested, hemorrhage was present in the lamina propria, and in some cases the entire mucosal epithelium was necrotic. Focal necrosis occurred in the hematopoietic tissue of the kidney where nuclei of glomerular cells were often swollen and occasionally necrotic. The ellipsoid system of the spleen was engorged with macrophagelike cells, focal necrosis was seen in the reticuloendothelial cells, and gills exhibited epithelial necrosis and sloughing; however, the liver and pancreas appeared normal.

Totland et al. (1988) found that in *V. salmonicida* infections, the blood was heavily laden with bacteria. They also detected bacteria in the cell membrane on the luminal side of endothelial cells of capillaries. Intracellular bacteria were later seen within endothelial cells.

Histopathology substantiates the causative relationship of *V. salmonicida* in CV, but pathogenesis of the disease is not fully understood because knowledge of factors responsible for its pathogenicity is limited.

Significance

During the 1980s, 80% of lost revenue from fish diseases on Norwegian fish farms was attributed to Hitra disease (Poppe et al. 1985). Since 1989, however, mortalities due to the disease have decreased (B. Hjeltnes, personal communication). That the disease is no longer confined to Norway and not only affects Atlantic salmon but to a lesser degree rainbow trout and cod, makes it an important disease where they are cultured.

ENTERIC REDMOUTH

Enteric redmouth is a systemic bacterial disease of trout that is caused by *Y. ruckeri,* a member of the family Enterobacteriaceae. It is a chronic to subacute infection that primarily affects farm-reared trout. The disease is also known as Hagerman redmouth disease, redmouth, salmonid blood spot, and in Norway it is called "yersiniosis."

Geographical Range and Species Susceptibility

Enteric redmouth was initially diagnosed in the early 1950s in cultured rainbow trout in Idaho and has since been found in most states where trout are grown. Bacterial isolates obtained from rainbow and brook trout in West Virginia in 1952 and from rainbow trout in Australia in 1963 appeared to be identical to *Y. ruckeri* (Bullock et al. 1978b). It is probable that detection of *Y. ruckeri* in various areas during the past 40 years may not be totally the result of dissemination from Idaho but likely because the organism is endemic to at least some of these areas. The presence of *Y. ruckeri* has been confirmed in many other countries including Australia, Bulgaria, Canada, Denmark, Finland, France, Germany, Great Britain (including Scotland), Italy, Norway, South Africa, and Switzerland (Austin and Austin 1987; Davies and Frerichs 1989; Stevenson et al. 1993). To date, it has not been reported in Japan's extensive trout and salmon farms.

Rainbow trout is the species most severely affected by ERM; however, all salmonids are susceptible to some degree, and chinook salmon is the most severely affected species in the Pacific Northwest (K. Amos, personal communication). Significant mortalities of Atlantic salmon have been reported on Norwegian farms, and eight salmonid and seven nonsalmonid species were identified by Stevenson et al. (1993) in which *Y. ruckeri* caused disease. Nonsalmonid species include the emerald shiner, fathead minnow, three species of whitefish, sturgeon, and turbot, but for the most part, mortalities among these species have been insignificant. The bacterium has also been isolated from numerous, apparently healthy fish, invertebrates (crawfish), mammals (muskrat), seagulls, a human, sewage, and river water (Busch 1983; Michel et al. 1986a; Stevenson et al. 1993). It is quite possible that *Y. ruckeri* is more common in the environment than was once thought.

Clinical Signs and Findings

Enteric redmouth is categorized into subacute, chronic, and latent phases, and clinical signs vary within each phase. In the subacute phase, fish often exhibit a red mouth, head, and jaw as a result of subcutaneous hemorrhaging (Ross et al. 1966; Busch 1983). Affected fish are dark, have inflammation and/or hemorrhaging at the base of fins and around the vent, exhibit unilateral or bilateral exophthalmia with orbital hemorrhage, and gills appear hemorrhaged near tips of the filaments (Figure 17.4). Fish are inappetant and sluggish with a tendency to accumulate downstream in raceways or along pond edges. Frerichs et al. (1985) described *Y. ruckeri* infections in Atlantic salmon in which there was no reddening of the mouth and opercula; therefore, *Y. ruckeri* should not be ruled out if the classic "redmouth" is absent.

Fuhrmann et al. (1984) reported that *Y. ruckeri* often produces internal pathological conditions that resemble "furunculosis." Petechia may be present in the muscles, visceral fat, intestines, liver surface, pancreas, pyloric cecae, and swim bladder (Figure 17.4). The lower intestine is flaccid, hemorrhaged, and inflamed and contains a thick, opaque, purulent material. The kidney and spleen may be enlarged.

As disease progresses into the chronic phase, exophthalamia increases (bilaterally or unilaterally), often causing the eye to rupture. Fins become frayed and eroded, gills are pale, the abdomen is distended, and there is an accumulation of bloody, ascitic fluid in the body cavity (Busch and Lingg 1975). Fish become dark, lethargic, and emaciated before death. As clinical signs disappear, mortality abates and survivors move into a latent phase in which a few darkly pigmented fish may be present.

Diagnosis

Enteric redmouth is diagnosed by isolation of *Y. ruckeri* from internal organs on general purpose bacteriological media (BHI or TSA) (Ross et al. 1966). Rodgers (1992) described a selective *Y. ruckeri* media (ROD) that differentiates the organism by its yellow color. Waltman and Shotts (1984) also described a selective media (Shotts-Waltman:

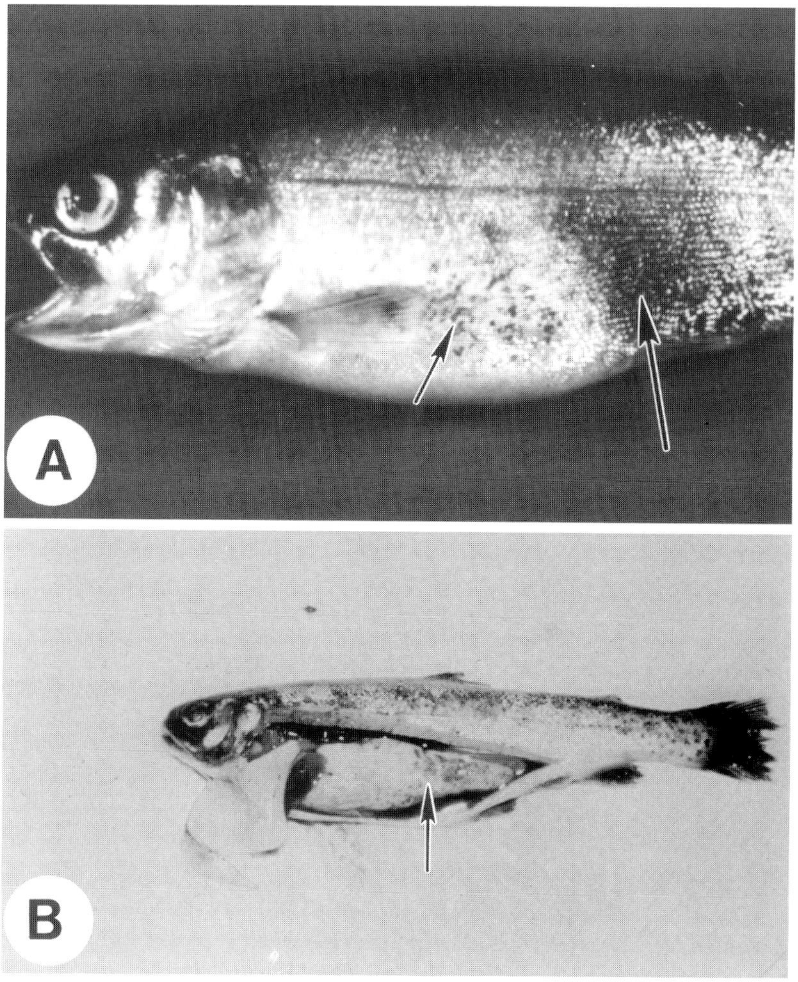

FIGURE 17.4. *Rainbow trout infected with* Yersinia ruckeri. (A) *Petechiae on the abdominal wall* (small arrow), *ecchymosis in posterior abdominal musculature* (large arrow), *hemorrhage in mandible and exophthalmia.* (B) *This fish has pale gills, mottled liver, and petechiae in the pyloric cecae area and in visceral adipose tissue* (arrow). *(Photographs courtesy of S. LaPatra, Clear Springs Food).*

SW) for *Y. ruckeri* on which it forms a green colony surrounded by a zone of hydrolysis. Hastings and Bruno (1985) evaluated the SW selective media and concluded that it alone was not reliable in isolating and identifying *Y. ruckeri* because of insufficient specificity; however, these selective media can be helpful in identification of *Y. ruckeri* from carrier fish, particularly if intestinal content or kidney material is used for pathogen isolation.

Amplification of *Y. ruckeri* DNA in PCR was used by Argenton et al. (1996) to diagnose enteric redmouth. The PCR assays detected three cell preparations of *Y. ruckeri* in seeded samples of serotype 01 and 02 and an unknown. The bacterium was also detected by PCR in infected trout kidney tissue after brief digestion with proteinase K. Because of speed, sensitivity, and specificity of PCR, Argenton et al. (1996) concluded that this system was preferable to conventional bacteriological diagnostic tests.

Bacterial Characteristics

Y. ruckeri grows aerobically and anaerobically at 9 to 37°C, with an optimum temperature range of 22 to 25°C at which colonies form in 24 to 48 hours. Isolates that grow at 37°C are generally avirulent to trout. The white to cream colored, translucent colonies are 1 to 2 mm in diameter, smooth, slightly convex, raised, and round with entire edges. Once the organism is isolated, it can be identified by conventional biochemical characteristics (Table 17.2) (Ewing et al. 1978; Stevenson and Daly 1982). It is a motile (peritrichous flagellation), Gram-negative, cytochrome oxidase negative rod that measures 1.0 × 1.0–3 µm. The most im-

TABLE 17.2. BIOCHEMICAL AND BIOPHYSICAL CHARACTERISTICS OF *YERSINIA RUCKERI*

Test/characteristic	% positive
Short, Gram-negative rods	100
Motility	82
Fermentation	100
Oxidase	0
Catalase	100
ß-galactosidase	100
Orinthine decarboxylase	100
Lysine	100
Arginine dihydrlase	0
Tryptophan deaminase	0
Urease	0
H_2S production	0
Indole	0
Citrate utilization	99
Methyl red	91
Voges-Proskauer	93
Gelatin hydrolysis	77
Casein hydrolysis	74
Tween 20, 80 hydrolysis	82
Growth on MacConkey agar	99
Nitrite reduction	99
Gas from glucose	8
Acid from	
Amygdalin	0
Arabinose	0
Fructose	100
Galactose	99
Glucose	100
Inositol	0
Lactose	0
Maltose	99
Mannitol	99
Mannose	100
Melibiose	0
Sorbitol	20
Sucrose	0
Trehalose	100

Source: Davies and Frerichs (1989).

portant biochemical reactions of *Y. ruckeri*, which are fairly homogeneous for the species, are as follows: it is positive for fermentative metabolism, production of catalase and ß-galactosidase, lysine and ornithine decarboxylation, methyl red test, nitrite reduction, and degradation of gelatin. It will grow on media containing up to 3% NaCl and it utilizes citrate; two characteristics that separate it from *Edwardsiella ictaluri*. *Y. ruckeri*'s inability to produce H_2S, indole, oxidase and phenylalanine deaminase, and its negative reaction for Voges-Proskauer, are significant features; however, some strains may vary in methyl red, Voges-Proskauer, lysine decarboxylase, arginine dihydrolase, and

TABLE 17.3. SEROLOGICAL GROUPINGS OF *YERSINIA RUCKERI*

Serological group (serovar)	Designation	% of isolates
I	Hagerman	59
I′	Salmonid blood spot	6
II	Oregon A	15
	Oregon B	8
III	Australian	6
IV	(Excluded)	0
V	COLORATO	3
VI	Ontario	3

Source: Stevenson et al. (1993).

lactose fermentation tests (Ross et al. 1966; Busch 1983). Fermentation of sorbitol has received some attention in discriminating between pathogenic strains (serotype I) and nonpathogenic strains (serotype II). Serotype I does not ferment sorbitol as does serotype II; consequently, Cipriano and Pyle (1985) developed a sorbitol-based media that can be used to distinguish them. Valtonen et al. (1992) questioned the accuracy of the sorbitol reaction test as a pathogenic indicator in Norwegian isolates. Austin and Austin (1987) stated that *Y. ruckeri* could be confused with *Hafnia alvei* if the API 20 system is used for identification; however, Grandis et al. (1988) showed that *H. alvei* is l-arabinose and l-rhamnose positive, whereas *Y. ruckeri* is uniformly negative for these characteristics.

Y. ruckeri can be positively identified by indirect fluorescent antibody test (IFAT) (Davies and Frerichs 1989), but one must be aware of the potential for cross-reactivity with other enteric bacteria such as *H. alvei*. Serological cross-reaction does not occur with other fish pathogens (*A. salmonicida*, *A. hydrophila*, *V. anguillarum*, or *R. salmoninarum*) that may be encountered in salmonids (Hansen and Lingg 1976). On the basis of agglutination patterns, *Y. ruckeri* was originally separated into two serotypes: serotype I, the Hagerman strain that is pathogenic and most commonly encountered in clinical diagnosis of diseased fish (Ross et al. 1966; Busch 1983), and serotype II, the Oregon strain that is normally not highly pathogenic (Bullock et al. 1978). Cipriano et al. (1986), however, did find that serotype II was pathogenic to chinook salmon. Currently, six whole-cell serotypes of *Y. ruckeri* are recognized (Table 17.3). Stevenson and Daly (1982) and Stevenson and Airdrie (1984) divided the species into five serological groups (I and

I′ through VI). They concluded that the Australian isolate (Bullock et al. 1977) should be included in serotype I (I′) and that two additional serotypes (V and VI) should be established. Daly et al. (1986) detected five strains of *Y. ruckeri* in Canada, most of which were in serotypes I and II.

Epizootiology

Although *Y. ruckeri* is generally considered an obligate pathogen, ERM epizootics are influenced by the fish's environment, and outbreaks are often allied with stress (Busch 1983), a fact that cannot be over emphasized. Hunter et al. (1980) noted that asymptomatic *Y. ruckeri* carrier steelhead trout did not transmit disease to uninfected fish unless stressed. When stressed at 25°C, the pathogen was transmitted from carrier to uninfected fish, but no deaths were reported in the recipients. Disease outbreaks can be precipitated by handling of apparently healthy trout and/or by presence of increased ammonia and metabolic waste in the water, which decreases oxygen levels (Bullock and Snieszko 1979). In a study to evaluate the effect of oxygen concentrations on *Y. ruckeri*–exposed rainbow trout, Caldwell and Hinshaw (1995) showed that a supersaturation of oxygen (150% of saturation) resulted in significantly higher mortality (17.9%) when fish were challenged with the pathogen, as opposed to a mortality of 12.6% at hypoxic (70% of saturation) and 10% at normoxic (100% of saturation) levels of dissolved oxygen. In view of these data, the normoxic condition is preferable.

Infected trout and salmon can become *Y. ruckeri* reservoirs. Busch and Lingg (1975) reported that 25% of rainbow trout checked 45 days after surviving an ERM epizootic had *Y. ruckeri* established in their lower intestines, making them asymptomatic carriers. When held at 14.5°C, these fish developed a recurrent ERM infection. It was demonstrated that *Y. ruckeri* carrier trout shed the organism in higher numbers on 36- to 40-day cycles and bacterial shedding preceded clinical signs and mortality by 3 to 5 days. Also, cyclic bacterial shedding and ensuing infections were influenced by water temperature and other water quality parameters; however, the full implication of this phenomenon is still not fully understood.

Y. ruckeri was found in water and bottom sediments throughout the year at two trout hatcheries. High-risk periods for ERM outbreaks at these hatcheries were related to stressful conditions and peak bacterial counts in the water (Romalde et al. 1994). These researchers reported that the pathogen was isolated from water of an effluent stream 15 km below the outfall of contaminated hatcheries.

Incubation time between exposure to *Y. ruckeri* and clinical ERM will vary inversely with water temperature, general health of fish, and presence of additional stressors (Ross et al. 1966; Busch 1983). At 15°C, incubation time is generally 5 to 7 days between exposure and first deaths, but if fish have a history of prior exposure, mortality may occur in 3 to 5 days. Onset of mortality may also depend on fish size and level of exposure. Enteric redmouth outbreaks usually occur only after fish have been exposed to large numbers of the pathogen, and deaths can continue for 30 to 60 days after initial onset of disease.

Enteric redmouth occurs primarily between April and September, when fish range in size from 5 to 200 g (Rodgers 1992). Larger fish can also be affected, but incidence of infection coincides with rise and fall of water temperature, poor water quality, overcrowding, and handling. Exposure to sublethal concentrations of copper (7 and 10 μg/L for 96 hours) significantly increases susceptibility of steelhead trout to ERM (Knittel 1981).

Using the *Y. ruckeri* selective media (ROD), Rodgers (1992) was able to detect very low numbers of the pathogen in intestines of carrier rainbow trout 6 to 8 weeks before infection actually occurred in the kidneys. On a farm where fish were severely stressed by frequent handling and poor water exchange, a significantly higher number of bacteria were found in the feces and kidneys of fish. Horizontal transmission occurs primarily through water from shedding fish to uninfected fish. Transmission of *Y. ruckeri* during spawning has not yet been proved; however, the organism can be isolated from brood stock at this time.

The effects of *Y. ruckeri* on trout depend on age and size of fish, their relative susceptibility, water temperature, and stress level. Naturally infected hatchery salmonids generally suffer less than 10% mortality during the subacute phase of disease, and recurrent infections in survivor populations result in low, chronic mortality during an extended period. In laboratory infectivity studies, mortality depends on the method of challenge. Intraperitoneal (IP) injection produces more consistent results, whereas waterborne exposures are inconsistent and can produce mortality ranging from near 0% to 100% (Busch and Lingg 1975).

In Europe, *Y. ruckeri*'s presence in imported bait fish would strongly suggest that they were the original disease source on that continent (Michel et al. 1986a). Sources other than fish may contribute to the spread of *Y. ruckeri.* Willumsen (1989) found the pathogen in the intestinal contents of seagulls during an ERM outbreak, and Stevenson and Daly (1982) reported the pathogen in muskrat's intestine. *Y. ruckeri* was also isolated from crayfish in a spring water supply. When the crayfish were eliminated, no more *Y. ruckeri* infections occurred, indicating the reality of a nonfish source (K. Amos, personal communication). *Y. ruckeri* may also survive in water and mud for an extended period. McDaniel (1972) reported that it could adapt to a normal aquatic saprophytic state and live for 2 months in mud.

Pathological Manifestations

Histopathology of ERM is characterized by bacterial colonization of capillaries of heavily vascularized tissues. Submucosal hemorrhage of the mouth, jaws, and under the head is most striking but is not always present. Gills develop telangiectasis; petechial hemorrhage, congestion, and edema occurs in the muscle, liver, kidney, spleen, and heart (Busch 1983). Inflammation occurs in the reticuloendothelial tissue, and necrotic foci develop in the liver with an accumulation of mononuclear cells. Macrophages phagocytose bacterial cells throughout the vascular system and hematopoietic tissue of the kidney. The spleen becomes necrotic, resulting in a loss of lymphoid tissue. The digestive tract is characteristically hemorrhaged, edematous, and necrotic with sloughing of the mucosa into the lumen. Owing to a septicemia and hemorrhaging, hematocrits, hemoglobin content, and blood protein are reduced to about half of normal (Quentel and Aldrin 1986).

Significance

Enteric redmouth is a significant disease in cold water aquaculture, particularly in rainbow trout. Its ability to cause high mortality in production size trout (which accounts for approximately 75% of investment cost) emphasizes its impact on aquaculture. In the past, ERM has accounted for up to 35% of overall losses in the trout industry at an annual cost of over $2.5 million in the Hagerman Valley (Busch 1978). In trout, detrimental effects of chronic *Y. ruckeri* infections in the form of reduced feeding, higher feed conversions, and retarded growth rates are unknown but could be economically significant. Also, the increase in its geographical range, either by dissemination or because of better surveillance and detection methods, has elevated the disease to one of international concern. The common practice of vaccinating trout for ERM has significantly reduced the disease to a minor problem in many areas.

BACTERIAL GILL DISEASE

Bacterial gill disease was named by Davis (1926) when he observed large numbers of long filamentous bacteria on clubbed gills of brook and rainbow trout. The term has since been used to describe infections caused by several different species of bacteria that affect fish gills. The principal etiological agent of BGD is *F. branchiophilum* (*branchiophila*) (Wakabayashi et al. 1989; von Graevenitz 1990; Bernardet et al. 1996). Also, Strohl and Tait (1978) described *Cytophaga aquatile* as a causative organism of BGD in trout and salmon, and Bernardet et al. (1996) determined that this species of bacteria should be *Flavobacterium aquatile,* which is the name used in the following discussion. Apparently, both organisms, *F. branchiophilum* and *F. aquatile,* can actually cause the same clinical disease, but *F. branchiophilum* infections appear to be more prevalent. Other bacteria can occasionally be involved with BGD, especially in nonsalmonids, but generally not as the primary pathogen. Bacterial gill disease is distinctly different from gill infections of *Flavobacterium columnare* (columnaris).

While recognizing that pathological gill changes can be caused by many factors, including nutrition, toxicants, fungi, and bacteria, only those gill infections of juvenile salmonids caused by the two filamentous flavobacteria species are discussed here.

Geographical Range and Species Susceptibility

Since first being described in the northeastern United States, BGD has been reported in Japan

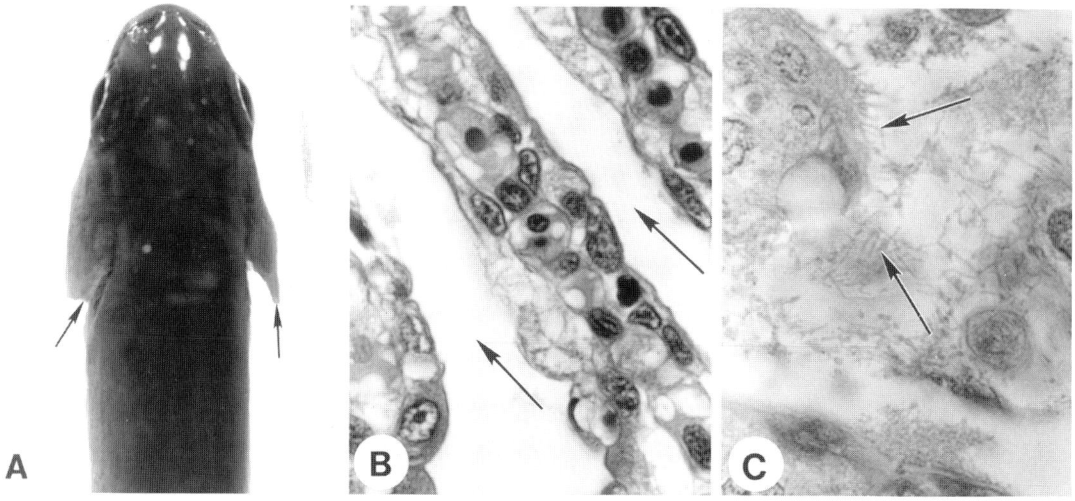

FIGURE 17.5. Bacterial gill disease of salmonids. (A) *Juvenile rainbow trout showing flared gill cover with some debris protruding from under one gill cover* (arrows). *(Photograph A courtesy of J. Ferguson.)* (B) *Clear lamellar trough* (arrow) *of healthy gill.* (C) *Mucus in the lamellar trough with long filamentous* Flavobacterium branchiophilum (arrows).

(Kimura et al. 1978), throughout North America (Wakabayashi et al. 1980; Ferguson et al. 1991), Hungary, and the Netherlands, and it likely occurs in cultured juvenile trout in most countries of the world. *F. aquatile,* described in Canada, is similar biochemically to gill isolates taken from numerous geographical locations, suggesting that it enjoys a wide geographical range (Strohl and Tait 1978).

Juveniles of many fish species are susceptible to BGD, but salmonids are most severely affected because of their method of culture. Bacterial gill disease has also been reported in carp, goldfish, catfish, eels, fathead minnows, and other fish species (Ostland et al. 1989). Whether the same bacterial species is responsible for BGD in all of these fish species is unclear.

Clinical Signs and Findings

Fish affected by BGD experience inappetence, move toward water inlets, and may ride high in the water. They become lethargic, sometimes spiraling, and moribund individuals will sink to the bottom. Affected fish have dark pigmentation, gills become very pale, gill covers are flared, and gills become so swollen and congested with mucus and debris that they protrude from beneath the operculum (Figure 17.5).

Diagnosis

Clinical BGD is diagnosed by microscopic examination of gill tissue wet mounts or by Gram staining smears (Figure 17.5). Large numbers of Gram-negative filamentous bacteria can be observed on the surface of gill intralamellar spaces, and they will often form clumps of long slender rods. Gill filaments begin swelling at the distal end where hyperplasia of the epithelium produces a "clubbed gill" condition that can become so acute that filaments will actually fuse (Figure 17.5). Huh and Wakabayashi (1989) compared light microscopy of wet mount material with IFAT for accuracy in detecting BGD. They found that IFAT was more accurate and resulted in identification of greater numbers of infected trout. Unless clinical disease was in progress, neither method detected *F. branchiophilum* on fish gills or in water. The organism is usually confined to the gill epithelium but some bacteria may be found within and beneath necrotic lamellar epithelial cells (Ostland et al. 1994).

F. branchiophilum and *F. aquatile* are cultured aerobically on media where both form yellow pigmented colonies, thus leading to the designation of "yellow pigmented bacteria" (YPB). Suspect *F. branchiophilum* colonies contain long filamentous bacteria and have refractile circular "cysts" (Ostland et al. 1994). Incubation at 12 to 18°C did not affect quantitative isolation whether incubation was 6 or 12 days. These bacteria have not been isolated from internal organs. Colonies of cultured *F. branchiophilum* are light yellow, round, translucent, mildly convex, smooth, and about 0.5 to 1 mm in diameter after being incubated at 18°C for 5 days (Wakabayashi et al. 1989).

Following bath exposure of rainbow trout to *F. branchiophilum*, an ELISA system was used by MacPhee et al. (1995a) to detect bacterium in crude gill extracts. The avidin–biotin system using polyclonal antiserum detected the pathogen when whole bacterial cell preparations, gill preparations spiked with bacteria, or extract of infected gills were tested. The authors proposed that the ELISA method is adaptable to field-collected gill samples but recognized that further antigenic specificity testing is required.

Bacterial Characteristics

Morphologically, *F. branchiophilum* is a Gram-negative rod that measures approximately 0.5 × 5–15 µm (average length, 8 µm), is nonmotile, does not glide or spread on agar, and is cytochrome oxidase positive. Cells are heavily fimbrianated (Ostland et al. 1994). Biochemical and biophysical characteristics are similar among all *F. branchiophilum* isolates (Table 17.4); however, variability in carbohydrate utilization has been noted. The organism does not grow on TSA agar unless the medium is diluted approximately 20-fold; it grows slowly at 10, 18, and 25°C on media containing up to 0.1% NaCl, but some strains will grow at 5 and 30°C but not at 37°C.

Antigenically, *F. branchiophilum* isolates from brook and rainbow trout, and Atlantic salmon in Ontario, Canada, were similar to those isolated from salmonids in Japan and Oregon (reference strains ATCC 3505 and ATCC 3506, respectively) (Ostland et al. 1994). All strains have similar mol% G + C of DNA of 30 to 32.

F. aquatile was isolated on 2% tryptone agar plates (Strohl and Tait 1978). Within 7 days, spreading colonies with pigmentation that varied from yellow to reddish orange appeared (Table 17.4).

TABLE 17. 4. CHARACTERISTICS OF *FLAVOBACTERIUM BRANCHIOPHILUM* AND *FLAVOBACTERIUM AQUATILE*, THE TWO FILAMENTOUS BACTERIA THAT HAVE BEEN ASSOCIATED WITH BACTERIAL GILL DISEASE OF TROUT

Test/characteristic	*F. branchiophilum*[a]	*F. aquatile*[a]
Yellow-red colony	+	+
Growth at 37°C	–	–
Motility	–	Gliding
Growth on nutrient agar	–	+
Hydrolysis of		
Gelatin	+	+
Casein	+	+
Chitin	–	+
Esculin	–	+
Starch	+	+
Indole from tryptone	–	+
Nitrate reduction	–	+
Binds Congo red	?	–
Catalase production	?	+
H_2S production	?	(+)
ONPG	?	+
Growth in		
0% NaCl	+	+
0.5% NaCl	–	+
1.0% NaCl	–	+
2.0% NaCl	–	+
Acid from		
Glucose	+	+
Fructose	+	+
Lactose	–	–
Sucrose	+	–
Maltose	+	–
Trehalose	+	(+)
Cellobiose	(+)	+
Arabinose	–	+
Xylose	–	+
Raffinose	(+)	±
Mannitol	–	+

Sources: Strohl and Tait (1978); Wakabayashi et al. (1989).
[a]+ = positive; – = negative; (+) = weakly positive.

Epizootiology

The epizootiology, and thus the etiological agent of BGD, is confusing because it has been associated with several different bacteria: *Flexibacter* sp., *Cytophaga* sp. (both of which are now flavobacteria), and *Flavobacterium* spp. Bullock (1972) implicated *Flavobacterium* sp. (*Cytophaga* sp.) in BGD but did not identify the species and could not create infections at will without severe environmental stress. Strohl and Tait (1978) described *Flavobacterium* (*Cytophaga*) *aquatile* that was isolated from diseased gills of salmon and trout in Michigan, and Wakabayashi et al. (1980) showed that organisms causing BGD in Japan, Oregon, and Hungary were all members of the genus *Flavobacterium*. Bacterial strains from Japan, the United States, and Hungary possess some identical antigens as well as slight differences; however, strains from the United States and Hungary were most similar (Huh and Wakabayashi 1989).

In general, experimental BGD transmission is unpredictable. *F. aquatile* has not been experimentally transmitted under any condition. *F. branchiophilum* has been successfully transmitted in the laboratory, but often only in conjunction with environmental stress. Rainbow trout were infected with *F. branchiophilum* by bath challenge; clinical disease developed in 24 hours and mortality in 48 hours (Ferguson et al. 1991). Lumsden et al. (1994) successfully transmitted BGD by immersion that resulted in a 23% to 41% survival rate in infected fish, compared with 49% to 70% in unexposed controls. Stressful conditions were not used in either of these studies.

It has been suggested that BGD may be incorrectly named because its development is so closely related to conditions in the culture environment and to gill epithelium injuries. It has been widely accepted that gills are usually injured by chemical or physical irritants in the water prior to bacterial colonization (Bullock 1972). Gill injuries will occur if water exchange is inadequate, ammonia levels are elevated, oxygen concentrations are decreased, or silt or excess feed is present in the water. These injuries will result in epithelial hyperplasia that makes gills more susceptible to microbial invasion by *Flavobacterium* and occasionally other bacteria.

Studies by Ferguson et al. (1991) and Lumsden et al. (1994) have shown that these injuries are not absolutely necessary for BGD to occur, and Speare et al. (1991a) presented a noninjurious scenario for BGD development. After an extensive study of 23 separate BGD outbreaks in rainbow trout, they concluded that "no other disease conditions, no gross errors in management, nor recent exposure to chemotherapeutics" preceded BGD. Although the causative organism was not definitively identified, it was thought to have been *F. branchiophilum*. During a 5-month monitoring regime before onset of a naturally occurring disease outbreak, gill morphology of examined fish remained unaltered, and it was proposed that *F. branchiophilum* could cause BGD without environmental stress or gill injury. Furthermore, MacPhee et al. (1995b) showed that BGD of trout is linked to the consumption of feed by the fish, rather than to environmental changes arising from feeding or water quality. They also suggested that alterations in the undisturbed layer on the gill may aid bacterial colonization, although this is secondary to feed consumption and waste excretion,.

Nevertheless, most BGD outbreaks have been associated with some management factor such as excessive feeding, poor feed quality, poor water quality, poor water circulation, or inadequate water flow. Wakabayashi and Iwado (1985) reported that fish infected with *F. branchiophilum* were more susceptible to hypoxia because the bacterium impaired respiratory functions. Noninfected fish consumed 251 to 289 mL of 0_2/kg · h, whereas oxygen consumption rates for infected fish at 2 and 5 days postinfection were 183 to 229 and 155 to 167 mL 0_2/kg · h, respectively. In comparing virulence of seven wild strain isolates and two ATCC strains of *F. branchiophilum,* Ostland et al. (1995) found that one strain was pathogenic to five species of salmonids and five species of shiner but that some isolates were avirulent. All isolates that colonized the gills had fimbria, but some were unable to produce pathology.

Historically, salmonid fry and fingerlings less than 5 cm in length are particularly susceptible to BGD, but larger fish can occasionally become infected. Adult rainbow and cutthroat trout and chinook salmon have suffered BGD outbreaks (R. Holt, Oregon State University, Corvallis, Oregon, personal communication) and Ferguson et al. (1991) successfully infected rainbow trout up to 3 years of age with *F. branchiophilum.* In view of these reports, BGD may be more of a problem in larger trout than previously realized.

Bacterial gill disease mortality among small fish has the potential to become subacute. In experimental BGD infection studies, mortality reached 39% to 80% after 13 days (Bullock 1972). Speare et al. (1991b) found that morbidity could increase from approximately 5% when disease was first detected to more than 80% within 24 to 48 hours. During this time, mortality rates rose to 20% per day, diminished by day 7, and only a few fish showed any clinical signs by days 10 to 14. Once juvenile trout have overcome an *F. branchiophilum* infection, disease may reoccur, particularly if environmental conditions are favorable. Bacterial gill disease generally occurs at 12 to 19°C (Heo et al. 1990).

Viability of *F. branchiophilum* in water is not fully known, but it is theorized that subclinically infected fish are a source of infection when they shed bacteria; however, trout reared entirely in well water have become infected (R. Holt, personal communication). Heo et al. (1990) detected *F. branchiophilum* in the water immediately before and during an epizootic, but not after disease had abated. Ostland et al. (1990) enumerated total bacteria and percentage of YPB on the gills of clinically healthy and BGD-infected rainbow trout. Healthy gills had 4.9×10^5 bacterial CFU/g of tissue, 15% of which were YPB. Severely BGD-infected fish had 3.9×10^6 CFU/g, of which 35% were filamentous YPB. Overall, these data suggest a strong association between the presence of filamentous yellow pigmented bacteria and BGD severity.

Pathological Manifestations

Bacterial gill disease usually results from gill injuries caused by environmental factors, but once the bacteria becomes established in the gills, it plays a major role in mortalities of affected fish. Bacterial gill disease generally begins with hyperplasia of the gill epithelium, which can be caused by mild but chronic toxins or water irritants. *F. branchiophilum* has large numbers of fimbriaelike surface structures that bridge the spaces between the bacterium and gill tissue and appear to be involved in attachment to the gill surface (Spear et al. 1991a). Bacterial colonization and lamellae fusion will begin on the distal end of filaments infected with BGD, whereas in "nutritional gill disease" (due to insufficient dietary pantothentic acid), lamellar and filament fusion begins at the proximal end of the filament and progresses distally. Speare et al. (1991a and 1991b) confirmed and expanded on earlier bacterial colonization observations of Kudo and Kimura 1983. They noted that colonization was immediately preceded by several gill changes detectable only at an ultrastructural level. The cause of these changes was not determined, but environmental conditions were not considered to be a factor. Microridges of superficial filament epithelium showed cytoplasmic blistering, and a slight irregularity of filament tips was noted, all of which would suggest mild hyperplasia. Explosive morbidity and acute mortality coincided with extensive bacterial proliferation on the lamellar surfaces, epithelial hydropic degeneration, necrosis, and edema (Figure 17.5). Lamellar fusion and epithelial hyperplasia were later detected in subacute (2 to 5 day) or chronic (7 to 14 day) infections (Spear et al. 1991a). *F. branchiophilum* does not appear to become systemic, even when injected IP, as it cannot be isolated at a later time from internal organs.

It has been suggested that *F. branchiophilum* physically occludes the gill surfaces and inhibits respiratory exchange (Wakabayashi and Iwado 1985). Speare et al. (1991b) challenged that suggestion because fish have large areas of underused gill surfaces and because massive numbers of bacteria did not substantially cover the tissue–water interface. Why BGD-infected fish behave in a manner that suggests a lack of sufficient oxygen has still not been adequately explained. Oxygen consumption data by Wakabayashi and Iwado (1985) support the theory that there is increased opercular movement to increase water flow across the gills to enable better oxygenation of the blood.

Trout infected with BGD develop a significant decrease in serum Na^+, Cl^-, and an osmolarity imbalance that results in hemo concentration (Byrne 1991). Fish with BGD exhibit increased respiration (tachybranchia) but are not hypoxic, therefore, the tachybranchia may not be a response to impaired oxygen exchange. It was suggested that fluctuations in blood components, other than those that are acid-base related, were responsible for death in BGD-infected fish. These circulatory disturbances result from a loss of blood electrolytes that triggers a fluid shift from extracellular to intracellular and leads to death. Also, Wakabayashi and Iwado (1985) concluded that a breakdown in gas exchange at the gills causes failure to provide enough oxygen to remove excess muscle lactate.

Significance

Bacterial gill disease is one of the most common diseases of juvenile salmonids. Because BGD is often a subacute condition with high mortality, it must be considered a potentially serious problem in trout and salmon seed and fingerling culture facilities. The Fish Pathology Laboratory at Ontario Veterinary College, University of Guelph, Ontario, Canada, reported that BGD accounted for approximately 21% of all samples submitted for diagnosis from fish farms (Spear and Ferguson 1989).

BACTERIAL COLDWATER DISEASE

Bacterial coldwater disease (BCWD) can cause low to moderate mortality among cultured salmonids

FIGURE 17.6. Rainbow trout with typical lesion on the caudal peduncle (arrow) *associated with bacterial coldwater disease* (Flavobacterium psychrophilum).

and occasionally other species. The disease was originally referred to as "peduncle disease" because of principal lesion location. Until recently, the causative agent of bacterial coldwater disease was named *Flexibacter psychrophila;* however, when reclassification of the *Flavobacterium–Flexibacteria–Cytophaga* complex was proposed by Bernardet et al. (1996), it was reclassified as *Flavobacterium psychrophilus,* which is synonymous with *Cytophaga psychrophila.* For clarification purposes the name *Flavobacterium psychrophilum* is used throughout this text (Bernardet and Grimont 1989; Holt 1993).

Geographical Range and Species Susceptibility

Although bacterial coldwater disease occurs throughout most trout- and salmon-growing regions of North America, it most seriously affects salmon culture in the northwestern United States and western Canada (Anderson and Conroy 1970). The disease also occurs in Japan (Wakabayashi et al. 1991), the United Kingdom (Austin 1992), many other European countries (Bernardet and Kerouault 1989; Wiklund et al. 1994), and Australia (Schmidtke and Carson 1995)

All salmonid species are suspected to be susceptible to *F. psychrophilum;* however, rainbow, brook, lake, and cutthroat trout and coho, sockeye, chum, and Atlantic salmon are known to be susceptible (Holt 1993). The bacterium was also reported in wild ayu in Japan (Iida and Mizokami 1996). *F. psychrophilum* can occasionally cause disease in nonsalmonids and has been detected in European eels, carp, tench, and crucian carp in Germany (Lehmann et al. 1991).

Clinical Signs and Findings

Clinical signs of BCWD vary with size and age of fish. In infected sac fry, the yolk membrane becomes eroded (Wood 1974), and older fish exhibit lethargy and sometimes a spiral swimming behavior (Kent et al. 1989). The caudal peduncle appears whitish and the skin becomes necrotic, detached, and sloughs, exposing underlying muscle (see Figure 17.6). In fish up to 1 year of age, the caudal peduncle lesion is the most typical clinical sign of BCWD. Muscle necrosis may continue to the point that a vertebral column becomes exposed and the tail is nearly separated from the body. Lesions may also appear dorsally, laterally, or on the isthmus. Fin and operculum hemorrhages and pale gills can occasionally be seen in infected fish. Darkly pigmented, moribund fish may be observed late in epizootics.

Internally, petechia may be present on the liver, pyloric cecae, adipose tissue, heart, swim bladder, and occasionally on the peritoneal lining. Bacterial coldwater disease survivors may develop vertebrae deformities and/or nervous disorders that cause them to swim in a spiral (Kent et al. 1989).

Diagnosis

Bacterial coldwater disease is diagnosed by recognition of characteristic clinical signs, detection of elongated vegetative cells in stained lesion smears, and isolation of *F. psychrophilum* on appropriate media. Cytophaga and tryptocase yeast extract glucose agar can be used for isolation of this pathogen, but a modified *Cytophaga* media, either broth or agar, is the usual culture substrate (Bullock et al. 1986). Cipriano et al. (1996d) failed to obtain *F. psychropilum* growth on tryptic soy, MacConkey, or triple sugar iron agar. Generally, the organism can be isolated from skin mucus, muscle lesions, and internal organs. On agar, *F. psychrophilum* either forms a 1- to 5-mm–diameter colony that is pale yellow, flat, and spreading with a thin irregular edge, or colonies that are smooth, entire, and yellow. Cultures may be a mixture of the two colony types (Holt 1993). Development of yellow colonies on cytophaga agar requires 14 days at 8°C (Iida and Mizokami 1996). The pathogen is strictly aerobic with an optimum growth temperature of about 15°C but will grow at temperatures from 4 to 25°C. Poor growth will occur at the temperature extremes, and no growth occurs at 30°C (Bernardet and Kerouault 1989; Holt et al. 1989). *F. psychrophilum* does not tolerate NaCl concentrations above 1%.

Bacterial Characteristics

On stained lesion smears, the bacterium is a slender, Gram-negative bacillus, with rounded ends, that measures 0.3 μm × 2 to 2.5 μm. In broth culture, cells may be several times this length with some cell branching (Holt 1993). *F. psychrophilum* is motile by gliding and does not produce fruiting bodies.

Biochemical properties do not vary greatly among different strains of *F. psychrophilum* (Table 17.5) (Bernardet and Grimont 1989; Holt 1993). Generally, the bacterium does not use carbohydrates but is proteolytic for casein, gelatin, albumin, and collagen. It is cytochrome oxidase positive by most accounts, weakly catalase positive, does not produce H_2S, and contains flexirubin pigments that are of diagnostic value. Some strains degrade elastin. Comparing *F. psychrophilum* isolates from three Pacific Northwest (United States) salmon hatcheries, Cipriano et al. (1996d) found very little biochemical diversity. It is rare for *F. psychrophilum* to be isolated in the Southern Hemisphere, but Schmidtke and Carson (1995) did isolate it from Atlantic salmon in Australia. Characteristics of the Australian isolate agreed with isolates from other parts of the world.

TABLE 17.5. BIOPHYSICAL AND BIOCHEMICAL CHARACTERISTICS OF 28 ISOLATES OF *FLAVOBACTERIUM PSYCHROPHILUM*

Test/characteristic	Reaction[a]
Growth in broth	
1% tryptone	+
1% Casamino acid	–
0.1% sodium lauryl sulfate	–
Nitrate reduction	–
Production of	
Ammonia	+
Hydrogen sulfide	–
Catalase	+
Cytochrome oxidase	–
Acetylmethyl carbinol	–
Indole	–
Degradation of	
Cellulose	–
Starch	–
Casein	+
Gelatin	+
Albumin	+
Elastin	+
Collagin	+
Chitin	–
Tyrosine	+
Carbohydrate oxidation	
Glucose	–
Lactose	–
Sucrose	–

Source: Holt (1993).
[a]+ = positive; – = negative.

Although most strains of *F. psychrophilum* are generally biophysically and biochemically similar, they are also serologically homogeneous. Holt (1993) reported that *F. psychrophilum* isolates from New Hampshire, Michigan, and Alaska were antigenically similar to Oregon strains. Also, Cipriano et al. (1996d) discovered almost no serological diversity in four isolates with which they worked. In spite of biochemical and serological homogeneity, these investigators did, however, show a diversity in ribotyping among the four isolates.

Currently, serum agglutination will positively identify *F. psychrophilum* (Holt 1993), but Lorenzen and Karas (1992) developed a rapid immunofluorescens diagnostic method using spleen imprints of diseased rainbow trout fry. A slight cross-reactivity with *F. columnare* may occur, but this can be corrected by adsorption with the "columnaris" organism. Bertolini et al. (1994) examined cell-free ECP by sodium dodecyl sulfate-polyacrylamide gel electrophoresis (SDS–PAGE) and proteases of 32, 86, 114, and 152 kD and found that isolates taken from ayu in Japan were similar to those taken from coho salmon in the United States.

Epizootiology

Bacterial coldwater disease normally occurs in spring when water temperatures range from 4 to 10°C, but Lehmann et al. (1991) noted that non-salmonid infections occurred at water temperatures of 8 to 12°C. Experimental *F. psychrophilum* infections were induced by Holt et al. (1989) in coho and chinook salmon and rainbow trout at temperatures of 3 to 15°C; as water temperatures rose past 15°C, disease severity decreased.

Bacterial coldwater disease affects salmonids ranging from yolk sac fry to yearlings; the younger the fish, the more severe the disease (Leek 1987). Mortality in yolk sac fry generally ranges from 30% to 50%, and as fish begin feeding, it is usually in the 20% range. In Australia, mortality among Atlantic salmon reached as high as 80% (Schmidtke and Carson 1995). Chronic infections may develop in yearling coho and other Pacific salmon during winter months, when typical peduncle lesions are seen and low grade mortality occurs. Infected fish can also be anemic, a condition that is often compounded by presence of an erythrocytic inclusion body syndrome virus (Leek 1987). Experimentally, *F. psychrophilum* virulence varies with bacterial strain. Mortality in coho salmon infected with different strains varied from 0% to 100% (Holt 1993). When Bertolini et al. (1994) compared the pathogenicity of 29 *F. psychrophilum* isolates and classified them into four protease groups, they noted that groups 1 and 2 were generally more virulent than groups 3 and 4.

Two separate physical manifestations may be noted after an active BCWD infection. Some fish may have spinal deformities, a problem first reported by Wood (1974) when he described infected fish that had compressed vertebrae, lordosis, and scoliosis. Other fish may display nervous disorders such as a spinning behavior, swelling in the posterior skull, and dark pigmentation on either side. *F. psychrophilum* can readily be isolated from brain tissue but less often from internal organs of these fish (Meyers 1989).

Although the natural reservoir for *F. psychrophilum* is not fully understood, it is accepted that adult coho and chinook salmon, and in all probability rainbow trout and other salmonids, serve as carriers of the bacterium (Bullock and Snieszko 1970). According to Cipriano et al. (1996d), *F. psychrophilum* could be isolated from kidney and mucus of chinook and coho salmon even though no significant mortality was occurring. Bacterial concentrations were as high as 3.0 $\times 10^6$ CFU/g of tissue. They found that by combining isolation results from both kidney and mucus, *F. psychrophilum* could be isolated from 66% to 88% of tested fish. Horizontal transmission is highly likely, although experimental transmission from fish to fish cannot be achieved unless the mucous layer and epithelium are injured. It is generally concluded that members of the *Flavobacterium* group are a normal part of trout and salmon skin microflora, and if conditions are favorable, these bacteria can become pathogenic.

The organism has been isolated from the spleen, kidney, ovarian fluid, and milt of mature fish; therefore, bacterial transmission may occur on eggs during spawning (Holt 1972). *F. psychrophilum* was detected in 20% to 76% of sexually mature female chinook and coho salmon in hatchery populations and in 0% to 66% of males at the same facility (Holt 1993). Whether anadromous salmon carry the organism through the marine maturation phase is not clear. Upon returning to fresh water, these fish have a low incidence of *F. psychrophilum* infection; however, the longer they remain in fresh water, the higher the infection incidence. This suggests that adults become either infected or reinfected upon return to fresh water.

Pathological Manifestations

Bacterial coldwater disease is a septicemia; however, the most obvious pathology is sloughing of necrotic skin on the caudal peduncle. Although bacterial cells can be found throughout most organs and tissues of infected fish, death is generally attributed to severe heart lesions (Wood and Yasutake 1956). There is usually very little inflammation associated with BCWD, but some researchers have noted mild mononuclear infiltration of macrophages in diseased tissues. It has been suggested that *F. psychrophilum* extracellular products cause characteristic epithelial lesions in fish injected with either live bacteria or cell-free extracellular products. Internal organ lesions were also found in fish injected with either live bacteria or crude cell-free substances.

Early pathogenicity studies of *F. psychrophilum* indicated that exotoxins were not involved in the disease process, but Pacha (1968) noted that proteases may be involved. Later, it was found that steelhead infected with *F. psychrophilum* ECP developed microscopic lesions similar to those that developed when fish were infected with live, virulent organisms. An immunohistochemical method was used by Evenson and Lorenzen (1996) to show that *F. psychrophilum* was localized in the monocyte–macrophage system, skin lesions, retina, and choroid gland of the eye. Dermal changes in superficial or deep ulcers extended to the subcutaneous tissue or the musculature included inflammatory cell infiltrates in which polymorphonuclear cells were contained in their cytoplasm. Inflammation was also noted in the retina.

Significance

Bacterial coldwater disease is significant in salmonid culture, particularly juvenile coho salmon in the northwestern region of North America. It is not only the most frequently isolated fish pathogen but also causes high mortality (Holt 1993).

BACTERIAL KIDNEY DISEASE

Bacterial kidney disease (BKD) is a chronic, rarely subacute, bacterial infection of salmonids caused by *R. salmoninarum* (Sanders and Fryer 1980). Bacterial kidney disease was initially described in Atlantic salmon in Scotland in 1930 and was called Dee disease (Smith 1964). It was reported for the first time in the United States in 1935 (Belding and Merill 1935). Bacterial kidney disease has also been called corynebacterial kidney disease and salmonid kidney disease.

Geographical Range and Species Susceptibility

Bacterial kidney disease exists in many fresh water and marine environments where trout and salmon occur, with the possible exception of Australia, New Zealand, and the former Soviet Union (Evelyn and Prosperi-Porta 1992; Fryer and Sanders 1981). There are also no reports of BKD in China. Severe BKD has been reported in Canada, Chile, England, France, Germany, Iceland, Italy, Japan, Spain, and the United States. Most outbreaks in the United States occur in the northwest, upper midwest, and northeast (Mangin 1991).

All salmonids are susceptible to *R. salmoninarum,* but brook trout and chinook salmon are considered the most severely affected species (Fryer and Sanders 1981; Evelyn 1993). Naturally occurring BKD outbreaks have taken place only in cultured salmonids, and a few nonsalmonids are experimentally susceptible. It is not unusual to find the disease organism in wild salmonids, but clinical infections in these environments are rare. Experimental infections have been established in sablefish (Bell et al. 1990), Pacific herring (Traxler and Bell 1988), shiner perch (Evelyn 1993), common shiner, and fathead minnows (Hicks et al. 1986).

Clinical Signs and Findings

External clinical signs of BKD vary among fish species but are most obvious in terminal stages of disease (Fryer and Sanders 1981). The most common early signs of infection are dark pigmentation, exophthalmia, hemorrhages at the base of fins, abdominal distension, and lethargic swimming. Occasionally, superficial blisters and ulcerative abscesses are present on the body surface, particularly in rainbow trout (Figure 17.7). A bloody acetic fluid may be present in the coelomic cavity and petechia may appear in muscle under the peritoneal lining. As BKD progresses, small grayish white, granulomatous lesions appear in the kidney and to a lesser extent in the liver and/or spleen, which contain leukocytes, cellular debris, macrophages with phagocytosed bacteria, and cell-free bacteria. As disease progresses, the number and size of granulomatous lesions increase in all internal organs and the kidney becomes swollen and has a "corrugated" surface (Figure 17.7). Although systemic BKD infections are normal, they may occasionally be present only in the eye or on the skin (Hoffman et al. 1984).

Diagnosis

Infections of *R. salmoninarum* are diagnosed by noting clinical signs and detection of Gram-positive encapsulated (halo) diplobacillus bacteria in the internal organs. The indirect fluorescent antibody test (IFAT) or ELISA are more sensitive than

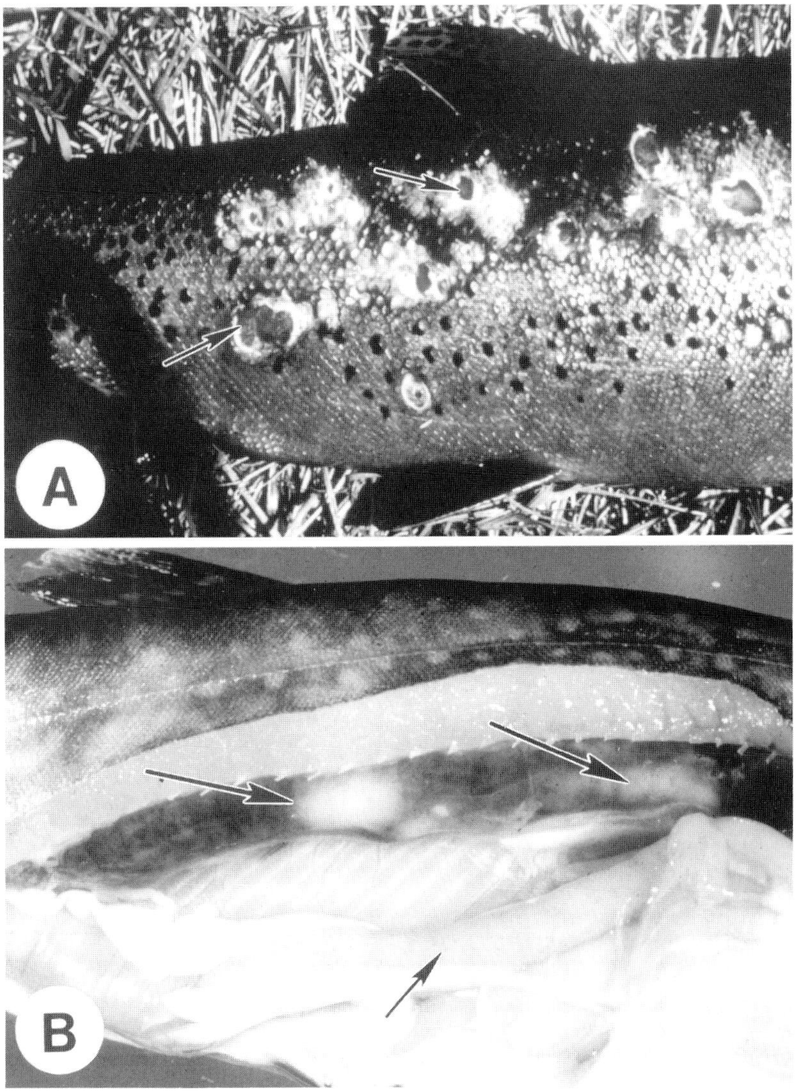

FIGURE 17.7. Bacterial kidney disease of salmonids. (A) Rainbow trout with epidermal, ulcerative lesions (arrow) *typical of chronic* Renibacterium salmoninarum *infection.* (B) *Granulomatous lesions* (large arrow) *in the swollen, corrugated kidney of brook trout. Note the petechia on the adipose tissue and intestine* (small arrow). *(Photographs courtesy of G. Camenisch.)*

the Gram stain and may be used on smears from infected kidney tissue to detect the presence of, and/or positively identify, the organism (Bullock et al. 1980b). Unless *R. salmoninarum* is present in abundance, Gram-stained smears are less precise than FAT or ELISA because pigment granules (melanin) can be confused with the pathogen. To enhance confidence in Gram-stained smears, they are compared to methylene blue and unstained smears. By comparing these slides, abundance of pigment granules that are brown and refractile in all preparations can be determined. Nevertheless, Gram staining to detect *R. salmoninarum* often results in a subjective analysis.

Bacterial Characteristics

R. salmoninarum is a Gram-positive bacillus (usually in pairs) that measures 0.3–0.5 µm × 0.6–1.0 µm (Young and Chapman 1978). It does not produce spores nor is it acid-fast or motile. Smith (1964) determined its optimum growth temperature to be 15°C, with growth slowed when temperature was reduced to 5°C or elevated to 22°C, and no growth at 37°C. *R. salmoninarum* is biochemically homogeneous (Table 17.6). The bacterium is proteolytic in litmus milk, produces catalase, and does not liquefy gelatin. It produces acid and alkaline phosphatase, caprylate esterase, glucosidase,

TABLE 17.6. BIOPHYSICAL AND BIOCHEMICAL CHARACTERISTICS OF *RENIBACTERIUM SALMONINARUM*

Characteristic	Reaction[a,b]
Cell morphology	Diplobacillus
Gram stain	+
Production of	
Acid phosphatase	+
Alkalin phosphatase	+
Caprylate esterase	+
Catalase	+
∂-glucosidase	+
ß-glucosidase	–
Leucine arylamidase	+
∂-mannosidase	+
Oxidase	–
Trypsinase	+
Degradation of	
Casein	+
Tributyrin	+
Tween 40	+
Tween 60	+
Tween 80	–
Acid from sugars	–
Growth at	
pH 7.8	+
0.25% bile salts	–
0.0001% crystal violet	+
0.001% methylene blue	–
0.00001% Nile blue	+
1% sodium chloride	+(poor)
Utilization of	
4MU-butyrate	+
4MU-heptanoate	+
4MU-laurate	+
4MU-nonanoate	+
4MU-oleate	+
4MU-palmitate	–
4MU-propionate	+

Sources: Goodfellow et al. (1985); Austin and Austin (1987).
[a]+ = positive; – = negative.
[b]Negative for production of butyrate esterase, chymotrypsinase, cystine arylamidase, ∂-fucosidase, ∂-galactosidase, ß-galactosidase, ß-glucosaminidase, ß-glucuronidase, myristate, esterase, and valine arylamidase. Negative for degradation of adenine, esculin, arbutin, chitin, chondroitin, DNA, elastin gelatin guanine, hyaluronic acid, hypoxanthine, lecithin, RNA, starch, tyrosine, and xanathine.

leucine arylamidase, a-mannosidase, and trypsinase. It degrades tributyrin, Tween 40 and 60, but not Tween 80. The organism grows at pH 7.8 (optimum), on 0.0001% crystal violet, 0.00001% Nile blue, and poorly in 1% sodium chloride (Goodfellow et al. 1985).

Isolates of *R. salmoninarum* from various parts of the world are serologically homogeneous when using polyclonal antisera (Bullock et al. 1974). However, when Arakawa et al. (1987) used monoclonal antibodies in an ELISA system, they found that in nine different *R. salmoninarum* isolates from the United States, Norway, and England, three serological groups were present. Supernate from one Norwegian clone reacted with all other isolates.

Isolation is the only method for detecting viable *R. salmoninarum,* but because of its fastidious nature, routine culture is difficult. The pathogen can be grown on general bacterial culture media, but Evelyn (1977) developed an improved medium, "kidney disease medium" (KDM-2), using peptone yeast extract agar supplemented with cysteine and bovine serum. Evelyn et al. (1989) improved *R. salmoninarum* growth on KDM-2 by introducing "cross feeding," whereby a nonfastidious feeder (nurse organism) is placed next to the BKD organism. Another innovation that further improves *R. salmoninlarum* growth on KDM-2 is addition to the medium of a small amount of KDM-2 broth in which *R. salmoninarum* had grown (supplemental–SKDM) (Evelyn et al. (1990).

At 15°C, the organism may require up to 20 days for appearance of unpigmented, creamy, shiny, smooth, raised, entire colonies measuring approximately 2 mm in diameter. Because the organism grows so slowly, precautions must be taken to ensure that culture media remains moist during incubation.

Detecting *R. salmoninarum* in subclinical carrier fish is challenging, but with improved FAT and more sensitive ELISA procedures, detection has become more accurate (Pascho and Mulcahy 1987). Elliott and Barila (1987) developed a membrane filtration–fluorescent antibody technique (MF–FAT) that detects less than 10^2 *R. salmoninarum* cells per milliliter of coelomic fluid. Lee (1989) detected less than 10^2 cells per gram of enzyme-digested kidney using FAT membrane filtrates. Treatment of kidney smears with the organic solvents xylene, acetone, or methanol for 30 to 60 minutes before FAT staining improved fluorescence intensity but did not alter detection of *R. salmoninarum* (Cvitanich 1994).

Combining MF–FAT with a culture method using SKDM can be of value in detecting *R. salmoninarum* carriers, particularly when carrier rate is low and no single test is conclusive (Teska et al. 1995). As part of a screening program to detect BKD, Griffiths et al. (1996) used kidney tissue and ovarian fluid from Atlantic salmon and compared pathogen detection by SKDM agar culture, FAT, ELISA, and

Western blot. Culture of kidney on SKDM agar identified the highest number of samples positive (34.5%) for *R. salmoninarum,* with lower sensitivity recorded by indirect IFAT (12%), ELISA (4%), and Western blot (1%). Assay of ovarian fluid showed that SKDM agar culture was best with 19.5% positive, compared with IFAT (7.5%) and ELISA (0%). In many instances, *R. salmoninarum* was detected in either kidney tissue or ovarian fluid but not both. These authors recommended using a combination of methods to increase reliability for detecting *R. salmoninarum.* By culturing ovarian fluid in SKDM broth and then applying Western blot assay, positive samples were increased by 32%.

Five methods for detecting subclinical *R. salmoninarum* in coho salmon were compared by Cipriano et al. (1985). They found that Gram stain and direct FAT were the least sensitive, and counterimmunoelectrophoresis the most sensitive. Sakai et al. (1987) modified an ELISA technique to a peroxidase–antiperoxidase procedure for detecting *R. salmoninarum* in coho salmon and found the modified method to be approximately 10 times more sensitive than FAT. The procedure also gave excellent results in a BKD survey. Eight different methods for detecting *R. salmoninarum* in supernatants from rainbow trout kidneys were compared by Sakai et al. (1989). They found that Gram stain and immunodiffusion were the least sensitive, FAT was very sensitive, but ELISA (direct and indirect) was able to detect the lowest concentration of bacterial cells when used in conjunction with a dot blot assay. The latter method could detect from 10^2 to 10^3 cells per gram of kidney tissue. In comparing a double-sandwich ELISA procedure with culture on SKDM-2 to detect *R. salmoninarum,* Gudmundsdóttir et al. (1993) sampled 1239 individual fish from 12 locations in Norway. Seven of the 12 groups tested negative by both methods, 5 groups tested positive by ELISA, and 4 were positive by culture. Within the 12 groups, a significantly higher number of fish were positive by ELISA than by culture. From these studies, it appears that no single method for detecting *R. salmoninarum* in carrier fish should be relied on but that a combination of culture, serological, or dot-blot methods should be used for most dependable results. This was further borne out by Jansson et al. (1996), who attempted to improve BKD detection, sensitivity, and reliability by comparing isolation on SKDM with commercial monoclonal and polyclonal preparations. By culture, 45% of fish were positive and by combining culture with polyclonal ELISA, 50% were positive. They determined that bacterial culture, polyclonal ELISA, and a commercially available monoclonal ELISA system were equally accurate in detecting *R. salmoninarum.*

The soluble, heat-stable 57-kD protein (p57 or F antigen) is released by *R. salmoninarum* into culture media and is considered to be a primary mediator of the bacteria's virulence (Bruno 1990). Griffiths et al. (1991) showed that in overtly infected Atlantic salmon, detection of soluble p57 by Western blot was significantly more sensitive and reliable than detection of whole cells by direct FAT in tissue homogenates. Whether the Western blot is applicable in detection of *R. salmoninarum* carrier fish with low level infections is not clear; however, about one third of epizootic survivors were seropositive for the p57 by this method. Indications are that ELISA is the most sensitive method for detecting soluble antigens in subclinical *R. salmoninarum* infections. However, p57 can cross-react with *Corynebacterium aquaticum* and *Carnobacterium piscicola,* two other Gram-positive bacteria (Bandin et al. 1993; Toranso et al. 1993). Recently, Wood et al. (1995) showed that a 60-kD protein may actually be responsible for this cross-reactivity and by using monoclonal antibody with p57, they were able to eliminate cross-reactivity. By use of rocket electrophoresis and cross adsorption analysis, Getchell et al. (1985) showed a common p57 antigen in seven isolates.

Genomic DNA was extracted from *C. aquaticum, C. piscicola,* and a Gram-negative *Pseudomonas maltophila* (Brown et al. 1995), all of which had reacted with polyclonal antisera for *R. salmoninarum* in an indirect FAT. The genomic DNA of these organisms was negative by PCR designed to amplify a segment of the gene encoding p57 of *R. salmoninarum.* These results further suggest that although *R. salmoninarum* antibodies reacted with antigens of these three bacteria, cross-reactivity is not associated with p57. To complement the high degree of specificity of *R. Salmoninarum,* Starliper (1996) showed that there is little genetic diversity in the North American isolates.

Epizootiology

Bacterial kidney disease is primarily a disease of hatchery fish, but it also occurs in wild salmonids, including anadromous populations. *R. salmoninarum* can cause overt infections or can be present only in the carrier state. In proceedings of the National Workshop on Bacterial Kidney Disease (Mangin 1991), it was noted that between 1986 and 1991, the pathogen was found at 63 locations in the United States where it caused no disease and at 97 locations where overt infections were present. Nine states reported asymptomatic presence of the pathogen, and 16 states reported its presence in either the carrier state or as overt disease. These data suggest that factors other than pathogen presence are related to BKD outbreaks and species dependency. The presence of very low numbers of *R. salmoninarum* in some brood rainbow trout without clinical disease may be explained by the possibility that unless the host becomes stressed, the pathogen remains dormant. Starliper and Teska (1995) found in a study of brook trout brood stock and their progeny that overall prevalence of *R. salmoninarum* decreased during a 3-year period. Although carrier rate ranged from 11.6% to 64.4% depending on type of detection method used and year of study, adults and progeny remained asymptomatic for the entire study.

Water quality may influence cumulative mortality and disease severity. A higher BKD incidence has been noted at hatcheries with soft water than at those using water with high total hardness; however, this relationship has not proved to be absolute (Warren 1963). Bacterial kidney disease has an adverse effect on salmon smolt survivability as they move from fresh water to salt water. During a 150-day period, coho salmon smolts experienced more than 17% cumulative mortality in seawater, compared with only 4% in a group of siblings held in fresh water (Fryer and Sanders 1981). Paterson et al. (1981) confirmed that heavy losses of Atlantic salmon smolts caused by BKD occurred as they became acclimated to seawater. Sanders et al. (1992) found that 20% of both hatchery released and wild salmonids migrating down the Columbia River were infected with *R. salmoninarum*. Prevalence in fish held in fresh water was 9%, compared with 46% prevalence in those held in salt water. Implications are that disease becomes more severe when salmon reach the marine environment but that after an initial epizootic, losses tend to decrease. In a survey of six chinook salmon hatcheries in the Columbia River and Snake River basins, Maule et al. (1996) found that prevalence of *R. salmoninarum* infections on the Snake River decreased from greater than 90% to less than 65% from the beginning to the end of a 5-year study; this decrease was attributed to initiation of health management practices on the hatcheries. It was noted that the increase in infection rate as fish migrated down the Snake River was not seen in the Columbia River. These variations may have resulted from differences in river conditions and distance from hatcheries to dams.

Bacterial kidney disease is affected by water temperature. Most epizootics in chinook and sockeye salmon occur during fall (12 to 18°C) and winter (8 to 11°C) and generally not in summer, but Smith (1964) observed that mortality accelerated in summer when temperatures rose from less than 8°C to greater than 8°C. Mortalities in cool water tend to be slow and chronic, compared with a more subacute mortality pattern in warmer water. Water temperature also influences time between exposure and death. At 11°C, death occurs 30 to 35 days postexposure, compared with 60 to 90 days at 7 to 10°C. Sanders et al. (1978) presented data concerning the relationship of temperature and mortality in BKD infections that conflicted with earlier reports. They indicated that at 6.7 to 12°C, 78% to 100% mortality occurred but that as temperatures rose to 20°C, mortality declined to a low of 8% to 14%, indicating the probability that factors other than temperature play a role in BKD-related mortality. Lovely et al. (1994) infected Atlantic salmon at 7 to 8°C and established *R. salmoninarum* carrier populations from which bacteria were isolated for up to 29 weeks, after which time only 4% of fish were carrier by immunoblot of the soluble *R. salmoninarum* antigen.

In addition to differential species susceptibility, various strains within species respond to BKD differently. Beacham and Evelyn (1992b) compared *R. salmoninarum* susceptibility in three chinook salmon strains and found substantial differences in mortality rates. Susceptibility in the most affected strain was seven times higher than in the least affected. Breeding and genetic manipulation of fish stocks to increase BKD resistance is a management technique that should be further exploited.

Several studies indicate that feed ingredients affect *R. salmoninarum* susceptibility. Nutritional studies involving Atlantic salmon indicate that insufficient amounts of dietary vitamin A, iron, zinc, iodine, and other minerals increased *R. salmoninarum* susceptibility (Paterson 1981). Woodall and

LaRoche (1964) noted that salmon receiving insufficient amounts of iodine were more susceptible to BKD, and when 4.5 mg of iodine per kg of feed was added, BKD prevalence was reduced from 95% to 3%. Addition of fluorine to feed at the same feeding rate also reduced prevalence of disease from 38% to 4% (Lall et al. 1985).

According to Evelyn et al. (1973), *R. salmoninarum* was first reported in anadromous salmonids in the United States in the early 1950s. There are indications that the disease is becoming more widespread, but this increase may be due to more sensitive detection methods than to an actual increase in geographical range. There has been a high BKD incidence reported in adult Pacific and Atlantic salmon when they return to spawn (Evelyn 1993). When salmon migrate from fresh water to seawater, infections continue to develop, with some fish succumbing to disease while at sea and those that survive returning to spawn and become a pathogen reservoir for the next generation (Banner et al. 1986).

Bacterial kidney disease also exists in wild or feral trout and salmon populations confined to fresh water. Mitchum et al. (1979) demonstrated BKD in wild populations of brook, brown, and rainbow trout in fresh water streams and when uninfected hatchery-reared trout were stocked into these waters *R. salmoninarum* was transmitted to the newly stocked fish. The newly infected rainbow trout died in 9 months or sooner after exposure. In Lake Michigan, coho salmon have been so affected by BKD that a detrimental impact on the salmon fishery has resulted (Hnath 1991). Souter et al. (1987) found *R. salmoninarum*–infected Arctic char and lake trout in the Northwest Territory of Canada, an area where no hatchery fish had been stocked. They postulated that this constituted a "natural" presence of the organism.

R. salmoninarum does not survive well outside the host; however, Austin and Rayment (1985) did isolate it for 28 days from spiked filter-sterilized river water at 15°C, and it is possible that the bacterium can survive undetected for longer periods. Balfry et al. (1996) showed that the organism survived for up to 1 week in seawater.

Although Koch's postulates for BKD were fulfilled in 1956 (Ordal and Earp 1956), horizontal transmission has been achieved with varying degrees of success. Wood and Wallis (1955) demonstrated *R. salmoninarum* transmission by feeding infected raw fish viscera to young salmon. It can also be transmitted by cohabitation in either fresh water or seawater. Bell et al. (1984) successfully transmitted *R. salmoninarum* from deliberately infected fish in cohabitation with naive fish. Recently, Balfry et al. (1996) showed that feces containing *R. salmoninarum* can significantly contribute to horizontal transmission via oral ingestion; fish naturally exposed to pathogen laden feces had 70% infectivity, compared with 98% in fish orally intubated with contaminated feces. Murray et al. (1992) successfully transmitted BKD to uninfected fish in two ways: chinook salmon were experimentally immersed in a solution containing 10^4 to 10^6 cells per milliliter for 15 to 30 minutes, and the pathogen was transmitted from injected fish to naive fish by cohabitation. Both procedures required 5 to 6 months for clinical disease to appear. *R. salmoninarum* transmission and infection is enhanced by any skin injury to the host (Hoffmann et al. 1984).

Vertical transmission of *R. salmoninarum* is important because the organism is uniquely adapted to this transmission method. It is small enough to enter the micropyle of the egg, which facilitates transmission; therefore, eggs may become infected during oogenesis or while they are surrounded by bacteria-contaminated coelomic fluid before, and during, the act of spawning (Bruno and Munro 1986; Lee and Evelyn 1989). *R. salmoninarum* can become a chronic infection, and as is the case in trout and Atlantic salmon, it can remain through multiple spawning periods or throughout the fish's life span, thus enhancing potential for vertical transmission. Allison (1958) first reported BKD in offspring from infected fish. Amos (1977) proved, however, that the pathogen could be transmitted from adults to offspring during spawning and that injection of adults with erythromycin prior to spawning prevented the pathogen from being transmitted to progeny. Bullock et al. (1978) further demonstrated vertical transmission of BKD, and Evelyn et al. (1984) also reported presence of *R. salmoninarum* within trout and salmon eggs, which further supports the case for vertical transmission.

Pathological Manifestations

Austin and Austin (1987) hypothesized that *R. salmoninarum* is a normal resident of the kidney and possibly the gastrointestinal tract of salmonids, where it remains dormant in low numbers. When the host is stressed, the pathogen will multiply in the kidney or break through the gut epithelial barrier and become systemic. This theory is based on the normally nonaggressive activity of the organism. It has also been postulated that an unidentified substance may occur in the kidneys of rainbow trout that inhibits pathogen growth, thus suppressing its ability to cause disease (Daly and Stevenson 1988). In fact, Hardie et al. (1996) showed that activated rainbow trout macrophages in vitro released H_2O_2, which aided in the inhibition of *R. salmoninarum*'s growth, suggesting that this event may be important in protective cell-mediated immunity against BKD.

R. salmoninarum infections are usually systemic and characterized by diffuse granulomatous inflammation lesions primarily in the kidney (Figure 17.7), but they can also occur in the liver, heart, spleen, and muscle. The granulomas are often large, particularly those in the kidney, with zones of caseous necrosis surrounded by epithelioid cells accompanied by lymphoid cell infiltration. These granulomas might or might not be encapsulated. Postorbital eye lesions are also characterized by granulomas with chronic inflammation and infiltration by macrophages and leukocytes (Hoffmann et al. 1984).

Chronologically, *R. salmoninarum* develops a septicemia within 4 to 11 days of experimental infection (Young and Chapman 1978). At about 18 days, bacteria will colonize on the outer surface of organs, and by day 25, they will have penetrated deeper into tissues where granulomas begin to form. These findings are congruous with long incubation periods between *R. salmoninarum* exposure and appearance of clinical disease and chronic mortality.

At 12°C, Bruno (1986) found that experimentally infected rainbow trout and Atlantic salmon exhibited initial pathological signs 10 days postinfection and suffered 75% and 80% mortality, respectively, by day 35. *R. salmoninarum* was present in the collecting ducts of kidneys and may have affected their function by causing sufficient damage to disrupt normal organ filtration. Death, therefore, was attributed to possible obliteration of normal kidney and liver structure owing to large granulomatous lesions. Death was also partly attributed to heart failure resulting from invasion of the myocardium by phagocytic cells containing *R. salmoninarum* and the release of hydrocytic and oxidizing enzymes from disrupted macrophages. In rainbow trout, Sami et al. (1992) described extended pathogenesis in which glomerulonephritis was associated with chronic BKD. Fifteen months postinfection, the following pathology was noted: fibrosis, adhesions in the Bowman's capsule, shrinkage or swelling of the capillaries, and degeneration of proximal tubules. In experimentally infected coho salmon, three different infection levels were determined by light microscopy and were characterized according to bacterial location in injured tissue (Flaño et al. 1996). Changes included destruction of hematopoietic tissue and plasmacytopoietic foci in renal and splenic tissue. Also, epithelioid and barrier cells appeared that were thought possibly to be a local defense response to the pathogen.

Wiens and Kaattari (1991), using monoclonal antibody, showed that the p57 could agglutinate salmonid leukocytes in vitro and whether or not it serves the same function in vivo is not known; however, it was suggested by Daly and Stevenson (1990) that p57 may aid *R. salmoninarum*'s ability to attach to host cells thus enabling intracellular invasion.

Significance

Because of BKD's wide occurrence in anadromous salmonid hatchery stock and wild trout populations, it is considered one of the most serious and controversial diseases of these fish. Although it seldom causes acute mortality, cumulative losses can be significant, and *R. salmoninarum*'s unique ability to be transmitted vertically within the egg, its intriguing epizootiology, and the absence of effective therapy emphasize its importance and economic impact on salmonid aquaculture.

PISCIRICKETTSIOSIS

Rickettsia have been observed in fish since 1939, but few instances of substantial infections have been reported (Almendras and Fuentealba 1997). During 1988, seawater cage cultured coho salmon in Chile experienced a significant epizootic that was eventually determined to be caused by *Piscirickettsia salmonis* and the disease called piscirickettsiosis or salmonid rickettsial septicemia (SRS) (Fryer et al. 1990; Cvitanich et al. 1991).

Geographical Range and Species Susceptibility

P. salmonis causes disease primarily in salmonids that are farmed in seawater cages (Fryer et al.

1992). The *P. salmonis* epizootics in Chile involved coho salmon, which is apparently the most susceptible species, but infections have also occurred in chinook and Atlantic salmon and rainbow trout (Garcés et al. 1991). Rickettsia infections have also been reported in Ireland (Rodger and Drinan 1993), Europe (Bravo 1994a), and British Columbia, Canada (Brocklebank et al. 1993); however, it has not been determined whether all of these rickettsialike organisms are the same. Although diseases caused by *P. salmonis* primarily affect marine fish, it has been found in freshwater coho salmon and rainbow trout (Bravo 1994b, Gaggero et al. 1995).

Clinical Signs

P. salmonis infected fish are anorexic, anemic, exophthalmic, lethargic, swim at the surface, and have pale gills and darkened skin coloration (Lannan and Fryer 1993). Skin lesions include hemorrhages on the operculum and around the anus, petechia on the abdomen, and shallow hemorrhagic ulcers on the body ranging in size from 0.5 to 2 cm in diameter. Firm, white nodules (1 cm in diameter) may also be present in the skin.

The most characteristic internal lesions are off-white to yellow subcapsular nodules that measure up to 2 cm in diameter throughout the liver (Cvitanich et al. 1991). The kidney becomes swollen, the spleen is enlarged, and the liver occasionally becomes yellow and mottled. Rodger and Drinan (1993) reported that in Ireland a rickettsialike organism infected Atlantic salmon and caused hemorrhages in the skeletal muscle. Other obvious pathology includes splenomegaly, congestion of the pyloric caeca, and paleness of the spleen and liver. Infected coho salmon and rainbow trout in fresh water had no skin ulcerations, but the abdominal cavity was distended with ascites (Gaggero et al. 1995). Petechiae and ecchymoses have been reported on serosal surfaces of the pyloric caeca, swim bladder, and hind gut of Atlantic salmon (Brocklebank et al. 1993).

Diagnosis

Diagnosis of SRS is usually by gross lesions and use of histochemical stains (H & E, Gram, Giemsa, acridine orange, Giménez, Machivello, and periodic acid-Schiff [PAS]) to detect the pathogen in tissue sections or smears (Almendras and Fuentealba 1997). The pathogen can be isolated in tissue culture for confirmation, but this process is time consuming. The most susceptible cell lines are CHSE-214 (chinook salmon embryo), CSE-119, CHH-1, and RTG-2 (rainbow trout gonad) at 15 to 18°C with up to 14 days required for cytopathic effect (CPE) to develop (Fryer et al. 1990). Cytopathic effect includes cell rounding and development of one or more vacuoles in the cytoplasm. An indirect FAT was developed for *P. salmonis* detection in blood films, tissue section, and smears from infected fish (Lannan et al. 1991). Presumably, this method could also be used to positively identify the pathogen in inoculated cell cultures. *P. salmonis* can also be detected and identified by immunohistochemistry with peroxidase–anti-peroxidase using rabbit anti–*P. salmonis* polyclonal antibody, ELISA, or PCR.

Pathogen Characteristics

The etiological agent of rickettsiallosis, *P. salmonis,* is an obligate intracellular bacterium (Fryer et al. 1992). The *P. salmonis* type strain (LF-89) has been deposited with the American Type Culture Collection as ATCC VR 1361. It is a weakly Gram-negative, pleomorphic, usually paired, coccoid cell that varies in diameter from 0.5 to 1.5 µm. The organism is also nonmotile, unencapsulated and PAS, acid fast, and Gieménez-negative but stains well with H & E, Giemsa, and methylene blue (Cvitanich et al. 1991).

Lannan and Fryer (1994) determined that extracellular survival of *P. salmonis* lasted a maximum of 3 weeks at 5°C and 1 week at 20°C when suspended in cell-free Eagle's minimum essential medium with 10% fetal bovine serum. Viability of *P. salmonis* in fresh water was destroyed almost immediately, which may help explain why SRS is seldom seen in freshwater fish.

Epizootiology

Piscirickettsiosis occurred from May to November in Chile, and cumulative losses at some farms reached 90%, with up to 40% mortality per month (Cvitanich et al. 1991). Disease first appeared about 6 to 12 weeks after fish were transferred from fresh water to salt water cages.

Experimental SRS transmission has been successful via IP injection, but cohabitation transmission experiments have yielded mixed results. Garcés et al. (1991) failed to observe transmission by cohabitation, but experimental transmission studies by Cvitanich et al. (1991) were successful. In naturally occurring outbreaks, SRS occurs about 2 weeks after smolts are transferred to seawater cages where they are exposed to infected fish. It was suggested that naturally occurring transmission is oral. Vertical transmission of *P. salmonis* was thought to have occurred in coho salmon when 98.3% of fingerlings from SRS-positive brood stock were positive, but only 27% of fingerlings from negative brood fish were positive (Almendras and Fuentealba 1997).

Two salmonid species (coho salmon and Atlantic salmon) were tested for *P. salmonis* susceptibility by Garcés et al. (1991). Although typical clinical disease developed only in the coho salmon, both species suffered nearly 100% mortality at the highest pathogen concentration. The organism was recovered in pure culture from infected fish of both species. More recently, an experimental challenge of 16- to 17-g coho salmon and rainbow trout was carried out in fresh water by Smith et al. (1996). After IP injection, clinical disease occurred in both species in the third week. At the highest pathogen concentration, cumulative mortality was 60% in the coho salmon and 28% in the rainbow trout.

Piscirickettsiosis is generally considered a disease of salmon in salt water, but Gaggero et al. (1995) indicated that coho salmon and rainbow trout developed the disease in fresh water when 60 to 90 days old. They noted that mortality usually occurred in fish 6 to 12 weeks after introduction to seawater cages. *P. salmonis* and BKD occur in the same fish in seawater cultured salmonids in Chile (Smith et al. 1996).

Pathological Manifestations

Hematopoietic tissue becomes necrotic with inflammation, and *P. salmonis* cells can be seen in cytoplasmic vacuoles in histological sections or in imprints and smears from the kidney, brain, skeletal muscle, skin, heart, intestine, gills, and ovary (Cvitanich et al. 1991; Lannan and Fryer 1993). Histopathological lesions show necrosis of intravascular thrombi (Branson and Diaz-Munoz 1991). Necrosis was most obvious in spleens, livers, and kidneys of infected Atlantic salmon in Ireland (Rodger and Drinan 1993). Actual pathogenesis of SRS has not been established.

Significance

When piscirickettsiosis first appeared in Chile, it killed approximately 1.5 million coho salmon and has continued to have a major economic impact on the salmon industry there. Its presence in other geographical areas has also caused concern wherever trout and salmon are cultured.

OTHER BACTERIAL DISEASES OF SALMONIDS

Several bacterial diseases that affect salmonids but are more important in other fish groups are discussed in greater depth in the following chapters: motile *Aeromonas* septicemia (*Aeromonas hydrophila*) and columnaris (*Flavobacterium columnare*) (Chapter 14), ulcer disease (atypical *Aeromonas salmonicida*) (Chapter 15), edwardsiellosis (*Edwardsiella tarda*) (Chapter 16), and mycobacteriosis (*Mycobacterium marinum*) (Chapter 18).

A disease of trout and Pacific salmon known as pseudokidney disease was recognized in the mid-1970s (Hiu et al. 1984). The causative agent of this disease was named *Lactobacillus piscicola* and reclassified as *Carnobacterium piscicola* (Wallbanks et al. 1990). The organism has been reported in North America and Europe (Hiu et al. 1984; Foott 1994). Primary susceptible species are coho and chinook salmon and rainbow and cutthroat trout but other marine and freshwater fish may also be susceptible under certain conditions. Fish older than 1 year are most often infected, but Michel et al. (1986) showed that very young salmonids and carp may also be affected by the organism. External clinical signs are abdominal distension, erythema at the base of fins, and subdermal blood blisters. Internally the liver, spleen, and kidney may be enlarged and the peritoneal cavity often contains ascitic fluid. Hemorrhages may occur in the testes, intestine, and musculature.

Isolation of *C. piscicola* is accomplished by streaking kidney or lesion material on TSA or BHI agar and incubating it aerobically at 15 to 24°C for 24 to 72 hours (Foott 1994). Colonies are pinpoint, opaque, entire, circular, and nonpigmented. The bacterium is a nonmotile, non–spore-forming, non–acid fast, Gram-positive rod or coccobacillus that measures 1.1–1.4 × 0.5–0.6 μm in size. As cultures age, the bacteria becomes Gram variable. The organism is negative for oxidase, catalase, urease, H_2S, nitrate reduction, and lactose fermentation (Hiu et al. 1984). It is positive for arginine dihydrolase and lactic acid production from glucose, maltose, mannitol, and sucrose. Disease caused by *C. piscicola* is usually associated with environmental or spawning-induced stress. The role of *C. piscicola* in fish diseases appears to be genuine, although experimental infections are not easily established.

MANAGEMENT OF SALMONID BACTERIAL DISEASES

The philosophical approach to health maintenance and control of salmonid bacterial diseases does not differ from that of nonsalmonids, but more management, chemotherapy, and vaccination options are available. To best control these diseases, a comprehensive program should include good management procedures, proper use of drugs, and vaccination when indicated.

Management

Some salmonid bacterial infections are caused by obligate pathogens and can produce clinical disease without adverse environmental influences, but many epizootics are stress related. In view of this fact, it is important to maintain a high standard of water quality and other associated environmental conditions because overall management practices are critical in controlling, preventing, and reducing effects of disease. To manage most bacterial trout and salmon diseases successfully, the water temperature should be maintained below 17°C, moderate stocking densities should be used, aeration should take place when needed to provide near 100% oxygen saturation, and an adequate water flow should be maintained to remove metabolites, uneaten feed, and solid waste. Eggs, fry, and fingerlings should be cultured in water that is fish free whenever possible.

Management of facilities where *A. salmonicida* is enzootic should include stocking fish strains that are more resistant to furunculosis. Destruction of *A. salmonicida*-infected fish, facility sanitation, and restrictions on egg use from infected brood stock have been successful in maintaining a disease rate of 2% on "health-controlled" farms in Sweden (Wichardt et al. 1989). High sanitation standards should be a top priority in prevention and/or control of bacterial disease. Newly arrived eggs should be disinfected to kill any surface-associated bacteria. In the United States, the most reliable disinfectants for this purpose are buffered polyvinyl pyrrolidon (PVP) iodophores (McFaddin 1969). Eyed eggs should be immersed in 25 to 200 mg of iodine/L for at least 10 minutes using well-oxygenated chlorine-free water, rinsed immediately, and put into flowing water incubators.

Because opinions concerning the etiological agent of coldwater vibriosis differ and because it is a multifactorial disease, its severity depends to a large extent on fish culture practices and environmental conditions (Hjeltnes et al. 1987a). H. Mitchell (personal communication) suggested that to minimize effects of CV, early disease detection is essential, and once confirmed, all handling of fish should be avoided. Moribund and dead fish should be promptly removed and buried, incinerated, or composted. Once surviving fish are removed from cages, nets should be cleaned and sanitized by drying in sunlight, and all clothing and/or equipment exposed to diseased fish should be disinfected. Also, effluence from slaughter areas should be disinfected to prevent large numbers of bacteria from being reintroduced into culture areas.

Enteric redmouth disease control dictates that ponds and raceways be kept clean, culture unit loading limits not be exceeded, the highest quality environmental conditions including adequate water flow be maintained, nets and other utensils be sanitized, and facilities be disinfected following fish removal. Also, removal of all possible *Y. ruckeri* carrier fish or invertebrates from the water supply is essential.

Reducing environmental stress, maintaining a high-quality water supply with adequate water flow, and using proper sanitation are the best methods for controlling bacterial gill disease (Snieszko 1981). A water supply should originate from a fish-free source, have a high oxygen concentration level and low levels of accumulated metabolites such as ammonia, be free of abrasive suspended solids, and be sufficient to accommodate the chosen fish density. Feed with excessive "fines" should not be used for BGD-susceptible fish, and uneaten feed should not be allowed to accumulate in tanks.

Increasing water temperatures above the optimum range is one way to prevent bacterial coldwater disease, but K. Amos (personal communication) also suggests that it is beneficial to use a substrate when incubating eggs, keep loading densities down for yolk sac fry incubation, minimize yolk sac activity, and maintain minimum flow rates. Also, coho salmon fry infections were less severe when fish were held in shallow, rather than deep, troughs. From these observations, it can be concluded that any procedures that reduce egg, skin, and fin abrasions of juvenile fish are helpful in reducing incidence and severity of *F. psychrophilum*.

Because of its complex epizootiological nature, BKD is one of the most difficult infectious fish diseases to manage. Control requires a variety of integrated management procedures that include reducing the possibility of vertical and horizontal transmission, chemoprophylaxis, chemotherapy when applicable, and environmental and population manipulation (Elliott et al. 1989). Owing to improved detection sensitivity to *R. salmoninarum* in brood stock, "BKD free certification" is now more feasible. In hatchery-reared salmonids, it is important to reduce potential vertical transmission during artificial spawning activities by culling and by segregating brood fish and eggs. Gametes from brood fish that are most likely to transmit *R. salmoninarum* to offspring should be destroyed. Each brood fish should be tested for *R. salmoninarum* at spawning, and their gametes should be held separately under refrigeration until BKD status of each adult is known. Gametes from BKD-positive fish are destroyed, and those of BKD-negative fish are fertilized. Pascho et al. (1991) suggested that using ELISA and FAT to test adults for *R. salmoninarum* in conjunction with segregation of eggs significantly reduces disease prevalence in chinook salmon to an acceptable level. It was also shown that BKD testing must be done when eggs are taken to get a correct measure of carrier status of brood stock (O'Halloran et al. 1995).

The National Workshop on Bacterial Kidney Disease addressed BKD management and recommended the following for implementation by the U.S. Fish and Wildlife Service (Mangin 1991):

1. There is no reason to destroy stocks of fish that are infected with *R. salmoninarum*.
2. Approved *R. salmoninarum* detection methods in fish include ELISA, FAT, and FAT with membrane filtration.
3. If *R. salmoninarum* is present at any brood stock facility, stocks should be managed to reduce infection severity, and suspect carrier fish should be stocked only in areas where the pathogen is already known to exist.
4. Eggs should not be taken from clinically diseased fish unless a specific program (such as endangered species) requires use of such eggs.
5. Asymptomatic infected fish should not be stocked in areas that are free of *R. salmoninarum*.
6. In the United States, no matter the locale, all BKD problems should be treated the same.

According to Almendras and Fuentealba (1997), effects of piscirickettsiosis can be reduced by decreasing stress during grading and handling, decreasing biomass and stocking density, and removal and proper disposal of dead and moribund infected fish. Also, proper disposal of blood and offal from harvested and dressed fish will eliminate an infection source.

Chemotherapy

In the United States, chemotherapy for furunculosis, vibriosis, ERM, CV, and other bacterial salmonid diseases is limited to feeding Terramycin and Romet-30. However, several other drugs are being used under Investigational New Animal Drug (INAD) exemption (see Table 5.3) (Schnick et al. 1989; Guide to Drug, Vaccine, and Pesticide Use in Aquaculture 1994). For furunculosis, vibriosis, and ERM, Terramycin is fed at a rate of 50 to 75 mg/kg of fish per day for 10 days, followed by a 21-day withdrawal period before harvest and marketing. Romet-30 is fed at 50 mg/kg of fish per day for 5 days with a 42 day withdrawal period. Before chemotherapy is applied, it is recommended that antibiotic sensitivity for a particular isolate of any bacterial pathogen be determined to detect antibiotic resistance. Antibiotics should never be used indiscriminately, for long durations, or at less-than-recommended concentrations because these practices lead to increased drug resistance (Inglis and Richards 1992).

Additional drugs to treat salmonid bacterial infections are approved in other countries but can be used only experimentally in the United States. Oxolinic acid is used extensively in Europe at 10 mg/kg body weight per day and several fluoroquinolones have been examined for their efficacy against furunculosis. Enrofloxacin was shown by Bowser et al. (1994) to significantly ($P < .05$) increase survival of experimentally *A. salmonicida*–infected rainbow trout when fed at rates of 1.25 to 2.5 mg/kg of fish daily for 10 days. During reduced feeding activity in winter, enrofloxacin fed at 10 mg/kg · d to lake trout and 5 mg/kg · d to Atlantic salmon for 20 days significantly reduced mortality due to furunculosis (Hsu et al. 1994). A second fluoroquinolone, difloxacin, was evaluated for controlling *A. salmonicida* in Atlantic salmon by Elston et al. (1995). Survival was significantly increased by feeding difloxacin at dosages of 2.5 to 5.0 mg/kg of fish daily for 5 days. No increased benefit was achieved by increasing medication for 10 days. Amoxicillin is being seriously considered for fish. Barnes et al. (1994) found that all *A. salmonicida* subsp. *salmonicida* tested in Scotland were sensitive to minimum inhibitory concentrations of 0.3 to 1.5 mg of amoxicillin per liter of water. Florfenicol (Aqua Flor) is approved to treat furunculosis in Canada, where it is fed at 10 mg/kg of body weight for 10 days.

Cipriano et al. (1992) showed that *A. salmonicida* resides in the mucus of carrier trout. The monitoring of an Atlantic salmon population for *A. salmonicida* via mucous analysis and subsequent oral treatment with oxytetracycline (77 mg/kg of fish for 10 d) and topical treatment with Chloramine-T (INAD exemption) (15 mg/L) in a 60-minute bath was described by Cipriano et al. (1996b and c). These treatments reduced the presence of *A. salmonicida* in the mucus by 90% to 100%, allowing affected fish to be stocked in accordance with established fish health regulations for salmonids in New England.

Terramycin is generally used to treat coldwater vibriosis. However, Poppe et al. (1986) used Terramycin, Tribrissen, or furazolidone (the latter two are not approved in the United States) in feed at 75 to 100 mg/kg of body weight per day for 10 days and found that furazolodone was most effective against CV. However, Bruno et al. (1986) held cumulative mortality to 3% with Terramycin. Mortality due to *V. salmonicida* in Atlantic salmon can be reduced in 3 to 4 days by feeding oxolinic acid at 75 to 90 mg/kg of body weight per day for 10 days. Because a variety of antibiotics have been used to control CV, multiple antibiotic resistance of *V. salmonicida* to oxytetracycline, sulfonamides, and oxolinic acid have emerged (Hjeltnes et al. 1987b).

Bacterial gill disease therapy is effective when applied expediently, as infected fish tend to recover rapidly after gill bacteria are reduced. Disinfectants containing quaternary ammonia, such as benzalkonium chlorides (Hyamine 1622, Hyamine 3500, or Roccal), can be added to water as a prolonged treatment. A 1 to 2 mg/L application of active benzalkonium chloride for 1 hour on 3 consecutive days is recommended; however, caution must be taken in soft water because the compound can be toxic to some trout. The herbicide Diquat has been used successfully in treating BGD at 2 to 4 mg/L of active ingredient. Potassium permanganate can also be used at 2 mg/L for 1 hour, but the safety margin is very narrow. Chloramine-T (INAD exemption) applied at 8 to 10 mg/L for 1 hour for 3 days is the most effective treatment for BGD (R. Holt, personal communication); however, 15 to 20 mg/L are required in stubborn cases in cold water (K. Amos, personal communication). Byrne et al. (1991) suggested that salt concentrations in water should be increased during BGD outbreaks or during chemotherapy treatment to compensate for loss of fish electrolytes. None of the aforementioned compounds are Food and Drug Administration (FDA) approved for therapeutic use against BGD if fish are destined for United States markets, but as noted, some are used under INAD exemption.

Serious outbreaks of BCWD are generally difficult to treat because infections are systemic and affected fish feed poorly, thus neutralizing any advantage of medicated feed (Holt 1993). Bullock and Snieszko (1970) and Schachte (1983) recommended bath treatments with water-soluble oxytetracycline at 10 to 50 mg/L or quaternary ammonium compounds at 2 mg/L while infections are confined to the skin. After disease becomes systemic, oxytetracycline in the diet at 50 to 75 mg/L of fish per day for 10 days is effective if fish are feeding. Amoxicillin is incorporated in the feed at 50 µg/kg · d for 10 days in Great Britain. Oxytetracycline is injected into adult fish at 50 mg/kg prior to spawning to prevent vertical transmission.

Chemoprophylaxis is used to reduce vertical transmission of *R. salmoninarum*. Adult female salmon are injected once or twice with 10 to 20 mg of erythromycin phosphate per kg of body weight at 3- to 4-week intervals prior to spawning, with a 10-day window between last injection and spawning (Amos 1977; Evelyn et al. 1986; Lee and Evelyn 1994). Erythromycin has also been widely used as an oral treatment at 4.5 g/45 kg (about 100 mg/kg) of body weight per day for 10 to 21 days for clinical BKD (Elliott et al. 1989). Moffitt (1992) found that erythromycin fed at 200 mg/kg of body weight per day for 21 days reduced mortality more than when lower dosages were fed for shorter periods.

Subsequently, Peters and Moffitt (1996) showed that an optimum erythromycin dosage for BKD-infected coho salmon was 100 mg/kg daily for 28 days. They noted that a higher dose was unpalatable, particularly late in the treatment regimen or when water temperature was higher than 10°C. Protection against *R. salmoninarum* lasted for about 3 months at 10°C and 2 months at 14°C before BKD mortality reoccurred. Experiments to treat *R. salmoninarum* with enrofloxacin (fluoroquinolone) by Hsu et al. (1994) showed that feeding the drug at 1.25 to 2.5 mg/kg of body weight per day for 10 days reduced mortality, and when fed up to 20 mg/kg · d, mortality was further reduced but palatability problems occurred with 100 mg/kg of fish. Overall, chemotherapy of clinically infected BKD fish is not overly successful because it does not totally eliminate *R. salmoninarum* from all treated fish and relapses can be expected following medication (Austin 1985).

Among drugs currently used for *P. salmonis,* oral application of the quinolones, oxolinic acid, and flumequine appear to be the most effective (Almendras and Fuentealba 1997). However, losses have gradually increased, which is in part due to antibiotic resistance of the pathogen. Injection of enrofloxacin has shown some efficacy.

Vaccination

Vaccination of salmonids has made major inroads in controlling and preventing several diseases of these fish (see Table 5.5). All salmonid diseases do not respond favorably to vaccination. In a treatise on vaccinations in Europe, Press and Lillehaug (1995) pointed out that application of modern molecular technology to fish vaccines has produced products that include purified virulence factors that have increased protection against certain pathogens and are being used in situations involving large numbers of fish. Commercial vaccines are now available for vibriosis, CV, ERM (yersiniosis), and furunculosis and are being used extensively throughout the world where salmonids are cultured with an excellent cost-to-benefit ratio. The highly successful sea-cage salmonid culture in Europe, the United Kingdom, the United States, and other places can in large part be credited to vaccination.

Generally, injectable furunculosis vaccines are most effective, but application on a production scale has been cost-prohibitive unless fish are individually valuable. Rodgers (1990) used a vaccine composed of whole cells and extracellular products to significantly enhance protection from a natural challenge of *A. salmonicida* in juvenile rainbow trout. Mortality of vaccinated fish was about 11%, compared with 37% for unvaccinated controls. Vaccinated fish also grew more rapidly than did nonvaccinates. Paterson et al. (1992), in laboratory and field trials, used a pelletized diet containing a dried, coated *A. salmonicida* culture preparation to vaccinate orally Atlantic salmon against furunculosis. Press et al. (1996) compared IP injection, immersion, and oral application of monovalent and trivalent vaccines and found the trivalent preparation to be the only one that led to high levels of specific antibody and possible immunological enhancement.

When Atlantic salmon were vaccinated by injection for furunculosis, they experienced a temporary immunosuppression that resulted in subclinical carrier fish (Inglis et al. 1996). The researchers found that feeding amoxicillin following vaccination improved survival to more than 80%, compared with 0% survival for nonmedicated fish. As a result of vaccination followed by drug treatment, a relative percent survival of 86% was achieved in these fish when challenged 4 months postvaccination.

Vaccination against vibriosis, particularly *V. anguillarum* and *V. ordalii,* has become an accepted practice for cultured salmon. Studies have also demonstrated its positive effect against vibriosis in other species as well, including ayu, eels, milkfish, and striped bass (Kawano et al. 1984; Tiecco et al. 1988; Rogers and Xu 1992). Initially, vaccination was accomplished by injection of a formalin-killed bacterin into salmon smolts before seawater transfer (Rohovec et al. 1975). Vaccination by immersion and/or spraying, however, has proved to be more efficient and cost-effective when done on a large scale (Amend and Johnson 1981). Bivalent vibrio vaccines containing *V. anguillarum* (types I and II) and *V. ordalii* (type III) (see Table 5.5) are commercially available. Vaccinating salmon at appropriate times against vibriosis can improve survival as much as 90%. In some vaccinations of coho salmon, mortalities were reduced from 52% in unvaccinated controls to 4% in fish vaccinated for 20 to 30 seconds by immersion in a preparation containing 1% vaccine. Horne et al. (1982) reduced *V. anguillarum* mortalities from 100% in unvaccinated rainbow trout to 53% in immersion vaccinated fish. Ayu and rainbow trout were also successfully vaccinated against *V. anguillarum* by immersion and IP injection by Muroga et al. (1995). Humoral agglutinating antibody resulted from both vaccination methods, and significant protection was demonstrated by immersion challenge.

Evidence exists that salmonid vaccination against *V. anguillarum* and *V. ordalii* not only significantly reduces mortality but vaccinated fish often have increased growth and lower feed conversion ratios (Håstein et al. 1980). Thorburn et al. (1987) noted that on Swedish marine net-cage farms, the decision to vaccinate depends on farm size and anticipated vibriosis risk. Vaccination against vibriosis is an appropriate management procedure when fish are to be exposed to an environment in which *V. anguillarum* or *V. ordalii* are indigenous. Fish that are to be transferred to salt water should be vaccinated against vibriosis 2 to 4 weeks prior to release. Salmonids can be vaccinated at any size greater than 2 g, but for best and most economical results, it should be done when fish are less than 200 fish/kg.

Vaccination of Atlantic salmon with formalin-killed whole-cell bacterins has shown promise and is used as the primary means for preventing *V. salmonicida* in the species. Before stocking Atlantic salmon into sea net cages, immersion vaccination reduced *V. salmonicida* mortality from 7.8% to 0.4% (Holm and Jørgensen 1987). When Lillehaug et al. (1990) vaccinated Atlantic salmon on Norwegian fish farms, CV mortality was reduced from 24.9% in nonvaccinated fish to 1.87% in vaccinated groups. Hjeltnes et al. (1989) showed that vaccination by injection afforded the most dependable protection against *V. salmonicida,* but double immersion was probably more practical and economical. Schroder et al. (1992) pointed out that vaccination of Atlantic salmon against *V. salmonicida* is a practical and beneficial management tool, even though protective immunity breaks down in 1.5 to 2 years.

Enteric redmouth was one of the first fish diseases to be managed with vaccine (Ross and Klontz 1965). A commercial ERM vaccine was introduced in 1976 and has since become an integral part of disease control in cultured salmonids in the United States, Great Britain, Scotland, Scandinavia, and other parts of Europe (Horne and Robertson 1987) (see Table 5.5). To protect against ERM, trout are vaccinated by immersion in a killed bacterin (Johnson et al. 1982a). The most cost effective fish size for vaccination is 4 to 4.5 g, but fish up to 200 g can be vaccinated. Trout smaller than 4 g can be vaccinated, but protection will not be as lasting; for 1.0-g fish, protection will last 4 months; 2.0 g, 6 months; and 4.0 g, 12 months (Johnson et al. 1982b). Fish should be either immersed for 30 seconds or sprayed with vaccine. A secondary immune response that acts as a booster vaccination followed exposure to living *Y. ruckeri* for up to 7 months after initial vaccination (Lamers and Muiswinkel 1984).

In a study of *Y. ruckeri* vaccinates, Tebbit et al. (1981) showed an 84% reduction in ERM mortalities, a 77% reduction in need for medication, and a 13.7% lower food conversion rate. These added values resulting from ERM vaccinations were confirmed by Horne and Robertson (1987). In a trout farm survey in the United Kingdom, approximately half the farms that vaccinated against ERM believed it failed to protect fish against disease. Failure was usually attributed, however, to poor condition of the fish and low water temperatures at time of vaccination (Rodgers 1991). A detrimental side effect of trout vaccination against ERM is that subclinical IPNV or infectious hematopoietic necrosis virus (IHNV) infections can be exacerbated into a clinical state with potential for mortality (Busch 1983).

Since the first experimental vaccination for BKD was reported by Evelyn (1971b), there have been attempts to examine vaccines potential for controlling this disease (Paterson et al. 1980a; McCarthy et al. 1984). Some successful vaccinations have been reported, but overall results have not been encouraging; consequently, no vaccine is currently available for BKD. It was found by Wood and Kaattari (1996), however, that the removal of p57 from *R. salmoninarum* cells enhanced its immunogenicity and resulted in a 20-fold increase in detectable antibody titers. Tests indicated that the antibody almost exclusively reacted with carbohydrate moieties on p57-negative cells, leading to the conclusion that removal of the virulence factors from *R. salmoninarum* enhances antibody response in fish and is another step toward vaccine development.

There has been very little effort to develop a vaccine for *F. branchiophilum*. It was noted by Heo et al. (1990) that survivors of BGD episodes were susceptible to subsequent pathogen challenge. On the other hand, Lumsden et al. (1994) found that fish that had been previously exposed to live *F. branchiophilum*, had received *F. branchiophilum*–specific serum intravenously, or those that had been bath vaccinated, experienced declining mortality in subsequent exposure, compared with unexposed controls. Also, fish that had been bath vaccinated three times by immersion were almost completely protected from experimental challenge. These researchers recognize the vaccination potential for BGD but emphasized that additional research is needed.

Trout and salmon vaccination against bacterial coldwater disease is under investigation and has shown some promise. Holt (1993) reported successful vaccination with a formalin-killed bacterin of *F. psychrophilum* by immersion and injection. Injection (with Freund's adjuvant) produced complete protection against challenge, compared with 43% mortality in nonvaccinated controls; however, immersion resulted in only 11% improved survival. Obach and Laurenci (1991) reported that 40-day-old posthatch rainbow trout were not protected by immersion vaccination with a heat-inactivated preparation of *F. psychrophilum*. Practical vaccination against BCWD is very difficult because yolk sac fry frequently contract the disease before immunity is possible unless it has been passively acquired from brood stock.

REFERENCES

Allen-Austin, D., B. Austin, and R. R. Colwell. 1984. Survival of *Aeromonas salmonicida* in river water. *FEMS Microbiological Letters* 21:143–146.

Allison, L. N. 1958. Multiple sulfa therapy of kidney disease among brook trout. *The Progressive Fish Culturist* 20:66–68.

Almendras, F. E., and I. C. Fuentealba. 1997. Salmonid rickettsial septicemia caused by *Piscirickettsia salmonis*: a review. *Diseases of Aquatic Organisms* 29:137–144.

Amend, D. F., and K. A. Johnson. 1981. Current status and future needs of *Vibrio anguillarum* bacterins. *Developments in Biological Standardization* 49:403–417.

Amos, K. 1977. The control of bacterial kidney disease in spring chinook salmon. Moscow, ID: University of Idaho. M.S. thesis.

Anderson, D. P. 1972. Virulence and persistence of rough and smooth forms of *Aeromonas salmonicida* inoculated into coho salmon (*Oncorhynchus kisutch*). *Journal of the Fisheries Research Board of Canada* 29:204–206.

Anderson, J. I. W., and D. A. Conroy. 1970. Vibrio disease in marine fishes. In: *A Symposium on Diseases of Fishes and Shellfishes*, edited by S. F. Snieszko, 266–272. Special publication no. 5. Bethesda, MD: American Fisheries Society.

Aoki, T., and F. I. Holland. 1985. The outer membrane proteins of the fish pathogens *Aeromonas hydrophila, Aeromonas salmonicida* and *Edwardsiella tarda. FEMS Microbiology Letters* 27:299–305.

Arakawa, C. K., J. E. Sanders, and J. L. Fryer. 1987. Production of monoclonal antibodies against *Renibacterium salmoninarum. Journal of Fish Diseases* 10:249–253.

Argenton, F., S. DeMas, C. Malocco, L. Dalla Valle, G. Giorgetti, and L. Colombo. 1996. Use of random DNA amplification to generate specific molecular probes for hybridization tests and PCR-based diagnosis of *Yersinia ruckeri. Diseases of Aquatic Organisms* 24:121–127.

Austin, B. 1985. Evaluation of antimicrobial compounds for the control of bacterial kidney disease in rainbow trout, *Salmo gairdneri* Richardson. *Journal of Fish Diseases* 8:209–220.

Austin, B. 1992. The recovery of *Cytophaga psychrophila* from two cases of rainbow trout (*Oncorhynchus mykiss* Waulbaum) fry syndrome in the UK. *Bulletin of the European Association of Fish Pathologist.* 12:207–208.

Austin, B., and D. A. Austin. 1987. *Bacterial Fish Pathogens: Diseases in Farmed and Wild Fish.* Chichester, UK: Ellis Horwood.

Austin, B., and J. Rayment. 1985. Epizootiology of *Renibacterium salmoninarum,* the causal agent of bacterial kidney disease in salmonid fish. *Journal of Fish Diseases* 8:505–509.

Austin, B., I. Bishop, C. Gray, B. Watt, and J. Dawes. 1986. Monoclonal antibody-based enzyme-linked immunosorbent assays for the rapid diagnosis of clinical cases of enteric redmouth and furunculosis in fish farms. *Journal of Fish Diseases* 9:469–474.

Balfry, S. K., L. J. Albright, and T. P. T. Evelyn. 1996. Horizontal transfer of *Renibacterium salmoninarum* among farmed salmonids via the fecal-oral route. *Diseases of Aquatic Organisms* 25:63–69.

Bandin, I., Y. Santos, J. L. Barja, and A. E. Toranzo. 1993. Detection of a common antigen among *Renibacterium salmoninarum, Corynebacterium aquaticum,* and

Carnobacterium pescicola by the western blot technique. *Journal of Aquatic Animal Health* 5:172–176.

Banner, C. R., J. J. Long, J. L. Fryer, and J. S. Rohovec. 1986. Occurrence of salmonid fish infected with *Renibacterium salmoninarum* in the Pacific Ocean. *Journal of Fish Diseases* 9:273–275.

Barnes, A. C., T. S. Hastings, and S. G. B. Amyes. 1994. Amoxicillin resistance of Scottish isolates of *Aeromonas salmonicida. Journal of Fish Diseases* 17:357–363.

Beacham, T. D., and T. P. T. Evelyn. 1992a. Population and genetic variation in resistance of chinook salmon to vibriosis, furunculosis, and bacterial kidney disease. *Journal of Aquatic Animal Health* 4:153–167.

Beacham, T. D., and T. P. T. Evelyn. 1992b. Population variation in resistance of pink salmon to vibriosis and furunculosis. *Journal of Aquatic Animal Health* 4:168–173.

Belding, D. L., and B. Merill. 1935. A preliminary report upon a hatchery disease of Salmonidae. *Transactions of the American Fisheries Society* 65:76–84.

Bell, G. R., D. A. Higgs, and G. S. Traxler. 1984. The effect of dietary ascorbate, zinc, and manganese on the development of experimentally induced bacterial kidney disease in sockeye salmon (*Oncorhynchus nerka*). *Aquaculture* 36:293–311.

Bell, G. R., R. W. Hoffmann, and L. L. Brown. 1990. Pathology of experimental infections of the sablefish, *Anoplopoma fimbria* (Pallas), with *Renibacterium salmoninarum,* the agent of bacterial kidney disease in salmonids. *Journal of Fish Diseases* 13:355–367.

Bergh, Ø., B. Hjeltnes, and A. B. Skiftesvik. 1997. Experimental infection of turbot *Scophthalmus maximus* and halibut *Hippoglossus hippoglossus* yolk sac larvae with *Aeromonas salmonicida* subsp. *salmonicida. Diseases of Aquatic Organisms* 29:13–20.

Bernardet, J.-F., and P. A. D. Grimont. 1989. Deoxyribonucleic acid relatedness and phenotypic characterization of *Flexibacter columnaris* sp., nov., nom. rev., *Flexibacter maritimus* Wakabyashi, Hikida and Masumura 1986. *International Journal of Systematic Bacteriology* 39:346–354.

Bernardet, J.-F., and B. Kerouault.1989. Phenotypic and genomic studies of *Cytophaga psychropila* isolated from diseased rainbow trout *Oncorhynchus mykiss* in France. *Applied and Environmental Microbiology* 55:1796–1800.

Bernardet, J.-F., P. Segers, M. Vancanneyt, F. Berthe, K. Kersters, and P. Vandamme. 1996. Cutting a Gordian knot: emended classification and description of the genus *Flavobacterium,* emended description of the family Flavobacteriaceae, and proposal of *Flavobacterium hydatis* nom. nov. (Basonym, *Cytophaga aquatilis* Strohl and Tait 1978). *International Journal of Systematic Bacteriology* 46:128–148.

Bertolini, J. M., H. Wakabayashi, V. G. Watral, M. J. Whipple, and J. S. Rohovec. 1994. Electrophoretic detection of proteases from selected strains of *Flexibacter psychrophilus* and assessment of their variability. *Journal of Aquatic Animal Health* 6:224–233.

Bowser, P. R., G. A. Wooster, and H. M. Hsu. 1994. Laboratory efficacy of enrofloxacin for the control of *Aeromonas salmonicida* infection in rainbow trout. *Journal of Aquatic Animal Health* 6:288–291.

Branson, E. J., and D. Nieto Diaz-Munoz. 1991. Description of a new disease condition occurring in farmed coho salmon, *Oncorhynchus kisutch* (Walbaum), in South America. *Journal of Fish Diseases* 14:147–156.

Bravo, S. 1994a. Piscirickettsiosis in freshwater. *Bulletin of the European Association of Fish Pathologist* 14:137–138.

Bravo, S. 1994b. First report of *Piscirickettsia salmonis* in freshwater. *FHS/AFS Newsletter* 22:6.

Bricknell, I. R., D. W. Bruno, and J. Stone. 1996. *Aeromonas salmonicida* infectivity studies in gold sing Wrasse, *Ctenolabrus rupestris* (L.). *Journal of Fish Diseases* 19:469–474.

Brocklebank, J. R., T. P. Evelyn, D. J. Spear, and R. D. Armstrong. 1993. Rickettsial septicemia in farmed Atlantic and chinook salmon in British Columbia: clinical presentation and experimental transmission. *Canadian Veterinary Journal* 34:745–748.

Brown, L. L., T. P. T. Evelyn, G. K. Iwama, W. S. Nelson, and R. P. Levine. 1995. Bacterial species other than *Renibacterium salmoninarum* cross-react with antisera against *R. salmoninarum* but are negative for the p57 gene of *R. salmoninarum* as detected by the polymerase chain reaction (PCR). *Diseases of Aquatic Organisms* 21:227–231.

Bruno, D. W. 1986. Histopathology of bacterial kidney disease in laboratory infected rainbow trout, *Salmo gairdneri,* Richardson, and Atlantic salmon, *Salmo salar* L. with reference to naturally infected fish. *Journal of Fish Diseases* 9:523–537.

Bruno, D. W. 1990. Presence of a saline extractable protein associated with virulent strains of the fish pathogen *Renibacterium salmoninarum. Bulletin of the European Association of Fish Pathologists* 10(2):8–10.

Bruno, D. W., and A. L. S. Munro. 1986. Observations on *Renibacterium salmoninarum* and the salmonid egg. *Diseases of Aquatic Organisms* 1:83–87.

Bruno, D. W., T. S. Håstings, and A. E. Ellis. 1986. Histopathology, bacteriology and experimental transmission of cold-water vibriosis in Atlantic salmon *Salmo salar. Diseases of Aquatic Organisms* 1:163–168.

Bullock, G. L. 1972. Studies on selected myxobacteria pathogenic for fishes and on bacterial gill disease in hatchery-reared salmonids. *Bureau of Sport Fisheries and Wildlife, Technical Paper* no. 60, 30 pp.

Bullock, G. L., and S. F. Snieszko. 1970. Fin rot, coldwater disease, and peduncle disease of salmonid fishes. *Fisheries Leaflet No. 462*. Washington, DC: U.S. Department of the Interior.

Bullock, G. L., and S. F. Snieszko. 1979. Enteric redmouth disease of salmonids. *Fish Disease Leaflet No. 57*. Washington, DC: U.S. Department of the Interior.

Bullock, G. L., and H. M. Stuckey. 1975. *Aeromonas salmonicida* detection of asymptomatically infected trout. *The Progressive Fish-Culturist* 37:237–239.

Bullock, G. L., H. M. Stuckey, and P. K. Chen. 1974. Corynebacterial kidney disease of salmonids: growth and serological studies on the causative bacterium. *Applied Microbiology* 28:811–814.

Bullock, G. L., H. M. Stuckey, and E. B. Shotts, Jr. 1977. Early records of North American and Australian outbreaks of enteric redmouth disease. *Fish Health News* 6(2):96.

Bullock, G. L., H. M. Stuckey, and D. Mulcahy. 1978a. Corynebacterial kidney disease: egg transmission following iodophore disinfection. *Fish Health News* 76:51–52.

Bullock, G. L., H. M. Stuckey, and E. B. Shotts, Jr. 1978b. Enteric redmouth bacterium: comparison of isolates from different geographic areas. *Journal of Fish Diseases* 1:351–356.

Bullock, G. L., B. R. Griffin, and H. M. Stuckey. 1980. Detection of *Corynebacterium salmoninus* by direct fluorescent antibody test. *Canadian Journal of Fisheries and Aquatic Sciences* 37:719–721.

Bullock, G. L., T. C. Hsu, and E. B. Shotts. 1986 Columnaris disease of fishes. *Fish Disease Leaflet no. 72*, United States Fish and Wildlife Service.

Busch, R. A. 1978. Enteric red mouth disease (Hagerman strain). *Marine Fisheries Review*, MFR Paper 1296, 40(3):42–51.

Busch, R. A. 1983. Enteric redmouth disease (*Yersinia ruckeri*). In: *Les Antigenes des Micro-orqanismes Pathogenes des Poissons*, edited by D. P. Anderson, M. Dorson, and Ph. Dubourget, 201–222. Lyon, France: Collection Fondation Marcel Meriux.

Busch, R. A., and J. A. Lingg. 1975. Establishment of an asymptomatic carrier state infection of enteric redmouth disease in rainbow trout (*Salmo gairdneri*). *Journal of the Fisheries Research Board of Canada* 32:2429–2432.

Byrne, P., H. W. Ferguson, J. S. Lumsden, and V. E. Ostland. 1991. Blood chemistry of bacterial gill disease in brook trout *Salvelinus fontinalis*. *Diseases of Aquatic Organisms* 10:1–6.

Caldwell, C. A., and J. M. Hinshaw. 1995. Tolerance of rainbow trout to dissolved oxygen supplementation and a *Yersinia ruckeri* challenge. *Journal of Aquatic Animal Health* 7:168–171.

Chart, H., and T. J. Trust. 1988. Characterization of the surface antigens of the marine fish pathogens, *Vibrio anguillarum* and *Vibrio ordalii*. *Canadian Journal Microbiology* 30:703–710.

Cipriano, R. C. 1982. Furunculosis in brook trout: infection by contact. *The Progressive Fish-Culturist* 44:12–14.

Cipriano, R. C., and J. B. Pyle. 1985. Development of a culture medium for determination of sorbitol utilization among strains of *Yersinia ruckeri*. *Microbios Letters* 28:79–82.

Cipriano, R. C., B. R. Griffin, and P. C. Lidgerding. 1981. *Aeromonas salmonicida*: relationship between extracellular growth products and isolate virulence. *Canadian Journal of Fisheries and Aquatic Sciences* 38:1322–1326.

Cipriano, R. C., C. E. Starliper, and J. H. Schachte. 1985. Comparative sensitivities of diagnostic procedures used to detect bacterial kidney disease in salmonid fishes. *Journal of Wildlife Diseases* 21:144–148.

Cipriano, R. C., W. B. Schill, S. W. Pyle, and R. Horner. 1986. An epizootic in chinook salmon (*Oncorhynchus tshwytscha*) caused by a sorbitol-positive serovar II strain of *Yersinia ruckeri*. *Journal of Wildlife Diseases* 22:488–492.

Cipriano, R. C., L. A. Ford, J. D. Teska, and L. L. Hale. 1992. Detection of *Aeromonas salmonicida* in mucus of salmonid fishes. *Journal of Aquatic Animal Health* 4:114–118.

Cipriano, R. C., G. L. Bullock, and A. Noble. 1996a. Nature of *Aeromonas salmonicida* carriage on asymptomatic rainbow trout maintained in a culture system with recirculating water and fluidized sand biofilters. *Journal of Aquatic Animal Health* 8:47–51.

Cipriano, R. C., L. A. Ford, J. T. Nelson, and B. N. Jensen. 1996b. Monitoring for early detection of *Aeromonas salmonicida* to enhance antibiotic therapy and control furunculosis in Atlantic salmon. *The Progressive Fish-Culturist* 58:203–208.

Cipriano, R. C., L. A. Ford, C. E. Starliper, J. D. Teska, J. T. Nelson, and B. N. Jensen. 1996c. Control of external *Aeromonas salmonicida*: topical disinfection of salmonids with Chloramine-T. *Journal of Aquatic Animal Health* 8:52-57.

Cipriano, R. C., W. B. Schill, J. D. Teska, and L. A. Ford. 1996d. Epizootiological study of bacterial cold-water disease in Pacific salmon and further characterization of the etiologic agent, *Flexibacter psychrophila*. *Journal of Aquatic Animal Health* 8:28–36.

Colwell, R. R., and D. J. Grimes. 1984. Vibrio diseases of marine fish populations. *Helgolander Wissenschaftliche Meeresuntersuchungen* 37:265–287.

Cvitanich, J. D. 1994. Improvements in the direct fluorescent antibody technique for the detection, identification and quantification of *Renibacterium salmoninarum* in salmonid kidney smears. *Journal of Aquatic Animal Health* 6:1–12.

Cvitanich, J. D., N. O. Garate, and C. E. Smith. 1991. The isolation of a rickettsia-like organism causing disease and mortality in Chilean salmonids and its confirmation by Koch's postulates. *Journal of Fish Diseases* 14:121–145.

Dalsgaard, I., O. Jurgens, and A. Mortensen. 1988. *Vibrio salmonicida* isolated from farmed Atlantic salmon in the Faroe Islands. *Bulletin of the European Association of Fish Pathologists* 8:53–54.

Daly, J. G., and R. M. W. Stevenson. 1988. Inhibitory effects of salmonid tissue on the growth of *Renibacterium salmoninarum*. *Diseases of Aquatic Organisms* 4:169–171.

Daly, J. G., and R. M. Stevenson. 1990. Characterization of the *Renibacterium salmoninarum* haemagglutinin. *Journal of General Microbiology* 136:949–953.

Daly, J. G., B. Lindvik, and R. M. W. Stevenson. 1986. Serological heterogeneity of recent isolates of *Yersenia ruckeri* from Ontario and British Columbia. *Diseases of Aquatic Organisms* 1:151–153.

Davies, R. L., and G. N. Frerichs. 1989. Morphological and biochemical differences among isolates of *Yersinia ruckeri* obtained from wide geographical areas. *Journal of Fish Diseases* 12:357–365.

Davis, H. S. 1926. A new gill disease of trout. *Transactions of the American Fisheries Society* 56:156–159.

Effendi, I., and B. Austin. 1994. Survival of the fish pathogen *Aeromonas salmonicida* in the marine environment. *Journal of Fish Diseases* 17:375–385.

Egidius, E., K. Andersen, E. Clausen, and J. Raa. 1981. Cold-water vibriosis or "Hitra disease" in Norwegian salmonid farming. *Journal of Fish Diseases* 4:353–354.

Egidius, E., O. Soleim, and K. Andersen. 1984. Further observations on cold-water vibriosis of Hitra disease. *Bulletin European Association of Fish Pathologists* 4(3):50–51.

Egidius, E., R. Wiik, K. Andersen, K. A. Hoff, and B. Hjeltnes. 1986. *Vibrio salmonicida* sp. nov., a new fish pathogen. *International Journal of Systematic Bacteriology* 36:518–520.

Ehlinger, N. F. 1977. Selective breeding of trout for resistance to furunculosis. *New York Fish and Game Journal* 24:26–36.

Elliott, D. G., and T. Y. Barila. 1987. Membrane filtration fluorescent antibody staining procedure for detecting and quantifying *Renibacterium salmoninarum* in coelomic fluid of chinook salmon (*Oncorhynchus tshawytscha*). *Canadian Journal of Fisheries and Aquatic Sciences* 44:206–210.

Elliott, D. G., R. J. Pascho, and G. L. Bullock. 1989. Developments in the control of bacterial kidney disease of salmonid fishes. *Diseases of Aquatic Organisms* 6:201–215.

Elston, R., A. S. Drum, and P. R. Bunnell. 1995. Efficacy of orally administered difloxacin for the treatment of Furunculosis in Atlantic salmon held in seawater. *Journal of Aquatic Animal Health* 7:22–28.

Espelid, S., K. O. Holm, K. Hjelmeland, and T. Jørgensen. 1988. Monoclonal antibodies against *Vibrio salmonicida*: the causative agent of coldwater vibriosis ("Hitra disease") in Atlantic salmon, *Salmo salar* L. *Journal of Fish Diseases* 11: 207–214.

Evelyn, T. P. T. 1971a. First records of vibriosis in Pacific salmon cultured in Canada, and taxonomic status of the responsible bacterium, *Vibrio anguillarum*. *Journal of the Fisheries Research Board of Canada* 28:517–525.

Evelyn, T. P. T. 1971b. Agglutinin response in sockeye salmon vaccinated intraperitoneally with a heat-killed preparation of the bacterium responsible for salmonid kidney disease. *Journal of Wildlife Diseases* 7:328–335.

Evelyn, T. P. T. 1977. An improved growth medium for the kidney bacterium and some notes in using the medium. *Bulletin Office International des Epizooties* 87:511–513.

Evelyn, T. P. T. 1993. Bacterial kidney disease—BKD. In: *Bacterial Diseases of Fish,* edited by V. Inglis, R. J. Roberts, and N. R. Bromage, 177–195. Oxford, UK: Blackwell Scientific Publishing.

Evelyn, T. P. T., and L. Prosperi-Porta. 1992. A new medium for growing the kidney disease bacterium: its performance relative to that of other currently used media. In: *Salmonid Diseases,* edited by T. Kimura, 143–150. Hakodate, Japan: Hokkaido University Press.

Evelyn, T. P. T., G. E. Hoskins, and G. R. Bell. 1973. First record of bacterial kidney disease in an apparently wild salmonid in British Columbia. *Journal of the Fisheries Research Board of Canada* 30:1578–1580.

Evelyn, T. P. T., J. E. Ketcheson, and L. Prosperi-Porta. 1984. Further evidence for the presence of *Renibacterium salmoninarum* in salmonid eggs and for the failure of povidine-iodine to reduce the intra-ovum infection in water-hardened eggs *Journal of Fish Diseases* 7:173–182.

Evelyn, T. P. T., G. R. Bell, L. Prosperi-Porta, and J. E. Ketcheson. 1989. A simple technique for accelerating the growth of the kidney disease bacterium *Renibacterium salmoninarum* on a commonly used culture medium (KDMl) *Diseases of Aquatic Organisms* 7:231–234.

Evelyn, T. P. T., L. Prosperi-Porta, and J. E. Ketcheson. 1990. Two new techniques for obtaining consistent results when growing *Renibacterium salmoninarum* on KDM2 culture medium, *Diseases of Aquatic Organisms* 9:209–212.

Evenberg, D., and B. Lugtenberg. 1982. Cell surface of the fish pathogenic bacterium *Aeromonas salmoni-*

cida. II. Purification and characterization of a major cell envelope protein related to autoagglutination, adhesion and virulence. *Biochemica et Biophysica Acta* 684:249–254.

Evensen, Ø., and E. Lorenzen. 1996. An immunohistochemical study of *Flexibacter psychrophilus* infection in experimentally and naturally infected rainbow trout (*Oncorhynchus mykiss*) fry. *Diseases of Aquatic Organism* 25:53–61.

Evensen, Ø., S. Espelid, and T. Håstein. 1991. Immunohistochemical identification of *Vibrio salmonicida* in stored tissue of Atlantic salmon *Salmo salar* from the first known outbreak of cold-water vibriosis ("Hitra disease"). *Diseases of Aquatic Organisms* 10:185–189.

Ewing, W. H., A. J. Ross, D. J. Brenner, and G. R. Fanning 1978. *Yersinia ruckeri* sp. nov., the redmouth (RM) bacterium. *International Journal of Systematic Bacteriology* 28:37–44.

Ferguson, H. W., V. E. Ostland, P. Byrne, and J. S. Lumsden. 1991. Experimental production of bacterial gill disease in trout by horizontal transmission and by bath challenge. *Journal of Aquatic Animal Health* 3:118–123.

Fjølstad, M., and A. L. Heyeroaas. 1985. Muscular and myocardial degeneration in cultured Atlantic salmon, *Salmo salar* L., suffering from "Hitra disease." *Journal of Fish Diseases* 8: 367–372.

Flaño, E., P. López-Fierro, B. Razquin, S. L. Kaatari, and A. Villena. 1996. Histopathology of the renal and splenic haemopoietic tissues of coho salmon *Oncorhyunchus kisutch* experimentally infected with *Renibacterium salmoninarum*. *Diseases of Aquatic Organisms* 24:107–115.

Foott, J. S. 1994. XV. Pseudokidney disease. In: *Bluebook: Suggested Procedures for the Detection and Identification of Certain Finfish and Shellfish Pathogens*, 4th ed., edited by J. C. Thoesen. Bethesda, MD: Fish Health Section/American Fisheries Society.

Frerichs, G. N., J. A. Stewart, and R. O. Collins. 1985. Atypical infection of rainbow trout, *Salmo gairdneri* Richardson, with *Yersinia ruckeri*. *Journal of Fish Diseases* 8:383–387.

Fryer J. L., and J. S. Rohovec. 1984. Principal bacterial diseases of cultured marine fish. *Helgolander Wissenschaftliche Meeresuntersuchungen* 37:533–545.

Fryer, J. L., and J. L. Sanders. 1981. Bacterial kidney disease of salmonid fish. *Annual Reviews in Microbiology* 35:273–298.

Fryer, J. L., C. N. Lannan, L. H. Garcés, J. J. Larenas, and P. A. Smith. 1990. Isolation of a rickettsiales-like organism from diseased coho salmon (*Oncorhynchus kisutch*) in Chile. *Fish Pathology* 25:107–114.

Fryer, J. L., C. N. Lannan, S. J. Giovanonni, and N. D. Wood. 1992. *Piscirickettsia salmonis* gen. nov., sp. nov., the causative agent of an epizootic disease in salmonid fishes. *International Journal of Systematic Bacteriology* 42:120–126.

Fuhrmann, H., K. H. Bohm, and H. J. Schlotfeldt. 1984. On the importance of enteric bacteria in the bacteriology of freshwater fish. *Bulletin of the European Association of Fish Pathologists* 4:42–46.

Gaggero, A., H. Castro, and A. M. Sandino. 1995. First isolation of *Piscirickettsia salmonis* from coho salmon, *Oncorhynchus kisutch* (Walbaum), and rainbow trout, *Oncorhynchus mykiss* (Walbaum), during the freshwater state of their life cycle. *Journal of Fish Diseases* 18:277–279.

Garcés, L. H., J. J. Larenas, P. A. Smith, S. Sandino, C. N. Lannan, and J. L. Fryer. 1991. Infectivity of rickettsia isolated from coho salmon *Oncorhynchus kisutch*. *Diseases of Aquatic Organisms* 11:93–97.

Getchell, R. G., J. S. Rohovec, and J. L. Fryer. 1985. Comparison of *Renibacterium salmoninarum* isolates by antigenic analysis, *Fish Pathology* 20:149–159.

Goodfellow, M., T. M. Embley, and B. Austin. 1985. Numerical taxonomy and emended description of *Renibacterium salmoninarum*. *Journal of General Microbiology* 131:2739–2752.

Grandis, S. A., P. J. Rrell, D. E. Flett, and R. M. W. Stevenson. 1988. Deoxyribonucleic acid relatedness of serovars of *Yersinia ruckeri*, the enteric redmouth bacterium. *International Journal of Systematic Bacteriology* 38:49–55.

Griffiths, S. G., G. Olivier, J. Fildes, and W. H. Lynch. 1991. Comparison of western blot, direct fluorescent antibody and drop-plate culture methods for the detection of *Renibacterium salmoninarum* in Atlantic salmon (*Salmo salar*). *Aquaculture* 89:117–129.

Griffiths, S. G., K. Liska, and W. H. Lynch. 1996. Comparison of kidney tissue and ovarian fluid from broodstock Atlantic salmon for detection of *Renibacterium salmoninarum*, and use of SKDM broth culture with Western blotting to increase detection in ovarian fluid. *Diseases of Aquatic Organisms* 24:3–9.

Grisez, L., M. Chair, P. Sorgeloos, and F. Ollevier. 1996. Mode of infection and spread of *Vibrio anguillarum* in turbot *Scophthalmus maximus* larvae after oral challenge through live feed. *Diseases of Aquatic Organisms* 26:181–187.

Groberg, W. J., Jr., J. S. Rohovec, and J. L. Fryer. 1983. The effects of water temperature on infection and antibody formation induced by *Vibrio anguillarum* in juvenile coho salmon (*Oncorhynchus kisutch*). *Journal of the World Mariculture Society* 14:240–248.

Gudmundsdottir, S., E. Benediktsdottir, and S. Helguson. 1993. Detection of *Renibacterium salmoninarum* in salmonid kidney samples: a comparison of results using double-sandwich ELISA and isolation on selective media. *Journal of Fish Diseases* 16:185–195.

Guide to drug, vaccine, and pesticide use in aquaculture. 1994. Prepared by the Federal Joint Subcommittee on Aquaculture, U.S. Department of Agriculture.

Hansen, C. B., and A. J. Lingg. 1976. Inert particle agglutination tests for detection of antibody to enteric redmouth bacterium. *Journal of the Fisheries Research Board of Canada* 33:2857–2860.

Harbell, S. C., H. O. Hodgins, and M. H. Schiewe. 1979. Studies on the pathogenesis of vibriosis in coho salmon *Oncorhynchus kisutch* (Walbaum). *Journal of Fish Diseases* 2:391–404.

Hardie, L. J., A. E. Ellis, and C. J. Secombes. 1996. In vitro activation of rainbow trout macrophages stimulates inhibition of *Renibacterium salmoninarum* growth concomitant with augmented generation of respiratory burst products. *Diseases of Aquatic Organism* 25:175–183.

Harrell, L. W. 1978. Vibriosis and current salmon vaccination procedures in Puget Sound, Washington. *Marine Fisheries Review* 40(3):24–25.

Håstein, T., F. Hallingstad, T. Refsti, and S. O. Roald. 1980. Recent experience of field vaccination trials against vibriosis in rainbow trout (*Salmo gairdneri*). In: *Fish Diseases,* edited by W. Ahne, 53–59. Third *COPRAQ* Session, Berlin: Springer-Verlag.

Håstein, T., and J. E. Smith. 1977. A study of *Vibrio anguillarum* from farms and wild fish using principal components analysis. *Journal of Fish Biology* 11:69–75.

Hastings, T. S., and D. W. Bruno. 1985. Enteric redmouth disease: survey in Scotland and evaluation of a new medium, Shotts-Waltman, for differentiating *Yersinia ruckeri. Bulletin of the European Association of Fish Pathologists* 54:2594–2597.

Heo, G-J., K. Kasai, and H. Wakabayashi. 1990. Occurrence of *Flavobacterium branchiophila* associated with bacterial gill disease at a trout hatchery. *Fish Pathology* 25:99–105.

Herman, R. L. 1968. Fish furunculosis 1952–1966. *Transactions of the American Fisheries Society* 97:221–230.

Hicks, B. D., J. G. Daly, and V. E. Ostland. 1986. Experimental infection of minnows with bacteria. *Proceedings of the Third Annual Meeting, Aquaculture Association of Canada.* Guelph, Ontario. Abstract.

Hiney, P. P., J. J. Kilmartin, and P. R. Smith. 1994. Detection of *Aeromonas salmonicida* in Atlantic salmon with asymptomatic furunculosis infections. *Diseases of Aquatic Organisms* 19:161–167.

Hirvelä-Koski, V., P. Koski, and H. Niiranen. 1994. Biochemical properties and drug resistance of *Aeromonas salmonicida* in Finland. *Diseases of Aquatic Organisms* 20:191–196.

Hiu, S. F., R. A. Holt, N. Sriranganathan, R. J. Seidler, and J. L. Fryer. 1984. *Lactobacillus piscicola,* a new species from salmonid fish. *International Journal of Systematic Bacteriology* 34:393–400.

Hjeltnes, B., and K. Julshamn. 1992. Concentrations of iron, copper, zinc and selenium in liver of Atlantic salmon *Salmo salar* infected with *Vibrio salmonicida. Diseases of Aquatic Organisms* 12:147–149.

Hjeltnes, B., and R. J. Roberts. 1993. Vibriosis. In: *Bacterial Diseases of Fish,* edited by V. Inglis, R. J. Roberts, and N. Bromage, 109–121, Oxford, UK: Blackwell Scientific Publications.

Hjeltnes, B., K. Andersen, H. M. Ellingsen, and E. Egidius. 1987a. Experimental studies on the pathogenicity of a *Vibrio* sp. isolated from Atlantic salmon, *Salmo salar* L., suffering from Hitra disease. *Journal of Fish Diseases* 10:21–27.

Hjeltnes, B., K. Andersen, and K. Egidius. 1987b. Multiple antibiotic resistance in *Vibrio salmonicida. Bulletin of the European Association of Fish Pathologists* 7:85.

Hjeltnes, B., K. Andersen, and Ellingsen. 1989. Vaccination against *Vibrio salmonicida*: The effect of different routs of administration and revaccination. *Aquaculture* 83:1–6.

Hjeltnes, B., Ø. Bergh, H. Wergeland, and J. C. Holm. 1995. Susceptibility of Atlantic cod *Gadus morhua,* halibut *Hippoglossus hippoglossus* and wrasse (Labridae) to *Aeromonas salmonicida* subsp. salmonicida and the possibility of transmission of furunculosis from farmed salmon *Salmo salar* to marine fish. *Diseases of Aquatic Organisms* 23:25–31.

Hnath, J. G. 1991. Bacterial kidney disease in Michigan. *Workshop on Bacterial Kidney Diseases,* Phoenix, Arizona, edited by S. Mangin, 46, November 20–21. Arlington, VA: U.S. Fish and Wildlife Service.

Hoffmann, R., W. Popp, and S. Van der Graaff. 1984. Atypical BKD predominantly causing ocular and skin lesions. *Bulletin of the European Association of Fish Pathologists* 4:7–9.

Holm, K. O., and T. Jørgensen. 1987. A successful "Hitra disease" or coldwater vibriosis. *Journal of Fish Diseases* 10:85–90.

Holm, K. O., E. Strom, K. Stensvag, J. Raa, and T. Jørgensen. 1985. Characteristics of a *Vibrio* sp. associated with the Hitra disease of Atlantic salmon in Norwegian fish farms. *Fish Pathology* 20:125–129.

Hollebeq, M-G., F. Bernadette, C. Bourmaud, and C. Michel. 1995. Spontaneous bactericidal and complement activities in serum of rainbow trout (*Oncorhynchus mykiss*) genetically selected for resistance or susceptibility to furunculosis. *Fish & Shellfish Immunology* 5:407–426.

Holt, R. A. 1972. Characterization and control of *Cytophaga psychrophila* (Borg) the causative agent of low temperature disease in young coho salmon (*Oncorhynchus kisutch*). Corvallis, OR: Oregon State University. M.S. thesis.

Holt, R. A. 1993. Bacterial cold-water disease. In: *Bacterial Diseases of Fish,* edited by V. Inglis, R. J. Roberts, and N. R. Bromage, 3–22. Oxford, UK: Blackwell Press.

Holt, R. A., A. Amandi, J. S. Rohovec, and J. L. Fryer. 1989. Relation of water temperature to bacterial cold-water disease in coho salmon, chinook salmon and rainbow trout. *Journal of Aquatic Animal Health* 1:97–101.

Horne, M. T., and D. A. Robertson. 1987. Economics of vaccination against enteric redmouth disease of salmonids. *Aquaculture and Fisheries Management* 18:131–137.

Horne, M. T., M. Tatner, S. McDerment, and C. Agius. 1982. Vaccination of rainbow trout, *Salmo gairdneri* Richardson, at low temperatures and the long-term persistence of protection. *Journal of Fish Diseases* 5:343–345.

Hsu, H.-M., G. A. Wooster, and P. R. Bowser. 1994. Efficacy of enrofloxacin for the treatment of salmonids with bacterial kidney disease, caused by *Renibacterium salmoninarum. Journal of Aquatic Animal Health* 6:220–223.

Huh, G-J., and H. Wakabayashi. 1989. Serological characteristics of *Flavobacterium branchiophila* isolated from gill diseases of freshwater fishes in Japan, USA, and Hungary. *Journal of Aquatic Animal Health* 1:142–147.

Hunter, V. A., M. D. Knittel, and J. L. Fryer. 1980. Stress-induced transmission of *Yersinia ruckeri* infection from carriers to recipient steelhead trout *Salmo gairdneri* Richardson. *Journal of Fish Diseases* 3:467–472.

Iida, Y., and A. Mizokami. 1996. Outbreaks of coldwater disease in wild ayu and pale chub. *Fish Pathology* 31:157–164.

Inglis, V., and R. H. Richards. 1992. Difficulties encountered in chemotherapy of furunculosis in Atlantic salmon (*Salmo salar* L.). In: *Salmonid Diseases,* edited by T. Kimura, 201–208. Hakodate, Japan: Hokkaido University Press.

Inglis, V., D. Robertson, K. Miller, K. D. Thompson, and R. H. Richards. 1996. Antibiotic protection against recrudescence of latent *Aeromonas salmonicida* during furunculosis vaccination. *Journal of Fish Diseases* 19:341–348.

Ishimaru, K., M. Akagawa-Matsushita, and K. Muroga. 1996. *Vibrio ichthyoenteri* sp. nov., a pathogen of Japanese flounder (*Paralichthys olivaceus*) larvae. *International Journal of Systematic Bacteriology* 46:155–159.

Jansson, E., T. Hongslo, J. Höglund, and O. Ljungberg. 1996. Comparative evaluation of bacterial culture and two ELISA techniques for the detection of *Renibacterium salmoninarum* antigens in salmonid kidney tissues. *Diseases of Aquatic Organisms* 27:197–206.

Jarp, J., A. G. Gjevre, A. B. Olsen, and T. Bruheim. 1995. Risk factors for furunculosis, infectious pancreatic necrosis and mortality in post-smolt of Atlantic salmon, *Salmo salar* L. *Journal of Fish Diseases* 18:67–78.

Johnson, K. A., J. K. Flynn, and D. F. Amend. 1982a. Onset of immunity in salmonid fry vaccinated by direct immersion in *Vibrio anguillarum* and *Yersinia ruckeri* bacterins. *Journal of Fish Diseases* 5:197–205.

Johnson, K. A., J. R. Flynn, and D. F. Amend. 1982b. Duration of immunity in salmonids vaccinated by direct immersion with *Yersinia ruckeri* and *Vibrio anguillarum* bacterins. *Journal of Fish Diseases* 5:207–213.

Jørgensen, T., K. Midling, S. Espelid, R. Nilsen, and K. Stensvag. 1989. *Vibrio salmonicida,* a pathogen in salmonids, also causes mortality in net-pen captured cod (*Gadus morhua*). *Bulletin of the European Association of Fish Pathologists* 9:42–44.

Kawano, K., T. Aoki, and T. Kitao. 1984. Duration of protection against vibriosis in ayu *Plecoglossus altivelis* vaccinated by immersion and oral administration with *Vibrio anguillarum. Bulletin of the Japanese Society of Scientific Fisheries* 50:771–774.

Kent, M. L., J. M. Groff, J. P. Morrison, W. T. Yasutake, and R. A. Holt. 1989. Spiral swimming behavior due to cranial and vertebral lesions associated with *Cytophaga psychrophila* infections in salmonid fishes. *Diseases of Aquatic Organisms* 6:11–16.

Kimura, N., H. Wakabayashi, and S. Kudo. 1978. Studies on bacterial gill disease in salmonids—I. Selection of bacterium transmitting gill disease. *Fish Pathology* 12:233–242.

Kingsbury, O. R. 1961. A possible control of furunculosis. *The Progressive Fish-Culturist* 23:136–137.

Klontz, G. W., W. T. Yasutake, and A. J. Ross. 1966. Bacterial diseases of the Salmonidae in the Western United States: pathogenesis of furunculosis in rainbow trout. *American Journal of Veterinary Research* 27:1455–1460.

Knappskog, D. H., O. M. Rødseth, E. Slinde, and C. Endresen. 1993. Immunochemical analysis of *Vibrio anguillarum* L., suffering from vibriosis. *Journal of Fish Diseases* 16:327–338.

Knittel, M. D. 1981. Susceptibility of steelhead trout *Salmo gairdneri* Richardson to redmouth infection *Yersinia ruckeri* following exposure to copper. *Journal of Fish Diseases* 4:33–40.

Kudo, S., and N. Kimura. 1983. Transmission electron microscopic studies on bacterial gill disease in rainbow trout fingerlings. *Japanese Journal of Ichthyology* 30(4):247–260.

Lall, S. P., W. D. Paterson, and N. J. Adams. 1985. Control of bacterial kidney disease in Atlantic salmon, *Salmo salar* L., by dietary modification. *Journal of Fish Diseases* 8:113–124.

Lamers, C. H. J., and W. B. Muiswinkel. 1984. Primary and secondary immune responses in carp (*Cyprinus carpio*) after administration of *Yersinia ruckeri* O-antigen. In: *Acuigrup Fish Diseases,* 119–127. Madrid: Editora ATP.

Lannan, C. N., and J. L. Fryer. 1993. *Piscirickettsia salmonis,* a major pathogen of salmonid fish in Chile. *Fisheries Research* 17:114–121.

Lannan, C. N., and J. L. Fryer. 1994. Extracellular survival of *Piscirickettsia salmonis. Journal of Fish Diseases* 17:545–548.

Lannan, C. N., S. A. Ewing, and J. L. Fryer. 1991. A fluorescent antibody test for detection of the rickettsia causing disease in Chilean salmonids. *Journal of Aquatic Animal Health* 3:229–234.

Larsen, J. L. 1983. *Vibrio anguillarum*: a comparative study of fish pathogenic, environmental, and reference strains. *Acta Veterinaria Scandinavica* 24:456–476.

Lee, E. G. H. 1989. Technique for enumeration of *Renibacterium salmoninarum* in fish kidney tissues. *Journal of Aquatic Animal Health* 1:25–28.

Lee, E. G. H., and T. P. T. Evelyn. 1989. Effect of *Renibacterium salmoninarum* levels in the ovarian fluid of spawning chinook salmon on the prevalence of the pathogen in their eggs and progeny. *Diseases of Aquatic Organisms* 7:179–184.

Lee, E. G. H. and T. P. T. Evelyn. 1994. Prevention of vertical transmission of the bacterial kidney disease agent *Renibacterium salmoninarum* by broodstock injection with erythromycin. *Diseases of Aquatic Organism* 18:1–4.

Leek, S. L. 1987. Viral erythrocytic inclusion body syndrome (EIBS) occurring in juvenile spring chinook salmon (*Oncorhynchus tshwytscha*) reared in fresh water. *Canadian Journal of Fisheries and Aquatic Sciences* 44:685–688.

Lehmann, J., D. Mock, F. J. Sturenberg, and J. F. Bernard 1991. First isolation of *Cytophaga psychrophilus* from a systemic disease in eel and cyprinids. *Diseases of Aquatic Organisms* 10:217–220.

Lillehaug, A. A. 1990. A field trial of vaccination against cold-water vibriosis in Atlantic salmon (*Salmo salar* L.). *Aquaculture* 84:1–12.

Lillehaug, A., R. H. Sorum, and A. Ramstad. 1990. Cross-protection after immunization of Atlantic salmon, *Salmo salar* L., against different strains of *Vibrio salmonicida. Journal of Fish Diseases* 13:519–523.

Lorenzen, E., and N. Karas. 1992. Detection of *Flexibacter psychrophilus* by immunofluorescence in fish suffering from fry mortality syndrome: a rapid diagnostic method. *Diseases of Aquatic Organisms* 13:231–234.

Lovely, J. E., C. Cabo, S. G. Griffiths, and W. H. Lynch. 1994. Detection of *Renibacterium salmoninarum* infection in asymptomatic Atlantic salmon. *Journal of Aquatic Animal Health* 6:126–132.

Lumsden, J. S., V. E. Ostland, D. D. MacPhee, J. Derksen, and H. W. Ferguson. 1994. Protection of rainbow trout from experimentally induced bacterial gill disease caused by *Flavobacterium branchiophilum. Journal of Aquatic Animal Health* 6:292–302.

MacPhee, D. D., V. E. Ostland, J. S. Lumsden, and H. W. Ferguson. 1995a. Development of an enzyme-linked immunosorbent assay (ELISA) to estimate the quantity of *Flavobacterium branchiophilum* on the gills of rainbow trout *Oncorhynchus mykiss. Diseases of Aquatic Organisms* 21:13–23.

MacPhee, D. D., V. E. Ostland, J. S. Lumsden, J. Derksen, and H. W. Ferguson. 1995b. Influence of feeding on the development of bacterial gill disease in rainbow trout *Oncorhynchus mykiss. Diseases of Aquatic Organisms* 21:163–170.

Mangin, S., Editor. 1991. *Workshop on Bacterial Kidney Diseases,* Phoenix, Arizona. Arlington, VA, November 20–21, U.S. Fish and Wildlife Service.

Maule, A. G., D. W. Rondorf, J. Beeman, and P. Haner. 1996. Incidence of *Renibacterium salmoninarum* infections in juvenile hatchery spring chinook salmon in the Columbia and Snake Rivers. *Journal of Aquatic Animal Health* 8:37–46.

McCarthy, D. H. 1975. Fish furunculosis. *Journal of the Institute of Fisheries Management* 6:13–18.

McCarthy, D. H. 1980. Some ecological aspects of the bacterial fish pathogen - *Aeromonas salmonicida.* In: *Aquatic Microbiology,* 299–324, Symposium of the Society of Applied Bacteriology, no. 6.

McCarthy, D. H., and C. T. Rawle. 1975. The rapid serological diagnosis of fish furunculosis caused by "smooth" and "rough" strains of *Aeromonas salmonicida. Journal of General Microbiology* 86:185–187.

McCarthy, D. H., and R. J. Roberts. 1980. Furunculosis of fish—The state of our knowledge. In: *Advances in Aquatic Microbiology,* edited by M. A. Droop and H. W. Janasch, 293–341. London: Academic Press.

McCarthy, D. H., T. R. Croy, and D. F. Amend. 1984. Immunization of rainbow trout, *Salmo gairdneri* Richardson, against bacterial kidney disease: preliminary efficacy evaluation. *Journal of Fish Diseases* 7:65–71.

McDaniel, D. W. 1972. *Hatchery Biologist Quarterly Report.* U.S. Fish and Wildlife Service, First Quarter.

McFaddin, T. W. 1969. Effective disinfection of trout eggs to prevent egg transmission of *Aeromonas liquefaciens. Journal of the Fisheries Research Board of Canada* 26:2311–2318.

McGraw, B. M. 1952. Furunculosis of fish. *U.S. Fish and Wildlife Service, Special Scientific Report Fisheries Series* 84, 87 pp.

Meyers, T. R. 1989. Apparent chronic bacterial myeloencephalitis in hatchery-reared juvenile coho salmon *Oncorhynchus kisutch* in Alaska. *Diseases of Aquatic Organisms* 6:217–219.

Michel, C., and A. Dubois-Darnaudpeys. 1980. Persistence of the virulence of *Aeromonas salmonicida* strains kept in river sediments. *Annual Rech Veterinary* 11:375–386.

Michel, C., B. Faivre, and P. A. de Kinkelin. 1986a. A clinical case of enteric redmouth in minnows (*Pimephales promelas*) imported in Europe as baitfish. *Bulletin of the European Association of Fish Pathologists* 6:97–99.

Michel, C., B. Faivre, and B. Kerouault. 1986b. Biochemical identification of *Lactobacillus piscicola* strains from France and Belgium. *Diseases of Aquatic Organisms* 2:27–30.

Mitchum, D. L., L. E. Sherman, and G. T. Baxter. 1979. Bacterial kidney disease in feral populations of brook trout (*Salvelinus fontinalis*), brown trout (*Salmo trutta*), and rainbow trout (*Salmo gairdneri*). *Journal of Fisheries Research Board of Canada* 36:1370–1376.

Moffitt, C. M. 1992. Survival of juvenile chinook salmon challenged with *Renibacterium salmoninarum* and administered oral doses of erythromycin thiocyanate for different durations. *Journal of Aquatic Animal Health* 4:119–125.

Mooney, J., E. Powell, C. Clabby, and R. Powell. 1995. Detection of *Aeromonas salmonicida* in wild Atlantic salmon using a specific DNA probe test. *Diseases of Aquatic Organisms* 21:131–135.

Munro, A. L. S., and T. S. Hastings. 1993. Furunculosis. In: *Bacterial Diseases of Fish*, edited by V. Inglis, R. J. Roberts, and N. R. Bromage, 122–142. Oxford, UK: Blackwell Scientific Publishers.

Munro, A. L. S., T. S. Hastings, A. E. Ellis, and J. Livorsidge. 1980. Studies on an ichthyotoxic material produced extracellularly by the furunculosis bacterium *Aeromonas salmonicida*. In: *Fish Diseases*, edited by W. Ahne, 98–106. Third *COPRAQ* Session, Berlin: Springer-Verlag.

Muroga, K. 1975. Studies on *Vibrio anguillarum* and *V. anguillarum* infection. *Journal of Faculty of Fisheries and Husbandry* (Hiroshima University) 14:101–205.

Muroga, K. 1992. Vibriosis of cultured fishes in Japan. In: *Salmonid Diseases*, edited by T. Kimura, 165–171. Hakodate, Japan: Hokkaido University Press.

Muroga, K., H. Yamanoi, Y. Hironaka, S. Yamamoto, M. Tatani, Y. Jo, S. Takahashi, and H. Hanada. 1984. Detection of *Vibrio anguillarum* from wild fingerlings of ayu *Plecoglossus altivelis*. *Bulletin of the Japanese Society of Scientific Fisheries* 50:591–596.

Muroga, K., A. Nakajima, T. Nakai, and H. Yamanoi. 1995. Humoral immunity in ayu, *Plecoglossus altivelis* immunized with *Vibrio anguillarum* by immersion method. In: *Diseases in Asian Aquaculture II*, edited by M. Shariff, J. R. Arthur and R. P. Subasinghe, 441–449. Manila, Philippines: Fish Health Section, Asian Fisheries Society.

Murray, C. B., T. P. T. Evelyn, T. D. Beachan, L. W. Barner, J. E. Retcheson, and L. Prosperi-Porta. 1992. Experimental induction of bacterial kidney disease in chinook salmon by immersion and cohabitation challenges. *Diseases of Aquatic Organisms* 12:91–96.

Nomura, T., M. Yoshimizu, and T. Kimura. 1992. An epidemiological study of furunculosis in salmon propagation. In *Salmonid Diseases,* edited by T. Kimura. Hakodate, Japan: Hokkaido University Press, 187 p.

Novotny, A. J., and L. W. Harrell. 1975. Vibriosis—a common disease of Pacific salmon cultured in marine waters of Washington. College of Agriculture, Washington State University, *Extension Bulletin,* 663:8.

Obach, A., and F. B. Laurencin. 1991. Vaccination of rainbow trout *Oncorhynchus mykiss* against the visceral form of coldwater disease. *Diseases of Aquatic Organisms* 12:13–15.

O'Halloran, J., K. Coombs, E. Carpenter, K. Whitman, and G. Johnson. 1995. Prescreening Atlantic salmon, *Salmo salar* L., broodstock for the presence of *Renibacterium salmoninarum*. *Journal of Fish Diseases* 18:83–85.

Olsson, J. C., A. Jöborn, A. Westerdahl, L. Bloombery, S. Kjelleberg, and P. L. Conway. 1996. Is the turbot, *Scophthalmus maximus* (L), intestine a portal of entry for the fish pathogen *Vibrio anguillarum*? *Journal of Fish Diseases* 19:225–234.

Ordal, E. J., and B. J. Earp. 1956. Cultivation and transmission of etiological agent of kidney disease in salmonid fishes. *Proceedings of the Society of Experimental Biological Medicine* 92:85–88.

Ostland, V. E., B. D. Hicks, and J. G. Daly. 1987. Furunculosis in baitfish and its transmission to salmonids. *Diseases of Aquatic Organisms* 2:163–166.

Ostland, V. E., H. W. Ferguson, and R. M. V. Stevenson 1989. Case report: bacterial gill disease in goldfish *Carassius auratus, Diseases of Aquatic Organisms* 6:179–184.

Ostland, V. E., H. W. Ferguson, J. F. Prescott, R. M. W. Stevenson, and I. K. Barker. 1990. Bacterial gill disease of salmonids; relationship between the severity of gill lesions and bacterial recovery. *Diseases of Aquatic Organisms* 9:5–14.

Ostland, V. E., J. S. Lumsden, D. D. MacPhee, and H. W. Ferguson. 1994. Characteristics of *Flavobacterium branchiophilum,* the cause of salmonid bacterial gill disease in Ontario. *Journal of Aquatic Animal Health* 6:13–26.

Ostland, V. E., D. D. MacPhee, J. S. Lumsden, and H. W. Ferguson. 1995. Virulence of *Flavobacterium bran-*

chiophilum in experimentally infected salmonids. *Journal of Fish Diseases* 18:249–262.

Pacha, R. E. 1968. Characteristics of *Cytophaga psychrophila* (Borg) isolated during outbreaks of bacterial cold-water disease. *Applied Microbiology* 16:97–101.

Pacha, R. E., and E. D. Kiehn. 1969. Characterization and relatedness of marine vibrios pathogenic to fish: physiology, serology and epidemiology. *Journal of Bacteriology* 100:1242–1247.

Pascho, R. J., and D. Mulcahy. 1987. Enzyme-linked immunosorbent assay for a soluble antigen of *Renibacterium salmoninarum*, the causative agent of bacterial kidney disease. *Canadian Journal of Fisheries and Aquatic Sciences* 44:183–191.

Pascho, R.J., D. G. Elliott, and J. M. Streufert. 1991. Brood stock segregation of spring chinook salmon *Oncorhynchus tshawytscha* by use of the enzyme linked immunosorbent assay (ELISA) and the fluorescent antibody technique (FAT) affects the prevalence and levels of *Renibacterium salmoninarum* infection in progeny. *Diseases of Aquatic Organisms* 12:25–40.

Paterson, W. D., D. Desautel, and J. M. Weber. 1980a. The immune response of Atlantic salmon, *Salmo salar* L., to the causative agent of bacterial kidney disease, *Renibacterium salmoninarum*. *Journal of Fish Diseases* 4:99–111.

Paterson, W. D., D. Douey, and D. Desautels. 1980b. Relationships between selected strains of typical and atypical *Aeromonas salmonicida, Aeromonas hydrophila* and *Haemophilus piscium*. *Canadian Journal of Microbiology* 26:588–598.

Paterson, W. D., S. P. Lall, and D. Desautels. 1981. Studies on bacterial kidney disease in Atlantic salmon (*Salmo salar*) in Canada. *Fish Pathology* 15:283–292.

Paterson, W. D., W. Parker, M. Poy, and M. T. Horne. 1992. Prevention of furunculosis using orally applied vaccines. In: *Salmonid Diseases,* edited by T. Kimura, 225–232. Hakodate, Japan: University of Hokkaido Press.

Pérez, M. J., A. I. G. Fernández, L. A. Rodriguez, and T. P. Nieto. 1996. Differential susceptibility to furunculosis of turbot and rainbow trout and release of the furunculosis agent from furunculosis-affected fish. *Diseases of Aquatic Organisms* 26:133–137.

Peters, K. K., and C. M. Moffitt. 1996. Optimal dosage of erythromycin thiocyanate in a new feed additive to control bacterial kidney disease. *Journal of Aquatic Animal Health* 8:229–240.

Piper, R. G., I. B. McElwain, L. E. Orme, J. P. McCraren, L. G. Fowler, and J. R. Leonard. 1982. *Fish Hatchery Management.* Washington, DC: U.S. Fish and Wildlife Service.

Popoff, M. 1984. *Aeromonas* Kluyer and Van Niel 1936. In: *Bergey's Manual of Systematic Bacteriology,* vol. I, edited by N. R. Krieg and J. G. Holt, 545–548. Baltimore, MD: Williams and Wilkins.

Poppe, T. T., T, Håstein, and R. Salte. 1985. "Hitra disease" (haemorragic syndrome) in Norwegian salmon farming: present status. In: *Fish and Shellfish Pathology,* edited by A. E. Ellis, 223–229. London: Academic Press.

Poppe, T. T., T. Håstein, A. Froslie, N. Koppang, and G. Norheim. 1986. Nutritional aspects of haemorrhagic syndrome ("Hitra disease") in farmed Atlantic salmon, *Salmo salar. Diseases of Aquatic Organisms* 1:155–162.

Press, C. M., and A. Lillihaug. 1995. Vaccination in European salmonid aquaculture: a review of practices and prospects. *British Veterinary Journal* 151:45–69.

Press, C. M., Ø. Evensen, L. J. Reitan, and T. Landsverk. 1996. Retention of furunculosis vaccine components in Atlantic salmon, *Salmo salar* L. following different routes of vaccine administration. *Journal of Fish Diseases* 19:215–224.

Quentel, C., and J. F. Aldrin. 1986. Blood changes in catheterized rainbow trout (*Salmo gairdneri*) intraperitoneally inoculated with *Yersinia ruckeri. Aquaculture* 53:169–185.

Ransom, D. P. 1978. Bacteriologic, immunologic and pathologic studies of *Vibrio* sp., pathogenic to salmonids. Corvallis, OR: Oregon State University. Ph.D. dissertation.

Ransom, D. P., C. N. Lannan, J. S. Rohovec, and J. L. Fryer. 1984. Comparison of histopathology caused by *Vibrio anguillarum* and *Vibrio ordalii* and three species of Pacific salmon. *Journal of Fish Diseases* 7:107–115.

Rodger, H. D., and E. M. Drinan. 1993. Observation of a rickettsia like organism in Atlantic salmon, *Salmo salar* L, in Ireland. *Journal of Fish Diseases* 16:361–369.

Rodgers, C. J. 1990. Immersion vaccination for control of fish furunculosis. *Diseases of Aquatic Organisms* 8:69–72.

Rodgers, C. J. 1991. The use of vaccination and antimicrobial agents for control of *Yersinia ruckeri. Journal of Fish Diseases* 14:291–301.

Rodgers, C. J. 1992. Development of a selective-differential medium for the isolation of *Yersinia ruckeri* and its application in epidemiological studies. *Journal of Fish Diseases* 15:243–254.

Rockey, D. D., J. L. Fryer, and J. S. Rohovec. 1988. Separation and in vivo analysis of two extracellular proteases and the T-hemolysin from *Aeromonas salmonicida. Diseases of Aquatic Organisms* 5:197–204.

Rogers, W. A., and D. Xu. 1992. Protective immunity induced by a commercial *Vibrio* vaccine in hybrid striped bass. *Journal of Aquatic Animal Health* 4:303–305.

Rohovec, J. S., R. L. Garrison, and J. L. Fryer. 1975. Immunization of fish from the control of vibriosis. *Third U.S.-Japan Meeting on Aquaculture,* Tokyo 105–112.

Romalde, J. L., B. Magariños, F. Pazos, A. Silv, and A. E. Toranzo. 1994. Incidence of *Yersinia ruckeri* in two farms in Galacia (NW) Spain during a one-year period. *Journal of Fish Diseases* 17:533–539.

Rose, A. S., A. E. Ellis, and A. L. S. Munro. 1990. The survival of *Aeromonas salmonicida* subsp. *salmonicida* in sea water. *Journal of Fish Diseases* 13:205–214.

Ross, A. J., and G. W. Klontz. 1965. Oral immunization of rainbow trout (*Salmo gairdneri*) against the etiologic agent of "redmouth disease." *Journal of the Fisheries Research Board of Canada* 22:713–719.

Ross, A. J., R. R. Rucker, and E. W. Ewing. 1966. Description of a bacterium associated with redmouth disease of rainbow trout (*Salmo gairdneri*). *Canadian Journal of Microbiology* 19:763–770.

Saeed, M. O. 1995. Association of *Vibrio harveyi* with mortalities in cultured marine fish in Kuwait. *Aquaculture* 136:21–29.

Sakai, M., G. Koyama, S. Atsuta, and M. Kobayashi. 1987. Detection of *Renibacterium salmoninarum* by a modified peroxidase-antiperoxidase (PAP) procedure. *Fish Pathology* 22:1–5.

Sakai, M., S. Atsuta, and M. Kobayashi. 1989. Comparison of methods used to detect *Renibacterium salmoninarum,* the causative agent of bacterial kidney disease. *Journal of Aquatic Animal Health 1:21–24.*

Sami, S., T. Fischer-Scherl, R. W. Hoffmann, and C. Pfeil-Putzien, 1992. Immune complex-mediated glomerulonephritis associated with bacterial kidney disease in the rainbow trout (*Oncorhynchyus mykiss*). *Veterinary Pathology* 29:169–174.

Sanders, J. E., and J. L. Fryer. 1980. *Renibacterium salmoninarum* gen. nov., sp. nov., the causative agent of bacterial kidney disease in salmonid fishes. *International Journal of Systematic Bacteriology* 30:496–502.

Sanders, J. E., K. S. Pilcher, and J. L. Fryer. 1978. Relation of water temperature to bacterial kidney disease in coho salmon (*Oncorhynchus kisutch*), sockeye salmon (*O. nerka*) and steelhead trout (*Salmo gairdneri*). *Journal of the Fisheries Research Board of Canada* 35:8–11.

Sanders, J. E., J. J. Long, C. K. Arakawa, J. L. Bartholomew, and J. S. Rohovec. 1992. Prevalence of *Renibacterium salmoninarum* among downstream migrating salmonids in the Columbia River. *Journal of Aquatic Animal Health* 4:72–75.

Santos, Y., F. Pazos, and A. E. Toranzo. 1996. Biochemical and serological analysis of *Vibrio anguillarum* related organisms. *Diseases of Aquatic Organisms* 26:67–73.

Santos, Y., F. Pazos, S. Nuñez, and A. E. Toranzo. 1997. Antigenic characterization of *Vibrio anguillarum* related organisms isolated from turbot and cod. *Diseases of Aquatic Organisms* 28:45–50.

Sawyer, E. S., R. G. Strout, and B. A. Coutermarsh. 1979. Comparative susceptibility of Atlantic (*Salmo salar*) and coho (*Oncorhynchus kisutch*) salmon to three strains of *Vibrio anguillarum* from the Maine–New Hampshire coast. *Journal of the Fisheries Research Board of Canada* 36:280–285.

Schachte, J. H. 1983. Coldwater disease, In: *A Guide to Integrated Fish Health Management in the Great Lakes Basin,* edited by F. P. Meyer, J. W. Warren, and T. G. Carey, 193–197. Special publication no. 83-2. Ann Arbor, MI: Great Lakes Fisheries Commission.

Schiewe, M. H. 1981. Taxonomic status of marine vibrios pathogenic for salmonid fish. *Developments in Biological Standards* 49:149–158.

Schiewe, M. H., J. H. Crosa, and E. J. Ordal. 1977. Deoxyribonucleic acid relationships among marine vibrios pathogenic to fish. *Canadian Journal of Microbiology* 23:954–958.

Schiewe, M. H., T. C. Trust, and J. H. Crosa. 1981. *Vibrio ordalii* sp. nov.: a causative agent of vibriosis in fish. *Current Microbiology* 6:343–348.

Schmidtke, L. M., and J. Carson. 1995. Characteristics of *Flexibacter psychrophilus* isolated from several fish species in France. *Diseases of Aquatic Organisms* 21:157–161.

Schnick, R. A., F. P. Meyer, and D. L. Gray. 1989. *A guide to approved chemicals in fish production and fishery resources management.* U.S. Fish and Wildlife Service and Arkansas Cooperative Extension Service MP241-5M-3-89RV, 17 pp.

Schroder, M. B., S. Espelid, and T. O. Jørgensen. 1992. Two serotypes of *Vibrio salmonicida* isolated from diseased cod (*Gadus morhua* L.); virulence, immunological studies and vaccination experiments. *Fish & Shellfish Immunology* 2:211–221.

Schubert, R. H. W. 1967. The taxonomy and nomenclature of the genus *Aeromonas* Kluyver and van Niel 1936, Part I. Suggestions on the taxonomy and nomenclature of the aerogenic *Aeromonas* species. *International Journal of Systematic Bacteriology* 17:23–37.

Shotts, E. B., Jr., and G. L. Bullock. 1975. Bacterial diseases of fishes: procedures for Gram negative pathogens. *Journal of Fisheries Research Board of Canada* 32:1243–1247.

Smith, I. W. 1964. The occurrence of pathology of Dee disease. *Department of Agriculture of Fisheries, Scottish Freshwater Salmon Fisheries Research* 34:1–12.

Smith, P. A., J. R. Contreras, L. H. Garces, J. J. Larenas, and S. Oynedel. 1996. Experimental challenge of coho salmon and rainbow trout with *Piscirickettsia salmonis. Journal of Aquatic Animal Health* 8:130–134.

Snieszko, S. F. 1981. Bacterial gill disease of freshwater fishes. *Fish Disease Leaflet no. 62,* U.S. Fish & Wildlife Service.

Souter, B. W., A. G. Dwilow, and K. Knight. 1987. *Renibacterium salmoninarum* in wild Arctic char *Salvelinus alpinus* and lake trout *S. namaycush* from the Northwest Territories, Canada, *Diseases of Aquatic Organisms* 3:151–154.

Speare, D. J., and H. W. Ferguson. 1989. Clinical features of bacterial gill disease of salmonids in Ontario. *Canadian Veterinary Journal* 30:882–887.

Speare, D. J., H. W. Ferguson, F. W. M. Beamish, J. A. Yager, and S. Yamashiro. 1991a. Pathology of bacterial gill disease: ultrastructure of branchial lesions. *Journal of Fish Diseases* 14:1–20.

Speare, D. J., H. W. Ferguson, F. W. M. Beamish, J. A. Yager, and S. Yamashiro. 1991b. Pathology of bacterial gill disease: sequential development of lesions during natural outbreaks of disease. *Journal of Fish Diseases* 14:21–32.

Starliper, C. E. 1996. Genetic diversity on North American isolates of *Renibacterium salmoninarum. Diseases of Aquatic Organisms.* 27: 207–213.

Starliper, C. E., and J. D. Teska. 1995. Relevance of *Renibacterium salmoninarum* in an asymptomatic carrier population of brook trout, *Salvelinus fontinalis* (Mitchell). *Journal of Fish Diseases* 18:383–387.

Stevenson, R. M. W., and D. W. Airdrie. 1984. Serological variation among *Yersinia ruckeri* strains. *Journal of Fish Diseases* 7:247–254.

Stevenson, R. M. W., and J. G. Daly. 1982. Biochemical and serological characteristics of Ontario isolates of *Yersinia ruckeri. Canadian Journal of Fisheries and Aquatic Sciences* 39:870–876.

Stevenson, R., D. Flett, and B. T. Raymond. 1993. Enteric redmouth (ERM) and other enterobacterial infections of fish. In: *Bacterial Diseases of Fish,* edited by V. Inglis, R. J. Roberts, and N. R. Bromage, 80–105. Oxford, UK: Blackwell Scientific Press.

Strohl, W. R., and L. R. Tait. 1978. *Cytophaga aquatilis* sp. nov., a facultative anaerobe isolated from the gill of freshwater fish. *International Journal Systematic Bacteriology* 28:293–303.

Tajima, K., M. Yoshimizu, S. Ezura, and T. Kimura. 1981. Causative organisms of vibriosis among pen cultured coho salmon (*Oncorhynchus kisutch*) in Japan. *Bulletin of the Japanese Society of Scientific Fisheries* 47:35–42.

Tajima, K., Y. Ejura, and T. Kimura. 1985. Studies on the taxonomy and serology of causative organisms of fish vibriosis. *Fish Pathology* 20:131–142.

Tebbit, G. L., J. D. Erickson, and R. B. Vande Water. 1981. Development and use of *Yersinia ruckeri* bacterins to control enteric redmouth disease. *Developments in Biological Standardization* 49:395–401.

Teska, J. D., A. Dawson, C. E. Starliper, and D. Tillinghast. 1995. A multiple-technique approach to investigating the presumptive low-level detection of *Renibacterium salmoninarum* at a broodstock hatchery in Maine. *Journal of Aquatic Animal Health* 7:251–256.

Thorburn, M. A. 1987. Factors influencing seasonal vibriosis mortality rates in Swedish pen-reared rainbow trout. *Aquaculture* 67:79–85.

Thorburn, M. A., T. E. Carpenter, and R. E. Plant. 1987. Perceived vibriosis risk by Swedish rainbow trout netpen farmers: Its effect on purchasing patterns and willingness-to-pay for vaccination. *Preventive Veterinary Medicine* 4: 419–434.

Tiecco, G., C. Sebastio, E. Francioso, G. Tantillo, and L. Corbari. 1988. Vaccination trials against "red plague" in eels. *Diseases of Aquatic Organisms* 4:105–107.

Titball, R. W., and C. B. Munn. 1981. Evidence for two haemolytic activities from *Aeromonas salmonicida. FEMS Microbiological Letters* 12:27–30.

Toranzo, A. E., J. L. Romalde, S. Numez, A. Figueras, and J. L. Barja. 1993. An epizootic in farmed, market-size rainbow trout in Spain caused by a strain of *Carnobacterium piscicola* of unusual virulence. *Diseases of Aquatic Organisms* 17:87–99.

Totland, G. K., A. Nylund, and K. O. Holm. 1988. An ultrastructural study of morphological changes in Atlantic salmon, *Salmo salar* L., during the development of cold water vibriosis. *Journal of Fish Disease* 11:1–3.

Traxler, G. S., and G. R. Bell. 1988. Pathogens associated with impounded Pacific herring *Clupea harenqus pallasi,* with emphasis on viral erythrocytic necrosis (VEN) and atypical *Aeromonas salmonicida. Diseases of Aquatic Organisms* 5:93–100.

Trust, T. J., A. G. Khouri, R. A. Austin, and L. D. Ashburner. 1980. First isolation in Australia of atypical *Aeromonas salmonicida. FEMS Microbiological Letters* 9:39–42.

Trust, T. J., E. E. Eshiguro, H. Chart, and W. W. Ray. 1983. Virulence properties of *Aeromonas salmonicida. Journal of the World Mariculture Society* 14:193–200.

Udey, L. R., and J. L. Fryer. 1978. Immunization of fish with bacterins of *Aeromonas salmonicida. Marine Fisheries Review* 40:12–17.

Umbreit, T. H., and M. R. Tripp. 1975. Characterization of the factors responsible for death of fish infected with *Vibrio anguillarum. Canadian Journal of Microbiology* 21:1272–1274.

Valtonen, E. T., P. Rintamaki, and M. Koskivaara. 1992. Occurrence and pathogenicity of *Yersinia ruckeri* at fish farms in northern and central Finland. *Journal of Fish Diseases* 15:163–171.

von Graevenitz, A. 1990. Revised nomenclature of *Campylobacter laridis, Enterobacter intermedium* and *"Flavobacterium branchiophila." International Journal of Systematic Bacteriology* 40:211.

Wakabayashi, H., and T. Iwado. 1985. Changes in glycogen, pyruvate and lactate in rainbow trout with bacterial gill disease. *Fish Pathology* 20:161–165.

Wakabayashi, H., S. Egusa, and J. L. Fryer. 1980. Characteristics of filamentous bacteria isolated from a gill disease of salmonids. *Canadian Journal of Fisheries and Aquatic Sciences* 37:1499–1504.

Wakabayashi, H., G. J. Huh, and N. Kimura. 1989. *Flavobacterium branchiophila* sp. nov., a causative agent of bacterial gill disease of freshwater fishes. *International Journal of Systematic Bacteriology* 39:213–216.

Wakabayashi, H., M. Horiuchi, T. Bunya, and G. Hoshiai. 1991. Outbreaks of cold-water disease in coho salmon in Japan. *Fish Pathology* 26:211–212.

Wallbanks, S., A. J. Martinez-Murcia, J. L. Fryer, B. A. Phillips, and M. D. Collins. 1990. 16S rRNA sequence determination for members of the genus *Carnobacterium* and related lactic acid bacteria and description of *Vagococcus salmoninarum* sp. Nov. *International Journal of Systematic Bacteriology* 40:224–230.

Waltman, W. D., and E. B. Shotts, Jr. 1984. A medium for the isolation and differentiation of *Yersinia ruckeri. Canadian Journal of Fisheries and Aquatic Sciences* 41:804–806.

Warren, J. W. 1963. Kidney disease of salmonid fishes and the analysis of hatchery waters. *The Progressive Fish-Culturist* 25:121–131.

Watkins, W. D., R. E. Wolke, and V. J. Cabelli. 1981. Pathogenicity of *Vibrio anguillarum* for juvenile winter flounder, Pseudopleuronectes americanus. *Canadian Journal of Fisheries and Aquatic Sciences* 38:1045–1051.

West, P. A., and J. V. Lee. 1982. Ecology of *Vibrio* species, including *Vibrio cholerae,* in natural waters of Kent, England. *Journal of Applied Bacteriology* 52:435–445.

Wichardt, U-P., N. Johansson, and O. Ljunberg. 1989. Occurrence and distribution of *Aeromonas salmonicida* infections on Swedish fish farms, 1951–1987. *Journal of Aquatic Animal Health* 1:187–196.

Wiens, G. D., and S. L. Kaattari. 1991. Monoclonal antibody characterization of a leukoagglutinin produced by *Renibacterium salmoninarum. Infection and Immunity* 59:631–637.

Wiik, R., K. Anderson, F. L. Daae, and F. A. Hoff. 1989. Virulence studies based on plasmid profiles of the fish pathogen *Vibrio salmonicida. Applied Environmental Microbiology* 55:819–825.

Wiklund, T., K. Kaas, L. Lömström, and I. Dalsgaard. 1994. Isolation of *Cytophaga psychrophila* (*Flexibacter psychrophilus*) from wild and farmed rainbow trout (*Oncorhynchus mykiss*) in Finland. *Bulletin of the European Association of Fish Pathologist.* 14:44–46.

Willumsen, B. 1989. Birds and wild fish as potential vectors of *Yersinia ruckeri. Journal of Fish Diseases* 12:275–277.

Wood, E. M., and W. T. Yasutake. 1956. Histopathology of fish III: Peduncle ("cold-water") disease. *The Progressive Fish-Culturist* 18:58–61.

Wood, J. W. 1974. *Diseases of Pacific salmon: their prevention and treatment,* 2nd ed. Olympia: State of Washington Department of Fisheries, Hatchery Division, 22 pp.

Wood, J. W., and J. Wallis. 1955. Kidney disease in adult chinook salmon and its transmission by feeding young chinook salmon. *Fisheries Commission of Oregon, Research Brochure* 6, 32 p.

Wood, P. A., and S. L. Kaattari. 1996. Enhanced immunogenicity of *Renibacterium salmoninarum* in chinook salmon after removal of the bacterial cell surface-associated 57 kDa protein. *Diseases of Aquatic Organism* 25:71–79.

Wood, P. A., G. D. Wiens, J. S. Rohovec, and D. D. Rockey. 1995. Identification of an immunologically cross-reactive 60-kilodalton *Renibacterium salmoninarum* protein distinct from p57: Implications for immunodiagnostics. *Journal of Aquatic Animal Health* 7:95–103.

Woodall, A. N., and G. Laroche. 1964. Nutrition of salmonid fishes, XI. Iodide requirements of chinook salmon. *Journal of Nutrition* 824:475–482.

Wooster, G. A., and P. R. Bowser. 1996. The aerobiological pathway of a fish pathogen: survival and dissemination of *Aeromonas salmonicida* in aerosols and its implications in fish health management. *Journal of the World Aquaculture Society* 27:7–14.

Young, C. L., and G. B. Chapman. 1978. Ultrastructural aspects of the causative agent and renal histopathology of bacterial kidney disease in brook trout (*Salvelinus fontinalis*). *Journal of the Fisheries Research Board of Canada* 35:1234–1248.

18 Striped Bass and Hybrid Bacterial Diseases

Many bacterial disease organisms associated with other aquaculture species also occur in striped bass and their hybrids (white bass). There appears to be no difference in disease susceptibility between genetically pure striped bass and hybrids. With expansion of striped bass culture and intensification of rearing methods (Harrell 1997), there has not been an increase in the number of diseases that affect these fish, but consequences of some infections have become more acute. In some instances, infectious bacterial diseases have limited and severely affected striped bass culture (Hawke 1996).

Most bacteria that infect striped bass, with several exceptions, are saprophytic, facultative, and opportunistic organisms that often cause debilitating infections following exogenous, inanimate predisposing factors. Species in several genera of bacteria, namely acid-fast staining (*Mycobacterium*) and *Photobacterium* (*Pasteurella)*, are the most serious, with Streptococcus being somewhat less so (Plumb 1997). Other bacterial pathogens of striped bass include species of *Aeromonas, Pseudomonas, Vibrio, Edwardsiella,* and *Flavobacteria.*

MYCOBACTERIOSIS AND NOCARDIOSIS

Acid-fast staining bacteria were first discovered as fish pathogens in carp in Europe during the latter part of the 19th century (Austin and Austin 1987). Mycobacteriosis was originally known as "fish tuberculosis" because the causative organism is taxonomically similar to that which causes tuberculosis and leprosy in humans. It was later suggested that the disease should more correctly be called fish "mycobacteriosis" because other than the organism's acid-fast staining characteristic and taxonomic classification, very little similarity exists between the fish infection and human tuberculosis. Members of two genera of acid-fast bacteria *Mycobacterium* (mycobacteriosis) and *Nocardia* (nocardiosis) cause similar fish infections; therefore, they are considered together here.

Mycobacteriosis is infrequently encountered in fish, but when present, it is usually chronic. The disease is caused by *Mycobacterium marinum, Mycobacterium fortuitum* (Van Duijn 1981), or *Mycobacterium chelonei* (Arakawa and Fryer 1984). Two species of *Nocardia* are known to cause nocardiosis in fish: *Nocardia asteroides* and *Nocardia kampachi* (Nigrelli and Vogel 1963; Conroy 1964) and are difficult to distinguish from mycobacteriosis.

Geographical Range and Species Susceptibility

Mycobacteriosis occurs in marine environments throughout the world and with increasing regularity in freshwater fishes. The disease is most severe in cultured or aquarium fish, but it does occur in wild populations, in which consequences are usually mild. Nigrelli and Vogel (1963) listed more than 150 marine and freshwater fish species, including salmonids and ornamentals, in which *M. fortuitum* and/or *M. marinum* had been documented. All teleosts, especially striped bass and salmonids, should be considered as possible hosts. Under certain conditions, cultured striped bass and hybrids are particularly susceptible to *M. marinum. M. chelonei* occurs in salmonids in Japan (Arakawa and Fryer 1984). *M. fortuitum* and *M. marinum* are also pathogenic to humans and other

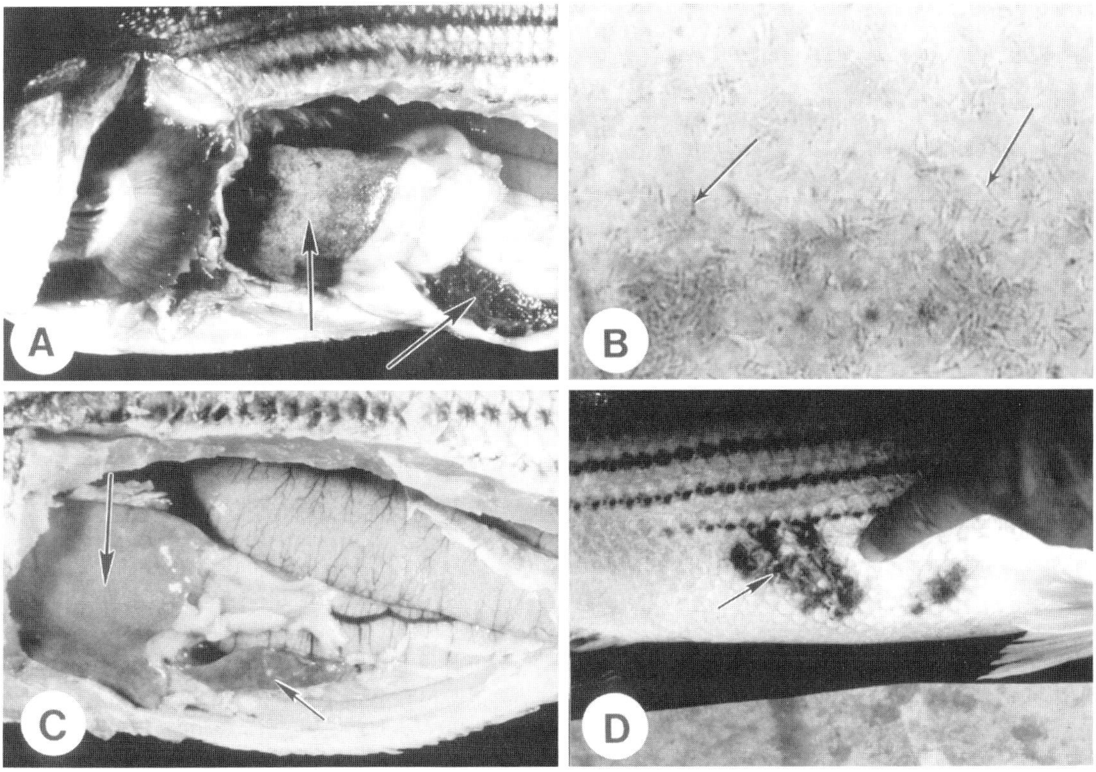

FIGURE 18.1. Diseases of striped bass. (A) Mycobacterium marinum *infection. Gills have pale areas at base as result of granuloma. The liver and spleen* (arrows) *have white granulomatous, rough surfaces.* (B) *Acid-fast staining mycobacterial cells* (arrows) *in a granuloma.* (C) *Striped bass infected with* Photobacterium damsela. *Note the pale, slightly mottled liver with granulomas* (large arrows) *and the pale, granulomatous spleen* (small arrows). *(Photographs A, B, and C courtesy of J. Hawke.)* (D) Aeromonas hydrophila *infection in the skin of striped bass* (arrow).

homotherms, frogs, snakes, and lizards (Goodfellow and Wayne 1982).

Nocardia sp. fish infections have been reported specifically from the United States, Argentina, Germany, Japan, and Taiwan (Post 1987; Chen et al. 1988). In all likelihood, it occurs worldwide in freshwater and saltwater fish but less frequently than mycobacteriosis. *Nocardia* infections have been documented in rainbow trout, brook trout, neon tetra, yellowtail, Formosa snakehead, giant gourami, and largemouth bass (Snieszko et al. 1964a; Chen 1992; Kitao et al. 1989), and ornamental fish.

Clinical Signs and Findings

Gross external clinical signs of mycobacteriosis vary depending on fish species (Van Duijn 1981). *M. marinum*–infected striped bass are lethargic, darkly pigmented, and emaciated with sunken abdomens and have occasional ulceration and hemorrhaging in the skin (Hedrick et al. 1957). Infected fish may also become anorexic, have grayish and irregular skin ulcerations (Figure 18.1), deformed vertebrae and mandibles, and exophthalmia that can result in loss of one or both eyes. Nodular tubercles that develop in the muscle appear externally as diffuse, light brown spots or swollen areas that can rupture. Scales develop lepidorthosis before being lost, and white streaks occur parallel to cartilaginous gill filament supports. Secondary sexual characteristics (hooked jaw and color changes) do not develop in infected adult Pacific salmon, and these fish may be smaller than normal, more darkly colored, and have undeveloped gonads. Diseased ornamental fish usually lose their bright coloration.

Internal gross pathology is more consistent among fish species; the most notable being a pale, granulomatous liver that has a granular surface that gives it and other organs a rather rough, sandpaper texture (Figure 18.1). Granulomas also develop in the spleen, heart, kidney, and mesenteries. The spleen and anterior kidney become extremely enlarged.

Clinical signs of *Nocardia* infections are similar to those of mycobacteriosis (Austin and Austin 1987). Fish swim in a rapid tail-chasing mode or are sluggish and become anorexic and emaciated and show abdominal distension. Fish may also lose scales; they frequently exhibit exophthalmia and have opaque eyes. Blood pools under the epithelium of the oral and opercular cavities, and multiple yellowish white nodules varying in size from 0.5 to 2.0 cm in diameter are scattered throughout the muscle, gill, heart, liver, spleen, ovary, and mesenteries.

Diagnosis

Mycobacteriosis is routinely diagnosed by detection of strongly acid-fast (Ziehl-Neelsen) staining (red) bacteria in smears from nodules or histological sections of the tubercles (see Figure 18.1). *Mycobacterium* spp. are Gram-positive, strongly acid-fast, nonmotile, pleomorphic rods that measure 0.25–0.35 × 1.5–2.0 μm (Wayne and Kubica 1986; Frerichs 1993).

Lowenstein-Jensen or Petriganis media provide the best growth for *Mycobacterium* spp. but *M. marinum, M. fortuitum,* and *M. chelonei* can be isolated on blood agar if inoculated plates are sealed to retain moisture. *M. marinum* forms colonies in 2 to 3 weeks when incubated at 25 to 30°C but will grow at 20 to 37°C (Hedrick et al. 1987). *M. fortuitum* forms colonies in about 7 days when incubated at 25°C and grows at 19 to 42°C, with an optimum range of 30 to 37°C (Van Duijn 1981). Colonies may be smooth, rough, moist, dry, raised, or flat depending on the media and age of culture. Material from skin and other tissues containing mixed bacterial species are treated with 0.3% Sepheran before primary culture, after which pure cultures can be maintained on Lowenstein-Jensen media.

Nocardia and *Mycobacterium* cause similar gross pathology and therefore can best be distinguished by culture; *Nocardia* spp. are more easily isolated than *Mycobacterium. Nocardia* spp. are Gram-positive, weakly acid-fast, nonmotile, long and branching rods that can be detected either in tissue sections or smears from granulomatous nodules (Chen et al. 1988). Colonies are irregular and rough; white, pinkish, orange, or yellow in color; and may require up to 21 days at 18 to 37°C for growth (Frerichs 1993).

Monoclonal antibodies (MABs) were developed by Adams et al. (1996) for identification of *M. marinum, M. fortuitum,* and *M. chelonei;* however, these MABs were not strain specific when used in a sandwich enzyme linked immunosorbent assay (ELISA) system. It is possible that these reagents could, however, be used to detect mycobacteria in tissues of infected fish, thus providing a rapid diagnostic method.

Bacterial Characteristics

M. marinum, M. fortuitum, and *M. chelonei* can be differentiated on the basis of several biophysical and biochemical properties (Table 18.1). *M. marinum* produces nicotinamidase and pyrazinamidase but *M. fortuitum* does not; *M. fortuitum* is positive for nitrate reductase, whereas *M. marinum* and *M. chelonei* are negative. *M. marinum* produces a yellowish orange pigment when exposed to light, which *M. fortuitum* lacks (Wheeler and Graham 1989).

Epizootiology

Mycobacteriosis in fish has been present for years, but recently, it has become one of the most serious infections to occur in intensive, recirculating striped bass (and hybrids) culture systems. The disease is usually chronic in intensively cultured striped bass, a condition that allows accumulation of bacteria to reach a point at which serious subacute infection occurs.

The bacterium source in recirculating systems is unknown. No mycobacteriosis problems have been reported in pond-cultured striped bass; however, Sakanari et al. (1983) found the pathogen in wild striped bass. Experimentally infected fish shed acid-fast bacteria into water, but little is actually known about its epizootiology in wild fish. It is likely that low-level mycobacterial infections in wild fish populations serve as reservoir pathogen sources for cultured fish. Months, or possibly years, may pass between naturally occurring expo-

TABLE 18.1. SELECTED BIOPHYSICAL AND BIOCHEMICAL CHARACERISTICS OF *MYCOBACTERIUM MARINUM*, *M. CHELONEI*, AND *NOCARDIA ASTEROIDES*

Characteristic	*M. marinum*[a]	*M. chelonei*[a]	*N. asteroides*[a]
Possesses mycolic acid	+	+	+
Colony morphology	Yellow to orange	Off white	?
Cell morphology	Long, branching rods	Long, rods	Polymorphic
Cell size (μm)	0.25–0.35 × 2	0.3–0.6 × 1–4	?
Acid fast	Strong	Strong	Weak
Growth at			
25°C	+	+	+
37°C	May adapt	–	+
Urease	+	+	+
Nitrate reduction*	–	+	+
Acid phosphatase	+	+	?
ß-Galactosidase	–	?	?
Acid phosphatase	+	+	?
Catalase	+	+	V
Peroxidase	+	?	?
Degrades Tween 80*	+	±	?
∂-Esterase	+	?	?
Photochromogenic	+	?	–
Growth on 5% NaCl	±	–	?
Glucose sole source of carbon	±	+	?
Pyruvate sole source of carbon	+	?	?
Nicotinamidase production*	+	–	?
Pyrazinamidase production*	+	–	?
Mol% G + C of DNA	62–70	61–65	64–72

Sources: Arakawa and Fryer (1984); Lechevalier (1986); Wayne and Kubica (1986).
[a]+ = positive; – = negative; V = variable; ? = unknown; and * = characteristics to separate *M. marinum* from *M. chelonei*.

sure to the bacterium and clinical disease. During an epidemiological study of Atlantic mackerel from the northeastern Atlantic Ocean, fish more than 2 years of age showed evidence of mycobacterial infection (MacKenzie 1988). This increased incidence of infection in older fish indicates a chronic disease; however, mycobacteriosis may occur in any age or size fish.

In the 1950s, ingestion of raw contaminated fish viscera was the probable source of mycobacteriosis in cultured salmon in the northwestern United States because at that time, raw fish was common in their diet (Ross 1970). Mycobacteria-free fish were found only in hatcheries where raw fish was not fed, and the bacteria disappeared from all hatcheries when raw fish diets were discontinued. This transmission theory was further substantiated when Chinabut et al. (1990) successfully transmitted mycobacteria in snakehead by feeding raw offal to naive fish. Conversely, in Australia, Ashburner (1977) reported *M. marinum* in fresh water cultured chinook salmon and he eliminated feed as a pathogen source of infection and presented evidence that the pathogen was vertically passed to the F_1 generation during spawning.

Juvenile rainbow trout and juvenile chinook salmon were experimentally infected with *M. chelonei* at 18°C; the trout suffered 20% to 52% mortality, whereas 98% of the salmon died within 10 days postinfection, indicating a difference in species susceptibility (Arakawa and Fryer 1984). They also found the prevalence of *Mycobacterium* to be 0% to 26% in wild juvenile coho, compared with 1.4% to 4.0% in chinook salmon. Mycobacteriosis was not diagnosed in Oregon salmon hatcheries between 1964 and 1981 (Fryer and Sanders 1981), but studies suggest that it was still present and would continue to occur throughout the anadromous salmon's life cycle.

Morbidity in striped bass populations may be low at any given time, but cumulative mortality can be high. Hedrick et al. (1987) reported that 50% of a *M. marinum*–infected yearling striped bass population died within months of being stocked into an intensive culture system; the greater the intensification, the more serious the disease became, and 80% of survivors were carriers. In closed recirculating systems where hybrid striped bass were experiencing chronic mortalities, 30% to 50% of randomly sampled fish had characteristic mycobacterial granulomas present in their internal organs (J. Newton, Auburn University, personal communication). Sakanari et al. (1983) found prevalence of mycobacteriosis in wild striped bass to be 25% to 68% in California and 46% in Oregon. Mortality in other farmed fish has varied from 35% in naturally infected populations of three-spot gourami in Columbia to 100% in experimentally infected pejerrey (Hatai et al. 1988). In Israel, cultured European seabass suffered 50% mortality from mycobacteriosis, but pathogen prevalence was 100% (Colorni 1992).

Nocardia may be a normal inhabitant of the environment (soil and water) or fish may serve as pathogen reservoirs, but whatever the source, it produces a slowly developing chronic infection. There is little information available on *Nocardia*-infected fish mortalities, but a documented case involving *N. asteroides*–infected Formosa snakehead cultured in a fresh water pond in Taiwan revealed that 20% of 30,000, 8- to 9-month-old (20 to 30 cm) fish were killed in 2 weeks (Kariya et al. 1968). Successful experimental transmission of nocardiosis has been inconsistent. Snieszko et al. (1964a) were unable to transmit *N. asteroides* orally from rainbow trout to other trout but were somewhat more successful with transmission by injection, requiring 1 to 3 months for disease to develop. Chen (1992) transmitted *N. asteroides* from largemouth bass to other individuals of the same species by intramuscular injection that resulted in characteristic granular nodules in the visceral organs and 100% mortality. Kusuda and Nakagawa (1978) reported successful transmission of the disease in yellowtail by injection or by smearing surface wounds with *Nocardia seriolae* (formerly *Nocardia kampachi*), suggesting that route of infection is more likely through epidermal injury than orally.

M. marinum has the potential to cause infection on human extremities that come in contact with infected fish; therefore, caution should be exercised by fish producers when handling striped bass or any other fish in facilities that may harbor the pathogen. Skin lesions caused by *M. marinum* are usually confined to hands, wrist, and forearms, where hard, raised, calcified, granulomas develop (Giavenni 1979). Occasionally, *M. marinum* has been known to be more invasive in humans, causing infection of the tendon sheaths, joints, and bone (Wolinski 1992). Fish isolates are able to adapt somewhat to the higher body temperatures of humans, but apparently the lower skin temperature of human extremities is more conducive to establishment of infection; therefore, it seldom becomes systemic unless an individual is otherwise debilitated (Frerichs 1993). Immunocompromised individuals, such as those suffering from human immune deficiency virus infections, may be particularly susceptible to the so-called atypical mycobacterial infections (Frerichs and Roberts 1989).

Individuals who clean marine fish for a living, handle certain cultured fish, or work with saltwater ornamental fish are at risk of contracting the disease. The presence of open scratches or wounds on the skin most likely enhances infection. *M. marinum* is also the causative agent of "swimming pool granuloma," a disease contracted by swimmers that appears as a cutaneous granuloma on elbows, fingers, knees, feet, and toes. *M. fortuitum* is an opportunistic pathogen in humans that may develop following superficial trauma or surgery, and because of its tolerance for higher temperatures, it can infect lungs, lymph nodes, and internal organs (Van Duijn 1981).

Pathological Manifestations

In fish, a mycobacteria infection results in proliferation of connective tissue but exhibits little inflammatory response other than granulomatous inflammation (Van Duijn 1981). Mycobacteriosis in teleosts is considered to be less cellular than is tuberculosis in mammals, and some refute the presence of Langerhans giant cells that are characteristic of tuberculosis (Nigrelli and Vogel 1963; Giavenni 1979). Timur et al. (1977), however, showed caseation, typical Langerhans cell production, and cell-mediated immunity in *M. marinum* infections in plaice. Large masses of bacteria were found in the visceral adipose tissue and hematopoietic tissue of the kidneys, spleens, and livers of young fish as well as in adult fish. Foci of bacteria that surrounded the intestinal tract of young fish disappeared in older fish, leaving large areas of caseous necrosis. The spleen, liver, and kidney had severe lesions with massive concentrations of acid-fast bacteria. Caseous necrosis also formed in the kidney.

In infected fish, bacterial metabolism is inhibited by the walling off of the pathogen by connective tissue that leads to the organism's death (Van Duijn 1981). The resulting tubercle, which may contain black pigmentation, becomes necrotic, and mineralization takes place, leading to cavitation. Acid-fast bacteria are usually present in young nodules but are absent in older granulomas.

In the early stages of a *Nocardia* largemouth bass infection, there was an acute, serous inflammation that resulted in production of an exudate containing cellular and bacterial debris that eventually became granulomatous (Chen 1992). The nodules consisted of necrotizing foci surrounded by epithelioid cells, fibroblasts, or fibrous encapsulation. The most characteristic tubercular nodule structures were bacillary masses within small cavities that were surrounded by concentric layers of fibrous tissue. Long, branching, filamentous, weakly acid-fast bacteria lay within these nodules.

Significance

In the southern and eastern United States where *M. marinum* has become established, its impact has been greatest in hybrid striped bass–white bass populations reared in recirculating culture systems. Because raw fish is no longer fed to salmon, mycobacteriosis is seldom a problem in today's cultured salmon populations, but it does occur in other species of seawater cultured fish. Mycobacteriosis continues to be a chronic disease problem in ornamental fish in home aquaria and/or in large public aquaria. Also, when an organism can be transmitted from a lower vertebrate to humans, the significance of that disease is immediately elevated. Nocardiosis does not appear to be a particularly significant disease because outbreaks have been sporadic and infectious incidence low.

PHOTOBACTERIOSIS (PASTEURELLOSIS)

Pasteurellosis, caused by *Pasteurella piscicida,* was first described in white perch and striped bass in the Chesapeake Bay of Virginia and Maryland (Snieszko et al. 1964b). Recently, the organism's name was changed to *Photobacterium damsela,* subspecies *piscicida* (Gauthier et al. 1995), and the disease it causes is now referred to as "photobacteriosis." The disease is a chronic to subacute, systemic infection of wild and cultured fish, and due to prominent white granulomatous lesions in the internal organs, it has been known as "pseudotuberculosis."

Geographical Range and Species Susceptibility

In the United States, *P. damsela* (subsp. *piscicida*) has been reported from the Chesapeake Bay area, Long Island Sound, New York, and the Gulf of Mexico (Snieszko 1964b; Robohm 1983; Hawke et al. 1987). Its known range outside of the United States now includes Japan, Norway, Taiwan, Mediterranean Sea region (Spain, Greece, and Malta), France, Italy, and Israel (Kimura and Kitao 1971; Tung et al. 1985; Toranzo et al. 1991; Baudin-Laurencin et al. 1991; Bakopoulos et al. 1997).

Photobacteriosis is primarily a disease of marine fish but does infrequently occur in fresh water. In addition to striped bass, fish known to be susceptible to *P. damsela* include yellowtail, red seabream, black seabream, striped jack, and gilthead seabream (Kimura and Kitao 1971; Kusuda and Yamaoka 1972; Toranzo et al. 1991; Nakai et al. 1992; Kitao 1993). There has been one report of the disease in snakehead in Taiwan; however, these fish had been fed raw marine fish products before infection (Tung et al. 1985).

Clinical Signs and Findings

Photobacteriosis can occur in a subacute or chronic form but is always a septicemia (Thune et al. 1993). In the acute form, clinical signs are loss of mobility and sinking in the water column, pale gills, dark pigmentation, and presence of discrete petechia at the base of fins and on the operculum (Robohm 1983; Hawke 1996). Internally, an enlarged spleen and mottled liver are evident, with other organs appearing normal (see Figure 18.1). In chronic infections, external lesions are similar; internally, however, small white miliary lesions may be present in the swollen spleen and kidney, giving rise to the name "pseudotuberculosis." Cut sections of the spleen may show white patches (granulomas).

Diagnosis

P. damsela can be isolated on ordinary bacteriological media (brain heart infusion [BHI]) or blood agar to which 0.5% to 4.0% NaCl (2% is optimum) has been added (Robohm 1983; Hawke 1996). It does not grow in salt-free media. Colonies of *P. damsela* are 1 to 2 mm in diameter, entire, convex, viscous, and opaque to translucent when incubated at 25°C for 48 to 72 hours.

Bacterial Characteristics

The organism is a Gram-negative, bipolar staining, nonmotile bacillus that is oxidase and catalase positive (Snieszko et al. 1964b; Jansen and Surgalla 1968; Hawke 1996). The pleomorphic cells measure 0.5 to 0.8 µm wide and 0.7 to 2.6 µm in length. *P. damsela* has an optimum growth temperature of 23 to 27°C and grows at 10 to 30°C but not at 37°C. *P. damsela* is generally considered to be a homogeneous species because isolates from various geographical locations are morphologically, physiologically, and biochemically similar (Magarinos et al. 1992; Bakopoulos et al. 1995; Hawke 1996) (Table 18.2).

P. damsela will survive less than 3 days in sterile brackish water, but the organism appears to be a normal inhabitant of the estuarine environment, where fish are most likely the natural host (Jansen and Surgalla 1968). Brief survival of *P. damsela* outside the host has been shown, but no carrier state has been defined; it was, therefore, suggested that fish species other than those that become clinically ill may be pathogen reservoirs (Toranzo et al. 1991).

All strains of *P. damsela* tested serologically by Robohm (1983) and Magarinos et al. (1992) were similar. Kimura and Kitao (1971) were unable to distinguish Japanese from American isolates serologically. Also, Kitao and Kimura (1974) identified the pathogen 100% of the time by using a direct fluorescent antibody technique (FAT) on impression smears from organs that exhibited characteristic white lesions. Mori et al. (1976) detected incipient infection in spleens or kidneys, or both, of yellowtail by using FAT and suggested that this technique could be used for detection of subclinical infections. It was determined by Hawke (1996) that the Aquarapid-Pp ELISA kit, developed to detect European isolates, was cross-reactive with the *P. damsela* isolate from Louisiana, providing further evidence of homogeneity. Kusuda et al. (1978), however, were able to differentiate *P. damsela* isolates by immunoelectrophoreses, which implies that more than one serological strain may be involved.

TABLE 18.2. BIOPHYSICAL AND BIOCHEMICAL CHARACTERISTICS OF *PHOTOBACTERIUM DAMSELA* SUBSP. *PISCICIDA*

Characteristic[a]	Reaction
Gram stain	– (bipolar)
Motility	–
Growth at	
10°C	–
15°C	+
25°C	+
35°C	–
Production of	
Catalase	+
Oxidase	+
Phenylalanine deaminase	–
Gluconidase	–
Arginine dihydrolase	+
2-3, butanediol dehydroginase	+
Nitrate reduction	–
Methyl red	+
Voges-Proskauer	+
Oxidation/fermentation glucose	+/+
Gas from glucose	–
Growth in	
0.0% NaCl	–
0.5% NaCl	+
3.0% NaCl	+
5.0% NaCl	–
Arginine dihydrolase	+
Acid from	
Glucose	+
Mannose	+
Galactose	+
Fructose	+
Maltose	(+)
Phospholipase	+
Sensitivity to	
Vibriostat 0/129	+
Novobiocin	+
Degradation of	
Arginine	+
Starch	+
Tween 80	–

Sources: Gauthier et al. (1995); Hawke (1996)

[a]Negative for indole production, citrate utilization, ß-galactosidase (ONPG), elastase, amylase, urease, tryptophane deaminase, lysine, and ornithine decarboxylase; negative for acid from sucrose, lactose, rhamnose, arabinose, amygdalin, melibiose, mannitol, inositol, sorbitol, and glycerol.

Epizootiology

The initial epizootic of *P. damsela* occurred in the Chesapeake Bay where a massive white perch and striped bass die-off took place (Snieszko et al.

1964b). The epizootic began in the lower Potomac River in June and spread into the Chesapeake Bay during July. At the time of the epizootic, the white perch population was high and the Bay and its tributaries were heavily polluted with organic material; conditions thought to have contributed to the disease outbreak (Sinderman 1970). During the year following the epizootic, commercial harvest of white perch in the Bay was reduced by almost half, leading to speculation that *P. damsela* had killed up to 50% of the population. No other wild fish episode of *P. damsela* has been as devastating as was the Chesapeake Bay epizootic.

P. damsela is currently the most serious infectious disease problem of the fledgling striped bass industry along the Gulf of Mexico coast (Hawke 1996). At some coastal Louisiana fish farms from autumn of 1990 through autumn of 1992, it was estimated that 30% to 80% mortality due to *P. damsela* occurred in cultured hybrid striped bass reared in water that ranged from 20 to 30°C (Hawke 1996). Previously, Hawke et al. (1987) reported mortality of about 80% in juvenile striped bass cultured in brackish water ponds in Alabama. Nakai et al. (1992) reported that in Japan, *P. damsela* was responsible for the loss of 34% of 10,000 sea-cage–reared juvenile striped jack. Mortality from *P. damsela* varies in cultured fish but is generally higher in young cultured yellowtail and black seabream.

Environmental conditions probably play a major role in determining the seriousness of photobacteriosis. Matsusato (1975) reported that disease incidence in yellowtail rose during the rainy season when salinity dropped below 30 ppt and water temperatures were optimum (25°C) for the pathogen. Photobacteriosis generally occurs in striped bass during spring and autumn when water temperatures are optimum for the pathogen.

Route of *P. damsela* transmission is unknown, but its short-lived nature in brackish water has led to speculation that transmission is probably fish to fish, even though a carrier or latent state has not yet been proved. In support of the theory that fish are the pathogen reservoir, Toranzo et al. (1982) demonstrated that *P. damsela* survived for less than 2 days in fresh water and less than 5 days in salt water. Dead fish are definitely a reservoir of *P. damsela* because Matsudka and Kamada (1995) showed that yellowtail shed 10^7 to 10^9 CFU/fish 10 minutes after death and for 5 days thereafter.

P. damsela can be highly pathogenic but varies with fish species. In experimental transmission studies with an immersion LD_{50} of about 687 CFU/mL, fish began to die on day 5 and mortality continued through day 10 postexposure (Hawke 1996). An injectable LD_{50} of about 100 CFU/mL was established for Formosa snakehead (Tung et al. 1985), but in experiments with other species, greater numbers of the organism were required to kill fish. Nakai et al. (1992) established an injectable LD_{50} of 1000 CFU/fish in striped jack and an LD_{50} of 10 million CFU/fish in red seabream.

Pathological Manifestations

The histopathology of photobacteriosis varies to some degree between fish species. According to Hawke (1996), photobacteriosis produced a generalized septicemia in experimentally infected hybrid striped bass. Most severe histopathology occurred in the spleen and kidney, with the liver being affected to a lesser degree. There was necrosis of the gills, spleen, and kidney. Macrophages laden with bacteria were seen in each of these tissues, but as was also noted by Wolke (1975), little inflammation was present. Bacterial counts in these organs and the blood 6 days postinfection ranged from $10^{7.7}$ to $10^{9.6}$ CFU/g of tissue or milliliter of blood. No histopathology was seen in the olfactory lamellae, brain, intestine, heart, or skin.

In wild fish populations, few pathological changes can be noted in fish with acute photobacteriosis, whereas a chronic infection is characterized by miliary lesions in the kidney and spleen (Wolke 1975; Robohm 1983). Also in the chronic form, necrotic lymphoid and peripheral blood cells collect in the spleen, and focal hepatocyte necrosis occurs in the liver. No inflammatory response occurred in infected white perch, but spleens of experimentally infected juvenile striped bass developed multiple foci of bacterial colonies and reduced densities of cells. In infected yellowtail, live bacteria were present in phagocytes that became swollen to the point that capillary blood flow was blocked (Kubota et al. 1970). In chronically infected yellowtail, granulomas formed in the spleen and kidney where bacteria were localized.

In naturally *P. damsela*–infected striped bass, there was extensive necrosis of splenic lymphoid tissue, coagulation necrosis, and karyorrhexis (Hawke et al. 1987). Similar histopathology was seen in the liver but to a lesser degree. Rod-shaped bacteria were observed in sinusoids and hepatic vessels of the livers, and large areas of this organ exhibited hyperplasia of reticuloendothelial cells lining the hepatic sinusoids. Inflammatory cellular accumulations were absent in striped bass.

Pathogenesis of *P. damsela* is unknown; however, Nakai et al. (1992) found that its extracellular products (ECPs) were as pathogenic to striped jack and red seabream as was the bacterium, suggesting that pathology is the result of these products. In support of this, Noya et al. (1995) found a reduction in circulating red blood cells in gilthead seabream injected with *P. damsela* ECPs. Severe lesions in the liver and gills further suggested the importance of toxins in the pathogenesis of *P. damsela*. Injection of ECPs also produced an inflammatory response that was similar to that associated with live bacteria, including lymphopenia, granulocytosis, an increase in peritoneal exudate cells, and degranulation of eosinophilic granular cells. As pathology stimulated by live bacteria progressed, it was noted that degenerating macrophages contained intact bacteria, causing Noya et al. (1995) to postulate that these macrophages played a major role in the dissemination of *P. damsela*. Death of infected hybrid striped bass can likely be attributed to respiratory failure resulting from a combination of gill epithelium and support tissue necrosis and congestion of the sinusoids and capillaries of secondary gill lamellae (Hawke 1996).

Significance

Photobacteriosis has become a major disease problem in some mariculture communities. It is currently posing a major threat to the striped bass cage culture industry in Louisiana, where it has caused closure of some operations and threatens others (Hawke 1996). Photobacteriosis is also a serious threat to yellowtail culture in Japan (Kitao 1993) and to culture of several fish species in the Mediterranean region.

OTHER BACTERIAL DISEASES OF STRIPED BASS

Other bacteria that cause mild to severe disease in cultured and wild striped bass discussed in detail in other chapters are *Aeromonas hydrophila* complex (motile *Aeromonas* septicemia [MAS]) and *Flavobacterium columnare* (*Flexibacter columnaris*) (Chapter 14), *Vibrio anguillarum* (Chapter 17), *Edwardsiella tarda* (Chapter 16), and *Streptococcus* sp. (Chapter 19).

Striped bass afflicted with motile *Aeromonas* septicemia will reduce or completely stop feeding and will swim lethargically at the surface. Although clinical signs are not specific, fish may develop mild to severe hemorrhage in the skin and fins, and the fins will also have pale, frayed margins (see Figure 18.1). As the result of fluid (edema), scales may protrude (lepidorthosis) and slough, and the eyes become exophthalmic. A cloudy, bloody fluid is often present in the body cavity, and internal organs may be hyperemic or pale depending on stage of infection.

The causative agent of columnaris in fresh water is *F. columnare* (*F. columnaris*). Columnaris occurs in both cultured and wild striped bass, but degree of severity is greater in cultured populations. Columnaris is usually confined to the body, fins, and/or gills, but will occasionally occur systemically. Whitish areas appear on the skin, and lost scales often expose underlying musculature. Fins are usually white and in various stages of fraying, and necrotic gill lesions appear pale. Lesion margins may be yellowish because of the presence of large numbers of bacteria. Although there is no published record of *Flexibacter maritimus* (Wakabayashi et al. 1986), the equivalent of "columnaris" in fresh water, infecting striped bass in salt water, there is no reason to believe it cannot.

Vibriosis can be a significant problem in saltwater cage-reared striped bass and in wild populations. Although several species of the genus *Vibrio* are implicated in the disease, the most common species is *Vibrio anguillarum*. These infections are usually related to stress, high stocking densities, handling, temperature shock, and/or poor water quality factors. In addition to lethargy, clinical signs of vibriosis in striped bass are hyperemia of the fins and skin, resulting in scale loss and development of ulcerated epidermal lesions at any location on the body including head and gill covers. The gills are often pale, and eyes may be hemorrhaged and exophthalmic. Internally, the body cavity may contain a bloody fluid, the liver may be pale and/or mottled, the spleen is usually swollen and dark red, and the kidney is often swollen and soft. The gastrointestinal tract is usually void of food, flaccid, and inflamed.

Streptococcus septicemia affects a variety of fish

species, but it is particularly serious in some striped bass culture operations (J. P. Hawke, Louisiana State University, personal communication). Although *Streptococcus* spp. has been reported in wild striped bass populations inhabiting brackish water, the infection is more significant in cultured fish reared in brackish water ponds and net cages, where the organism apparently occurs naturally. The disease is usually chronic, but on occasion can be subacute. Clinical signs of *Streptococcus* spp. infection are not particularly specific in striped bass, but fish are generally darker than normal, exhibit erratic, spiral swimming, and often display body curvature. They often have either bilateral or unilateral exophthalmia with hemorrhage in the iris. Hemorrhages develop at the base of fins, in scale pockets, and on the operculum and mouth, and ulcerative lesions occasionally occur. A bloody fluid is present in the intestine, which is also hyperemic; the liver is pale; and the spleen is dark and greatly enlarged. Mortality of intensively cultured striped bass can be high, especially when water temperatures are 25 to 30°C. The handling and moving of fish seems to trigger overt disease where *Streptococcus* is endemic. Abrasions, loss of scales, and other skin injuries and environmental stressors are also important precursors to *Streptococcus* spp. infections in some fish species (Chang and Plumb 1996).

The genus *Enterococcus* is a relatively new taxonomic group infecting fish that was previously included in *Streptococcus* (Mundt 1986). *Enterococcus* infections have been noted in a few intensive freshwater striped bass units when water temperatures were 28 to 32°C (J. P. Hawke, personal communication). The causative organism is *Enterococcus faecium* (formerly *Streptococcus faecium*). When infected with *Enterococcus*, adult and juvenile striped bass develop a septicemia with hemorrhages in the scale pockets, and the eyes become swollen, hemorrhaged, or cloudy. *Enterococcus* is a Gram-positive, nonmotile cocci, but more ovoid than round. The organisms are ß hemolytic (Lancefield's Group D) and grow at 45°C, in 40% bile, and in 6.5% NaCl, characteristics that separate them from *Streptococcus* spp. The full impact of *E. faecium* on striped bass is still unknown, but judging from recent epizootics, the potential is notable.

Edwardsiella tarda is a common pathogen in some cultured fish species and will occasionally infect cultured striped bass. Herman and Bullock (1986) reported that 4- to 5-cm freshwater cultured juvenile striped bass were infected with *E. tarda* and the pathogen was transmitted to naive juveniles by waterborne exposure. The fish became moribund, swam lethargically at the surface, displayed pale gills, and had a slightly discolored area in the cranium. Histologically, epithelium of experimentally infected fish was necrotic and fins were frayed. Numerous abscesses were present in the kidney, accompanied by necrosis of the trunk kidney hematopoietic tissue. A significant *E. tarda* infection was found in wild adult striped bass in the Chesapeake Bay (A. Baya, University of Maryland, personal communication). The most notable clinical signs in these fish were numerous irregular, coalescing, hemorrhagic areas on the skin and fins and some ulcerations that emitted an unpleasant odor.

Several additional bacteria, which might or might not be serious pathogens, have been known to cause disease in striped bass. Baya et al. (1990) isolated one such bacteria that belongs to the genus *Moraxella*. The organism is a short, Gram-negative rod that often appears in pairs; it exhibits bipolar bodies, is cytochrome oxidase positive, nonfermentative in glucose, and does not produce acid from most carbohydrates. Fish affected with the bacterium had large hemorrhagic lesions on the dorsolateral body surface, and scales had been lost. Internally, the liver was enlarged and pale to mottled. The swim bladder was inflamed and adhesions connected the liver to the body wall. The role of *Moraxella* in the disease process was somewhat speculative because a viral agent, tentatively named striped bass reovirus, was isolated from these fish and also because gills of afflicted fish were heavily parasitized with *Trichodina* and *Ergasilus*.

Baya et al. (1991) isolated a *Carnobacterium*-like organism from moribund and dead striped bass and other fish in the Chesapeake Bay of Maryland. The fish had been stressed before infection occurred, and no clinical signs were described. The organism was similar to *Carnobacterium piscicola*, which is a Gram-positive bacillus that tolerates salinities from 0% to 6% and has a growth temperature range of 10 to 37°C. It is easily isolated on BHI media or trypticase soy agar (TSA). Toranzo et al. (1993) were unable to kill striped bass with *C. piscicola* by injection, but the organism did cause mild infiltration of the liver, hemorrhage in the kidney, and inflammation of the meninges. The investigators did, however, kill about 35% of rainbow trout injected with 4.5 ×

10^6 cells. These observations led to the conclusion that *C. piscicola* possesses low virulence to striped bass and produces only moderate injury to internal organs. However, infected striped bass can be carriers of *C. piscicola* for at least 2 months, which suggests the possibility that infected fish would be more susceptible to invasion by secondary pathogens or environmental stress.

Baya et al. (1992) isolated a bacterium, identified as *Corynebacterium aquaticum,* from the brain of striped bass exhibiting exophthalmia. This is the first report that *C. aquaticum,* a normal waterborne organism, is pathogenic to fish. The LD_{50} of the organism in striped bass was 1.0×10^5 colony forming units. Experimentally infected fish developed hemorrhaging in most internal organs, with it being most severe in the brain and eyes. It should also be noted that there is the possibility that *C. aquaticum* may be infectious to homeothermic animals.

MANAGEMENT OF STRIPED BASS BACTERIAL DISEASES

Chemical and drug availability for controlling infectious diseases of striped bass is extremely limited; therefore, management, good health maintenance, and disease prevention are keys to successful culture.

Management

To reduce the potential for catastrophic disease outbreaks in striped bass populations, the highest water quality possible should be maintained. This is best accomplished by using prudent stocking densities, adequate flow of fresh water through intensive culture units, removal of metabolites from recirculating water, supplemental aeration in flow-through and pond culture systems, and use of high-quality feed. Striped bass are sensitive to improper handling; therefore, gentle netting and handling is essential. Prophylactic chemotherapy during or after handling and/or moving of fish will reduce external parasite numbers and reduce the possibility of secondary bacterial infections.

In cultured striped bass, the absence of raw fish products in diets will usually disrupt the cycle of a mycobacteriosis infection; however, recent striped bass infections in a recirculating system were not associated with a raw fish diet. Once fish become infected with the organism, it is extremely difficult to treat, and it has been suggested by some that infected fish populations be destroyed, buried, and the facility sterilized (Lebovitz 1980). Fish should definitely be removed from contaminated facilities, the system cleaned thoroughly with an acid wash to remove as much organic material from water lines as possible, and the facility disinfected with 200 mg/L of calcium hypochlorite (HTH), chlorine dioxide, or phenolic compounds. Recirculating aquaculture systems should be equipped with ultraviolet and/or ozone water sterilization systems; however, water treatment rates and actual value of these units when acid-fast bacteria are present have not been studied in detail. According to Kitao (1993), avoidance of overcrowding and good management may help prevent photobacteriosis and other bacterial diseases.

Chemotherapy

Although no Food and Drug Administration (FDA)-approved drugs are available for striped bass, some drugs and chemicals have been successfully used prophylactically or in chemotherapy to treat clinical bacterial infections. Prophylaxis that have been used include baths in NaCl (0.5% to 3% for various periods) and potassium permanganate (2 to 5 mg/L for 1 hour to indefinitely). Clinical, systemic bacterial infections are usually treated with medicated feed containing oxytetracycline at a rate of 2.5 to 3.5 g/45 k of fish per day for 10 days; however, this is not FDA approved. Xu and Rogers (1993) determined that when oxytetracycline was injected intraperitoneally into striped bass, the drug was cleared from the muscle in 32 days, and when applied in feed clearance, time was reduced. Romet-30 fed at a rate of 50 mg/k of fish per day is also effective against most systemic bacterial infections. Romet-30 might or might not be effective against *Streptococcus.*

Recent efforts to control mycobacterial infections in striped bass culture populations have shown that some isolates are sensitive to Terramycin, whereas others are resistant to most antibiotics (J. Hawke, personal communication). The long-term antibiotic therapy required to keep the disease in remission is not economically feasible.

Application of medicated feed after disease occurs has been successful in controlling photobacteriosis. Ampicillin fed at 10 and 100 mg/kg of body weight per day reduced mortality in yellowtails to

30% and 0%, respectively, compared with a 100% mortality of nonmedicated controls (Kusuda and Inoue 1976). In Taiwan, when an antibiotic was fed to cultured snakehead infected with *P. damsela*, only a 30% loss was noted (Tung et al. 1985). Hawke et al. (1987) fed oxytetracycline at 50 to 150 mg/kg of body weight of striped bass per day with only a slight reduction in mortality. However, Nakai et al. (1992) controlled a *P. damsela* infection in striped jack by adding oxytetracycline and oxolinic acid to the feed.

Vaccination

Fish vaccination is becoming another tool that can be used to combat bacterial infections; however, few experiments have involved striped bass. Experimental vaccination has been shown to be effective in prevention of *Photobacterium*, especially in Japan, and this method of treatment is expected to become a management tool of the future for this disease. *Photobacterium* vaccines that contain killed whole cell bacterins can be effectively delivered by intraperitoneal injection, immersion, spray, or oral feeding (Fukuda and Kusuda 1981). These researchers reported survival rates of 60% to 88% with oral and immersion treatments and 100% when using injection and spray vaccination. Fukuda and Kusuda (1985) also found that lipopolysaccharide (LPS) preparations delivered by immersion or spray provided better protection than did whole cell preparations. Early vaccinations of yellowtail showed a response to an intraperitoneal injection of formalin-killed *P. damsela* in Freund's complete adjuvant. Kusuda et al. (1988) demonstrated a high degree of protection with an injectable ribosomal vaccine prepared from *P. damsela*. Vaccination by immersion with a live attenuated preparation has been attempted, but to date, no commercial immunogenic preparations have been developed. Rogers and Xu (1992) reported successful vaccination of striped bass against vibriosis. Vaccination shows some promise as a preventive treatment for *V. anguillarum*.

REFERENCES

Adams, A., K. D. Thompson, J. McEwan, S.C. Chen, and R. H. Richards. 1996. Development of monoclonal antibodies to *Mycobacterium* spp. isolated from chevron snakehead and Siamese fighting fish. *Journal of Aquatic Animal Health* 8:208–215.

Ashburner, L. D. 1977. Mycobacteriosis in hatchery-confined chinook salmon (*Oncorhynchus tshawytscha* Walbaum) in Australia. *Journal of Fish Biology* 10:523–528.

Arakawa, C. K., and J. L. Fryer. 1984. Isolation and characterization of a new subspecies of *Mycobacterium chelonei* infectious for salmonid fish. *Helgolander Wissenschaftliche Meeresuntersuchungen.* 37:329–342.

Austin, B., and D. A. Austin. 1987. *Bacterial Fish Pathogens: Diseases in Farmed and Wild Fish*, Chichester, UK: Ellis Horwood.

Bakopoulos, V., A. Adams, and R. H. Richards. 1995. Some biochemical properties and antibiotic sensitivities of *Pasteurella piscicida* isolated in Greece and comparison with strains from Japan, France and Italy. *Journal of Fish Diseases* 18:1–7.

Bakopoulos, V., Z. Peric, H. Dodger, A. Adams, and G. K. Iowama. 1997. First report of fish pasteurellosis from Malta. *Journal of Aquatic Animal Health* 9:26–33.

Baudin-Laurencin, F., J. F. Pepin, and J. C. Raymond. 1991. First observation of an epizootic pasteurellosis in farmed and wild fish of the French Mediterranean coast [abstract]. *5th International Conference of the European Association of Fish Pathologists.* Budapest, Hungary.

Baya, A., A. E. Toranzo, S. Nunez, J. L. Barja, and F. M. Hetrick. 1990. Association of a *Moraxella* sp. and a Reo-like virus with mortalities of striped bass, *Morone saxatilis*. In: *Pathology in Marine Science*, edited by F. O. Perkins and T. C. Cheng, 91–99. New York: Academic Press.

Baya, A. M., A. E. Toranzo, B. Lupiani, T. Li, B. S. Roberson, and F. M. Hetrick. 1991. Biochemical and serological characterization of *Carnobacterium* spp. isolated from farmed and natural populations of striped bass and catfish. *Applied and Environmental Microbiology* 57:3114–3120.

Baya, A. M., and seven others. 1992. Phenotypic and pathobiological properties of *Corynebacterium aquaticum* isolated from diseased striped bass. *Diseases of Aquatic Organisms* 14:115–126.

Chang, P. H., and J. A. Plumb. 1996. Effects of salinity on *Streptococcus* infection on Nile tilapia, *Oreochromis niloticus*. *Journal of Applied Aquaculture.* 6:39–45.

Chen, S-C. 1992. The study on the pathogenicity of *Nocardia asteroides* to largemouth bass *Micropterus salmoides* Lacepede, *Fish Pathology* 27:1–5.

Chen, S-C., M-C. Tung, and W-C. Tsai. 1988. An epizootic in Formosa snake-head fish (*Chana maculata* Lacepede), caused by *Nocardia asteroides* in fresh water pond in southern Taiwan. *COA Fisheries Series No. 15, Fish Disease Research* (IX) 6:42–48.

Chinabut, S., C. Limsuwan, and P. Chanaratchakool. 1990. Mycobacteriosis in snakehead, *Chana striatus* (Fowler). *Journal of Fish Diseases* 13:531–535.

Colorni, A. 1992. A systemic mycobacteriosis in the European sea bass *Dicentrarchus labrax* cultured in Eilat (Red Sea). *The Israeli Journal of Aquaculture—Bamidgeh* 44(3):75–81.

Conroy, D. A. 1964. Notes on the incidence of piscine tuberculosis in Argentina. *The Progressive Fish-Culturist* 26:89–90.

Frerichs, G. N. 1993. Mycobacteriosis: Nocardiosis. In: *Bacterial Diseases of Fish,* edited by V. Inglis, R. J. Roberts, and N. R. Bromage, 219–233. Oxford: Blackwell Scientific Press.

Frerichs, G. N., and R. J. Roberts. 1989. The bacteriology of teleosts. In: *Fish Pathology,* 2nd ed., edited by R. J. Roberts, 289–319. London: Bailliere-Tindall.

Fryer, J. L., and J. E. Sanders. 1981. Bacterial kidney disease in salmonid fish. *Annual Reviews in Microbiology* 35:273–298.

Fukuda, Y., and R. Kusuda. 1981. Efficacy of vaccination for pseudotuberculosis in cultured yellowtail by various administration. *Bulletin of the Japanese Society of Scientific Fisheries* 47:147–150.

Fukuda, Y., and R. Kusuda. 1985. Vaccination of yellowtail against pseudotuberculosis. *Fish Pathology* 20:421–425.

Gauthier, G., B. LaFay, R. Ruimy, V. Breittmayer, J. L. Nicolas, M. Gauthier, and R. Christen. 1995. Small subunit rRNA sequences and whole DNA relatedness concur for the reassignment of *Pasteurella piscicida* (Snieszko et al.) Jansen and Surgalla to the genus *Photobacterium* as *Photobacterium damsela* subsp. *piscicida* comb. nov. *International Journal of Systematic Bacteriology* 54(1):139–144.

Giavenni, R. 1979. Alcuni aspetti zoonosici delle micobatteriosi di origine Ittica. *Revista Italiano Piscicultura Ittiopathologia* 14:123–126.

Goodfellow, M., and L. G. Wayne. 1982. Taxonomy and nomenclature. In: *The Biology of the Mycobacteria,* edited by C. Ratledge and J. L. Stanfor, 471–521. New York: Academic Press.

Harrell, R. M., Editor. 1997. *Striped Bass and Other Morone Culture. Developments in Aquaculture and Fisheries Science,* vol. 30. Amsterdam: Elsevier.

Hatai, K., O. Lawhavinit, S. Kubota, K. Toda, and N. Suzuki. 1988. Pathogenicity of *Mycobacterium* sp. isolated from pejerrey *Odonthestes banariensis. Fish Pathology* 23:155–159.

Hawke, J. P. 1996. Importance of a siderophore in the pathogenesis and virulence of *Photobacterium damsela* subsp. *piscicida* in hybrid striped bass (*Morone saxatilis* X *Morone chrysops*). Baton Rouge, LA: Louisiana State University. Ph.D. dissertation,.

Hawke, J. P., S. M. Plakas, R. V. Minton, R. M. McPhearson, T. G. Snider, and A. M. Guarino. 1987. Fish pasteurellosis of cultured striped bass (*Morone saxatilis*) in coastal Alabama. *Aquaculture* 65:193–204.

Hedrick, R. P., T. McDowell, and J. Groff. 1987. Mycobacteriosis in cultured striped bass from California. *Journal of Wildlife Diseases* 23:391–395.

Herman, R. L., and G. L. Bullock. 1986. Pathology caused by the bacterium *Edwardsiella tarda* in striped bass *Morone saxatilis. Transactions of the American Fisheries Society* 115:232–235.

Jansen, W. A., and M. J. Surgalla. 1968. Morphology, physiology, and serology of a *Pasteurella* species pathogenic for white perch (*Roccus americanus*). *Journal of Bacteriology* 96:1606–1610.

Kariya, T., S. Kubota, Y. Nakamura, and R. Kira. 1968. Nocardial infection in cultured yellowtail (*Seriola guinguiradiata* and *S. purpurascens*). I. Bacteriological study. *Fish Pathology* 3:16–23.

Kimura, M., and T. Kitao. 1971. On the etiological agent of "bacterial tuberculosis" of *Seriola. Fish Pathology* 6:8–14.

Kitao, T. 1993. Pasteurellosis. In *Bacterial Diseases of Fish,* edited by V. Inglis, R. J. Roberts, and N. R. Bromage, 159–165. Oxford: Blackwell Press.

Kitao, T., and M. Kimura. 1974. Rapid diagnosis of pseudotuberculosis in yellowtail by means of fluorescent antibody technique. *Bulletin of the Japanese Society of Fisheries* 40:889–893.

Kitao, T., L. Ruangpan, and M. Fukudome. 1989. Isolation and classification of a *Nocardia* species from diseased giant gorami *Osphronemus gourami. Journal of Aquatic Animal Health* 1:154–162.

Kubota, S., M. Kimura, and S. Egusa. 1970. Studies of a bacterial tuberculosis of the yellowtail, I. Symptomatology and histopathology. *Fish Pathology* 4:111–118.

Kusuda, R., and K. Inoue. 1976. Studies on the application of ampicillin for pseudotuberculosis of cultured yellowtail. I. In vitro studies on sensitivity, development of drug-resistance, and reversion of acquired drug-resistance of *Pasteurella piscicida. Bulletin of the Japanese Society of Scientific Fisheries* 42:969–973.

Kusuda, R., and A. Nakagawa. 1978. Nocardial infection of cultured yellowtail. *Fish Pathology* 13:25–31.

Kusuda, R., and M. Yamaoka. 1972. Etiological studies on bacterial pseudotuberculosis in cultured yellowtail with *Pasteurella piscicida* as the causative agent—I. On morphological and biochemical properties. *Nippon Suisan Gakkaish* 38:1325–1332.

Kusuda, R., K. Kawai, and T. Matsui. 1978. Etiological studies on bacterial pseudotuberculosis in cultured yellowtails with *Pasteurella piscicida* as the causative agent—II. On the serological properties. *Fish Pathology* 13:79–83.

Kusuda, R., M. Ninomiya, M. Hamacuchi, and A.

Muroaka. 1988. The efficacy of ribosomal vaccine prepared from *Pasteurella piscicida* against pseudotuberculosis in cultured yellowtail. *Fish Pathology* 23:191–196.

Lebovitz, L. 1980. Fish tuberculosis (Mycobacteriosis). *Journal of American Veterinary Medical Association* 176:415.

Lechevalier, H. A. 1986. Nocardioforms. In: *Bergey's Manual of Systematic Bacteriology,* vol. 2, edited by P. A. Sneath, N. S. Naiv, M. E. Sharpe, and J. G. Holt, 1458–1465. Baltimore, MD: Williams & Wilkins.

MacKenzie, K. 1988. Presumptive mycobacteriosis in northeast Atlantic mackerel, *Scomber scombrus. Journal of Fish Biology* 32:263–275.

Magarinos, B., J. L. Romalde, I. Bandin, B. Fouz, and A. E. Toranzo. 1992. Phenotypic, antigenic, and molecular characterization of *Pasteurella piscicida* strains isolated from fish. *Applied and Environmental Microbiology* 58:3316–3322.

Matsudka, S., and S. Kamada 1995. Discharge of *Pasteurella piscicida* cells from experimentally infected yellowtail. *Fish Pathology* 30:221–225

Matsusato, T. 1975. Bacterial tuberculosis of cultured yellowtail. *Special Publication of Fisheries Agencies of the Japanese Government,* Japanese Sea Regional Fisheries Research Laboratory, 115–118.

Mori, M., T. Kitao, and M. Kimura. 1976. A field survey by means of the direct fluorescent antibody technique for diagnosis of pseudotuberculosis in yellowtail. *Fish Pathology* 11:11–16.

Mundt, J. O. 1986. Enterococci. In: *Bergey's Manual of Systematic Bacteriology,* edited by H. A. Smeatin, 1063–1065. Baltimore, MD: Williams & Wilkins.

Nakai, T., N. Fujiie, K. Muroga, M. Arimoto, Y. Mizuto, and S. Matsuoka. 1992. *Pasteurella piscicida* infection in hatchery-reared juvenile striped jack. *Fish Pathology* 27:103–108.

Nigrelli, R. F., and H. Vogel. 1963. Spontaneous tuberculosis in fishes and in other cold-blooded vertebrates with special reference to *Mycobacterium fortuitum* Cruz from fish and human lesions. *Zoologica* 48:130–143.

Noya, M., B. Magarinos, A. E. Toranzo, and J. Lamas. 1995. Sequential pathology of experimental pasteurellosis in gilthead seabream *Sparus aurata.* A light- and electronmicroscopic study. *Diseases of Aquatic Organisms* 21:177–186.

Plumb, J. A. 1997. Infectious Diseases of Striped Bass. In: *Striped Bass and Other Morone Culture,* edited by R. M. Harrell, 271–313. Developments in Aquaculture and Fisheries Science, vol. 30. Amsterdam: Elsevier.

Post, G. 1987. *Textbook of Fish Diseases.* Neptune, NJ: T. F. H. Publication.

Robohm, R. A. 1983. *Pasteurella piscicida.* In: *Antigens of Fish Pathogens,* edited by D. P. Anderson, M. Dorson, and Ph. Dubourget, 161–175. Lyon, France: Collection Fondation Marcel Merieux.

Rogers, W. A., and D. Xu. 1992. Protective immunity induced by a commercial vibrio vaccine in hybrid striped bass. *Journal of Aquatic Animal Health* 4:303–305.

Ross, A. J. 1970. Mycobacteriosis among salmonid fishes. In: *A Symposium on Diseases of Fishes and Shellfishes,* edited by S. F. Snieszko, 279–283. Special Publication no. 5, Washington, DC: American Fisheries Society.

Sakanari, J. A., C. A. Reilly, and M. Moser. 1983. Tubercular lesions in Pacific coast populations of striped bass. *Transactions of the American Fisheries Society* 112:565–566.

Sinderman, C. J. 1970. *Principal Diseases of Marine Fish and Shellfish.* New York: Academic Press.

Snieszko, S. F., G. L. Bullock, C. E. Dunbar, and L. L. Pettijohn. 1964a. Nocardial infection in hatchery-reared fingerling rainbow trout (*Salmo gairdneri*). *Journal of Bacteriology* 88:1809–1813.

Snieszko, S. F., G. L. Bullock, E. Hollis, and J. G. Boone. 1964b. *Pasteurella* sp. from an epizootic of white perch (*Roccus americanus*) in Chesapeake Bay tidewater areas. *Journal of Bacteriology* 88:1814–1815

Thune, R. L., L. A. Stanley, and R. K. Cooper. 1993. Pathogenesis of gram-negative bacterial infections in warmwater fish. *Annual Review of Fish Diseases* 3:37–68.

Timure, G., R. J. Roberts, and A. McQueen. 1977. The experimental pathogenesis of focal tuberculosis in the plaice (*Pleuronectes platessa* L.). *Journal of Comparative Pathology* 87:83–87.

Toranzo, A. E., J. L. Barja, and F. M. Hetrick. 1982. Survival of *Vibrio anguillarum* and *Pasteurella piscicida* in estuarine and fresh waters. *Bulletin of the European Association of Fish Pathologists* 3:43–45.

Toranzo, A. E., S. Barreiro, J. F. Casal, A. Figueras, B. Magarinos, and J. L. Barja. 1991. Pasteurellosis in cultured gilthead seabream (*Sparus aurata*): first report in Spain. *Aquaculture* 99:1–15.

Toranzo, A. E., B. Novoa, A. M. Baya, F. M. Hetrick, J. L. Barja, and A. Figueras. 1993. Histopathology of rainbow trout, *Oncorhynchus mykiss* (Walbaum), and striped bass, *Morone saxatilis* (Walbaum), experimentally infected with *Carnobacterium piscicola. Journal of Fish Diseases* 16:261–267.

Tung, M-C., S-S. Tsai, L-F. Ho, S-H. Huang, and S-C. Chen. 1985. An acute septicemic infection of *Pasteurella* organism in pond-cultured Formosa snakehead fish (*Chana maculata* Lacepede) in Taiwan. *Fish Pathology* 20:143–148.

Van Duijn, C. 1981. Tuberculosis in fishes. *Journal of Small Animal Practices.* 22:391–411.

Wakabayashi, H., M. Hikida, and K. Masumura. 1986. *Flexibacter maritimus* sp. nov., a pathogen of marine fishes. *International Journal of Systematic Bacteriology* 36:396–398.

Wayne, L. G., and S. Kubica. 1986. Genus *Mycobacterium.* In: *Bergey's Manual of Systematic Bacteriology,* vol. 2, edited by P. H. Sneath, N. S. Mair, M. E. Sharpe, and J. G. Holt, 1436–1457. Baltimore, MD: Williams & Wilkins.

Wheeler, A. P., and B. S. Graham. 1989. Saturday conference: atypical mycobacterial infections. *Southern Medical Journal* 82:1250.

Wolinski, E. 1992. Mycobacterial diseases other than tuberculosis. *Clinical Infectious Diseases* 15:1–12.

Wolke, R. E. 1975. Pathology of bacterial and fungal diseases affecting fish. In: *The Pathology of Fishes,* edited by R. Ribelin and G. Migaki, 33–116. Madison, WI: University of Wisconsin Press.

Xu, D., and W. A. Rogers. 1993. Oxytetracycline residue in hybrid striped bass muscle. *Journal of the World Aquaculture Society* 24:466–472.

19 Tilapia Bacterial Diseases

No known infectious bacterial agent is specific for tilapia. Extensive or moderately intensive cultured tilapia reared in a normal warm water habitat seldom contract serious infectious disease. Under certain conditions, however, such as stress due to low water temperature, improper handling, exposure to poor water quality, or high density accompanied by high feeding rates (causing poor water quality), these fish can actually be more severely affected by some bacterial pathogens than are other fish species. Tilapia culture has become increasingly intensive, expanding into temperate and colder climates where it is more difficult to maintain a proper environment artificially and where, consequently, infectious diseases have become a more serious problem.

With the popularity of intensively rearing tilapia in closed recirculating and raceway systems, infections of *Streptococcus* spp. and *Enterococcus* sp. have become the most serious infectious diseases of these fish. Other bacterial diseases that affect tilapia include motile *Aeromonas* septicemia (MAS), *Pseudomonas* septicemia, vibriosis, columnaris, and edwardsiellosis (Roberts and Sommerville 1982).

STREPTOCOCCOSIS (*STREPTOCOCCUS* AND *ENTEROCOCCUS*)

Reports of *Streptococcus* spp. fish infections date back to the mid-1950s (Hoshina et al. 1958), but the first case involving tilapia was reported by Wu (1970). Hubert (1989) stated that "undoubtedly this disease (streptococcosis) represents a real danger to farmers (fish) engaged in warm water aquaculture," and this would seem to be most applicable to intensively cultured tilapia, in which *Streptococcus* spp. has emerged as a serious infectious disease. The disease may be subacute, but more often, it manifests itself as a chronic condition. Austin and Austin (1987) listed seven species of *Streptococcus* (*S. agalactia, S. dysqalactiae, S. equi, S. equisimilis, S. faecium, S. pyogenes,* and *S. zooepidemicus*) that at various times have been pathogenic to fish, but often infections are caused by unspeciated streptococci that differ from other known members of the genus. Also, *Streptococcus iniae* has become a very important pathogen of tilapia.

Geographical Range and Species Susceptibility

Infections of *Streptococcus* have been reported in fish from North America (Canada and the United States), several Central American countries, Saudi Arabia, Japan, South Africa, Israel, Great Britain, and Norway (Hoshina et al. 1958; Robinson and Meyer 1966; Barham et al. 1979; Kitao et al. 1981; Kawahara and Kusuda 1987; Hubert 1989; Al-Harbe 1994). Members of the genus *Enterococcus* have been described in Japan and the United States and are becoming more widely found. These pathogens occur in both fresh water and marine environments.

Tilapia and a variety of other fishes are affected by streptococci. Kitao (1993) listed 22 fish species that are naturally susceptible to *Streptococcus* spp. The most severely affected culture species are Mozambique tilapia, Nile tilapia, yellowtail, eels, ayu, rainbow trout, and striped bass or their hy-

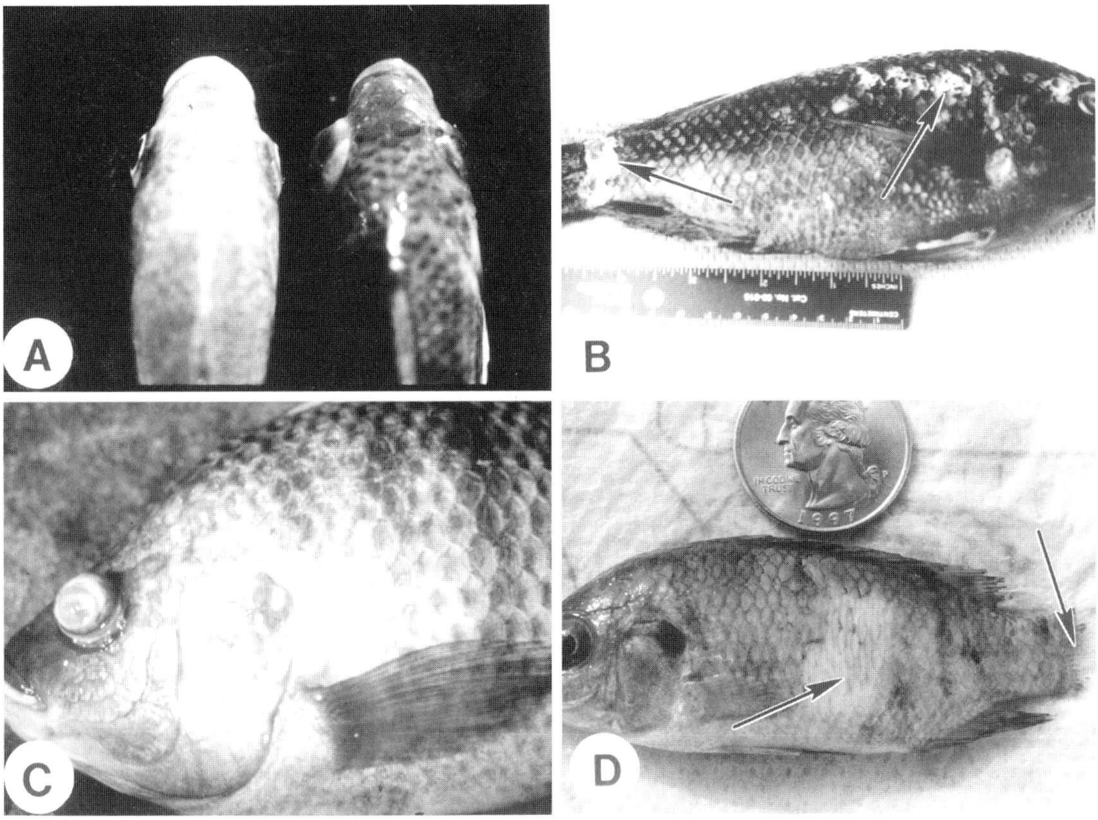

FIGURE 19.1. Diseased tilapia. (A) *Tilapia with* Streptococcus *infection. Note the exophthalmia of the fish on the right.* (B) *Tilapia with skin infection of* Aeromonas hydrophila (arrows). (C) *Opaque and exophthalmic eye of tilapia with* Edwardsiella tarda *infection.* (D) *Columnaris infection on body and caudal fin* (arrows) *of tilapia.*

brids. Other less-susceptible species include cultured channel catfish and golden shiners. Plumb et al. (1974) isolated *Streptococcus* from about 10 wild marine fish species in estuaries of the northern Gulf of Mexico: Gulf menhaden, silver seatrout, pinfish, Atlantic croaker, hardhead catfish, and others. *Enterococcus seriolicida* occurs in yellowtail, eels, and other fish species in Japan. According to Ferguson et al. (1994), the ornamental species zebra danio and pearl danio may also be naturally susceptible to *Streptococcus* spp. Clearly, this group of bacteria has no host specificity.

Clinical Signs and Findings

Clinical signs of streptococci in tilapia are not particularly specific, but, generally, these fish are darkly pigmented, lethargic, exhibit erratic and spiral swimming, and have curved bodies (Figure 19.1) (Plumb et al. 1974; Kitao 1993; Chang and Plumb 1996). Infected fish also exhibit abdominal distension, exophthalmic eyes that have hemorrhage and opaque corneas, and diffused hemorrhage that occurs in the operculum, skin, and base of fins. These epidermal lesions are usually more superficial than those associated with the aeromonads or vibriosis but will occasionally become necrotic, bloody ulcers. Bloody mucoid fluid in the lower intestine exudes from the anus. Internal findings include a bloody, sometimes gelatinous, exudate in the abdominal cavity, pale livers, hyperemic digestive tract, and a greatly enlarged, nearly black spleen. The lower gut is flaccid and hyperemic. Clinical signs of streptococcosis in other fish species are similar. *Enterococcus* sp. infections produce a host response similar to, but clinically indistinguishable from, that of *Streptococcus* spp.

Diagnosis

Isolation of *Streptococcus* spp. on media is preferable when diagnosing the disease, but presumptive diagnosis can be made by detection of Gram-positive coccus (sometimes ovoid) in histological sections or smears from infected tissues. In infected fish, bacteria may be single or paired or form short chains of two to six cells. Streptococci can be isolated on Todd-Hewitt, brain heart infusion (BHI), or trypticase soy agar (TSA) media. An addition of 5% blood and 10 mg of colistin and nalidixic acid per liter (to discourage other bacteria) to the agar will enhance and expedite isolation (Kitao et al. 1981). Inoculated media should be incubated at 20 to 30°C; small yellowish to gray, translucent, rounded, slightly raised, pinpoint or pinhead size colonies will appear in 24 to 48 hours. When streptococci are grown on media (agar or broth), they have a greater tendency to form chains of up to seven or eight cells. These bacteria are nonmotile, noncapsulated, and nonsporeformers.

Group B *Streptococcus* spp. can be presumptively identified by a positive CAMP test and lack of hemolysis on sheep blood agar (Facklam 1980). Many *Streptococcus* spp. isolates from fish, especially group B, do not conform biochemically to described species (Table 19.1). The Group D species can be identified by hemolysis and biochemical characteristics.

Bacterial Characteristics

Several species of *Streptococcus* have been isolated from tilapia and other fish, with the most common being an unspeciated nonhemolytic isolate, *S. iniae, S. faecalis, S. faecium, S. agalactia. Streptococcus faecalis, S. faecium,* and *S. agalactia,* are ∂ or ß hemolytic and are in Lancefield group D, whereas nonhemolytic isolates are in group B. *S. iniae* cannot be grouped by the Lancefield antigen method. Some previously identified *Streptococcus* spp. fish isolates in Japan have been renamed *Enterococcus serolocida* (Kusuda et al. 1991).

Group D ∂ and ß hemolytic streptococci conform to the enterococcus group that grows at 45°C, in 40% bile and 6.5% NaCl, and at pH 9.6 (Kusuda et al. 1991). In Japan, *Streptococcus* spp. isolates made from rainbow trout, yellowtail, and eel in the 1970s appear to also fit into this group. The cells are more ovoid than round but do occur in chains. Similar isolates came from rainbow trout in South Africa (Boomker et al. 1979) and hybrid striped bass in fresh water systems in the United States (J. P. Hawke, personal communication). A ∂ hemolytic strain isolated from eel and yellowtail in Japan does not react with the Lancefield group D antisera, but it has other characteristics similar to the *Enterococcus* group (see Table 19.1) (Kitao 1982). Whether this isolate is the same as the one from striped bass in the United States is not clear.

S. iniae has been implicated in tilapia diseases in Israel, the United States, Taiwan, and Japan, as well as in other fish species in Japan (Kawahara and Kusuda 1987; Eldar et al. 1994; Perera et al. (1994). Perera et al. (1994) reported that *S. iniae* was responsible for a chronic mortality of hybrid tilapia (Nile X blue tilapia) on a Texas fish farm. Two species of *Streptococcus* (*Streptococcus shiloi* and *Streptococcus dificile*) were reported to cause meningoencephalitis in tilapia that resulted in great economic losses in Israel during the middle to late 1980s (Eldar et al. 1994). Using DNA–DNA hybridization, Elder et al. (1995) showed that *S. shiloi* from different fish species in Israel was synonymous with *S. iniae* from tilapia in Texas and Taiwan and similar to streptococci found in Japan; therefore, *S. iniae* could be cosmopolitan in distribution.

Epizootiology

Although *Streptococcus* spp. infections occur in fresh or salt water grown tilapia, generally infections are more prevalent and more severe in marine and brackish water fish. In vitro *Streptococcus* salinity tolerance supports this phenomenon (Rasheed and Plumb 1984; Baya et al. 1990), but most isolates from freshwater fish do not grow in media salinity above 3% (Kitao et al. 1981). It appears that streptococci isolates adapt to salinity levels in various ecosystems and that they are generally less virulent when found in an ecosystem that differs from their origin. Tilapia in water with 15 to 30 ppt salinity at 25 and 30°C are more susceptible to *Streptococcus* than when in fresh water at the same temperature. Little difference in susceptibility is noted at 20°C (Chang and Plumb 1996).

Tilapia susceptibility to *Streptococcus* is unquestionably associated with environmental stress, skin injury, scale loss, and other factors associated with intensive aquaculture (Chang and Plumb 1996). According to J. P. Hawke (personal commu-

TABLE 19.1. CHARACTERISTICS OF MOST FREQUENTLY ISOLATED *STREPTOCOCCUS* SPP. AND *ENTEROCOCCUS* SP. FROM FISH

Test/Characteristic	*S. iniae*[a]	*S. faecalis*[a]	*S. faecium*[a]	*S. agalactia*[a]	*Group B*[a]	*E. seriolicida*[a]
Cell morphology	Spherical envelope long chains	Ovoid pairs or short chains	Sherical pairs or short chains	?	Spherical short chains	Ovoid short chains
Motility	–	–	–	–	–	–
Growth at						
10°C	+	+	+	?	+	+
45°C	–	+	+	–	–	+
Growth in						
6.5% NaCl	–	+	+	–	–	+
0.4% tellurite	?	?	?	?	?	–
pH 9.6	–	+	+	–	–	+
Tetrazolium	?	?	±	?	?	+
0.1% methylene blue	?	?	?	?	–	+
40% bile esculin	–	+	+	+	–	+
Arabinose	–	?	+	?	–	–
Salacin	+	?	?	?	?	±
Trehalose	+	?	?	?	–	+
Catalase	–	–	–	–	–	–
Starch	+	–	–	–	–	+
Esculin hydrolysis	+	+	+	–	–	+
Hemolysis	α/ß	α/ß	?	–	–	γ
H_2S	–	–	–	–	–	–
Arginine hydrolysis	?	+	+	+	±	+
Hippurate hydrolysis	–	+	?	+	+	–
Voges-Proskauer	–	?	?	+	+	+
Acid production from						
D-xylose	?	?	?	?	?	–
L-Rhamnose	?	?	?	?	?	–
Sucrose	?	?	?	?	±	–
Lactose	–	?	?	?	–	–
Melibiose	?	?	+	?	?	–
Raffinose	–	?	?	–	–	–
Glycerol	?	?	?	?	?	–
Adonitol	?	?	?	?	?	–
Sorbitol	–	?	–	–	–	+
Glucose	+	+	+	+	+	+
Mannitol	+	?	–	?	?	+
Lancefield grouping	None	D	D	B	B	Not D
Mol% G + C DNA	32.9	33.5–38	38.3–39	34	?	44

Sources: Rotta (1984); Kusuda et al. (1991).
[a]+ = positive; – = negative; ? = unknown.

nication), infestations of *Gyrodactylus,* a monogenetic trematode, may also enhance *Streptococcus* infections.

Mortality of fish infected with *Streptococcus* may be low to high depending on other circumstances. Under culture conditions, mortality has been as high as 75% in naturally infected tilapia. In Japan, it was demonstrated that *Streptococcus* spp. remains in seawater and mud year-round near rearing facilities but that higher numbers of bacteria were present during the summer. Kitao et al. (1979) found, however, that isolation of *Streptococcus* from mud was easier during autumn and winter. In naturally occurring infections, *Streptococcus* transmission is thought to be by contact, but experimentally, transmission occurs by immersion in water containing the pathogen, injection, or cohabitation. Immersion transmission is enhanced by injury to the epithelium or by stressful environmental conditions. Chang and Plumb (1996) had difficulty establishing *Streptococcus* spp. infections in Nile tilapia or channel catfish unless skin was scarified before exposure. In contrast, Ferguson et al. (1994) easily produced infection and high mortality in unstressed ornamental fish and rainbow trout that were exposed to *Streptococcus* spp.

Epizootics in wild fish populations may be associated with environmental conditions. Plumb et al. (1974) noted that fish kills along the Gulf of Mexico coast during a period of very low rainfall were confined to bays with restricted flow to the open Gulf. This combination likely interfered with flushing of the bays, creating poor water quality conditions and causing fish to become stressed. A similar phenomenon was noted in the Chesapeake Bay, Maryland, involving a *Streptococcus* spp. kill (Baya et al. 1990). In the United States and Japan, most epizootics occur during late summer through autumn (Kitao 1993).

Streptococcus spp. can occur with other opportunistic bacterial infections. *Streptococcus* and *Aeromonas sobria* were implicated in mortality of adult Nile tilapia in the Philippines when approximately 1200, 800, and 500 individuals died during successive weeks (Yambot 1996). Also, dual infections of *Edwardsiella tarda* and *Streptococcus* have been noted in intensive tilapia culture systems.

From late 1995 through mid-1996, there were reports of *Streptococcus* spp. infections being transmitted to humans from infected tilapia as a result of puncture wounds or cuts acquired while cleaning farm-reared tilapia purchased live from a fish market (Invasive Infection with *Streptococcus iniae* in Ontario, 1995–1996, 1996). The bacteria in these human infections were identified as *S. iniae* and caused major concern for the commercial tilapia industry, particularly in North America. There has been very little conclusive evidence that the organism isolated from affected humans is the same in all respects as the isolates taken from fish.

Pathological Manifestations

In tilapia, *Streptococcus* spp. affects the spleen, liver, eye, and in some cases the kidneys; a general bacteremia occurs, and the eye may display severe granulomatous inflammation (Miyazaki et al. 1984). Also, infiltration of bacteria-laden macrophages and granulomas in infected lesions of the epicardium, peritoneum, stomach, intestine, brain, ovary, testes, and capsules of the liver and spleen were noted. Rasheed et al. (1985) found that spleens of infected Gulf killifish were about 10 times larger than normal, the splenic pulp was severely congested, some cells were necrotic, and it appeared that phagocytosed bacteria multiplied in the macrophages. Livers of these fish were edematous, the sinusoids were dilated, and hepatocytes were atrophied and focally necrotic.

Pathogenesis of *Streptococcus* spp. is thought to be facilitated by exotoxins. Kimura and Kusuda (1979) reported that when yellowtail were injected with a cell-free culture media in which *Streptococcus* spp. was grown, susceptibility increased upon subsequent exposure to the bacterium. They also found that the kidney and spleen were more conducive to high numbers of bacteria than were the blood, liver, or intestines.

In a detailed pathology study of *Enterococcus* sp. in farmed turbot, the disease was described as having two forms: focal and generalized (Nieto et al. 1995). Pathology in the focal form was characterized by exophthalmia, muscular hemorrhages, acute branchitis, and suppurative inflammation of periorbital tissues, eyeball, and brain meninges. In the generalized form, similar lesions were seen, but more extensive hemorrhage, ulceration and purulent inflammation of the skin, and desquamative enteritis of the intestine were reported. Necrosis of the spleen and kidney was also present.

Significance

Streptococcus spp. has become a major disease of cultured tilapia in North America, South America, Asia, and the Middle East and could be a limiting factor in the expansion of tilapia culture. *Streptococcus* can also cause serious problems in hybrid striped bass culture populations, particularly along coastal areas. The disease is potentially serious in Japanese mariculture, particularly in yellowtail and eel populations. In 1989, there were 484 reported *Streptococcus* incidences in cultured yellowtail in Japan (Kitao 1993). Its potential for human infections is uncertain but worthy of concern.

OTHER BACTERIAL TILAPIA DISEASES

In tilapia, MAS is associated with several different species of bacteria, the most common of which is *Aeromonas hydrophila* (*liquifaciens, punctata*); however, *A. sobria* and other related species do occasionally occur (Plumb 1997). Tilapia infected with MAS lose their equilibrium, swim lethargically, gasp at the surface, and generally display the same clinical signs as other fish species (see Figure 19.1).

Although MAS is not uncommon in cultured tilapia, there are few published "case reports" of its occurrence (Figure 19.1). Recently, *A. hydrophila* was shown to cause "eye disease" of cage cultured Nile tilapia in the Philippines (Yambot and Inglis 1994).

In tilapia, *Pseudomonas fluorescens* will occasionally produce clinical signs and pathology similar to *A. hydrophila* and can cause significant mortality (Duremdez and Lio-Po 1985). In Japan, this bacterium caused a disease in Nile tilapia characterized by hemorrhagic lesions in the gonads and the ovaries in particular (Miyashita 1984). The infection occurred primarily in winter and spring when water temperatures were 15 to 20°C; resulting mortalities were 0.2% to 0.3% per day.

Plesiomonas shigelloides is becoming more frequently reported as an opportunistic fish pathogen, but little is known about its pathological capability (Desrina 1994). Faisal and Popp (1987) reported that *P. shigelloides* was responsible for 30% to 60% mortality among overcrowded 4-week-old Nile tilapia. The disease could be transmitted only by injection to 6-week-old tilapia but not to 9-month-old fish.

The epizootiology of vibriosis in tilapia is similar to that of MAS in the respect that both usually occur as secondary infections. In salt water, *Vibrio anguillarum* or *Vibrio vulnificus* are the usual organisms involved, whereas in fresh water, *Vibrio mimicus* or *Vibrio cholerae* are found; *Vibrio parahaemolyticus* can occur in either environment. Vibriosis in tilapia is usually mild to chronic, and clinical signs do not differ significantly from those for MAS. Mortality of vibrio-infected tilapia is usually chronic with relatively low daily losses, but cumulative mortality can become serious. Hubert (1989) stated that *Vibrio parahaemolyticus* infections were not spontaneous but were fulminating septicemias that occurred in market-size tilapia 2 to 3 days after handling and transfer to wintering ponds.

Sakata (1988) reported that as a result of a vibriosis infection, Nile tilapia suffered 10% to 20% mortality after transfer to salt water pens at 18 to 20°C. Decreasing water temperatures, coupled with high salinity, are considered to be compounding stressors on tilapia populations (Balarin and Haller 1982).

Columnaris has not been frequently reported in cultured tilapia, even though the pathogen is ubiquitous in fresh water and is not species specific. The presence of frayed fins is the most frequently observed clinical sign of disease (see Figure 19.1), and infected fish will swim lethargically or float at the surface. Primary cause of death is attributed to injury to skin, fins, and gills. Any type of physical injury or environmentally induced stress can precipitate these infections. Marzouk and Bakeer (1991) showed that Nile tilapia were more susceptible to columnaris infections and that a higher mortality resulted when pH was either very acidic or alkaline. Amin et al. (1988) isolated *Flavobacterium columnare* from cultured Nile tilapia that had gill lesions. In this study, pathogenicity varied in seven different isolates, but infection could not be established with any isolate unless fish were stressed or injured.

Edwardsiellosis (*E. tarda*) affects tilapia cultured under high density and other stressful conditions in either fresh water or marine environments. Tilapia infected with *E. tarda* swim lethargically on their sides at the surface; have an enlarged abdominal area; and swollen, opaque and hemorrhaged eyes (Figure 19.1); and possibly some discrete inflammation or skin discoloration. Internally, focal areas of necrosis are present, hemorrhage and gas filled cavities can occur in the muscle, bloody fluid can accumulate in the body cavity, the liver is often pale but mottled, spleen is dark red and swollen, kidney swollen and soft, and intestine can be inflamed and is usually void of food.

In Japan, an *E. tarda* infection was reported in Nile tilapia with mortalities of 0.2% to 0.3% per day when water temperatures were 20 to 30°C (Miyashita 1984). Infected fish had small white nodules in the spleen and abscesses in the swim bladder. In some *E. tarda*–infected tilapia, the eyes are severely affected with exophthalmia and opaqueness. *Edwardsiella* occurs most often in tilapia that are reared in intensive culture systems with marginal water quality, high organic load, and high fish density and can occur in conjunction with *A. hydrophila* or *Streptococcus* spp. In Egypt, an *E. tarda* infection was identified in Nile tilapia that were reared in ponds that received domestic waste water (Badran 1993).

Mycobacteriosis (fish tuberculosis) was detected in tilapia in Central Africa (Roberts and Sommerville 1982). These organisms can present problems in tilapia populations similar to those in intensively reared striped bass.

A rickettsialike organism was detected in diseased Nile tilapia in Taiwan (Chen et al. 1994). Affected fish were pale and lethargic, displayed hemorrhages and ulcers on the skin, had varying degrees of ascites and enlarged spleens and kidneys, and livers were marked with distinct white nodules of varying sizes. A Gram-negative rickettsialike organism appeared in inclusions or within host cell cytoplasmic vacuoles. The organism was cultured in CHSE-214 (chinook salmon embryo) cells and experimentally transmitted to naive tilapia. The full pathological implication of these rickettsialike organisms in tilapia is not known.

MANAGEMENT OF TILAPIA BACTERIAL DISEASES

As tilapia culture systems become more intensive, the need for environmental control, water-quality stability through management, stress reduction on fish, and other sound health maintenance procedures become more essential.

Management

When starting a tilapia culture operation, great care should be taken to make certain the facility is "clean" of wild fish that can harbor pathogens. Sound management procedures should be implemented to eliminate accumulation of detritus, waste, and dead fish. When fish are introduced, only specific pathogen free stocks should be used if possible. To reduce potential for catastrophic disease outbreaks, water quality should be maintained at the highest possible level, and reduction of environmentally induced stress should receive top priority. Prudent stocking densities, adequate water exchange in intensive culture units, removal of metabolites from recirculating water, supplemental aeration, and feeding high quality diets are a means to that end. Although tilapia feed does not generally contain raw fish products, diets of other fish species sometimes do, and this should not be practiced because it is possible to contract *Streptococcus* or other pathogenic bacteria through this source if the fish used in the diet are infected (Minami 1979). Also if disease does occur, infected fish and dead carcasses should be promptly removed from the culture unit.

The use of prophylactic chemotherapy such as salt (NaCl or CaCl) or potassium permanganate, during or after handling, aids in healing minor skin abrasions and reduces external parasites and the possibility of contracting secondary bacterial infections.

Disinfection of water with ultraviolet light, and ozone will help reduce bacterial populations in recirculating water or open water supplies. Sanitation by routine sterilization of nets, buckets, and other utensils will reduce accidental cross contamination of culture units. Sterilization can be accomplished by dipping utensils and boots into 200 mg of chlorine per liter or in 100 mg of a quaternary ammonium compound per liter. Iodine at 1000 mg/L can also be used as a disinfectant and may be safer to use than chlorine or the quaternary ammonia compounds. Thoroughly rinsing nets and utensils in fresh water is essential before disinfected items come into contact with fish. Nets, seines, and the like should be thoroughly dried, preferably in the sun, to kill most pathogens. Anectodal data indicate that reducing fish density, changing handling practices, improving water quality, and adopting other health maintenance procedures will reduce effects of *Streptococcus* spp. infections. Also, maintaining brood stock on the premises will provide seed fish with a known disease history.

Chemotherapeutics

Currently there are no chemotherapeutics with approved labels for treating any tilapia disease in the United States; it is encouraging, however, that some chemicals and drugs are being considered and may be approved by the Food and Drug Admininstration (FDA) for use on cultured tilapia in the near future (R. Schnick, personal communication).

Some drugs and chemicals have been successfully used prophylactically or in chemotherapy for bacterial infections of tilapia (see Table 5.1). Prophylaxis includes salt baths (NaCl or CaCl at 0.5 to 3%) for dip or prolonged treatments and/or potassium permanganate (5 to 10 mg/L) for 1 hour or 2 to 5 mg/L indefinitely.

Systemic bacterial infections of tilapia are usually treated with Terramycin incorporated into feed to provide 2.5 to 3.5 g/45 kg (50 to 75 mg/kg) of fish per day for 14 days. Sulfadimethoxazine-ormetoprim (Romet-30) fed at a rate of 2.5 to 3.5 g/45 kg (50 to 75 mg/kg) of fish per day for 5 days is also effective against most systemic bacterial infections but may present a palatability or toxicity problem to some fish species. Erythromycin is effective against Gram-positive bacteria, and its use is currently under FDA consideration for use in *Streptococcus* spp.–infected tilapia (R. Schnick, personal communication). If approved, medication level would probably be 50 or 100 mg/kg of fish per day for 10 days. When bacterial infections are confined to the skin (columnaris or external motile aeromonad infections), potassium permanganate at previously noted rates may be used.

Vaccination

Few immunological or vaccination experiments have involved tilapia. In Nile tilapia, protective immunity to *A. hydrophila* was demonstrated by intraperitoneal injection with vaccine (Rungpan et al. 1986). They reported 53% to 61% protection within 1 week postvaccination and 100% protection in 2 weeks. Mass vaccination of tilapia by injection is impractical except possibly for brood stock. Some autogenous vaccines are used in striped bass and tilapia to prevent *Streptococcus* spp., but results are inconclusive. Vaccines have been used experimentally to prevent streptococcus infections in Japan (Sakai et al. 1987).

REFERENCES

Al-Harbe, A. H. 1994. First isolation of *Streptococcus* sp. from hybrid tilapia (*Oreochromis niloticus* x *O. aureus*) in Saudi Arabia. *Aquaculture* 128:195–201.

Amin, N. E., I. S. Abdallah, M. Faisal, M. E.-S. Easa, T. Alaway, and S. H. Alyan. 1988. Columnaris infection among cultured Nile tilapia *Oreochromis niloticus*. *Antonie van Leeuwenhoek* 54:509–520.

Austin, B., and D. A. Austin. 1987. *Bacterial Fish Pathogens: Disease in Farmed and Wild Fish*. Chichester, UK: Ellis Horwood.

Badran, A. F. 1993. An outbreak of edwardsiellosis among Nile tilapia (*Oreochromis niloticus*) reared in ponds supplied with domestic wastewater. *Zagazig Veterinary Journal* 21:771–777.

Balarin, J. D., and R. D. Haller. 1982. Intensive culture of tilapia in tanks, raceways and cages. In: *Recent Advances in Aquaculture*, edited by J. F. Muir and R. J. Roberts, 266–355. London: Croom Helro.

Barham, W. T., H. Schoonbee, and G. L. Smit. 1979. The occurrence of *Aeromonas* and *Streptococcus* in rainbow trout (*Salmo gairdneri*). *Journal of Fish Biology* 15: 457–460.

Baya, A. M., B. Lupiani, F. M. Hetrick, B. S. Roberson, R. Lukacovic, E. May, and C. Poukish. 1990. Association of a *Streptococcus* sp. with fish mortalities in the Chesapeake Bay and its tributaries. *Journal of Fish Diseases* 13:251–253.

Boomker, J., G. D. Imes, Jr., C. M. Cameron, T. W. Nause, and H. J. Schoonbee. 1979. Trout mortalities as a result of *Streptococcus* infection. *Onderstepoort Journal of Veterinary Research* 46:71–78.

Chang, P. H., and J. A. Plumb. 1996. Effects of salinity on *Streptococcus* infection of Nile tilapia, *Oreochromis niloticus*. *Journal of Applied Aquaculture*. 6:39–45.

Chen, S. C., and seven others. 1994. Systematic granulomas caused by a rickettsia-like organism in Nile tilapia, *Oreochromis niloticus* (L.), from southern Taiwan. *Journal of Fish Diseases* 17:591–599.

Desrina. 1994. Pathogenicity of *Plesiomonas shigelloides* in channel catfish (*Ictalurus punctatus*). Auburn, AL: Auburn University. M.S. thesis.

Duremdez, R. C., and G. D. Lio-Po. 1985. Studies on the causative organism of *Sarotheradon niloticus* (Linnaeus) fry mortalities-2: isolation, identification and characterization of *Pseudomonas fluorescence*. *Fish Pathology* 20:115–123.

Eldar, A., Y. Bejerano, and H. Bercovier. 1994. *Streptococcus shiloi* and *Streptococcus dificile:* two new streptococcal species causing a meningoencephalitis in fish. *Current Microbiology* 28:139–143.

Eldar, A., P. Frelier, L. Assenta, P. W. Varner, S. Lawhon, and H. Bervovier. 1995. *Streptococcus shiloi,* the name for an agent causing septicemic infection in fish is a junior synonym of *Streptococcus iniae*. *International Journal of Systematic Bacteriology* 45:840–842.

Facklam, R. R. 1980. Streptococci and Aerococci. In: *Manual of Clinical Microbiology,* 3rd ed., edited by E. H. Lennette, 88–110. Washington, DC: American Society for Microbiology.

Faisal, M., and W. Popp. 1987. *Plesiomonas shigelloides*: a pathogen for the Nile tilapia, *Oreochromis niloticus*. *Journal of the Egyptian Veterinary Medical Association* 47:63–70.

Ferguson, H. W., J. A. Morales, and V. E. Ostland. 1994. Streptococcosis in aquarium fish. *Diseases of Aquatic Organisms* 19:1–6

Hoshina, T., T. Sano, and Y. Morimoto. 1958. A streptococcus pathogenic to fish. *Journal of Tokyo University Fisheries* 44:57–68.

Hubert, R. M. 1989. Bacterial diseases in warmwater aquaculture. In: *Fish Culture in Warm Water Systems: Problems and Trends,* edited by M. Shilo and S. Sarig, 197–194. Boca Raton, FL: CRC Press Inc.

Invasive infection with *Streptococcus iniae* in Ontario, 1995–1996. 1996. *Morbidity and Mortality Weekly Report* 45(30):650.

Kawahara, E., and R. Kusuda. 1987. Direct fluorescent antibody technique for differentiation between alpha and beta-hemolytic *Streptococcus* spp. *Fish Pathology* 22:77–82.

Kimura, H., and R. Kusuda. 1979. Studies on the pathogenesis of streptococcal infection in culture yellowtails *Seriola* spp.: effect of the cell free culture on experimental streptococcal infection. *Journal of Fish Diseases* 2:501–510.

Kitao, T. 1982. The method for detection of *Streptococcus* sp., causative bacteria of streptococcal disease of culture yellowtail (*Seriola guingueradiata*): especially their cultural, biochemical and serological properties. *Fish Pathology* 17:17–26.

Kitao, T. 1993. Streptococcal infections. In: *Bacterial Diseases of Fish,* edited by V. Inglis, R. J. Roberts, and N. R. Bromage, 196–210. London: Blackwell Press.

Kitao, T., T. Aoki, and K. Iwata. 1979. Epidemiological study on streptococcosis of cultured yellowtail (*Seriola guingueradiata*)—Distribution of *Streptococcus* sp. in seawater and muds around yellowtail farms. *Bulletin of the Japanese Society of Scientific Fisheries* 45:567–572.

Kitao, T., T. Aoki, and R. Sakoh. 1981. Epizootic caused by ß-haemolytic *Streptococcus* species in cultured freshwater fish. *Fish Pathology* 15:301–307.

Kusuda, R., K. Kawai, F. Salati, C. R. Banner, and J. L. Fryer. 1991. *Enterococcus seriolicida* sp. Nov., a fish pathogen. *International Journal of Systematic Bacteriology* 41:406–409.

Marzouk, M. S. M., and A. Bakeer. 1991. Effect of water pH on *Flexibacter columnaris* infection in Nile tilapia. *Egyptian Journal of Comparative Pathology and Clinical Pathology* 4:227–235.

Minami, T. 1979. *Streptococcus* sp., pathogenic to cultured yellowtail, isolated from fishes for diets. *Fish Pathology* 14:15–19.

Miyashita, T. 1984. *Pseudomonas fluorescens* and *Edwardsiella tarda* isolated from diseased tilapia. *Fish Pathology* 19:45–50.

Miyazaki, T., S. S. Kubota, N. Kaige, and T. Miyashita. 1984. A histopathological study of streptococcal disease in tilapia. *Fish Pathology* 19:167–172.

Nieto, J. M., S. Devesa, I. Quinoga, and A. E. Toranzo. 1995. Pathology of *Enterococcus* sp. infection in farmed turbot, *Scophthalmus maximus* L. *Journal of Fish Diseases* 18:21–30.

Perera, R. P., S. K. Johnson, M. D. Collins, and D. H. Lewis. 1994. *Streptococcus iniae* associated with mortality of *Tilapia nilotica* x *T. aurea* hybrids. *Journal of Aquatic Animal Health* 6:335–340.

Plumb, J. A. 1997. Infectious diseases of tilapia. *Tilapia Aquacultures in the Americas,* edited by B. A. Costa-Pierce and J. Rakocy, Vol. 1, 212–228. World Aquaculture Society.

Plumb, J. A., J. H. Schachte, J. L. Gaines, W. Peltier, and B. Carroll. 1974. *Streptococcus* sp. from marine fishes along the Alabama and northwest Florida coast of the Gulf of Mexico. *Transactions of the American Fisheries Society* 103:358–361.

Rasheed, V. M., and J. A. Plumb. 1984. Pathogenicity of a non-haemolytic group B *Streptococcus* sp. in Gulf killifish (*Fundulus grandis* Baird and Girard). *Aquaculture* 37:97–105.

Rasheed, V. M., C, Limsuwan, and J. A. Plumb. 1985. Histopathology of bullminnows, *Fundulus grandis* Baird and Girard, infected with a non-haemolytic group B *Streptococcus* sp. *Journal of Fish Diseases* 8:65–74.

Roberts, R. J., and C. Sommerville. 1982. Diseases of tilapias. In: *The Biology and Culture of Tilapias,* edited by S. V. Pullin and R. H. Lowe-McConnel, 247–263. Manila, Philippines: International Center for Living Aquatic Resources Management.

Robinson, J. A., and F. P. Meyer. 1966. Streptococcal fish pathogen. *Journal of Bacteriology* 92:512.

Ruangpan, L., T. Kitao, and Y. Yoshida. 1986. Protective efficacy of *Aeromonas hydrophila* vaccines in Nile tilapia. *Veterinary Immunology and Immunopathology* 12:345–350.

Rotta, J. 1984. Pyogenic hemolytic streptococci. In *Bergey's Manual of Systematic Bacteriology,* vol. 2. edited by P. H. A. Sneath, N. S. Mair, M. E. Sharpe, and J. G. Holt, 1047– 1065. Baltimore, MD: Williams & Wilkins.

Sakai, M., R. Kubota, S. Atsuta, and M. Kobayashi. 1987. Vaccination of rainbow trout, *Salmo gairdneri* against beta-hemolytic streptococcal disease. *Bulletin of the Japanese Society of Scientific Fisheries* 53:1373–1376.

Sakata, T. 1988. Characteristics of *Vibrio vulnificus* isolated from diseased tilapia (*Saratherodon niloticus*). *Fish Pathology* 23:33–40.

Wu, S. Y. 1970. New bacterial disease of Tilapia. *FAO Fish Culture Bulletin* 2:14.

Yambot, P. 1996. *Streptococcus* spp. and/or *Aeromonas* spp. associated with fish kill in the Nile tilapia (*Oreochromis niloticus*) breeders in the Philippines [abstract]. *World Aquaculture, Bangkok, Thailand* 96.

Yambot, A. V., and V. Inglis. 1994. *Aeromonas hydrophila* isolated from Nile tilapia (*Orechromis niloticus* L.) with "eye disease" [abstract]. *International Symposium on Aquatic Animal Health, Fish Health Section/American Fisheries Society.* Seattle, WA.

20 Miscellaneous Bacterial Diseases

Some bacterial fish diseases cannot conveniently be categorized according to host because they do not affect one fish species or group more than another. One of those diseases is "marine flexibacteriosis" caused by *Flexibacter maritimus* (Wakabayashi et al. 1986).

MARINE FLEXIBACTERIOSIS

The marine fish disease caused by *F. maritimus* has not been given a common name. Because of its similarity to columnaris of freshwater fishes, it is referred to here as marine flexibacteriosis. The disease named "black batch disease" of Dover sole is also caused by this same organism (Bernardet et al. 1990).

Geographical Range and Species Susceptibility

To date, *F. maritimus* has been reported to infect marine fish in Japan, the Pacific coast of North America, the United Kingdom, Spain, and Tasmania. A variety of susceptible fish species include white seabass, Pacific sardine, northern anchovy, chinook salmon, rainbow trout, turbot, seabream, greenback flounder, olive flounder, striped trumpeter, and Dover sole (Bernardet et al. 1990; Pazos et al. 1993; Chen et al. 1995; Magariños et al. 1995; Handlinger et al. 1997). A similar disease was described in Atlantic salmon by Frelier et al. (1994).

Clinical Signs

Affected fish become lethargic, white to pinkish lesions develop on the skin, scale loss occurs, and underlying muscle is exposed (Figure 20.1). Fins become frayed, and white necrotic lesions form on the gills (Chen et al. 1995). Clinical signs in Atlantic salmon, rainbow trout, and greenback flounder were consistent among the species, with eroded skin lesions being the most prominent sign noted (Handlinger et al. 1997).

Diagnosis

Marine flexibacteriosis is diagnosed by detection of long, thin flexing bacterial rods in wet mount scrapings from body, fin, or gill lesions at 400×. Phase contrast microscopy is advantageous in detecting the bacteria.

Bacterial Characteristics

The Gram-negative cells range in size of 0.5 μm in diameter to 2 to 30 μm in length. *F. maritimus* can be cultured on seawater–Hsu-Shotts (SW-HS) media made with 50% seawater and supplemented with neomycin sulfate (4 μg/mL) and polymyxin B (200 IU/mL) (Baxa et al. 1986). The 50% salt water can be made by adding 18.7 g/L of sea salts to the media. Optimum incubation temperature is 22 to 35°C. Colonies of *F. maritimus* are rhizoid and white or light tan. These bacteria grow at 15 but not at 37°C, require at least 33% seawater, with maximum growth occurring in broth made with 66% to 100% seawater (Chen et al. 1995). Cysts or microcysts are not formed by cultured *F. maritimus* (Chen et al. 1995), but Baxa et al. (1986) and Kent et al. (1988) did report microcysts by marine gliding bacteria.

Key biochemical characteristics of *F. maritimus*

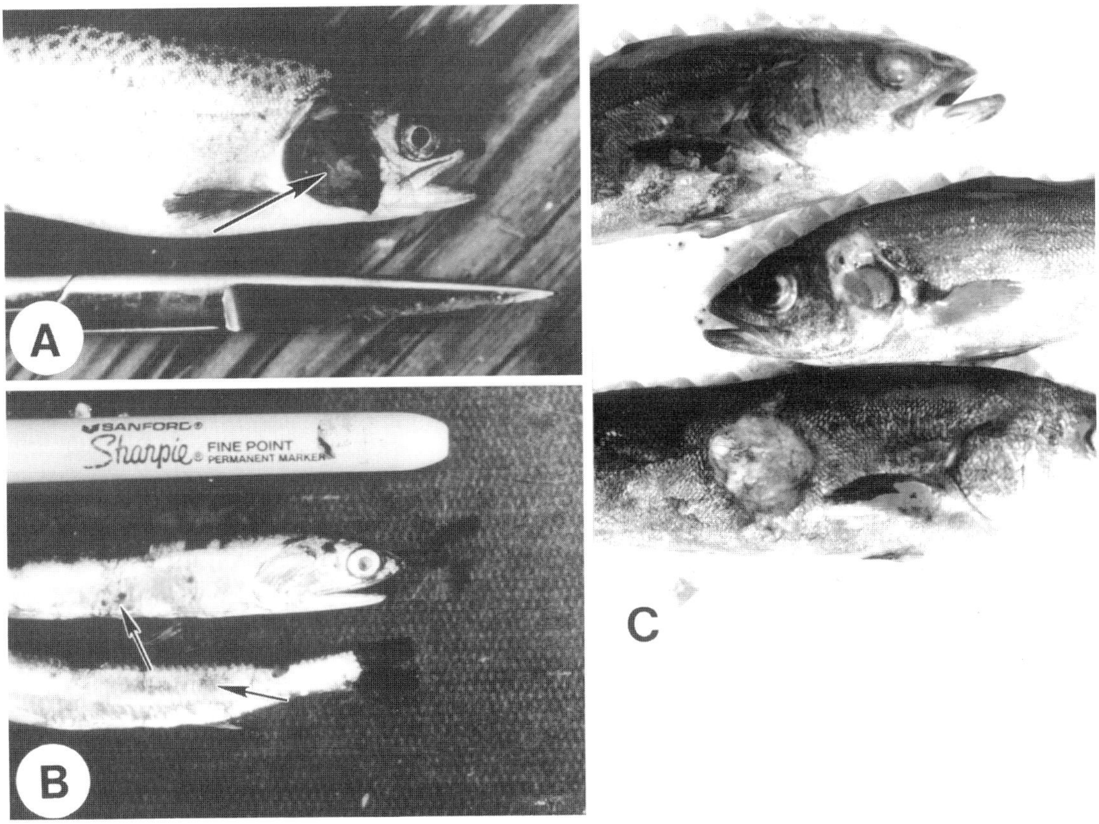

Figure 20.1. Marine flexibacteriosis (Flexibacter maritimus). (A) *Chinook salmon with a filamentous bacterial* (F. maritimus) *induced lesion on the gill* (arrow). (B) *A Northern anchovy with hemorrhagic and necrotic lesions of* F. maritimus *on the body* (arrows). (C) *White seabass with ulcerative body lesions caused by* F. maritimus. *(Photographs courtesy of M. Chen, California Department of Fish and Game.)*

are that it is negative for flexirubin pigments and starch hydrolysis and positive for Congo red, catalase, cytochrome oxidase, hydrolysis of gelatin, and tyrosine (see Table 14.1). Also, the bacteria's lack of 0% salt tolerance is important.

Epizootiology

F. maritimus is a pathogen that affects only marine fish and appears capable of infecting a wide variety of cultured species. To date, most epizootics have involved marine cage-cultured fish.

In marine flexibacteriosis–infected anchovies and sardines, trauma and skin abrasions are thought to be important precursors of disease (Chen et al. 1995). Net cage–reared white seabass, which are normally aggressive feeders, developed the disease following an interruption of feeding due to climatic conditions and/or mechanical failure. These researchers also reported gill lesions in caged chinook salmon down-current from cages of *F. maritimus*–infected bait fish. They theorized that gill lesions found in the salmon occurred when pieces of infected tissue from the bait fish lodged in their gills. Chen et al. (1995) also speculated that a trichodina gill infestation may have predisposed the salmon to *F. maritimus*. Although salmon culture in this coastal area has been discontinued for various reasons, *F. maritimus* being one of them, successful culture of the more resistant white seabass continues.

Fish appear to become infected via colonization of *F. maritimus* in the skin mucus (Magariños et al. 1995), as the bacterium was noted to adhere strongly to the mucus and skin of three fish species (turbot, seabream, and seabass) examined.

Pathological Manifestations

Long, thin, basophilic bacteria were seen in sectioned skin and muscle lesions taken from northern anchovy and white seabass (Chen et al. 1995). The bacteria extended into subdermal connective tissue and produced congestion and hemorrhage. A loss of epidermis was noted in the ulcerated tissue, and mats of bacteria extended into the dermis and subdermal layers. Bacteria also colonized the scale pockets but did not produce scale loss or ulceration. Infiltration and inflammatory cells in the affected areas were mild or absent. The earliest lesions to develop in salmonids and nonsalmonids were fragmentation and degeneration of epithelium with infiltration of amorphous proteinlike materials, congestion, and hemorrhage of the superficial dermis (Handlinger et al. 1997).

Significance

Marine flexibacteriosis is emerging as a significant pathogen of a variety of cage-cultured marine fishes. In some instances, it has contributed to the discontinuance of culture of certain highly susceptible species.

Management

Management recommendations for rearing white seabass suggested by Chen et al. (1995) include placement of culture cages of seabass as far from live bait pens as possible and provision of frequent and sufficient feeding to avoid antagonism and cannibalism. Antibiotics in the feed should be considered when marine flexibacteriosis occurs. The *F. maritimus* isolates from white seabass were sensitive to Terramycin and Romet-30.

REFERENCES

Baxa, D. V., K. Kawai, and R. Kusuda. 1986. Characteristics of gliding bacteria isolated from diseased cultured flounder, *Paralichthys olivaceus*. *Fish Pathology* 21:251–258.

Bernardet, J. F., A. C. Campbell, and J. A. Buswell. 1990. *Flexibacter maritimus* is the agent of "black patch necrosis" in Dover sole of Scotland. *Diseases of Aquatic Organisms* 8:233–237.

Chen, M. F., D. Henry-Ford, and J. M. Groff. 1995. Isolation and characterization of *Flexibacter maritimus* from marine fishes in California. *Journal of Aquatic Animal Health* 7:318–326.

Frelier, P. F., R. A. Elston, J. K. Loy, and C. Mincher. 1994. Macroscopic and microscopic features of ulcerative stomatitis in farmed Atlantic salmon *Salmo salar*. *Diseases of Aquatic Organisms* 18:227–231.

Handlinger, J., M. Soltani, and S. Percival. 1997. The pathology of *Flexibacter maritimus* in aquaculture species in Tasmania, Australia. *Journal of Fish Diseases* 20:159–168.

Kent, M. L., C. F. Dungan, R. A. Elston, and R. A. Holt. 1988. *Cytophaga* sp. (Cytophagales) infection in seawater pen-reared Atlantic salmon *Salmo salar*. *Diseases of Aquatic Organisms* 4:173–179.

Magariños, B., F. Pazos, Y. Santos, J. L. Romalde, and A. Toranzo. 1995. Response of *Pasteurella piscicida* and *Flexibacter maritimus* to skin mucus of marine fish. *Diseases of Aquatic Organisms* 21:103–108.

Pazos, F., Y. Santos, S. Núñez, and A. E. Toranzo. 1993. Characterization of *Flexibacter maritimus* isolated in northwest of Spain [abstract]. *Proceedings of the 6th International Conference of the European Association of Fish Pathologist*, Brest, France.

Wakabayashi, H., M. Hikida, and K. Masumura. 1986. *Flexibacter maritimus* sp. Nov., a pathogen of marine fishes. *International Journal of Systematic Bacteriology* 36:396–398.

Index

Italicized numbers followed by *fig.* or *tab.* indicate pages that contain figures or tables.

Q

R